I0066518

Genomics: Principles and Analysis

Genomics: Principles and Analysis

Edited by Jamie Spooner

New York

Published by Syrawood Publishing House,
750 Third Avenue, 9th Floor,
New York, NY 10017, USA
www.syrawoodpublishinghouse.com

Genomics: Principles and Analysis
Edited by Jamie Spooner

International Standard Book Number: 978-1-68286-712-9 (Hardback)

Cataloging-in-Publication Data

Genomics : principles and analysis / edited by Jamie Spooner.
p. cm.
Includes bibliographical references and index.
ISBN 978-1-68286-712-9
1. Genomics. 2. Genetics. 3. Genomes. I. Spooner, Jamie.
QH447 .G46 2019
572.86--dc23

TABLE OF CONTENTS

PREFACE

A genome is the complete genetic code of an organism. The science of genomics deals with the study of genomes along with their function, structure, evolution and manipulation. High throughput DNA sequencing and bioinformatics have advanced the understanding of this field. Some of the research dimensions of genomics are in the areas of structural and functional genomics, epigenomics, metagenomics, etc. Genomics has contributed significantly to the fields of medicine, biotechnology and conservation. From theories to research to practical applications, all contemporary topics of relevance to this field have been included in this book. Molecular biologists, geneticists, biochemists and students will find this book full of crucial and unexplored concepts.

This book unites the global concepts and researches in an organized manner for a comprehensive understanding of the subject. It is a ripe text for all researchers, students, scientists or anyone else who is interested in acquiring a better knowledge of this dynamic field.

I extend my sincere thanks to the contributors for such eloquent research chapters. Finally, I thank my family for being a source of support and help.

Editor

1

Expression-based clustering of CAZyme encoding genes of *Aspergillus niger*

Birgit S. Gruben[1,2†], Miia R. Mäkelä[1,3,4†], Joanna E. Kowalczyk[1,3], Miaomiao Zhou[1,5], Isabelle Benoit-Gelber[1,2,3,6] and Ronald P. De Vries[1,2,3*]

Abstract

Background: The *Aspergillus niger* genome contains a large repertoire of genes encoding carbohydrate active enzymes (CAZymes) that are targeted to plant polysaccharide degradation enabling *A. niger* to grow on a wide range of plant biomass substrates. Which genes need to be activated in certain environmental conditions depends on the composition of the available substrate. Previous studies have demonstrated the involvement of a number of transcriptional regulators in plant biomass degradation and have identified sets of target genes for each regulator. In this study, a broad transcriptional analysis was performed of the *A. niger* genes encoding (putative) plant polysaccharide degrading enzymes. Microarray data focusing on the initial response of *A. niger* to the presence of plant biomass related carbon sources were analyzed of a wild-type strain N402 that was grown on a large range of carbon sources and of the regulatory mutant strains Δ*xlnR*, Δ*araR*, Δ*amyR*, Δ*rhaR* and Δ*galX* that were grown on their specific inducing compounds.

Results: The cluster analysis of the expression data revealed several groups of co-regulated genes, which goes beyond the traditionally described co-regulated gene sets. Additional putative target genes of the selected regulators were identified, based on their expression profile. Notably, in several cases the expression profile puts questions on the function assignment of uncharacterized genes that was based on homology searches, highlighting the need for more extensive biochemical studies into the substrate specificity of enzymes encoded by these non-characterized genes. The data also revealed sets of genes that were upregulated in the regulatory mutants, suggesting interaction between the regulatory systems and a therefore even more complex overall regulatory network than has been reported so far.

Conclusions: Expression profiling on a large number of substrates provides better insight in the complex regulatory systems that drive the conversion of plant biomass by fungi. In addition, the data provides additional evidence in favor of and against the similarity-based functions assigned to uncharacterized genes.

Keywords: Transcriptional regulators, Plant biomass degradation, CAZy genes, XlnR, AmyR, GalX, AraR, RhaR, *Aspergillus niger*

Background

Aspergillus niger is a saprobic fungus that degrades a broad range of plant polysaccharides. Its genome encodes a versatile set of polysaccharide degrading enzymes [1, 2], which can be classified into families of glycoside hydrolases (GHs), polysaccharide lyases (PLs), carbohydrate esterases (CEs) and auxiliary activities (AAs) according to the CAZy (Carbohydrate-Active Enzymes) database (www.cazy.org; [3]). The classification is based on amino acid sequence and structural similarity. Among the 176 genes of *A. niger* CBS513.88 [4] that are predicted to encode CAZymes involved in plant biomass degradation less than half have been biochemically characterized, while the others have been assigned to CAZy families merely based on homology to functionally characterized genes.

In addition to the production of a wide variety of CAZyme encoding genes, the efficient depolymerization of the polysaccharides present in plant biomass requires

* Correspondence: r.devries@westerdijkinstitute.nl
†Equal contributors
[1]Fungal Physiology, Westerdijk Fungal Biodiversity Institute, Uppsalalaan 8, 3584, CT, Utrecht, The Netherlands
[2]Microbiology, Utrecht University, Padualaan 8, 3584, CH, Utrecht, The Netherlands
Full list of author information is available at the end of the article

a fine-tuned regulatory system. The expression of fungal CAZy genes have been shown to be controlled by multiple transcriptional regulators, most of which belong to fungi specific Zn_2Cys_6zinc binuclear family of transcriptional factors [5]. In *A. niger*, several regulators related to plant polysaccharide degradation have been identified [6]. These include XlnR [7], AraR [1], AmyR [8], InuR [9], RhaR [10], ManR/ClrB [11, 12], ClrA [13], GalX [14] and GaaR [15] that have been reported as transcriptional activators of CAZymes (Table 1). These regulators respond to mono- and small oligosaccharides that act as inducers (Table 1) [16], but so far, a limited set of target genes of these regulators have been identified. While some genes can be controlled by a single regulator, coregulation of several CAZyme encoding genes has been described in *Aspergillus* species.

AmyR, a transcriptional regulator that controls the genes involved in starch degradation, was the first well-studied regulator in several *Aspergillus* species [17, 18]. In Aspergilli, AmyR is induced by maltose and regulates genes encoding α-amylases, glucoamylase and α-glucosidases all of which are involved in depolymerization of starch, the major storage polysaccharide in plants [6]. In addition, AmyR has been shown to have a broader physiological role in *A. niger* by controlling some of the genes encoding D-glucose and D-galactose releasing enzymes, i.e. β-glucosidases, and α- and β-galactosidases [8]. Also, D-glucose or its metabolic product has been suggested to have a possible role as the inducer of the AmyR system in *A. niger*.

XlnR has an important role in biomass degradation by controlling the expression of genes encoding enzymes that degrade xylan, cellulose and xyloglucan, which are the most abundant polysaccharides in nature [19–21]. The *xlnR* gene has also been shown to be present in almost in all filamentous ascomycete fungi [22]. The range of genes regulated by XlnR include genes encoding endoxylanase, β-xylosidase, α-glucuronidase, acetylxylan esterase, arabinoxylan arabinofuranohydrolase, feruloyl esterase, α- and β-galactosidases, endoglucanase and cellobiohydrolase, as well as *aglB* and *lacA* genes that encode enzymes putatively involved in xyloglucan or galactomannan degradation [23].

A homolog of XlnR, AraR, is a transcriptional regulator induced by L-arabinose and its degradation product, L-arabitol [22]. These monomers are building blocks of arabinan present in side chains of arabinoxylan and pectin. Two arabinan hydrolyzing enzymes produced by *A. niger*, α-L-arabinofuranohydrolases A and B, are controlled by AraR [22]. In addition, AraR controls the expression of the genes involved in L-arabinose catabolism. AraR and XlnR also co-regulate genes from pentose catabolic pathway and pentose phosphate pathway [24].

The expression of the genes encoding inulinases and invertase, which hydrolyze plant storage polymer inulin, is controlled by the transcriptional regulator InuR in *A. niger* [9]. Inulinolytic enzyme encoding genes are also inducted by sucrose, and moreover, the repertoire of the genes regulated by InuR has been suggested to include other genes related to degradation of inulin and sucrose.

Several plant polysaccharides, e.g. xylan, pectin and glucogalactomannan, include D-galactose, which is released by fungal α- and β-galactosidases and endogalactanases. While a galactose-related regulator GalR was reported to be unique for *Aspergillus nidulans* [25], it has also been found in related species of *Aspergillus* section *nidulantes* [26]. In contrast, GalX is more generally present in *Aspergillus* species. In *A. niger*, GalX regulates the expression of the genes from oxido-reductive pathway for D-galactose catabolism [14].

In addition to AraR, the other pectinolytic regulators described from *A. niger* are RhaR [10] and GaaR [15]. RhaR, induced by a metabolic conversion product of L-rhamnose, influences the degradation of rhamnogalacturonan I part of

Table 1 Transcriptional activators involved in plant polysaccharide degradation and/or sugar catabolism in *A. niger*

Regulator[a]	Inducer	Function	Reference
AmyR	D-glucose or maltose	Starch degradation	[8]
AraR	L-arabitol	Arabinan degradation, Pentose Catabolic Pathway	[22, 66]
ClrA	cellulose[b]	Degradation of cellulose	[13]
GaaR	2-keto-3-deoxy-L-galactonate	Degradation of polygalacturonic acid and more complex pectins, transport of D-galacturonic acid, D-galacturonic acid catabolism	[29, 47]
GalX	D-galactose or derivative	D-galactose catabolism	[15]
InuR	sucrose	Inulin and sucrose degradation	[9, 43]
ManR/ClrB		Galactomannan and cellulose degradation	[11, 12]
RhaR	L-rhamnose derivative	Rhamnogalacturonan I degradation, L-rhamnose catabolism	[10, 27]
XlnR	D-xylose	Xylan, xyloglucan and cellulose degradation, Pentose Catabolic Pathway	[7, 21, 22]

[a]The *A. niger* deletion strains of the underlined regulators were used in this study
[b]Based on data from *Neurospora crassa* [45]

pectin by controlling several genes involved in L-rhamnose release and catabolism [10, 27], as well as a L-rhamnose transporter [28]. The more recently described transcriptional regulator GaaR is induced by 2-keto-3-deoxy-L-galactonate, a metabolic conversion product of D-galacturonic acid, and involved in the release of galacturonic acid from polygalacturonic acid and more complex pectins, as well as transport of galacturonic acid and induction of the galacturonic acid catabolic genes [15, 29].

Other plant-biomass related transcriptional regulators described in *Aspergillus* species include the regulator of mannan degrading enzymes, ManR/ClrB, which was first described in *Aspergillus oryzae,* with a role in galactomannan and cellulose degradation [11, 12]. In *A. oryzae,* ManR/ClrB is induced by the disaccharide mannobiose, but not mannose [11, 12]. Furthermore, the genomes of Aspergilli possess various homologs of plant-polysaccharide related regulators from other fungal species, and the gene expression studies have also indicated the presence of several additional regulators involved in this process, including e.g. those responding to the presence of ferulic acid, glucuronic acid and galacturonic acid [6].

The aim of this study was to evaluate co-regulation/co-expression of characterized and putative CAZymes to gain more insight in the function of uncharacterized CAZyme encoding genes in plant biomass utilization and to identify new targets of transcriptional regulators. The focus of the study was on the initial response of *A. niger* to the presence of a carbon source. For this, microarray data were analyzed of *A. niger* N402 (wild type) that was grown on a set of 23 carbon sources (including eight monosaccharides, two oligosaccharides, 11 polysaccharides, a crude plant biomass substrate and ferulic acid), and of regulatory mutant strains (Δ*xlnR*, Δ*araR*, Δ*amyR*, Δ*rhaR* and Δ*galX*) that were grown on their specific inducing compounds. Hierarchical clustering of the expression data revealed several gene clusters that appear to be under control of the same regulators.

Results and discussion

Microarray data were analyzed of *A. niger* N402 that was grown on 23 carbon sources (Tables 2, 3) and of the regulatory mutants Δ*xlnR*, Δ*araR*, Δ*amyR*, Δ*rhaR* and Δ*galX* that were grown on their inducing compounds (Tables 1, 3). The mycelial samples were collected after 2 h or 4 h (for N402 and Δ*amyR* on D-maltose) exposure to the carbon source of interest. Thus, this study focused on the initial response to the presence of a carbon source to avoid changes due to carbon source limitation or changes in the polymeric substrates. Although we can therefore not exclude that genes that were not expressed may have been induced after longer incubation times or on substrates that were not included in this analysis, it provides a detailed understanding of the initial response of *A. niger* to the presence of plant-biomass related carbon sources.

Clustering the expression profiles of *A. niger* (putative) CAZyme encoding genes that are related to plant polysaccharide degradation by Pearson correlation resulted in nine clusters, A-I (Additional file 1). After the initial clustering analysis, genes with a signal value below 50 under all growth conditions were removed from the analysis. These genes were considered not to be significantly expressed (Additional files 2 and 3). The genes that were significantly expressed (signal value >50) at least under one condition are shown in Tables 4, 5, 6, 7, 8, 9, 10, 11. In addition, the fold-changes of the significantly expressed genes between N402 and the regulatory mutant strains were determined (Tables 4, 5, 6, 7, 8, 9, 10, 11). Negative fold-changes indicate genes for which the expression is lower in the mutant than in the wild type strain, while positive fold-changes indicate higher expression in the mutant than in the wild type strain. If the negative fold-change is larger than 2.5, we consider this gene under control of the respective regulator.

AraR and XlnR regulated genes involved in cellulose, xyloglucan, xylan and arabinan degradation cluster together based on their expression profile

The genes of cluster B were specifically induced on L-arabinose, D-xylose and/or polygalacturonic acid. This cluster can be divided into sub-clusters B-1 and B-2 that contain seven and ten significantly expressed genes, respectively (Additional file 1, Table 4). The highest expression for genes of sub-cluster B-1 was detected on L-arabinose, except for *axhA* that was also induced on L-arabinose, but was higher expressed on polygalacturonic acid. The *axhA* gene encodes an arabinoxylan arabinofuranohydrolase and is specific for arabinoxylan degradation [30]. The high expression of this and other genes of cluster B on polygalacturonic acid may be due to impurity of the substrate (Table 3). The expression of XlnR-regulated genes has been shown to decrease with increasing concentrations of D-xylose due to carbon catabolite repression [31]. Small traces of D-xylose and L-arabinose in the polygalacturonic acid substrate may therefore lead to higher expression of the xylanolytic, arabinanolytic and cellulolytic genes than on 25 mM of D-xylose or L-arabinose used in our study. One gene of sub-cluster B-1 has been characterized as an endoglucanase (*eglB*), which has activity towards cellulose [21]. The other significantly expressed genes of this cluster encode an arabinofuranosidase (*abfA*), two putative endoxylanases (*xlnC* and *xynA*), a putative α-galactosidase (*aglD*) and a putative β-endogalactanase (An03g01050) (Table 4).

The highest expression level of these genes was found on L-arabinose (Additional file 1). Regulation of two of these genes, *abfA* and *axhA*, is controlled by the

Table 2 *A. niger* strains used in this study

Strain	Genotype	Reference
A. niger N402	*cspA1*	[67]
FP-304 (Δ*rhaR*)	*cspA1*, Δ*kusA::amdS+*, *pyrA5*, Δ*rhaR::pyrA+*	[10]
UU-A101.1 (Δ*amyR*)	*cspA1*, Δ*argB*, *pyrA6*, *leuA1*, *nicA1*, Δ*amyR::argB+*	[8]
FP-306 (Δ*galX*)	*cspA1*, Δ*kusA::amdS+*, *pyrA5*, Δ*galX::pyrA+*	[14]
UU-A033.21 (Δ*araR*)	*cspA1*, *pyrA6*, *nicA1*, *leuA1*, Δ*argB*::pIM2101 Δ*araR::argB+*	[22]
UU-A062.10 (Δ*xlnR*)	*cspA1*, *pyrA6*, Δ*argB*, *nicA1*, *leuA*, *pyrA6*, Δ*xlnR::pyrA+*	[22]

transcriptional activator AraR [23] that is induced by L-arabitol, a metabolic product of L-arabinose [32]. Co-regulation of AraR-regulated arabinanolytic genes (i.e. *abfA*, *abfB* and *abnA*) has been suggested previously [22, 33] and *abfA* has been shown to be controlled by GaaR [15, 34]. A previous principal component analysis (PCA) clustering of the pectinolytic genes has been shown to result in one cluster which contained *abfA*, *abfB*, *abnA* and *lacA* [35], which matches a more resent hierarchical clustering of the expression of pectinolytic genes in wild type and *gaaR* deletion mutant strains resulting in a cluster containing *abfA, abfB, abfC, lacA, lacB* and An03g01620 [15]. However, in this study, the *abfB*, *abnA* and *lacA* genes were separated from *abfA*, which indicates that *abfA* has a distinct expression profile from the other genes. This is likely due to the large set of carbon sources that were tested in our study, which

Table 3 Composition, purity and concentration of the carbon sources used in this study

Substrate	Conc.	Company	Purity	Composition
D-glucose	25 mM	Sigma-Aldrich	≥99.5%	–
D-fructose	25 mM	Sigma-Aldrich	≥99%	–
D-galactose	25 mM	Sigma-Aldrich	≥99%	–
L-arabinose	25 mM	Sigma-Aldrich	≥99%	–
D-xylose	25 mM	Sigma-Aldrich	≥99%	–
D-mannose	25 mM	Sigma-Aldrich	≥99%	–
D-galacturonic acid	25 mM	Sigma-Aldrich	≥99%	–
L-rhamnose	25 mM	Sigma-Aldrich	≥99%	–
Maltose	25 mM	Sigma-Aldrich	≥99%	–
Sucrose	25 mM	Sigma-Aldrich	≥99%	–
Ferulic acid	25 mM	Sigma-Aldrich	≥99%	–
Inulin from chicory	1%	Sigma-Aldrich	ns	ns
Cellulose	1%	Sigma-Aldrich	ns	ns
Guar gum	1%	Sigma-Aldrich	ns	ns
Xyloglucan	1%	Sigma-Aldrich	ns	ns
Xylan (from beechwood)	1%	Sigma-Aldrich	>90%	>90% D-xylose residues
Polygalacturonic acid (from apples)	1%	Sigma-Aldrich	≥85%	ns
Apple pectin	1%	Sigma-Aldrich	ns	ns
Galactan (from potato)	1%	Megazyme	ns	Gal: Ara: Rha: GalA = 88: 2: 3: 7
Debranched 1,5-α-L-arabinan (from sugar beet)	1%	Megazyme	~95%	Ara: Gal: Rha: GalA = 88: 4: 2: 6
Rhamnogalacturonan I (from potato)	1%	Megazyme	>97%	GalA: Rha: Ara: Xyl: Gal: Os = 62: 20: 3.3: 1: 12: 1.7
Mannan (ivory nut)	1%	Megazyme	>98%	99% Mannan, Ara + Xyl traces
Citrus pulp	1%	ns	ns	Glc: GalA: Ara: Gal: Xyl: Man: Rha = 39: 35: 11: 7: 4: 4: 1
Sugar beet pulp	1%	ns	ns	Glc: Ara: GalA: Gal: Xyl: Man: Rha = 33: 28: 26: 7: 2: 2: 1
Soy bean hulls	1%	ns	ns	Glc: GalA: Xyl: Ara: Man: Gal: Rha = 49: 16: 15: 8: 7: 4: 1

Gal D-galactose, *Ara* L-arabinose, *Rha* L-rhamnose, *GalA* D-galacturonic acid, *Xyl* D-xylose, *Man* D-mannose, *Os* other sugars. ns = not specified

Table 4 Significantly expressed *A. niger* N402 genes from cluster B in the expression profiling tree

	Gene[a]	Enzyme[b]	CAZy family	Up/Down-regulated[c]					Regulated by		
				Δ*xlnR*	Δ*araR*	Δ*rhaR*	Δ*galX*	Δ*amyR*	This study	Literature	Reference
B-1	An03g01050	GLN	GH5	4.8	**−74.6**	3.1	29.9	6.3	AraR	nr	[38]
	An14g01800 (*aglD*)	AGL	GH27	−1.5	**−97.4**	–	–	–	AraR	nr	[38]
	An01g00330 (*abfA*)	ABF	GH51	1.0	**−113.9**	−1.7	4.6	38.3	AraR	AraR, GaaR	[15, 22, 34, 68]
	An03g00940 (*xlnC/xynA*)	XLN	GH10	**−17.6**	**−27.6**	–	–	–	AraR, XlnR	ClrA, XlnR	[13, 21, 38]
	An03g00960 (*axhA*)	AXH	GH62	**−17.3**	**−28.5**	–	–	–	AraR, XlnR	AraR, XlnR	[21, 22, 30]
	An16g06800 (*eglB*)	EGL	GH5	4.1	**−6.2**	−1.3	−1.1	1.0	AraR	nr	[38]
	An15g04550 (*xynA*)	XLN	GH11	–	**−6.6**	–	–	–	AraR	nr	[38]
B-2	An12g05010 (*axeA*)	AXE	CE1	**−51.5**	1.5	–	–	–	XlnR	XlnR	[21, 38]
	An01g09960 (*xlnD/xynD*)	BXL	GH3	**−502.6**	1.6	−1.8	2.1	5.9	XlnR	XlnR	[21, 69]
	An08g01900	BXL	GH43	**−4.0**	1.2	–	–	–	XlnR	nr	[38]
	An11g02100	BGL	GH1	**−5.3**	2.0	1.0	2.1	2.0	XlnR	nr	[38]
	An14g05800 (*aguA*)	AGU	GH67	**−32.5**	2.1	**−3.7**	1.5	1.5	RhaR, XlnR	XlnR	[21, 70]
	An09g00120 (*faeA*)	FAE	SF7[d]	**−78.3**	−1.5	–	–	–	XlnR	XlnR	[21, 36]
	An08g01760	CBH	GH6	**−3.8**	–	–	–	–	XlnR	nr	[38]
	An09g03300 (*axlA/xylS*)	AXL	GH31	**−10.4**	**−6.3**	1.3	3.8	4.3	AraR, XlnR	nr	[41]
	An01g00780 (*xlnB/xynB*)	XLN	GH11	**−40.7**	**−5.6**	1.7	1.1	−2.2	AraR, XlnR	XlnR	[21, 38]

[a]Genes with expression value of <50 in all studied *A. niger* N402 cultures are not included in the table
[b]Enzyme codes are provided in Additional file 2
[c]Fold-change between *A. niger* N402 and the regulatory mutants grown on their relevant carbon source. - = expression value <50 in both N402 and regulatory mutant strain. Negative fold-changes >2.5 were considered as proof of regulator function and are depicted in bold
[d]Sub-family (SF) classification of fungal FAEs according to Dilokpimol et al. [49]
nr not reported

provides a more detailed view of the expression of these genes than has been published previously, and also reveals the complexity of the expression of plant-biomass related genes. It should be noted that in nature, fungi are confronted with mixtures of carbon sources, and therefore likely activate a combination of the gene sets we observed in response to pure substrates.

Similar expression profiles for the other genes in this sub-cluster (*eglB, xlnC, aglD, xynA* and An03g01050) suggest that they are also regulated by AraR. This is supported by the reduced expression of these genes in the Δ*araR* strain on L-arabinose compared to N402 (Table 4). The *axhA* and *xlnC* genes are also regulated by XlnR [21], which was confirmed in our analysis, as these genes were down-regulated in the Δ*xlnR* strain. In addition, *xlnC* has been reported to be ClrA-regulated [13]. Thus, our results indicate a broader role for AraR as some of the genes related to cellulose (*eglB*), galactomannan (*aglD, mndA*), pectin (*lacA, lacB, xghA*), xyloglucan (*axlA*) and xylan (*gbgA, xlnB, xlnC,* An01g01320) degradation were significantly down-regulated in the Δ*araR* strain.

The genes of sub-cluster B-2 were significantly down-regulated in the Δ*xlnR* strain (Table 4), thus suggesting that they are controlled by XlnR. Indeed, five of these genes (*axeA, xlnD, aguA, faeA* and *xlnB*) have previously been shown to be regulated by XlnR [21, 36, 37]. The highest expression for most genes of this sub-cluster was detected on D-xylose, except for *aguA* and An11g02100 that were higher expressed on polygalacturonic acid, and *axlA* and An16g00540 that were higher expressed on L-arabinose (Additional file 1). High expression of *axlA* on D-xylose has previously been reported [9, 37]. This gene encodes a putative α-xylosidase that is suggested to have a role in xyloglucan degradation [38]. An16g00540 encodes an α-L-fucosidase, which also has a putative role in xyloglucan hydrolysis, indicating co-regulation of some of the genes involved in this process. An11g02100 and An08g01760 encode the cellulolytic enzymes β-glucosidase and cellobiohydrolase, respectively. This is in line with the previous finding that XlnR is a regulator of xylanolytic, xyloglucanolytic and cellulolytic genes [21]. The co-regulation of AraR- and XlnR-regulated genes in cluster B that are involved in cellulose, xyloglucan, xylan and arabinan degradation supports combined action of regulators. Co-regulation of these genes is an efficient strategy for polysaccharide degradation, since L-arabinose, D-xylose and D-glucose often co-occur in plant cell wall polysaccharides.

Expression of pectinolytic genes involved in degradation of the pectin main chains were clustered

Cluster C contains 28 significantly upregulated genes of which most are pectin backbone hydrolyzing genes, mainly from CAZy families GH28 (several types of pectin

Table 5 Significantly expressed *A. niger* N402 genes from cluster C in the expression profiling tree

	Gene[a]	Enzyme[b]	CAZy family	Up/Down-regulated[c]					Regulated by		
				Δ*xlnR*	Δ*araR*	Δ*rhaR*	Δ*galX*	Δ*amyR*	This study	Literature	Reference
C-1	An01g13610 (*amyD*)	AMY	GH13	1.2	–	–	1.3	–	nd	nr	[43]
	An18g04810 (*rgxC*)	RGX	GH28	–	–	**−17.1**	–	–	RhaR	GaaR, RhaR	[15, 34, 40]
	An06g00290 (*lacC*)	LAC	GH35	1.3	−1.7	**−20.2**	1.6	1.5	RhaR	AmyR, AraR, GaaR, RhaR	[15, 34, 40]
C-2	An11g04040 (*pgxA*)	PGX	GH28	–	–	–	–	–	nd	GaaR	[15, 34, 40]
	An03g06740 (*pgxB*)	PGX	GH28	–	–	–	–	–	nd	GaaR	[15, 34, 40]
	An02g12450 (*pgxC*)	PGX	GH28	−1.3	**−3.2**	–	–	–	RhaR	AraR, GaaR	[15, 34, 40]
	An12g07500 (*pgaX*)	PGX	GH28	–	–	–	–	–	nd	GaaR	[15, 34, 40]
	An02g00140 (*xynB*)	BXL	GH43	–	2.3	–	1.6	1.9	nd	GaaR	[34, 38]
	An03g06310 (*pmeA*)	PME	CE8	–	–	–	–	–	nd	GaaR, AraR, RhaR	[15, 34, 71]
	An14g04370 (*pelA*)	PEL	PL1	1.2	−1.2	−1.5	1.0	1.0	nd	GaaR	[15, 34, 72, 73]
	An14g01130 (*rglA*)	RGL	PL4	–	–	–	–	–	nd	GaaR	[15, 34, 35]
C-3	An18g05620 (*agdF*)	AGD	GH31	–	–	**−3.4**	2.7	3.2	RhaR	nr	[41]
	An09g02160 (*rgaeA*)	RGAE	CE12	–	–	**−22.1**	2.3	–	RhaR	GaaR, RhaR	[10, 34, 74]
	An04g09360 (*rgaeB*)	RGAE	CE12	–	1.5	**−38.1**	2.0	–	RhaR	GaaR, RhaR	[10, 34, 39]
	An01g14650 (*rgxA*)	RGX	GH28	–	–	**−26.4**	–	–	RhaR	RhaR	[10, 40]
	An01g06620	RHA	GH78	−1.2	1.3	**−90.2**	−1.1	2.8	RhaR	RhaR	[10, 38]
	An12g05700	RHA	GH78	–	–	**−132.0**	–	4.8	RhaR	GaaR, RhaR	[10, 34, 38]
	An14g02920 (*urhgA*)	URH	GH105	–	2.3	**−50.2**	2.1	3.6	RhaR	AraR, GaaR, RhaR	[10, 34, 39]
	An07g00240	RHA	GH78	2.0	–	**−17.0**	–	–	RhaR	RhaR	[10, 38]
	An10g00290	RHA	GH78	–	–	**−8.2**	–	–	RhaR	RhaR	[10, 38]
	An01g14600	XLN	GH11	–	–	**−13.5**	–	–	RhaR	nr	[38]
	An03g02080 (*rgxB*)	RGX	GH28	–	–	**−137.7**	–	–	RhaR	RhaR	[40]
	An04g09070	RHA	GH78	**−2.8**	1.1	**−40.4**	1.1	2.0	RhaR, XlnR	RhaR	[10, 38]

[a]Genes with expression value of <50 in all studied *A. niger* N402 cultures are not included in the table
[b]Enzyme codes are provided in Additional file 2
[c]Fold-change between *A. niger* N402 and the regulatory mutants grown on their relevant carbon source. - = expression value <50 in both N402 and regulatory mutant strain. Negative fold-changes >2.5 were considered as proof of regulator function and are depicted in bold
nd not detected, *nr* not reported

hydrolases) and GH78 (α-rhamnosidases) (Table 5). It can be divided into the sub-clusters C-1, C-2 and C-3 (Additional file 1). Sub-cluster C-3 contains 12 significantly expressed genes, of which 10 have been shown to be regulated by RhaR and are specifically induced on L-rhamnose [10]. The other two genes of this cluster, *agdF* and An01g14600, were also specifically induced on L-rhamnose and down-regulated in the Δ*rhaR* strain suggesting that they are also under control of this regulator (Table 5). However, our results suggest a broader role for RhaR, since in addition to its target genes of cluster C, some other genes were identified that were down-regulated in the Δ*rhaR* strain, such as *aguA*, *aglC* and *mndA*.

Notably, the *agdF* gene has previously been assigned to encode a putative enzyme of the starch degrading GH31 family [38]. Our data does not support a function in starch degradation as, in addition to induction on L-rhamnose, this gene was significantly up-regulated in the Δ*amyR* strain (Table 5), while the opposite would be expected for a starch-related gene. The expression profile of An01g14600, which encodes a putative enzyme of the GH11 endoxylanase family, is unexpected as no link between this family and rhamnogalacturonan degradation has been described. Therefore, our data suggests the involvement of *agdF* and An01g14600 in rhamnogalacturonan degradation, although their enzymatic function is unclear at this point. A high expression level on L-rhamnose has been previously reported for *rgaeB*, *rgxA*, *rgxB*, *urhgA* and *rglB* [39, 40]. In our analysis, *rgaeB* appears to have a slightly different expression profile from the other genes of sub-cluster C-3 as it is located in a separate branch of the hierarchal cluster (Additional file 1). The inclusion of the L-rhamnose and D-galacturonic acid mixture data enabled us to evaluate the co-operation of these two sugars as inducers by comparing them to the individual sugar cultivations. Interestingly, despite the dominant role for galacturonic acid and GaaR in regulation of

Table 6 Significantly expressed *A. niger* N402 genes from cluster D in the expression profiling tree

	Gene[a]	Enzyme[b]	CAZy family	Up/Down-regulated[c]					Regulated by		
				Δ*xlnR*	Δ*araR*	Δ*rhaR*	Δ*galX*	Δ*amyR*	This study	Literature	Reference
D-1	An02g01400 (*abnB*)	ABN	GH43	−1.6	1.4	1.6	−1.1	-	nd	nr	[39]
	An11g07660 (*exgD*)	EXG	GH5	−1.2	1.0	1.2	−1.3	−1.3	nd	nr	[38]
	An04g09890 (*agsA*)	AGS	GH13	1.4	-	-	−2.1	−1.9	nd	nr	[41, 75]
	An18g04800	RHA	GH78	-	-	1.7	1.9	-	nd	GaaR	[34, 38]
	An15g01890	BGL	GH3	1.4	1.4	1.0	−1.2	1.2	nd	nr	[38]
	An13g02110	AFC	GH29	-	-	-	-	-	nd	nr	[38]
	An01g01340	UGH	GH88	-	-	-	-	-	nd	AraR	[34, 38]
	An14g05340 (*urhgB*)	URH	GH105	-	-	-	-	-	nd	GaaR, RhaR	[34, 39]
D-2	An04g06930 (*amyC*)	AMY	GH13	-	1.0	1.2	6.2	**−3.4**	AmyR	AmyR	[38]
	An03g06550 (*glaA*)	GLA	GH15	−2.3	**−4.6**	5.9	9.1	**−44.4**	AmyR, AraR	AmyR	[8, 38, 76, 77]
	An11g03340 (*aamA*)	AMY	GH13	4.3	**−9.2**	12.8	38.4	**−35.8**	AmyR	AmyR	[41, 42]
	An04g06920 (*agdA*)	AGD	GH31	−1.1	−1.2	4.3	6.7	**−42.7**	AmyR	AmyR	[8, 38, 78]
	An12g02460 (*agtB*)	AGT	GH13	-	-	-	-	**−3.8**	AmyR	nr	[41]
	An03g05530	XG-EGL	GH12	-	-	−1.8	−1.9	**−3.0**	AmyR	nr	[4]

[a]Genes with expression value of <50 in all studied *A. niger* N402 cultures are not included in the table
[b]Enzyme codes are provided in Additional file 2
[c]Fold-change between *A. niger* N402 and the regulatory mutants grown on their relevant carbon source. - = expression value <50 in both N402 and regulatory mutant strain. Negative fold-changes >2.5 were considered proof of regulator function and are depicted in bold
nd = not detected; nr = not reported

pectinolytic genes [16, 34], the mixture of L-rhamnose and D-galacturonic acid clusters more closely with L-rhamnose than with D-galacturonic acid in our analysis. This may indicate that the induction by L-rhamnose is more discriminative than the induction by D-galacturonic acid in distinguishing genes by expression pattern.

Sub-cluster C-1 contains three significantly expressed genes, two of which are regulated by RhaR on L-rhamnose and by GaaR: *lacC* and *rgxC* (Table 5) [10, 15]. The *lacC* and *rgxC* genes were previously reported to be expressed on D-galacturonic acid, polygalacturonic acid and L-rhamnose, in contrast to the genes of sub-cluster C-3 that were specifically induced on L-rhamnose [39]. High expression of *lacC* and *rgxC* on galactan could be due to the small traces of D-galacturonic acid and L-rhamnose in the substrate (Table 3). The *lacC* has also been reported to be under control of AraR [34] and AmyR [8], but it was not observed to be down-regulated in the Δ*amyR* strain in our study (Table 5). The third gene of the sub-cluster C-1, *amyD*, has been classified as an α-amylase [38], but its expression was not detected on D-maltose in *A. niger* N402 [41]. In our study, the gene was expressed on D-galacturonic acid, polygalacturonic acid and the mixture of D-galacturonic acid and L-rhamnose (Additional file 1). A role for *amyD* in starch degradation is therefore doubtful.

Table 7 Significantly expressed *A. niger* N402 genes from cluster E in the expression profiling tree

	Gene[a]	Enzyme[b]	CAZy family	Up/Down-regulated[c]					Regulated by		
				Δ*xlnR*	Δ*araR*	Δ*rhaR*	Δ*galX*	Δ*amyR*	This study	Literature	Reference
E-1	An15g04900 (*eglD*)	LPMO	AA9	2.1	−1.2	−1.7	−1.6	−1.8	nd	nr	[38]
	An03g00190 (*pelB*)	PEL	PL1	-	-	-	1.2	1.0	nd	nr	[73]
E-2	An07g09760	BGL	GH3	-	-	-	-	-	nd	nr	[38]
	An08g01100	EXG	GH5	1.6	1.1	1.1	-	3.8	nd	nr	[38]
	An11g03200 (*inuA/inuB*)	INU	GH32	-	-	-	-	-	nd	InuR	[8, 79]
	An12g08280 (*inuE/inu1*)	INX	GH32	2.9	2.0	1.9	5.2	4.1	nd	InuR	[8]
	An02g04900 (*pgaB*)	PGA	GH28	2.6	−2.3	-	2.2	−2.1	nd	GaaR	[15, 34, 80]
	An08g11070 (*sucA/suc1*)	SUC	GH32	-	-	-	-	-	nd	InuR	[9, 81, 82]

[a]Genes with expression value of <50 in all studied *A. niger* N402 cultures are not included in the table
[b]Enzyme codes are provided in Additional file 2
[c]Fold-change between *A. niger* N402 and the regulatory mutants grown on their relevant carbon source. - = expression value <50 in both N402 and regulatory mutant strain
nd not detected, *nr* not reported

Table 8 Significantly expressed *A. niger* N402 genes from cluster F in the expression profiling tree

				Up/Down-regulated[c]					Regulated by		
	Gene[a]	Enzyme[b]	CAZy family	Δ*xlnR*	Δ*araR*	Δ*rhaR*	Δ*galX*	Δ*amyR*	This study	Literature	Reference
F-1	An02g00610	GUS	GH2	11.7	-	-	-	-	nd	nr	[38]
	An09g03070 (*agsE*)	AGS	GH13	1.6	1.1	−2.0	**−11.0**	**−5.8**	AmyR, GalX	nr	[43, 75]
	An16g02730 (*abnD*)	ABN	GH43	3.6	-	-	-	**−2.5**	AmyR	GaaR, RhaR	[15, 34, 39]
	An09g00260/An09g00270 (*aglC*)	AGL	GH36	1.3	1.0	**−3.5**	1.3	1.3	RhaR	AmyR	[8, 44]
F-2	An09g05350	FAE	SF9[d]	1.5	-	-	-	−1.1	nd	nr	[4]

[a]Genes with expression value of <50 in all studied *A. niger* N402 cultures are not included in the table
[b]Enzyme codes are provided in Additional file 2
[c]Fold-change between *A. niger* N402 and the regulatory mutants grown on their relevant carbon source. - = expression value <50 in both N402 and regulatory mutant strain. Negative fold-changes >2.5 were considered proof of regulator function and are depicted in bold
[d]Sub-family (SF) classification of fungal FAEs according to Dilokpimol et al. [49]
nd not detected, *nr* not reported

The pectinolytic genes in sub-cluster C-2 are involved in the degradation of homogalacturonan (Table 5). These genes are not regulated by RhaR but were induced on D-galacturonic acid and polygalacturonic acid in this study (Additional file 1) and most of them are under control of GaaR [15, 34]. The significantly expressed genes of sub-cluster C-2 include four exopolygalacturonases (*pgxA*, *pgxB*, *pgxC* and *pgaX*), a pectin methyl esterase (*pmeA*), a pectin lyase (*pelA*), and rhamnogalacturonan lyase (*rglA*) (Table 5), all of which have been shown to be GaaR-

Table 9 Significantly expressed *A. niger* N402 genes from cluster G in the expression profiling tree

				Up/Down-regulated[c]					Regulated by		
	Gene[a]	Enzyme[b]	CAZy family	Δ*xlnR*	Δ*araR*	Δ*rhaR*	Δ*galX*	Δ*amyR*	This study	Literature	Reference
G-1	An17g00520	BGL	GH3	-	1.6	-	-	1.5	nd	nr	[38]
	An11g06540 (*mndA*)	MND	GH2	4.1	**−3.0**	**−2.8**	1.2	2.6	AraR, RhaR	nr	[4]
	An03g03740 (*blg4*)	BGL	GH1	4.1	1.0	−1.2	5.8	9.6	nd	nr	[38]
	An12g01850 (*mndB*)	MND	GH2	**−2.5**	−1.5	1.0	1.9	13.4	XlnR	nr	[38]
	An02g07590	na	GH3	-	1.6	1.0	1.3	-	nd	nr	[38]
	An09g05880 (*agdE*)	AGD	GH31	1.0	1.2	1.6	1.3	−1.1	nd	nr	[41]
G-2	An09g01190 (*abnA*)	ABN	GH43	5.5	−2.4	1.8	1.4	13.1	nd	AraR, GaaR	[34, 83]
	An01g10350 (*lacB*)	LAC	GH35	1.9	**−2.7**	1.3	3.5	12.9	AraR	AraR, GaaR	[15, 34, 38]
	An18g05940 (*galA*)	GAL	GH53	-	-	-	-	-	nd	AraR, GaaR, RhaR	[34, 84]
	An01g12150 (*lacA*)	LAC	GH35	**−8.9**	**−6.3**	1.4	6.4	13.6	AraR, XlnR	AraR, XlnR	[23, 34, 85]
	An08g10780 (*gbgA*)	BXL	GH43	−1.6	**−29.7**	2.8	11.7	5.2	AraR	AraR	[34, 39]
	An08g01710 (*abfC*)	ABF	GH51	1.8	**−8.7**	1.8	7.7	31.1	AraR	AraR, GaaR	[15, 34, 39]
	An15g02300 (*abfB*)	ABF	GH54	2.2	**−6.2**	1.3	18.8	159.4	AraR	AraR	[34, 46]
G-3	An01g01320	AGL	GH27	-	**−3.8**	-	-	-	AraR	nr	[38]
	An17g00300 (*xarB*)	BXL/ABF	GH3	**−2.8**	−1.9	1.2	1.2	1.0	XlnR	nr	[47]
	An01g04880 (*axlB*)	AXL	GH31	−2.2	1.0	1.8	1.8	1.2	nd	nr	[41]
	An16g02760 (*afcA*)	AFC	GH95	**−5.2**	2.2	−1.3	1.7	5.9	XlnR	nr	[38]
	An01g03340 (*xgeA*)	XG-EGL	GH12	−1.1	−2.1	-	-	-	nd	nr	[38]
	An04g09690 (*pmeB*)	PME	CE8	-	-	-	-	-	nd	AraR, GaaR	[15, 34, 39]
	An04g09700 (*xghA*)	XGH	GH28	-	**−4.6**	-	-	-	AraR	AraR, GaaR	[34, 86]
	An01g11520 (*pgaI*)	PGA	GH28	-	-	-	-	-	nd	GaaR	[15, 34, 50]
	An19g00270 (*pelD*)	PEL	PL1	-	-	-	-	-	nd	AraR, GaaR, RhaR	[15, 34, 72, 87]

[a]Genes with expression value of <50 in all studied *A. niger* N402 cultures are not included in the table
[b]Enzyme codes are provided in Additional file 2
[c]Fold-change between *A. niger* N402 and the regulatory mutants grown on their relevant carbon source. - = expression value <50 in both N402 and regulatory mutant strain. Negative fold-changes >2.5 were considered proof of regulator function and are depicted in bold
na no assigned or predicted function, *nd* not detected, *nr* not reported

Table 10 Significantly expressed *A. niger* N402 genes from cluster H in the expression profiling tree

				Up/Down-regulated[c]					Regulated by		
	Gene[a]	Enzyme[b]	CAZy family	Δ*xlnR*	Δ*araR*	Δ*rhaR*	Δ*galX*	Δ*amyR*	This study	Literature	Reference
H-1	An04g03170	BGL	GH1	–	–	–	2.3	3.3	nd	nr	[38]
	An14g01770	BGL	GH3	1.0	1.5	−1.2	1.0	1.5	nd	AmyR	[8, 38]
	An11g00200 (*bg1M)*	BGL	GH3	–	3.7	–	–	2.2	nd	nr	[38]
	An12g02550 (*faeC*)	FAE	CE1	–	–	–	–	–	nd	nr	[39]
	An12g10390 (*faeB*)	FAE	SF1[d]	–	2.4	**−2.8**	1.1	6.5	RhaR	AraR, GaaR, RhaR	[10, 34, 84]
H-2	An15g04570	LPMO	AA9	–	**−2.7**	1.3	1.3	–	AraR	nr	[38]
	An14g04200 (*rhgB*)	RHG	GH28	–	–	–	–	–	nd	nr	[88]
	An16g09090	na	GH3	−1.1	1.2	−1.1	1.2	2.0	nd	nr	[38]
	An18g03570 (*bglA/bgl1*)	BGL	GH3	**−13.1**	2.9	−2.4	1.5	101.3	XlnR	nr	[53]
	An16g02100	EGL	GH5	−1.4	−1.1	1.3	1.3	1.7	nd	nr	[38]
	An04g02700	AGL	GH36	–	–	1.2	1.6	–	nd	AmyR	[8, 38]
	An18g04100 (*gp43*)	EXG	GH5	–	**−5.7**	−1.5	6.0	1.8	AraR	nr	[38]
	An01g06120 (*gdbA*)	GDB	GH13	−1.2	1.0	−2.1	1.9	6.3	nd	nr	[41]
	An01g10930 (*agdB*)	AGD	GH31	2.2	3.4	1.0	1.9	1.3	nd	AmyR	[8, 41]
	An11g03120 (*xynD*)	BXL	GH43	−1.7	4.5	1.2	4.9	22.5	nd	nr	[38]
	An06g00170 (*aglA*)	AGL	GH27	–	**−4.2**	−1.6	6.5	–	AraR	AmyR	[8, 89, 90]
	An02g11150 (*aglB*)	AGL	GH27	−1.5	1.1	2.5	9.6	5.5	nd	XlnR	[23]
	An02g13240 (*agdC*)	AGD	GH13	1.0	1.1	−1.2	2.8	2.4	nd	nr	[41]
	An05g02410	GUS	GH2	–	1.7	1.9	4.5	3.9	nd	nr	[38]
	An14g04190 (*gbeA*)	GBE	GH13	−1.7	1.3	−1.6	2.9	6.1	nd	nr	[41]

[a]Genes with expression value of <50 in all studied *A. niger* N402 cultures are not included in the table
[b]Enzyme codes are provided in Additional file 2
[c]Fold-change between *A. niger* N402 and the regulatory mutants grown on their relevant carbon source. - = expression value <50 in both N402 and regulatory mutant strain. Negative fold-changes >2.5 were considered proof of regulator function and are depicted in bold
[d]Sub-family (SF) classification of fungal FAEs according to Dilokpimol et al. [49]
na no assigned or predicted function, *nd* not detected, *nr* not reported

Table 11 Significantly expressed *A. niger* N402 genes from clusters A and I in the expression profiling tree

				Up/Down-regulated[c]					Regulated by		
	Gene[a]	Enzyme[b]	CAZy family	Δ*xlnR*	Δ*araR*	Δ*rhaR*	Δ*galX*	Δ*amyR*	This study	Literature	Reference
A	An15g00320 (*sucB*)	SUC	GH32	1.5	−1.3	−2.1	–	1.7	nd	InuR	[9, 52]
	An15g07160 (*pelF*)	PEL	PL1	1.5	1.4	−2.1	−1.1	1.7	nd	GaaR	[15, 35, 72]
I	An08g05230	LPMO	AA9	–	–	**−2.9**	−1.2	1.3	RhaR	nr	[38]
	An14g02670	LPMO	AA9	−1.1	−1.7	−1.6	1.0	–	nd	XlnR	[21, 38]
	An03g05380	XG-EGL	GH12	–	–	–	–	–	nd	nr	[38]
	An10g00870 (*plyA*)	PLY	PL1	–	–	–	–	–	nd	nr	[91]
	An02g10550 (*abnC*)	ABN	GH43	**−27.8**	15.2	1.4	−1.8	49.2	XlnR	nr	[39]
	An15g03550	ABN	GH43	–	10.0	–	–	6.0	nd	nr	[38]
	An07g07630	BGL	GH3	–	2.9	–	–	3.1	nd	nr	[4]
	An11g06080	BGL	GH3	–	2.0	1.4	2.4	3.8	nd	nr	[4]

[a]Genes with expression value of <50 in all studied *A. niger* N402 cultures are not included in the table
[b]Enzyme codes are provided in Additional file 2
[c]Fold-change between *A. niger* N402 and the regulatory mutants grown on their relevant carbon source. - = expression value <50 in both N402 and regulatory mutant strain. Negative fold-changes >2.5 were considered proof of regulator function and are depicted in bold
nd not detected, *nr* not reported

regulated [15]. In addition, regulation by AraR has been reported for *pgxC*, and by AraR and RhaR for *pmeA* [34]. Also, gene An02g00140, which encodes a putative β-xylosidase, showed significant expression (Table 5). The expression profiles of *pelA*, *pmeA* and *pgaX* genes were previously shown to cluster and these genes were suggested to play a major role in the initial degradation of pectin [35]. This is also supported by the results reported from sugar beet pectin [15]. In line with our results, strong induction on D-galacturonic acid and polygalacturonic acid has been reported for *pgxB*, *pgxC* and *pgaX*, while lower expression has been observed for *pgxA* on these substrates [15, 40]. The *pelA* gene was well expressed on all tested substrates, but its highest expression was detected on polygalacturonic acid (Additional file 1). In agreement with the previous studies [15, 39], the *rglA* gene was expressed on D-galacturonic acid, polygalacturonic acid and galactan, but not on L-rhamnose. The GaaR-regulated *pmeA* gene [15] was slightly induced on D-galacturonic acid and polygalacturonic acid in our study and that of de Vries et al. [35]. In contrast to the results of Kowalczyk et al. [34], the regulation of *pmeA* by AraR or RhaR was not detected. The function of five out of eight putative α-rhamnosidase encoding genes (i.e. An01g06620, An12g05700, An07g00240, An10g00290 and An04g09070) in sub-cluster C-2 is supported by our analysis as they are specifically induced on L-rhamnose and are under control of RhaR [10, 38]. In addition, An12g05700 is controlled by GaaR and RhaR, and An18g04800 by GaaR [34].

The pectinolytic genes found in cluster C were expressed on L-rhamnose, D-galacturonic acid and/or polygalacturonic acid, suggesting that these genes encode initial pectin degrading enzymes. Pectinolytic genes that showed no significant, or constitutive expression, may be induced on pectin-related substrates after longer incubation times. Expression of *plyA*, *pgaII*, *pgaB*, *pgaD*, *pgaE*, *pelB*, *pelC* and *pelF* was low or not significant on all substrates in our study. However, expression of these genes on D-galacturonic acid, polygalacturonic acid and sugar beet pectin has been reported to increase in time [35] and *pgaB*, *pgaE* and *pelF* have been shown to be regulated by GaaR [15, 34].

Constitutively expressed genes clustered with genes involved in starch degradation

In cluster D, sub-cluster D-1 contains nine significantly expressed genes encoding enzymes from different GH families, while in sub-cluster D-2 six genes are present that mainly encode enzymes from GH families assigned to starch degradation (GH13, 15 and 31) (Table 6). The genes of sub-cluster D-1 were not down-regulated in any of the tested regulatory mutant strains, indicating that they are not regulated by these transcriptional activators (Table 6). They show a relatively distant separation from each other, and most showed low, but similar expression levels on all substrates (Additional file 1) indicating that the genes in sub-cluster D-1 are likely constitutively expressed. Indeed, the *abnB* gene, present in sub-cluster D-1, was previously reported to be constitutively expressed on D-fructose, D-xylose, sorbitol, L-rhamnose, D-galacturonic acid, polygalacturonic acid and sugar beet pectin [39].

The sub-cluster D-2 contains genes that are involved in starch degradation and are down-regulated in the Δ*amyR* strain. Two *glaA* and *agdA* genes, encoding a glucoamylase and an α-glucosidase, respectively [38, 41], showed high expression on all substrates, while the highest expression levels were detected in N402 on maltose (Additional file 1), in line with the previous study [41]. Gene *aamA*, which encodes an acid α-amylase [42], has also been reported to be highly expressed on maltose [41], but was expressed at a much lower level in our study. For this gene, significant expression was also detected on L-arabinose, polygalacturonic acid and sugar beet pulp (Additional file 1). The similar expression patterns and the down-regulation of *glaA*, *agdA* and *aamA* genes in the Δ*amyR* strain (Table 6) indicates their co-regulation by AmyR, as has been suggested by Yuan et al. [41]. All three genes were up-regulated in the Δ*galX* mutant on D-galactose to a higher level than the expression on maltose in N402 (Additional file 1). The α-amylase gene *amyC* was also most highly expressed on D-galactose in the Δ*galX* mutant. Like *glaA*, *agdA* and *aamA*, expression of this gene was reported to be reduced in the Δ*amyR* strain [41]. However, the expression profile of *amyC* in our study differs from the other three amylolytic genes, because a similar expression level of this gene was found on D-maltose, L-rhamnose and guar gum, making its induction on D-maltose less specific (Additional file 1). In a previous study, expression of *amyC* was similar on D-xylose and D-maltose after 2 h of incubation, but the gene was not expressed after 8 h on xylose, while its expression on maltose was still detected [41].

Low expression for *agtB* encoding a putative 4-α-glucanotransferase was detected on all substrates, with only significant expression levels and down-regulation in the Δ*amyR* strain (Additional file 1, Table 6). This data is in contrast with a previous study [41], where expression was only detected after 8 h on D-maltose and *agtB* was reported to be AmyR independent. Co-expression of *agtB* and *agsC*, encoding a putative α-glucan synthase, has previously been observed [41]. Even though *agsC* was not significantly expressed in our study (Additional file 2), it did cluster with *agtB* in our initial correlation analysis (Additional files 1 and 2).

An03g05530 is also found in sub-cluster D-2, even though its highest expression level was detected on L-rhamnose and D-galacturonic acid. However, this gene is significantly down-regulated in the Δ*amyR* strain, which may explain its presence in sub-cluster D-2.

InuR-regulated inulinolytic genes were co-expressed on sucrose and inulin

Cluster E contains eight significantly expressed genes that have relatively distant positions in the expression profile tree (Additional file 1). Sub-cluster E-1 consists of only *eglD* and *pelB* encoding a putative LPMO and a pectin lyase, respectively (Table 7), that showed a low overall expression. While this is in contrast to the reported lack of expression for *pelB* in *A. niger* cultures on sugar beet pectin, galacturonic acid, rhamnose and xylose [39], the low expression we observed may indicate that expression levels of *pelB* are always around the detection cut-off. Sub-cluster E-2 contains six genes that were expressed on guar gum, inulin, sugar beet pulp and/or sucrose (Table 7, Additional file 1). High expression on inulin and to a lesser extent on guar gum was observed for a putative exo-inulinase encoding gene *inuE*, which clustered with an endo-inulinase encoding *inuA*, but expression levels of the latter gene were much lower. In addition to *inuE* and *inuA*, sub-cluster E-2 contains the extracellular inulinolytic gene *sucA*. These genes were all regulated by InuR, and co-regulation and expression on sucrose and inulin was previously reported for these genes [43]. The more distant position of *sucA* in the expression profile tree can be explained by its relative expression levels on sucrose, inulin and sugar beet pulp, the latter resulting in the highest expression for *sucA*. An08g01100 and to a lesser extent An07g09760 were specifically induced on guar gum, but are located close to *inuE* and *inuA* in the expression profile tree (Additional file 1). The correlation analysis also demonstrated which substrates are most similar when the expression of all the tested genes was taken into account. Guar gum was most closely related to inulin, sucrose and sugar beet pulp. The sugar beet pulp used in this study contains significant amounts of sucrose (data not shown), which explains the clustering of this substrate with sucrose and inulin. Our results suggest that guar gum may also contain some traces of sucrose, even though this was not reported by the supplier.

Other inulinolytic genes described for *A. niger*, i.e. *sucB*, *sucC* and *inuQ*, were not present in cluster E. Absence of expression of the intracellular invertase encoding *sucC* gene, and *inuQ*, which was described to be a pseudogene, confirmed a previous study [44]. The other intracellular invertase encoding gene, *sucB*, was reported to have an overall low expression on other substrates than sucrose and inulin [44], which was also confirmed by our study.

Only five significantly expressed genes are positioned in cluster F (Additional file 1, Table 8), with only one gene, An09g05350, in sub-cluster F-2. It was expressed on D-glucose, D-fructose, D-maltose and rhamnogalacturonan. The four genes that form sub-cluster F-1 differ in their expression profile, and therefore the reason for the clustering of these genes may be that they did not fit into any of the other clusters. It should be noted that the genes of cluster F are distantly separated from each other within the expression profiling tree (Additional file 1). A putative α-glucan synthase encoding gene (*agsE*) showed high expression levels on all substrates in N402, which confirms a previous study [41]. However, expression of this gene was strongly reduced in the Δ*amyR* strain (Table 8), which was not observed in the study of Yuan et al. [41]. The opposite was found for α-galactosidase encoding *aglC* that has been reported to be under control of AmyR [41], while our study only detected significant down-regulation in the Δ*rhaR* strain. Expression of endoarabinanase encoding *abnD* was previously reported to be constitutive [39], but more recently it was shown to be GaaR-dependent on D-galacturonic acid and GaaR and RhaR-dependent on sugar beet pectin [15, 34]. However, we only detected significant expression levels of *abnD* on D-maltose in N402 and down-regulation in the Δ*amyR* strain, suggesting control by this regulator.

Genes related to degradation of pectin side chains cluster separately from those acting on the pectin main chain

Most of the significantly expressed genes of cluster G (Table 9) were highly expressed on D-galacturonic acid and polygalacturonic acid (Additional file 1). The difference between these genes and D-galacturonic and polygalacturonic acid induced genes of cluster C is that the cluster G genes are less specifically induced on D-galacturonic acid and polygalacturonic acid, as they also show high expression levels on other carbon sources. Cluster G, the largest cluster detected with 23 genes, can be divided into the sub-clusters G-1, G-2, and G-3 (Additional file 1).

Expression of some of the genes in cluster G has been previously analyzed on D-fructose, L-rhamnose, D-xylose, sorbitol, D-galacturonic acid, polygalacturonic acid and sugar beet pectin [15, 39, 40]. Specific induction has been observed for *pmeB*, *xghA*, *pgaI*, *abfB*, *abfC*, *lacA*, *lacB*, *galA* and *abnA* on D-galacturonic acid, polygalacturonic acid and sugar beet pectin [15, 45], and all these genes have been shown to be GaaR-controlled, except *lacA* and *abfB* [15, 34]. Furthermore, the *abfB* and *abfC* genes were also highly expressed on D-xylose [39, 46]. In our study, induction of these genes on D-galacturonic acid and polygalacturonic acid was also observed. In addition, *abfB*, *abfC*, *lacA*, *lacB*, *galA* and *abnA*, all members of sub-cluster G-2, were highly expressed on galactan (Additional file 1). Co-regulation of *abfB*, *abnA* and *galA* was suggested previously [33, 35], but in our study only *abnA* and *galA* fall in the same cluster, while the expression profile of *abfA* is different.

High expression for most of the sub-cluster G-2 genes, except *lacA* and *galA*, was observed on arabinan, while high expression on L-arabinose was observed for *abfB*, *abfC* and *lacA*, all of which were down-regulated in the Δ*araR* strain. The genes of this sub-cluster all encode enzymes that could be involved in the degradation of the pectinolytic side chains, suggesting a strong link between function and expression.

High expression levels of the genes of sub-cluster G-1 were detected on polygalacturonic acid, but to a much lower extent than for the sub-cluster G-2 genes. The highest expression for three genes of sub-cluster G-1, *mndA*, *mndB* and *bgl4*, was found on mannan. The *mndA* gene encodes a β-mannosidase [44], involved in mannan degradation, while *mndB* and *bgl4* encode a putative β-mannosidase and β-glucosidase, respectively. Their co-expression with *mndA* supports these functions as both activities are needed for complete degradation of galactoglucomannan. However, these genes were not inducted by mannose. This is in line with the ManR/ClrB regulator from *A. oryzae* induced by mannobiose, but not by mannose [11, 12]. The highest expression for the other genes of this sub-cluster, *agdE*, An17g00520 and An02g07590, was detected on polygalacturonic acid.

The highest expression levels of all the genes of sub-cluster G-3 were found on polygalacturonic acid. The GaaR, AraR and RhaR-regulated *pelD* gene [15, 34] was specifically induced on polygalacturonic acid, in contrast to a previous study where this gene was reported to be non-expressed [39]. Expression of *pgaI*, which is under control of GaaR [15], and *pmeB* and *xghA*, which are under control of GaaR and AraR [15, 34], has previously been reported on D-galacturonic acid and polygalacturonic acid [39], which was confirmed in our study. Three genes of sub-cluster G-3, *xarB*, *axlB* and *afcA*, which encode a putative bi-functional xylosidase/arabinofuranosidase [47], an α-glucosidase and an α-fucosidase, respectively, were down-regulated in the Δ*xlnR* strain, suggesting control by XlnR. One gene of this sub-cluster, *xghA*, was down-regulated in the Δ*araR* strain, suggesting regulation by AraR in line with Kowalczyk et al. [34]. As mentioned earlier, the polygalacturonic acid specific induction of arabinanolytic and xylanolytic genes may be due to impurity of the substrate with small traces of D-xylose and L-arabinose.

Cluster H contains a diverse set of genes that are expressed on a broad range of substrates

Six and 15 significantly expressed genes form sub-cluster H-1 and H-2, respectively (Additional file 1, Table 10). The *faeB* gene was expressed at a basal level on L-rhamnose, D-xylose, sorbitol, D-fructose D-galacturonic acid, polygalacturonic acid and sugar beet pectin, while *faeC*, which is also found in this sub-cluster, was not expressed on these substrates [39]. The genes of sub-cluster H-1 were all specifically induced on ferulic acid. While induction of *faeB* on ferulic acid has previously been reported [48], the *faeC* was also induced on this substrate, suggesting co-regulation of these two feruloyl esterase encoding genes, which was confirmed by a recent study [49]. Interestingly, the other genes of the sub-cluster H-1 specifically induced on ferulic acid encode putative β-glucosidases (An04g03170, An14g01770 and *bgm1*) and a putative LPMO (An15g04570).

The genes of sub-cluster H-2 were expressed at a constant level on most carbon sources tested, but showed low expression on D-glucose, D-fructose, sucrose and sugar beet pulp (Additional file 1). As mentioned before, the sugar beet pulp used in this analysis contains sucrose (data not shown). These genes may therefore be under strong carbon catabolite repression. Binding sites for CreA have been found in the promoter regions of all these genes [38], and low overall expression of a putative α-glucosidase encoding *agdC* has previously been described [41]. Another α-glucosidase encoding gene, *agdB*, has been reported to be strongly induced on D-maltose and down-regulated in the Δ*amyR* strain [41]. Our study, however, revealed that this gene was highly expressed on most carbon sources tested and no down-regulation in the Δ*amyR* strain was observed (Table 10).

Two clusters of putatively not co-expressed genes were detected

In clusters A and I, only a small number of genes (two and eight, respectively) were significantly expressed (Additional file 1, Table 11). Furthermore, the genes in clusters A and F share no specific trends in their expression profiles and are relatively distantly separated from each other within the expression profiling tree (Additional file 1), and are probably not co-expressed.

In cluster A, the significantly expressed genes, *sucB* and *pelF*, encode enzymes from CAZy families GH32 and PL1, respectively (Table 11, Additional file 1). The overall expression of these genes was very low on all substrates. Furthermore, the genes were not significantly down- or up-regulated in the studied regulatory mutant strains, indicating that these genes are not regulated by any of these transcriptional activators. The low overall expression of *pelF*, a gene encoding a putative pectin lyase, has been reported previously [39] and it has been shown to be regulated by GaaR [15]. Notably, *pelF* did not cluster with any of the other pectinolytic genes in our data. In contrast, *pelF* clustered distantly with the other pectinolytic genes in a previous study [35], which, however, included a smaller set of genes and a more focused set of growth conditions that may explain the differences with our study. In addition, the previous hierarchical clustering suggested induction of *pelF* during starvation or derepressed conditions [15]. Gene *sucB* encodes an intracellular invertase with transfructosylation activity [50, 51]. Its expression profile was distinct from other inulinolytic genes (Additional file 1). The *sucB* gene has been reported to be under control of the inulinolytic regulator InuR [52] and to be constitutively expressed at low level [43]. In our study, significant expression of *sucB* was found on inulin, which supports regulation by InuR. In addition, *sucB* expression was observed on D-maltose in the Δ*amyR* strain (Additional file 1). This suggests interaction between AmyR and InuR, similarly as was described for XlnR and AraR in *A. niger* [22, 53].

All genes in cluster I were expressed at low level on D-maltose and sugar beet pulp in the N402 strain. Some of these genes (i.e. An15g03550, *abnC*, An07g07630 and An11g06080) were up-regulated in the Δ*amyR* strain. The highest expressed gene of this cluster was a putative endoarabinanase encoding *abnC*, which was highly expressed on all the tested substrates except D-maltose and sugar beet pulp (Additional file 1). Expression levels of this gene have previously been reported to be elevated after 24 h on D-fructose, L-rhamnose, sorbitol, D-xylose and D-galacturonic acid [39]. The *abnC* gene was significantly down-regulated in the Δ*xlnR* strain on D-xylose, which indicates that this gene is regulated by XlnR (Table 11). The *abnC* gene and An15g03550, both encoding putative endoarabinanases from family GH43, were highly expressed on galactan, while An15g03550 was also highly expressed on mannan. The highest expression levels of An08g05230 and An14g02670 encoding putative LPMOs from family GH61, An03g05380 encoding putative xyloglucan-active endoglucanase and *plyA* encoding putative pectate lyase were detected on arabinan (Additional file 1).

Upregulation of genes in regulatory mutants suggests interaction between the different regulatory systems

While the down-regulation of gene expression in *A. niger* regulatory mutants compared to the wild type strain can be taken as evidence of control by this regulator, we surprisingly also found a significant number of genes for which the expression in a regulatory mutant was higher than in the wild type. While in most cases this was a moderate increase (less than 3-fold,), for 46 genes the difference was higher and 13 of these had fold-changes >10. The largest set of strongly upregulated genes was observed in the *amyR* mutant on maltose. Interestingly, this seems to especially affect L-arabinose related genes as the fold-change for *abfA*, *abfB*, *abfC*, *abnA*, *abnC* and An159g3550 (putative ABN) was 38, 160, 31, 13, 49 and 6, respectively. In addition, *bglA* was also 100-fold upregulated. Antagonistic interactions between regulators have been observed before, in particular for the two pentose-related regulators XlnR and AraR [54]. However, more recently, this was also observed for three pectinolytic regulators, GaaR, AraR and RhaR [34], suggesting that this is more common phenomenon has been so far considered. The nature of the antagonistic interaction and whether this is a direct or indirect is not clear at this point and requires further study.

Conclusions

This study aimed to reveal co-expression patterns of plant biomass polysaccharide degradation related genes from *A. niger*, using a more global approach than is usually performed by including a wide range of carbon sources, as well as five regulatory mutants, thus generating an unprecedented view of this system. The broader range of substrates revealed the highly complex expression patterns of these CAZy genes, and demonstrated that the focused analyses of the transcriptional regulators involved in this process that have been identified so far only revealed initial indications of the overall regulatory system. In fact, many of the genes tested in this study were shown to be under control of more than one regulator (Fig. 1a). Interestingly, the role of the regulators appears to be less linked to a specific polysaccharide when the genes encoding a certain enzyme activity and the regulators that act on them were combined (Fig. 1b). This could imply that the role of the enzymes may in fact be broader than currently assumed. E.g. the role of BXL in removing xylose from xylogalacturonan could explain the influence of GaaR on the expression of some BXL-encoding genes.

Previous studies in *T. reesei* and *N. crassa* also addressed induction of CAZy genes under different conditions and in some cases by using deletion mutants of plant biomass related regulators [55–58]. However, these studies, similar to previous studies in *Aspergillus*, used a limited number of carbon sources and did not compare different regulatory mutants. It can therefore be expected that also in these studies the complexity of the regulatory network driving plant biomass degradation is underestimated. More detailed studies in *A. niger* as well as other fungi are needed to fully grasp the organization of the regulatory network and reveal the differences between fungal species.

Methods

Strains, media and culture conditions

The *A. niger* strains used in this study are listed in Table 2. Strains were grown at 30 °C on minimal medium (MM) or complete medium (CM) [51] either or not containing 1.5% agar. Liquid cultures were grown on a rotary shaker at 250 rpm. Pre-cultures for RNA isolation were grown for 16 h in 1 L Erlenmeyer flasks that

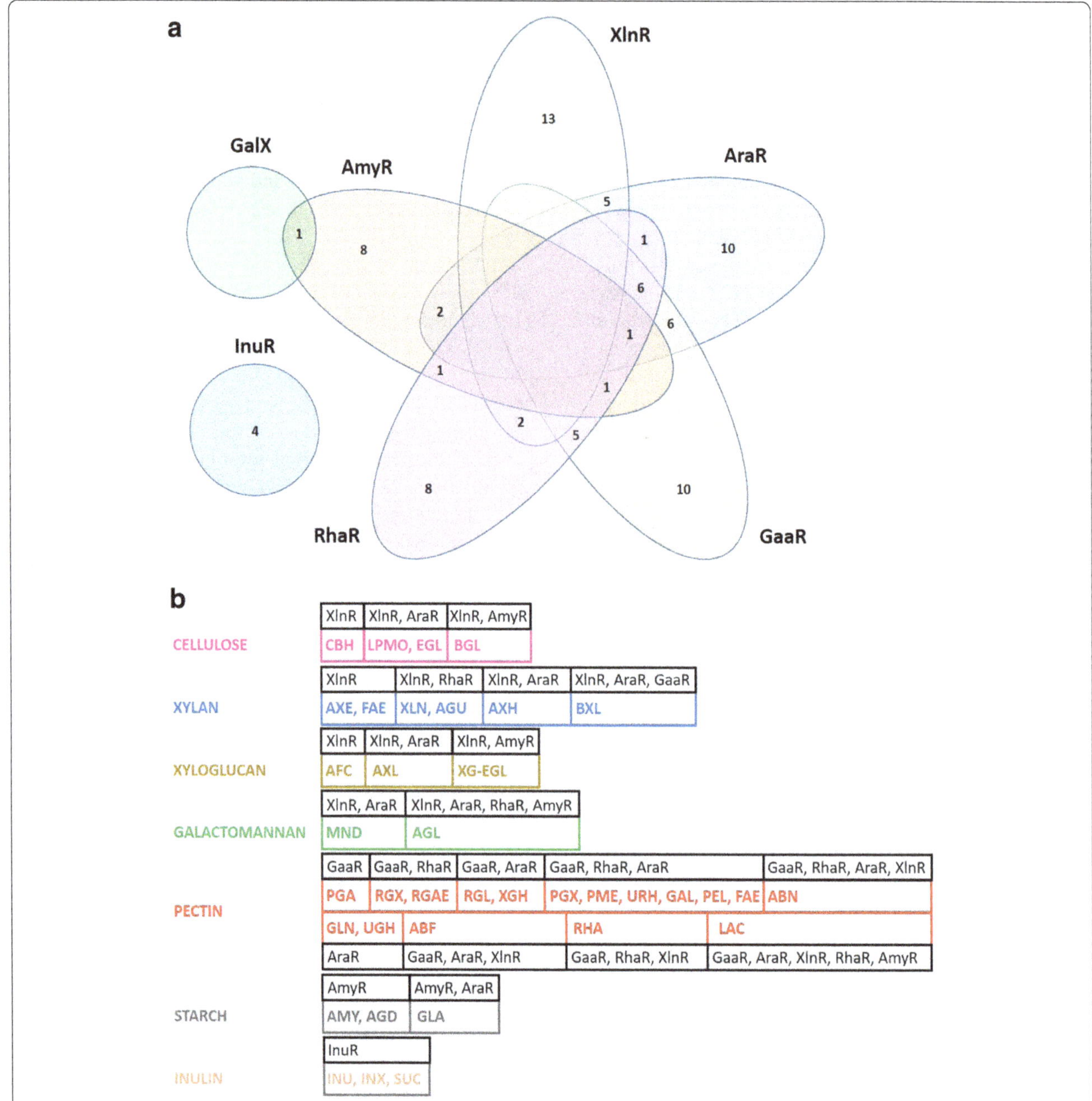

Fig. 1 Global analysis of the expression profiles of CAZy genes related to plant polysaccharide degradation. **a** Number of genes under control of one or more regulators are indicated in a VENN diagram. **b** Comparison of the influence of regulators on enzyme activities linked to the polysaccharide they act on. Regulatory effects on individual genes encoding the same enzyme activity were combined in the boxes

contained 250 ml CM supplemented with 2% D-fructose. Mycelium was washed with MM and 1 g (wet weight) aliquots were transferred for 2 h to 250 ml Erlenmeyer flasks containing 50 ml MM supplemented with 25 mM mono- or disaccharide or ferulic acid, or mixture of 25 mM L-rhamnose and 25 mM D-galacturonic acid, or 1% polysaccharide or complex plant biomass (Table 3). The only exceptions were D-maltose cultures of N402 and Δ*amyR* strains that were incubated for 4 h and for which 1% maltose was used. These data originate from a different study [8], but were included to help with the grouping of the genes and assess the AmyR effect. Mycelium was harvested by vacuum filtration, dried between towels and frozen in liquid nitrogen. While N402 liquid cultures were performed on all carbon sources listed in Table 3 as well as on the mixture of L-rhamnose and D-galacturonic acid, the regulatory mutant strains Δ*xlnR*, Δ*araR*, Δ*amyR*, Δ*rhaR* and Δ*galX* were grown on D-xylose, L-arabinose, maltose, L-rhamnose and D-galactose, respectively, and L-rhamnose and D-galacturonic acid. All cultures were performed as biological duplicates.

Microarray processing

RNA isolation and microarray hybridization were performed as described previously [59]. In brief, RNA for microarray analysis was extracted using TRIzol reagent (Invitrogen) and purified using TRIzol® Plus RNA Purification Kit (Sigma-Aldrich) according to the instructions of the manufacturer. The concentration of RNA was calculated from the absorbance at 260 nm in a spectrophotometer (Biochrom Libra S22). The quality of the RNA was analyzed with an Agilent 2100 Bioanalyzer using a RNA6000 LabChip kit (Agilent Technology). Microarray hybridization using the Affymetrix GeneChips *A. niger* Genome Array was performed at GenomeScan (Leiden, The Netherlands).

Transcriptome analysis

Microarray data was analyzed using the Bioconductor tool package version 2.8 (http://www.bioconductor.org/) together with house-made Perl (version .5.0) and Python (version 3.0) scripts. Probe intensities were normalized for background by the robust multi-array average (RMA) method [60] using the R statistical language and environment [61]. This method makes use of only perfect match (PM) probes.

Normalization was processed by the quantiles algorithm. The median polish summary method [62] was used to calculate the gene expression values. Further statistical analyses were performed with the CyberT tool package using multiple testing (http://cybert.ics.uci.edu/). BayesAnova and paired BayesT-test tests were performed on each gene through pairing carbon sources, PPDE (Posterior Probability of Differential Expression) analysis and multiple hypothesis testing correction are performed on the *p*-values [63]. Adjusted cut off value of $p < 0.05$ was used to determine the statistical significance of gene expression difference. Reproducibility of the replicates was verified by PCA analysis (Additional file 4). Genome scale PCA analysis was performed with the gene expression values of the different samples. The PCA was generated using R (v3.40) statistical language and environment, the PCA function from FactoMineR package (v1.35) and plotted using ggplot2 package (v 2.2.1). Replicates are plotted using the same color. Due to the large amount of data, the calculation of the matrix was not possible.

Gene expression clustering, visualization and annotation

Hierarchical clusters were made using complete linkage with the normalized expression data from selected CAZyme encoding genes by calculating the Pearson correlation distances [64]. Clusters were set manually based on the branch-length differences of the gene-tree. The genes were selected based on the annotation of the CAZy families and their (putative) role in plant biomass degradation. Clusters and expression correlation profiles were visualized by Genesis [65]. Genes with an expression value <50 were colored dark blue, the ones >1000 were colored red and the values ≥50 and ≤1000 were colored by a gradient of these 2 colors.

Gene functional annotations were based on previous study [1]. When the data of this study suggested a different function, this was verified by performing phylogenetic analysis of the CAZy family this gene belongs to. The phylogeny analysis was performed using all the *A. niger* genes of the corresponding family together with all functionally characterized fungal members of that family, which allowed us to verify to which activity this gene clustered.

Additional files

Additional file 1: Expression profiling tree containing 168 *A. niger* genes encoding putative CAZymes (www.cazy.org). Clusters A-I can be distinguished. (PDF 1780 kb)

Additional file 2: Expression of selected CAZy genes. (XLSX 215 kb)

Additional file 3: Significantly and not significantly expressed genes encoding CAZymes in *A. niger* CBS513.88 in this study. (XLSX 19 kb)

Additional file 4: PCA analysis of the gene expression values of the biological duplicate samples revealing the reproducibility of the duplicates. (PDF 90 kb)

Abbreviations

AA: Auxiliary activity; CAZy: Carbohydrate-active enzyme; CE: Carbohydrate esterase; CM: Complete medium; GH: Glycoside hydrolase; MM: Minimal medium; PL: Polysaccharide lyase; RMA: Robust multi-array average

Acknowledgements

Not applicable.

Authors' contributions

BSG performed part of the experiments and participated in the analysis and writing of the manuscript. MRM participated in the analysis and writing of the manuscript. JK analyzed part of the data. MZ performed the bioinformatics analysis. IBG performed part of the experiments. RPdV designed the study and participated in the analysis and writing of the manuscript. All authors read and approved the final manuscript.

Funding

BS and IBG were supported by a grant of the Dutch Technology Foundation STW, Applied Science division of NWO and the Technology Program of the Ministry of Economic Affairs UGC 07938 to RPdV. MZ was supported by a grant from the Netherlands Organisation for Scientific Research (NWO) and the Netherlands Genomics Initiative 93,511,035 to RPdV.

Competing interests

The authors declare that they have no competing interests.

Author details

[1]Fungal Physiology, Westerdijk Fungal Biodiversity Institute, Uppsalalaan 8, 3584, CT, Utrecht, The Netherlands. [2]Microbiology, Utrecht University, Padualaan 8, 3584, CH, Utrecht, The Netherlands. [3]Fungal Molecular Physiology, Utrecht University, Uppsalalaan 8, 3584, CT, Utrecht, The Netherlands. [4]Department of Food and Environmental Sciences, Division of Microbiology and Biotechnology, Viikki Biocenter 1, University of Helsinki, Helsinki, Finland. [5]Current affiliation: ATGM, Avans University of Applied Sciences, Lovensdijkstraat 61–63, 4818, AJ, Breda, The Netherlands. [6]Current affiliation: Center for Structural and Functional Genomics, Concordia University, 7141 Sherbrooke St. W, Montreal QC, Canada.

References

1. Benoit I, Culleton H, Zhou M, DiFalco M, Aguilar-Osorio G, Battaglia E, et al. Closely related fungi employ diverse enzymatic strategies to degrade plant biomass. Biotechnol Biofuels. 2015;8.
2. de Vries RP, Visser J. *Aspergillus* enzymes involved in degradation of plant cell wall polysaccharides. Microbiol Mol Biol Rev. 2001;65:497–522.
3. Lombard V, Ramulu HG, Drula E, Coutinho PM, Henrissat B. The carbohydrate-active enzymes database (CAZy) in 2013. Nucleic Acids Res. 2014;42:D495.
4. Pel HJ, de Winde JH, Archer DB, Dyer PS, Hofmann G, Schaap PJ, et al. Genome sequencing and analysis of the versatile cell factory *Aspergillus niger* CBS 513.88. Nat Biotech. 2007;25:221–31.
5. Todd R, Zhao M, Ohm RA, Leeggangers H, Visser L, de Vries R. Prevalence of transcription factors in ascomycete and basidiomycete fungi. BMC Genomics. 2014;15:214.
6. Kowalczyk JE, Benoit I, de Vries RP. Regulation of plant biomass utilization in *Aspergillus*. Adv Appl Microbiol. 2014;88:31–56.
7. van Peij NNME, Visser J, de Graaff LH. Isolation and analysis of *xlnR*, encoding a transcriptional activator co-ordinating xylanolytic expression in *Aspergillus niger*. Mol Microbiol. 1998;27:131–42.
8. vanKuyk PA, Benen JAE, Wösten HAB, Visser J, de Vries RPA. Broader role for AmyR in *Aspergillus niger*: regulation of the utilisation of D-glucose or D-galactose containing oligo- and polysaccharides. Appl Microbiol Biotechnol. 2012;93:285–93.
9. Yuan X-L, Roubos JA, van den Hondel CAMJJ, Ram AFJ. Identification of InuR, a new Zn(II)2Cys6 transcriptional activator involved in the regulation of inulinolytic genes in *Aspergillus niger*. Mol Gen Genomics. 2008;279:11–26.
10. Gruben BS, Zhou M, Wiebenga A, Ballering J, Overkamp KM, Punt PJ, et al. *Aspergillus niger* RhaR, a regulator involved in L-rhamnose release and catabolism. Appl Microbiol Biotechnol. 2014;98:5531–40.
11. Ogawa M, Kobayashi T, Koyama Y. ManR, a transcriptional regulator of the β-mannan utilization system, controls the cellulose utilization system in *Aspergillus oryzae*. Biosci Biotechnol Biochem. 2013;77:426–9.
12. Ogawa M, Kobayashi T, Koyama Y. ManR, a novel Zn(II)2Cys6 transcriptional activator, controls the β-mannan utilization system in *Aspergillus oryzae*. Fungal Genet Biol. 2012;49:987–95.
13. Raulo R, Kokolski M, Archer DB. The roles of the zinc finger transcription factors XlnR, ClrA and ClrB in the breakdown of lignocellulose by *Aspergillus niger*. AMB Express. 2016;6:5.
14. Gruben BS, Zhou M, de Vries RP. GalX regulates the D-galactose oxido-reductive pathway in *Aspergillus niger*. FEBS Lett. 2012;586:3980–5.
15. Alazi E, Niu J, Kowalczyk JE, Peng M, Aguilar Pontes MV, van Kan JAL, et al. The transcriptional activator GaaR of *Aspergillus niger* is required for release and utilization of D-galacturonic acid from pectin. FEBS Lett. 2016;590:1804–15.
16. de Vries RP. Regulation of *Aspergillus* genes encoding plant cell wall polysaccharide-degrading enzymes; relevance for industrial production. Appl Microbiol Biotechnol. 2003;61:10–20.
17. Gomi K, Akeno T, Minetoki T, Ozeki K, Kumagai C, Okazaki N, et al. Molecular cloning and characterization of a transcriptional activator gene, *amyR*, involved in the amylolytic gene expression in *Aspergillus oryzae*. Biosci Biotechnol Biochem. 2000;64:816–27.
18. Tani S, Katsuyama Y, Hayashi T, Suzuki H, Kato M, Gomi K, et al. Characterization of the *amyR* gene encoding a transcriptional activator for the amylase genes in *Aspergillus nidulans*. Curr Genet. 2001;39:10–5.
19. Gielkens MMC, Dekkers E, Visser J, de Graaff LH. Two cellobiohydrolase-encoding genes from *Aspergillus niger* require D-xylose and the xylanolytic transcriptional activator XlnR for their expression. Appl Environ Microbiol. 1999;65:4340–5.
20. Hasper AA, Dekkers E, van Mil M, van de Vondervoort PJI, de Graaff LH. EglC, a new endoglucanase from *Aspergillus niger* with major activity towards xyloglucan. Appl Environ Microbiol. 2002;68:1556–60.
21. van Peij NNME, Gielkens MMC, de Vries RP, Visser J, de Graaff LH. The transcriptional activator XlnR regulates both xylanolytic and endoglucanase gene expression in *Aspergillus niger*. Appl Environ Microbiol. 1998;64:3615–9.
22. Battaglia E, Visser L, Nijssen A, van Veluw GJ, Wösten HAB, de Vries RP. Analysis of regulation of pentose utilisation in *Aspergillus niger* reveals evolutionary adaptations in Eurotiales. Stud Mycol. 2011;69:31–8.
23. de Vries RP, van den Broeck HC, Dekkers E, Manzanares P, de Graaff LH, Visser J. Differential expression of three α-galactosidase genes and a single β-galactosidase gene from *Aspergillus niger*. Appl Environ Microbiol. 1999;65: 2453–60.
24. Battaglia E, Zhou M, de Vries RP. The transcriptional activators AraR and XlnR from *Aspergillus niger* regulate expression of pentose catabolic and pentose phosphate pathway genes. Res Microbiol. 2014;165:531–40.
25. Christensen U, Gruben BS, Madrid S, Mulder H, Nikolaev I, de Vries RP. Unique regulatory mechanism for D-galactose utilization in *Aspergillus nidulans*. Appl Environ Microbiol. 2011;77.
26. de Vries R, Riley R, Wiebenga A, Aguilar-Osorio G, Amillis S, Akemi Uchima C, et al. Comparative genomics reveals high biological diversity and specific adaptations in the industrially and medically important fungal genus *Aspergillus*. Genome Biol. 2017;18:28.
27. Khosravi C, Kun RS, Visser J, Aguilar-Pontes MV, de Vries RP, Battaglia E. In vivo functional analysis of L-rhamnose metabolic pathway in *Aspergillus niger*: a tool to identify the potential inducer of RhaR. BMC Microbiol. 2017;17:214.
28. Sloothaak J, Odoni DI, dos Santos VAP M, Schaap PJ, Tamayo-Ramos JA. Identification of a novel L-rhamnose uptake transporter in the filamentous fungus *Aspergillus niger*. PLoS Genet. 2016;12:e1006468.
29. Alazi E, Khosravi C, Homan T, du Pré S, Arentshorst M, Di Falco M, et al.. The pathway intermediate 2-keto-3-deoxy-L-galactonate mediates the induction of genes involved in D-galacturonic acid release and catabolism. FEBS Lett. 2017;591:1408–18.
30. Gielkens M, Visser J, de Graaff L. Arabinoxylan degradation by fungi: characterization of the arabinoxylan-arabinofuranohydrolase encoding genes from *Aspergillus niger* and *Aspergillus tubingensis*. Curr Genet. 1997;31:22–9.
31. de Vries RP, Visser J, de Graaff LH. CreA modulates the XlnR-induced expression on xylose of *Aspergillus niger* genes involved in xylan degradation. Res Microbiol. 1999;150:281–5.
32. de Vries RP, Flipphi MJA, Witteveen CFB, Visser J. Characterization of an *Aspergillus nidulans* L-arabitol dehydrogenase mutant. FEMS Microbiol Lett. 1994;123:83–90.
33. Flipphi MJA, Visser J, van der Veen P, de Graaff LH. Arabinase gene expression in *Aspergillus niger*: indications for coordinated regulation. Microbiology. 1994;140:2673–82.
34. Kowalczyk J, Lubbers R, Peng M, Battaglia E, Visser J, de Vries R. Combinational control of gene expression in *Aspergillus niger* grown on complex pectins. Sci Rep. 2017;7:12356.
35. de Vries RP, Jansen J, Aguilar G, Pařenicová L, Joosten V, Wülfert F, et al. Expression profiling of pectinolytic genes from *Aspergillus niger*. FEBS Lett. 2002;530:41–7.
36. de Vries RP, Visser J. Regulation of the feruloyl esterase (*faeA*) gene from *Aspergillus niger*. Appl Environ Microbiol. 1999;65:5500–3.
37. Jørgensen TR, Goosen T, van den Hondel CA, Ram AFJ, Iversen JJL. Transcriptomic comparison of *Aspergillus niger* growing on two different sugars reveals coordinated regulation of the secretory pathway. BMC Genomics. 2009;10:44.
38. Coutinho P, Andersen M, Kolenova K, VanKuyk P, Benoit I, Gruben B, et al. Post-genomic insights into the plant polysaccharide degradation potential of *Aspergillus nidulans* and comparison to *Aspergillus niger* and *Aspergillus oryzae*. Fungal Genet Biol. 2009;46:S161–9.
39. Martens-Uzunova ES, Schaap PJ. Assessment of the pectin degrading enzyme network of *Aspergillus niger* by functional genomics. Fungal Genet Biol. 2009;46:S170–9.
40. Martens-Uzunova ES, Zandleven JS, Benen JAE, Awad H, Kools HJ, Beldman G, et al. A new group of exo-acting family 28 glycoside hydrolases of *Aspergillus niger* that are involved in pectin degradation. Biochem J. 2006; 400:43–52.

41. Yuan X-L, van der Kaaij RM, van den Hondel CAMJJ, Punt PJ, van der Maarel MJEC, Dijkhuizen L, et al. *Aspergillus niger* genome-wide analysis reveals a large number of novel α-glucan acting enzymes with unexpected expression profiles. Mol Gen Genomics. 2008;279:545–61.
42. Korman D, Bayliss F, Barnett C, Carmona C, Kodama K, Royer T, et al. Cloning, characterization, and expression of two alpha-amylase genes from *Aspergillus niger* var. *awamori*. Curr Genet. 1990;17:203–12.
43. Yuan X-L, Goosen C, Kools H, van den Maarel MJEC, van den Hondel CAMJJ, Dijkhuizen L, et al. Database mining and transcriptional analysis of genes encoding inulin-modifying enzymes of *Aspergillus niger*. Microbiology. 2006; 152:3061–73.
44. Ademark P, de Vries RP, Hägglund P, Stålbrand H, Visser J. Cloning and characterization of *Aspergillus niger* genes encoding an α-galactosidase and a β-mannosidase involved in galactomannan degradation. Eur J Biochem. 2001;268:2982–90.
45. Tian C, Beeson WT, Iavarone AT, Sun J, Marletta MA, Cate JHD, et al. Systems analysis of plant cell wall degradation by the model filamentous fungus *Neurospora crassa*. Proc Natl Acad Sci U S A. 2009;106:22157–62.
46. Flipphi MJA, van Heuvel M, van der Veen P, Visser J, de Graaff LH. Cloning and characterization of the *abfB* gene coding for the major α-L-arabinofuranosidase (ABF B) of *Aspergillus niger*. Curr Genet. 1993;24:525–32.
47. de Souza WR, de Gouvea PF, Savoldi M, Malavazi I, de Souza Bernardes LA, Goldman MHS, et al. Transcriptome analysis of *Aspergillus niger* grown on sugarcane bagasse. Biotechnol Biofuels. 2011;4:40.
48. de Vries RP, VanKuyk PA, Kester HCM, Visser J. The *Aspergillus niger faeB* gene encodes a second feruloyl esterase involved in pectin and xylan degradation and is specifically induced in the presence of aromatic compounds. Biochem J. 2002;363:377–86.
49. Dilokpimol A, Mäkelä MR, Aguilar-Pontes MV, Benoit-Gelber I, Hildén KS, de Vries RP. Diversity of fungal feruloyl esterases: updated phylogenetic classification, properties, and industrial applications. Biotechnol Biofuels. 2016;9:231.
50. Bussink HJD, Brouwer KB, de Graaff LH, Kester HCM, Visser J. Identification and characterization of a second polygalacturonase gene of *Aspergillus niger*. Curr Genet. 1991;20:301–7.
51. de Vries RP, Frisvad JC, van de Vondervoort PJI, Burgers K, Kuijpers AFA, Samson RA, et al. *Aspergillus vadensis*, a new species of the group of black Aspergilli. Antonie Van Leeuwenhoek. 2005;87:195–203.
52. Goosen C, Yuan X-L, van Munster JM, Ram AFJ, van der Maarel MJEC, Dijkhuizen L. Molecular and biochemical characterization of a novel intracellular invertase from *Aspergillus niger* with transfructosylating activity. Eukaryot Cell. 2007;6:674–81.
53. Siegel D, Marton I, Dekel M, Bravdo B-A, He S, Withers SG, et al. Cloning, expression, characterization, and nucleophile identification of family 3, *Aspergillus niger* β-glucosidase. J Biol Chem. 2000;275:4973–80.
54. de Groot MJL, van de Vondervoort PJI, de Vries RP, VanKuyk PA, Ruijter GJG, Visser J. Isolation and characterization of two specific regulatory *Aspergillus niger* mutants shows antagonistic regulation of arabinan and xylan metabolism. Microbiology. 2003;149:1183–91.
55. Li J, Lin L, Li H, Tian C. Transcriptional comparison of the filamentous fungus *Neurospora crassa* growing on three major monosaccharides D-glucose, D-xylose and L-arabinose. Biotechnol Biofuels. 2014;7:31.
56. dos Santos CL, de Paula RG, Antoniêto ACC, Persinoti GF, Silva-Rocha R, Silva RN. Understanding the role of the master regulator xyr1 in *Trichoderma reesei* by global transcriptional analysis. Front Microbiol. 2016;7:175.
57. Sun J, Tian C, Diamond S, Glass NL. Deciphering transcriptional regulatory mechanisms associated with hemicellulose degradation in *Neurospora crassa*. Eukaryot Cell. 2012;11:482–93.
58. Benz JP, Chau BH, Zheng D, Bauer S, Glass NL, Somerville CRA. Comparative systems analysis of polysaccharide-elicited responses in *Neurospora crassa* reveals carbon source-specific cellular adaptations. Mol Microbiol. 2014;91: 275–99.
59. Benoit I, Zhou M, Vivas Duarte A, Downes D, Todd RB, Kloezen W, Post H, Heck AJ, Maarten Altelaar AF, de Vries RP. Spatial differentiation of gene expression in *Aspergillus niger* colony grown for sugar beet pulp utilization. Sci Rep. 2015;5:13592.
60. Irizarry RA, Hobbs B, Collin F, Beazer-Barclay YD, Antonellis KJ, Scherf U, et al. Exploration, normalization, and summaries of high density oligonucleotide array probe level data. Biostatistics. 2003;4:249.
61. Core Team R. R: a language and environment for statistical computing. Vienna: R Foundation for Statistical Computing; 2014. http://www.R-project.org/.
62. Bolstad BM. Comparing the effects of background, normalization and summarization on gene expression estimates. 2002. http://bmbolstad.com/stuff/components.pdf. Accessed June 2017.
63. Baldi P, Long ADA. Bayesian framework for the analysis of microarray expression data: regularized t-test and statistical inferences of gene changes. Bioinformatics. 2001;17:509–19.
64. gibbons FD, Roth FP. Judging the quality of gene expression-based clustering methods using gene annotation. Genome Res. 2002;12:1574–81.
65. Sturn A, Quackenbush J, Trajanoski Z. Genesis: cluster analysis of microarray data. Bioinformatics. 2002;18:207.
66. de Groot MJL, van den Dool C, Wösten HAB, Levisson M, VanKuyk PA, Ruijter GJG, et al. Regulation of the pentose catabolic pathway of *Aspergillus niger*. Food Technol Biotechnol. 2007;45:134–8.
67. Bos CJ, Debets AJM, Swart K, Huybers A, Kobus G, Slakhorst SM. Genetic analysis and the construction of master strains for assignment of genes to six linkage groups in *Aspergillus niger*. Curr Genet. 1988;14:437–43.
68. Flipphi MJA, Visser J, van der Veen P, de Graaff LH. Cloning of the *Aspergillus niger* gene encoding α-l-arabinofuranosidase a. Appl Microbiol Biotechnol. 1993;39:335–40.
69. van Peij NN, Brinkmann J, Vrsanská M, Visser J, de Graaff LH. β-Xylosidase activity, encoded by *xlnD*, is essential for complete hydrolysis of xylan by *Aspergillus niger* but not for induction of the xylanolytic enzyme spectrum. Eur J Biochem. 1997;245:164–73.
70. de Vries R, van de Vondervoort P, Hendriks L, van de Belt M, Visser J. Regulation of the α-glucuronidase-encoding gene (*aguA*) from *Aspergillus niger*. Mol Gen Genomics. 2002;268:96–102.
71. Khanh NQ, Ruttkowski E, Leidinger K, Albrecht H, Gottschalk M. Characterization and expression of a genomic pectin methyl esterase-encoding gene in *Aspergillus niger*. Gene. 1991;106:71–7.
72. Harmsen JAM, Kusters-van Someren MA, Visser J. Cloning and expression of a second *Aspergillus niger* pectin lyase gene (*pelA*): indications of a pectin lyase gene family in *A. niger*. Curr Genet. 1990;18:161–6.
73. Kusters-van Someren MA, Harmsen JAM, Kester HCM, Visser J. Structure of the *Aspergillus niger pelA* gene and its expression in *Aspergillus niger* and *Aspergillus nidulans*. Curr Genet. 1991;20:293–9.
74. de Vries RP, Kester HCM, Poulsen CH, Benen JAE, Visser J. Synergy between enzymes from *Aspergillus* involved in the degradation of plant cell wall polysaccharides. Carbohydr Res. 2000;327:401–10.
75. Damveld RA, VanKuyk PA, Arentshorst M, Klis FM, van den Hondel CAMJJ, Ram AFJ. Expression of *agsA*, one of five 1,3-α-D-glucan synthase-encoding genes in *Aspergillus niger*, is induced in response to cell wall stress. Fungal Genet Biol. 2005;42:165–77.
76. Boel E, Hansen MT, Hjort I, Høegh I, Fiil NP. Two different types of intervening sequences in the glucoamylase gene from *Aspergillus niger*. EMBO J. 1984;3:1581–5.
77. Fowler T, Berka RM, Ward M. Regulation of the *glaA* gene of *Aspergillus niger*. Curr Genet. 1990;18:537–45.
78. Nakamura A, Nishimura I, Yokoyama A, Lee D-G, Hidaka M, Masaki H, et al. Cloning and sequencing of an α-glucosidase gene from *Aspergillus niger* and its expression in *A. nidulans*. J Biotechnol. 1997;53:75–84.
79. Ohta K, Akimoto H, Matsuda S, Toshimitsu D, Nakamura T. Molecular cloning and sequence analysis of two endoinulinase genes from *Aspergillus niger*. Biosci Biotechnol Biochem. 1998;62:1731–8.
80. Parenicová L, Benen JA, Kester HC, Visser J. *pgaA* and *pgaB* encode two constitutively expressed endopolygalacturonases of *Aspergillus niger*. Biochem J. 2000;345:637–44.
81. Bergès T, Barreau C, Peberdy JF, Boddy LM. Cloning of an *Aspergillus niger* invertase gene by expression in *Trichoderma reesei*. Curr Genet. 1993;24:53–9.
82. Somiari RI, Brzeski H, Tate R, Bieleck S, Polak J. Cloning and sequencing of an *Aspergillus niger* gene coding for β-fructofuranosidase. Biotechnol Lett. 1997;19:1243–7.
83. Flipphi M, Panneman H, van der Veen P, Visser J, de Graaff L. Molecular cloning, expression and structure of the endo-1,5-α-L-arabinase gene of *Aspergillus niger*. Appl Microbiol Biotechnol. 1993;40:318–26.
84. de Vries RP, Pařenicová L, Hinz SWA, Kester HCM, Beldman G, Benen JAE, et al. The β-1,4-endogalactanase a gene from *Aspergillus niger* is specifically induced on arabinose and galacturonic acid and plays an important role in the degradation of pectic hairy regions. Eur J Biochem. 2002;269:4985–93.

85. Kumar V, Ramakrishnan S, Teeri T, Knowles J, Hartley B. *Saccharomyces cerevisiae* cells secreting an *Aspergillus niger* β-galactosidase grow on whey permeate. Biotechnology. 1992;10:82–5.
86. van der Vlugt-Bergmans CJB, Meeuwsen PJA, Voragen AGJ, van Ooyen AJJ. Endo-xylogalacturonan hydrolase, a novel pectinolytic enzyme. Appl Environ Microbiol. 2000;66:36–41.
87. Gysler C, Harmsen J, Kester H, Visser J, Heim J. Isolation and structure of the pectin lyase D-encoding gene from *Aspergillus niger*. Gene. 1990;30:101–8.
88. Suykerbuyk ME, Kester HC, Schaap PJ, Stam H, Musters W, Visser J. Cloning and characterization of two rhamnogalacturonan hydrolase genes from *Aspergillus niger*. Appl Environ Microbiol. 1997;63:2507–15.
89. den Herder IF, Mateo Rosell AM, van Zuilen CM, Punt PJ, van den Hondel CAMJJ. Cloning and expression of a member of the *Aspergillus niger* gene family encoding α-galactosidase. Mol Gen Genet. 1992;233:404–10.
90. Kulik N, Weignerová L, Filipi T, Pompach P, Novák P, Mrázek H, et al. The α-galactosidase type a gene *aglA* from *Aspergillus niger* encodes a fully functional α-N-acetylgalactosaminidase. Glycobiology. 2010;20:1410–9.
91. Benen JAE, Kester HCM, Pařenicová L, Visser J. Characterization of *Aspergillus niger* pectate lyase a. Biochemistry. 2000;39:15563–9.

2

Comparative transcriptome analysis of soybean response to bean pyralid larvae

Weiying Zeng, Zudong Sun*, Zhaoyan Cai, Huaizhu Chen, Zhenguang Lai, Shouzhen Yang and Xiangmin Tang

Abstract

Background: Soybean is one of most important oilseed crop worldwide, however, its production is often limited by many insect pests. Bean pyralid is one of the major soybean leaf-feeding insects in China. To explore the defense mechanisms of soybean resistance to bean pyralid, the comparative transcriptome sequencing was completed between the leaves infested with bean pyralid larvae and no worm of soybean (Gantai-2-2 and Wan82–178) on the Illumina HiSeq™ 2000 platform.

Results: In total, we identified 1744 differentially expressed genes (DEGs) in the leaves of Gantai-2-2 (1064) and Wan82–178 (680) fed by bean pyralid for 48 h, compared to 0 h. Interestingly, 315 DEGs were shared by Gantai-2-2 and Wan82–178, while 749 and 365 DEGs specifically identified in Gantai-2-2 and Wan82–178, respectively. When comparing Gantai-2-2 with Wan82–178, 605 DEGs were identified at 0 h feeding, and 468 DEGs were identified at 48 h feeding. Gene Ontology (GO) annotation analysis revealed that the DEGs were mainly involved in the metabolic process, single-organism process, cellular process, responses to stimulus, catalytic activities and binding. Pathway analysis showed that most of the DEGs were associated with the plant-pathogen interaction, phenylpropanoid biosynthesis, phenylalanine metabolism, flavonoid biosynthesis, peroxisome, plant hormone signal transduction, terpenoid backbone biosynthesis, and so on. Finally, we used qRT-PCR to validate the expression patterns of several genes and the results showed an excellent agreement with deep sequencing.

Conclusions: According to the comparative transcriptome analysis results and related literature reports, we concluded that the response to bean pyralid feeding might be related to the disturbed functions and metabolism pathways of some key DEGs, such as DEGs involved in the ROS removal system, plant hormone metabolism, intracellular signal transduction pathways, secondary metabolism, transcription factors, biotic and abiotic stresses. We speculated that these genes may have played an important role in synthesizing substances to resist insect attacks in soybean. Our results provide a valuable resource of soybean defense genes that will benefit other studies in this field.

Keywords: Soybean, Bean pyralid, Transcriptome sequencing, Differentially expressed genes (DEGs)

Background

Soybean (*Glycine max* (L.) Merr.) is the largest oil crop worldwide, and is widely used in the production of food, feed, industrial products and other sideline fields [1]. However, there have been large increases in soybean production costs due to pests. Bean pyralid (*Lamprosema indicate* (Fabricius)) is one of the major leaf-feeding insects that affects soybean crops in central and southern China, the larvae lurk inside soybean leaves, cause leaf curling and feed on leaf tissues, affecting photosynthesis. Therefore, the plants cannot grow normally [2]. So bean pyralid is different from other leaf-feeding insects with chewing mouthparts which cause holes or incisions by means of encroachment [3]. In the soybean-producing areas of southern China, many generations of bean pyralids may appear in 1 year. In serious pest-damaged years, only veins and petioles will be left, causing serious yield losses [4]. Sun et al. and Long et al. evaluated rolled leaflet number and larva number could be used as an evaluation index for bean pyralid in soybean, and screened the highly resistant line Gantai-2-2 and the highly sensitive line Wan 82–178 [4–6]. Two indicators of resistance to bean pyralid, rolled leaflet

* Correspondence: sunzudong639@163.com
Guangxi Academy of Agricultural Sciences, Nanning, Guangxi 530007, China

number and rolled leaflet percentage, were a significantly positively correlated with the pubescence angle, length on leaf blade, angle on petiole and a significantly negatively correlated with the pubescence density on leaf blade, but on correlation with pubescence tip shape was observed [7]. Xing et al. and Li et al. showed that soybean resistance to bean pyralid accords with two or three major genes and polygene, 81–92% of the phenotypic variation was accounted for by additive quantitative trait locus (QTL) (27–43%), epistatic QTL pairs (5–13%) and collective unmapped minor QTL (38–58%) [8–10]. The contents of soluble sugar, superoxide dismutase (SOD), polyphenol oxidase (PPO), jasmonate (JA) and abscisic acid (ABA) are significantly increased after bean pyralid feeding [11]. However, the results of comparative transcriptome research which has focused on soybeans' resistance to bean pyralid has not yet been made available. This is the first study of soybean transcriptome in response to bean pyralid feeding.

Transcriptome sequencing has become an important method for gene expression analysis, differentially expressed genes (DEGs) selection, functional gene mining, and genetic evolution analysis. The soybean genome was released in 2010 [12]. Based on the soybean genome data and transcriptome technology, can be better to examine all the transcription reactions, structural functions, and transcriptional regulation of resistant soybean varieties at the overall level. In the present paper, we tried to find important DEGs and metabolism pathways might related to the soybean in response to bean pyralid larvae through the comparative transcrptome analysis between the leaves infested with bean pyralid larvae and no worm of soybean using the Illumina HiSeq™ 2000 platform. Our results provide a valuable resource of soybean defense genes that will benefit other studies in this field.

Results

Transcriptome sequencing and sequence alignment

An Illumina HiSeq™2000 sequencer was employed to analyze the comparative transcriptome of eight samples of Gantai-2-2 and Wan82–178 leaves that bean pyralid had been feeding on 0 h and 48 h. The original image data obtained by sequencing base-calling were the original sequence reads. Each read in the Solexa paired-end (PE) sequencing was 100 bp in length. There were 45.88 G original data sets produced during sequencing. The mean sequencing depth was 5.67. After the raw data were trimmed, 442,422,398 clean reads were obtained. The clean/raw read rates of the eight samples ranged from 95.30% to 97.11%. All clean reads were matched to the soybean reference genome by BWA software, allowing two base mismatches. The mapped genome reads ranged from 42,474,863 to 44,489,050 sets, genome map rates ranged from 77.23% to 78.95%, and unique match rates ranged from 72.86% to 75.22%. The expressed genes ranged from 50,283 to 53,739 (Table 1). To estimate whether the sequencing depth was sufficient for transcriptome coverage, the sequencing saturation in the eight cDNA libraries was analyzed. The results showed that most genes became saturated when the amount of PE reads was 20 M (200 × 100 kb) (Fig. 1), which indicated that the overall quality of sequencing saturation in the eight cDNA libraries was high and that the sequencing *amount* covered the vast majority of expressed genes.

Correlation analysis of samples

The correlation of gene expression levels among samples is a key criterion to test whether the experiments are reliable and whether the chosen samples are reasonable. If one sample is highly similar to another one, the correlation value between them is very close to 1. We calculated the correlation value between each of two samples based on the FPKM results. According to the standard that Encode plan recommends, the square of the correlation value (R^2) should be ≥0.92 (under an ideal experimental environment with reasonable samples). Our results showed that the R^2 of all repetitions were >0.95 (Fig. 2), which signified that our experimental samples and results were satisfactory and reliable.

Table 1 Number of reads sequenced and mapped to soybean genome

	HRK0–1	HRK0–2	HRK48–1	HRK48–2	HSK0–1	HSK0–2	HSK48–1	HSK48–2	Sum
Raw reads	56,776,842	56,776,934	56,776,686	56,777,138	56,777,168	56,777,236	59,048,232	59,048,102	458,758,338
Clean reads	54,999,276	55,134,342	54,619,378	54,911,596	54,809,878	55,085,486	56,273,358	56,589,084	442,422,398
≥Q20(%)	97.22	97.45	97.51	97.55	97.17	97.57	97.69	97.68	–
Clean reads/Raw reads (%)	96.87	97.11	96.20	96.71	96.54	97.02	95.30	95.84	–
Genome map rates (%)	77.23	78.07	78.25	78.28	78.32	78.95	78.52	78.62	–
Unique Match (%)	72.86	74.01	74.94	75.22	74.50	74.46	74.76	74.89	–
Expressed gene	50,283	51,399	51,935	53,739	51,469	51,794	51,835	53,032	70,016

HRK represented the highly resistant line Gantai-2-2; *HSK* represented the highly susceptible line Wan82–178; numbers 0 and 48 represented the processing time; and −1 and −2 represented repetitions 1 and 2, respectively. Sequence length was 2 × 100 bp, length of each read was 100 bp using double end sequencing

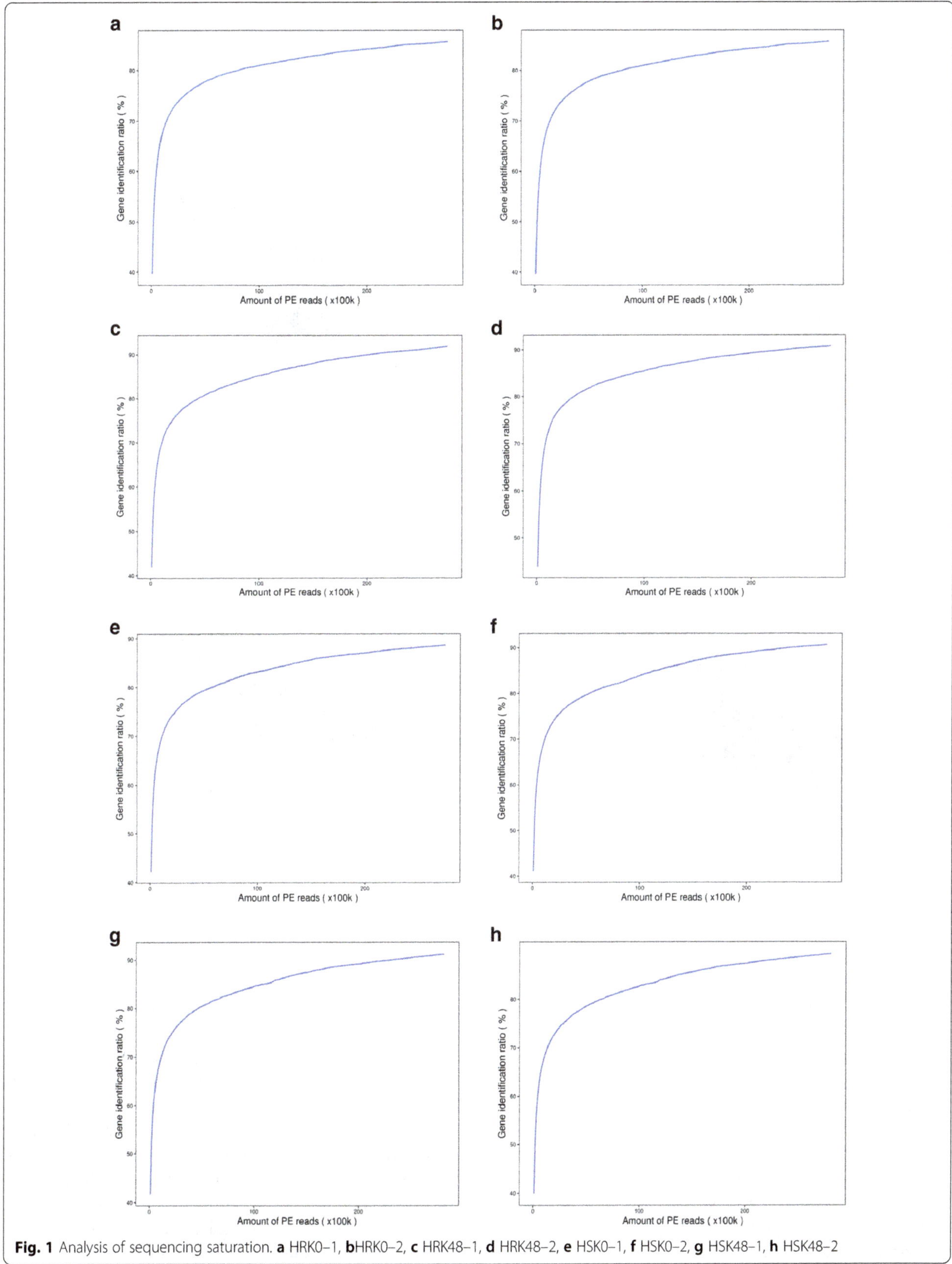

Fig. 1 Analysis of sequencing saturation. **a** HRK0–1, **b**HRK0–2, **c** HRK48–1, **d** HRK48–2, **e** HSK0–1, **f** HSK0–2, **g** HSK48–1, **h** HSK48–2

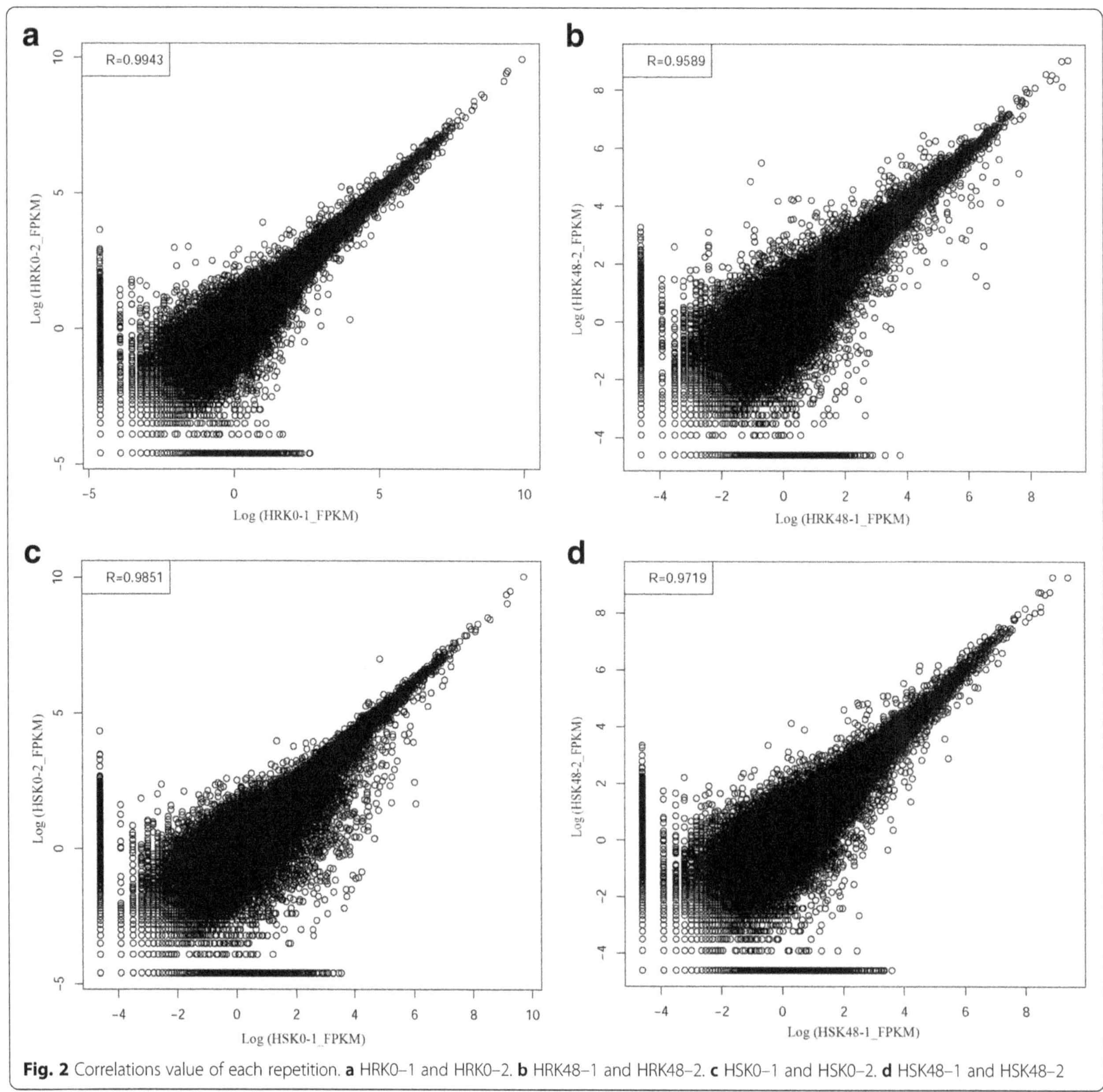

Fig. 2 Correlations value of each repetition. **a** HRK0–1 and HRK0–2. **b** HRK48–1 and HRK48–2. **c** HSK0–1 and HSK0–2. **d** HSK48–1 and HSK48–2

Screening of differentially expressed genes (DEGs)

Noiseq, DESeq2 and edgeR methods were used to screen DEGs (Fig. 3). The results showed that the Noiseq method can screen DEGs between two groups, showing a good performance when comparing it to other differential expression methods, such as edgeR and DESeq2. Noiseq maintains good True Positive and False Positive rates when the sequencing depth is increased, whereas most other methods show poor performance. Furthermore, Noiseq models the noise distribution from the actual data to better adapt to the size of the data set and be more effective in controlling the rate of false discoveries. Therefore, the Noiseq method was used to screen the DEGs.

As a result, under bean pyralid larvae feeding for 48 h, 1064 DEGs were identified in the Gantai-2-2, of which 894 DEGs were up-regulated and 170 DEGs were down-regulated compared to 0 h (Additional file 1: Table S1). Additionally, 680 DEGs were identified in Wan82–178, of which 495 DEGs were up-regulated and 185 DEGs were down-regulated (Additional file 2: Table S2). After being induced with bean pyralid larvae, the number of up-regulated genes was significantly higher than that of down-regulated genes. These results indicated that most of the genes were activate and a few genes were inhibited after insect feeding.

To screen the constitutive defense genes, the highly resistant line and highly sensitive line were compared at

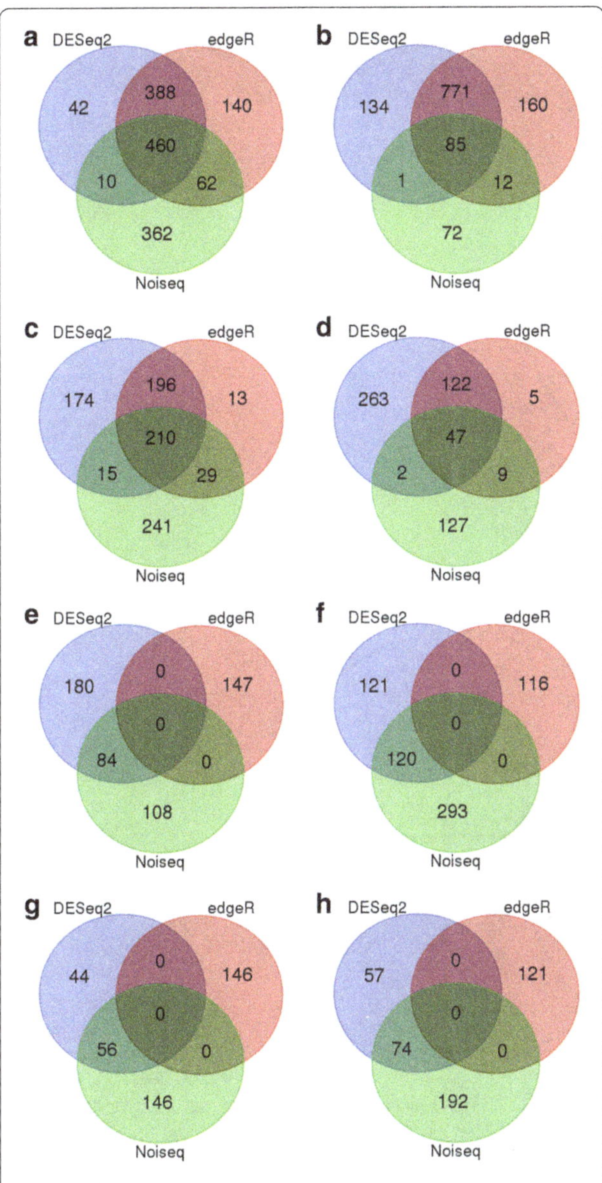

Fig. 3 The DEGs were screened by Noiseq, DESeq2 and edgeR. **a** HRK48/HRK0_UP In total, 894, 900 and 1050 up-regulated DEGs were identified by Noiseq, DESeq2 and edgeR, respectively. 460 DEGs were identified under the three methods, 62 DEGs were identified under both Noiseq and edgeR, 388 DEGs were identified under both DESeq2 and edgeR, 10 DEGs were identified under both Noiseq and DESeq2. **b** HRK48/HRK0_DOWN In total, 170, 991 and 1028 down-regulated DEGs were identified by Noiseq, DESeq2 and edgeR, respectively. 85 DEGs were identified under the three methods, 12 DEGs were identified under both Noiseq and edgeR, 771 DEGs were identified under both DESeq2 and edgeR, 1 DEGs were identified under both Noiseq and DESeq2. **c** HSK48/HSK0_UP In total, 495, 595 and 448 up-regulated DEGs were identified by Noiseq, DESeq2 and edgeR, respectively. 210 DEGs were identified under the three methods, 29 DEGs were identified under both Noiseq and edgeR, 196 DEGs were identified under both DESeq2 and edgeR, 15 DEGs were identified under both Noiseq and DESeq2. **d** HSK48/HSK0_DOWN In total, 185, 434 and 183 down-regulated DEGs were identified by Noiseq, DESeq2 and edgeR, respectively. 47 DEGs were identified under the three methods, 9 DEGs were identified under both Noiseq and edgeR, 122 DEGs were identified under both DESeq2 and edgeR, 2 DEGs were identified under both Noiseq and DESeq2. **e** HRK0/HSK0_UP In total, 192, 264 and 147 up-regulated DEGs were identified by Noiseq, DESeq2 and edgeR, respectively. 84 DEGs were identified under both Noiseq and DESeq2. **f** HRK0/HSK0_DOWN In total, 413, 241 and 116 down-regulated DEGs were identified by Noiseq, DESeq2 and edgeR, respectively. 120 DEGs were identified under both Noiseq and DESeq2. **g** HRK48/HSK48_UP In total, 202, 100 and 146 up-regulated DEGs were identified by Noiseq, DESeq2 and edgeR, respectively. 56 DEGs were identified under both Noiseq and DESeq2. **h** HRK48/HSK48_DOWN In total, 266, 131 and 121 down-regulated DEGs were identified by Noiseq, DESeq2 and edgeR, respectively. 74 DEGs were identified under both Noiseq and DESeq2

different feeding times. The results showed that 605 DEGs were identified in Gantai-2-2 at 0 h of feeding, of which 192 DEGs were up-regulated and 413 DEGs were down-regulated (Additional file 3: Table S3), compared to Wan82–178. And at 48 h feeding, 468 DEGs were identified in Gantai-2-2, of which 202 DEGs were up-regulated and 266 DEGs were down-regulated (Additional file 4: Table S4), compared to Wan82–178.

The DEGs were further divided into three categories. The first category was the "DEGs with non-bean pyralid-induced genotype", and there were 605 DEGs in total. This class of genes was the "DEGs identified in Gantai-2-2 compared to Wan 82-178 before bean pyralid feeding induction", in which 52 DEGs were always up-regulated and 83 DEGs were always down-regulated at 0 h and 48 h. In addition, 9 DEGs were up-regulated at 0 h but down-regulated at 48 h, and 2 DEGs were down-regulated at 0 h but up-regulated at 48 h, whereas the other 459 DEGs displayed no changes at 48 h (Fig. 4a). The second category was the "bean pyralid-induced DEGs that appeared in both materials". This category included 315 DEGs, which mainly displayed an up-regulated trend, with 274 DEGs were up-regulated and 31 DEGs were down-regulated in the two materials. A total of 8 DEGs were down-regulated in Gantai-2-2 but up-regulated in Wan82–178, 2 DEGs up-regulated in Gantai-2-2 but down-regulated in Wan82–178 (Fig. 4b). The third type was the "bean pyralid-induced genotype DEGs", which consisted of a total of 1114 DEGs, of which 749 DEGs were only expressed in Gantai-2-2 and 365 DEGs were only expressed in Wan82–178.

Gene Ontology (GO) annotation analysis of the DEGs

To further analyze the cellular components, molecular functions and biological processes of the DEGs, GO annotation analysis was performed on all of the identified DEGs. The results showed that, under bean pyralid larvae feeding for 48 h, 572 DEGs (53.76%) of Gantai-2-2

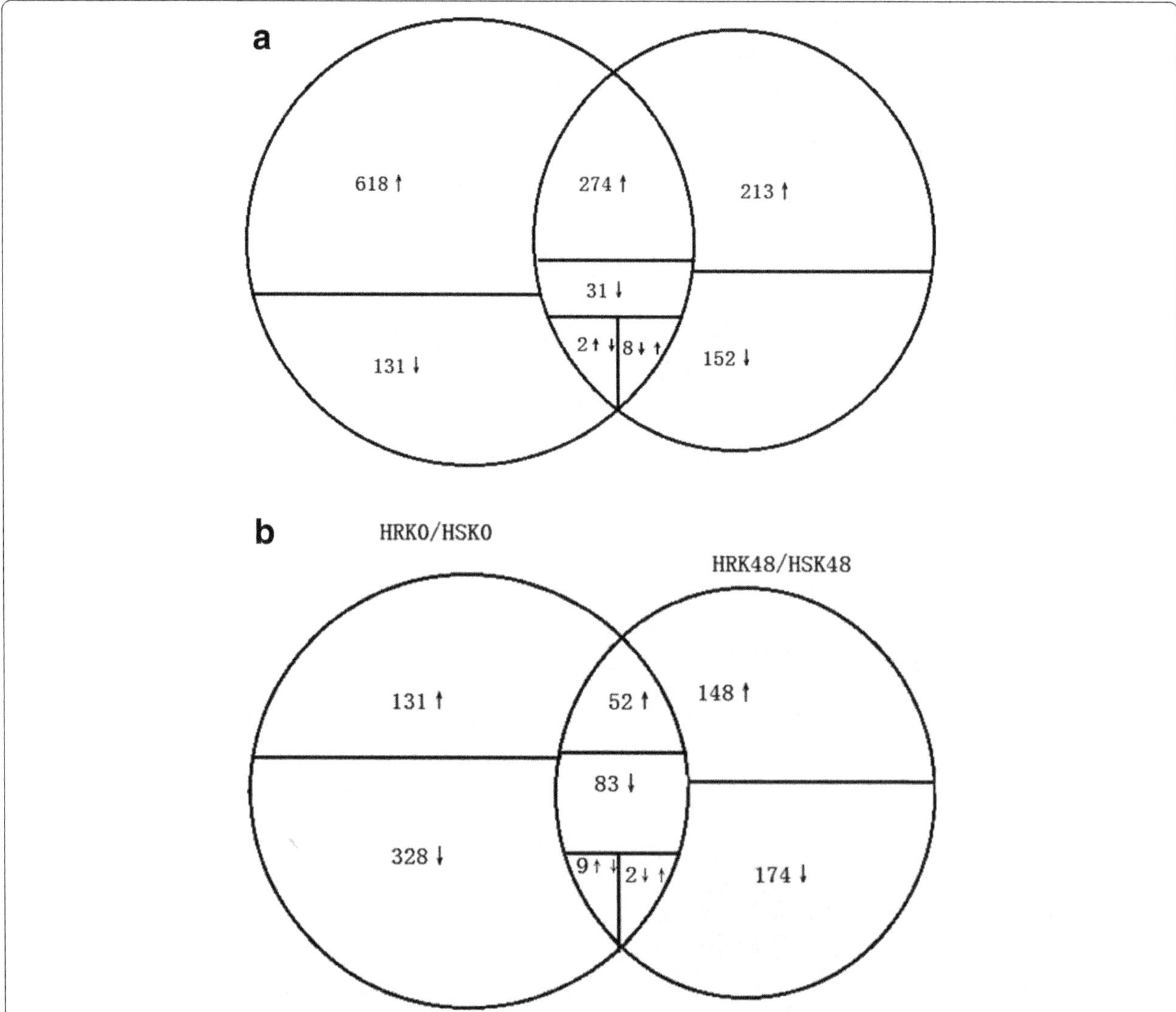

Fig. 4 Venn diagram of the distribution of DEGs. **a** HRK48/HRK0 and HSK48/HSK0. **b** HRK0/HSK0 and HRK48/HSK48. The circles are proportional to the number of genes identified in each treatment. The overlapping regions indicate the number of common genes. The ↑ indicate up-regulated, ↓ indicate down-regulated, ↑↓ indicate up-regulated in HRK48/HRK0 or HRK0/HSK0 but down-regulated in HSK48/HSK0 or HRK48/HSK48, ↓↑ indicate up-regulated in HSK48/HSK0 or HRK48/HSK48 but down-regulated in HRK48/HRK0 or HRK0/HSK0

were annotated to 42 functional groups, including 18 biological processes, 12 cellular components and 12 molecular functions compared to 0 h (Fig. 5a, Additional file 5: Table S5). Under bean pyralid larvae feeding for 48 h, 378 DEGs (55.59%) of Wan82–178 were annotated to 41 functional groups, including 18 biological processes, 11 cellular components and 12 molecular functions compared to 0 h (Fig. 5b, Additional file 6: Table S6). When comparing Gantai-2-2 with Wan82–178 at 0 h feeding, 285 DEGs (47.11%) were annotated to 39 functional groups, including 17 biological processes, 12 cellular components and 10 molecular functions (Fig. 5c, Additional file 7: Table S7). When comparing Gantai-2-2 with Wan82–178 at 48 h feeding, 240 DEGs (51.28%) were annotated to 34 functional groups, including 15 biological processes, 9 cellular components and 10 molecular functions (Fig. 5d, Additional file 8: Table S8).

Among GO annotations of the DEGs in the above four groups, the largest common functional groups were metabolic process, cellular process, single-organism process, responses to stimuli, catalytic activities and binding. These results indicated that when the soybean was subjected to bean pyralid larvae feeding, the defense systems in the plants would immediately respond to the stimuli, appropriately increase metabolic activities in vivo and produce defense substances, such as defendant enzymes, and proteinase inhibitors, thereby enhancing the activities of various enzymes to promote defense.

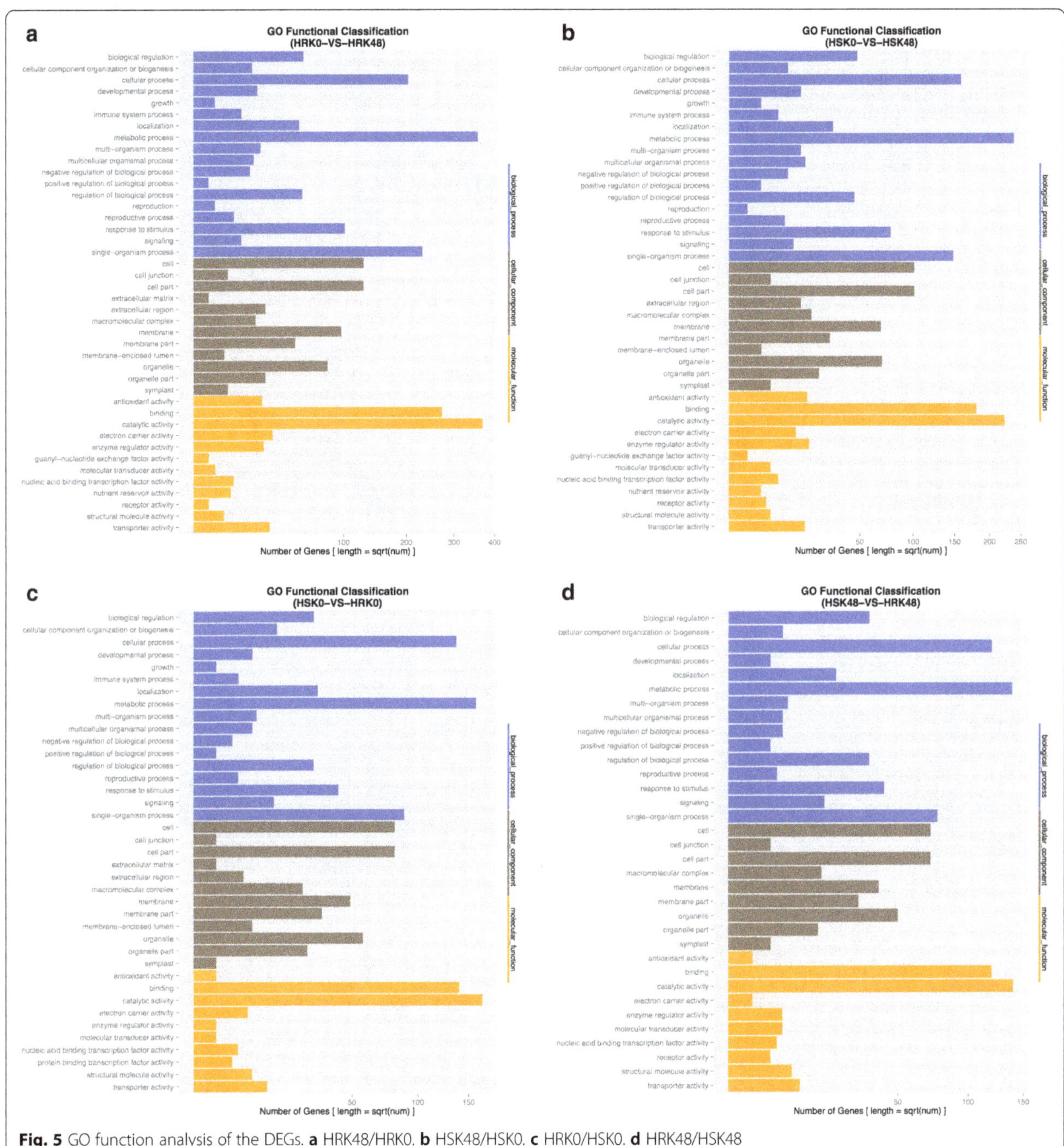

Fig. 5 GO function analysis of the DEGs. **a** HRK48/HRK0. **b** HSK48/HSK0. **c** HRK0/HSK0. **d** HRK48/HSK48

Pathway analysis of the DEGs

Pathway-based analysis was performed using the Kyoto Encyclopedia of Genes and Genomes (KEGG) pathway database. The results showed that 614 DEGs (57.71%) of Gantai-2-2 were assigned to 256 pathways under bean pyralid larvae feeding for 48 h, compared to 0 h. Which mainly included the metabolic pathways (263, 42.83%), biosynthesis of secondary metabolites (206, 33.55%), microbial metabolism in a diverse environment (85, 13.84%), flavonoid biosynthesis (52, 8.47%), phenylpropanoid biosynthesis (47, 7.65%) and phenylalanine metabolism (33, 5.37%) (Additional file 9: Table S9, Fig. 6a). And 380 DEGs (55.88%) of Wan82–178 were assigned to 224 pathways under bean pyralid larvae feeding for 48 h, compared to 0 h, which mainly included the metabolic pathways (147, 38.68%), biosynthesis of secondary metabolites (102, 26.84%), phenylpropanoid biosynthesis (27, 7.11%), phenylalanine metabolism (22, 5.79%) and flavonoid biosynthesis (14, 3.68%) (Additional file 10: Table S10, Fig. 6b). When comparing Gantai-2-2 with

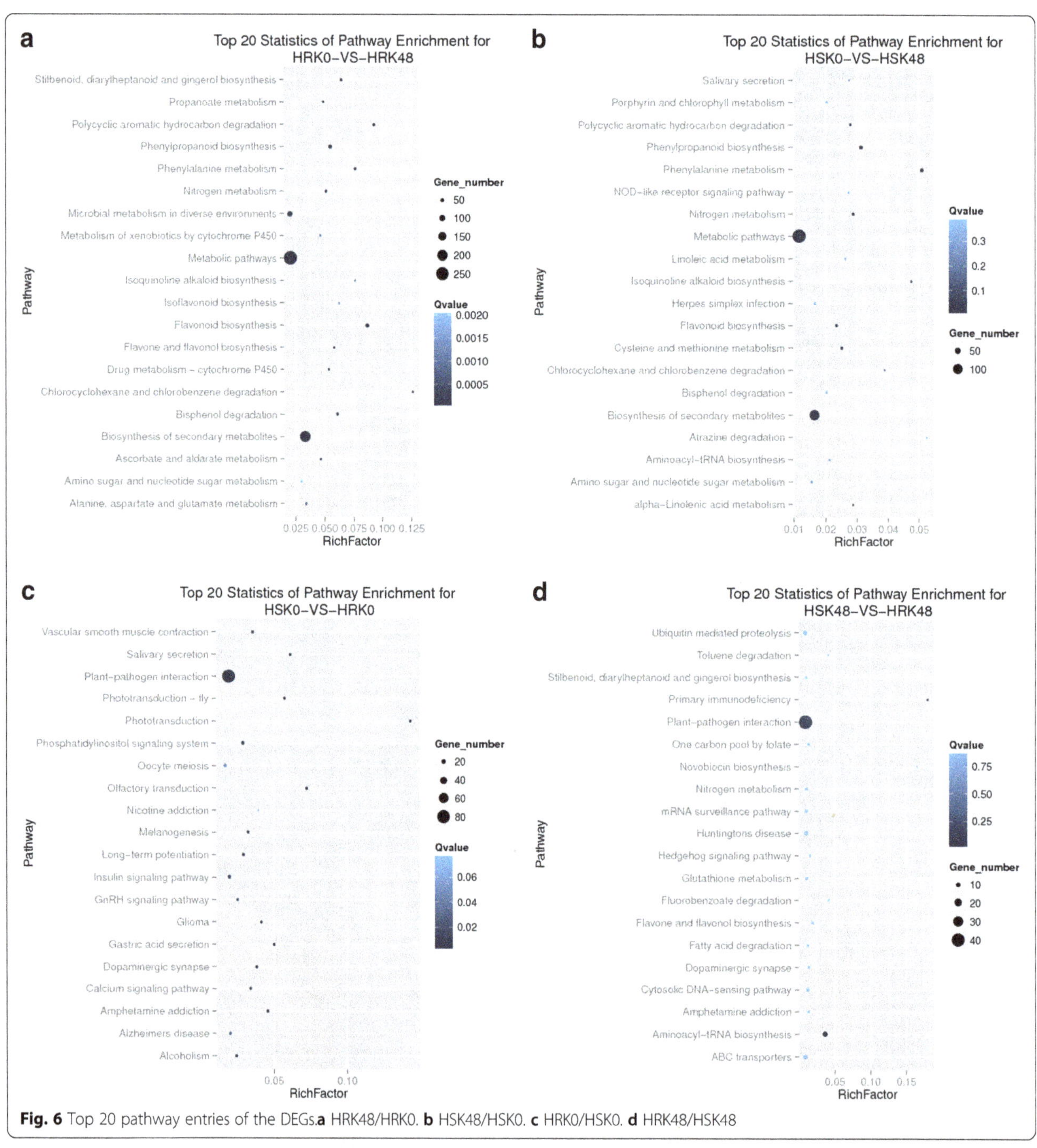

Fig. 6 Top 20 pathway entries of the DEGs.**a** HRK48/HRK0. **b** HSK48/HSK0. **c** HRK0/HSK0. **d** HRK48/HSK48

Wan82–178 at 0 h feeding, 367 DEGs (62.15%) were assigned to 208 pathways, which mainly included plant-pathogen interaction (84, 22.89%), insulin signaling pathway (14, 3.81%), phosphatidylinositol signaling system (13, 3.54%), and oocyte meiosis (13, 3.54%) (Additional file 11: Table S11, Fig. 6c). When comparing Gantai-2-2 with Wan82–178 at 48 h feeding, 209 DEGs (63.68%) were assigned to 209 pathways, which mainly included plant-pathogen interaction (42, 14.09%), aminoacyl-tRNA biosynthesis (12, 4.03%), ABC transporters (10, 3.36%) and huntington's disease (9, 3.02%) (Additional file 12: Table S12, Fig. 6d).

Pathway analysis showed that differentially expressed of these genes in the metabolic pathways after induced by bean pyralid might be related to the resistance, such as plant-pathogen interactions, phenylpropanoid biosynthesis, phenylalanine metabolism, flavonoid biosynthesis, peroxisome, plant hormone signal transduction, terpenoid backbone biosynthesis, and so on, and that they played a defensive role against insect stress.

R language was used to conduct a super geometric algorithm. The result showed that the DEGs belonged to the bean pyralid-induced DEGs that appeared in both materials, mainly involving the global and overview maps, biosynthesis of the other secondary metabolites, carbohydrate metabolism and amino acid metabolism (Fig. 7a). A total of 146 DEGs were identified in Gantai-2-2 compared to Wan 82–178 before and after bean pyralid feeding that were mainly involved environmental adaptation, translation, global and overview maps and signal transduction (Fig. 7b).

Analysis of DEGs potentially related to anti-bean pyralid in soybean

Man pathway cluster analysis was employed to identify the pathway classification of the DEGs. The results showed that some important DEGs were categorized into ROS removal, hormone metabolism, signaling, stress, secondary metabolism and cell wall (Tables 2 and 3).

Under biotic and abiotic stresses, a great number of ROS in plants will be produced, and the cell structure is then destroyed. Plants often remove ROS in vivo through the production of a reactive oxygen scavenging agent, in order to alleviate damage to the plants caused by oxidative stress [13]. ROS removal system mainly includes enzymes, such as POD and PPO, and low molecular weight antioxidants, such as ferredoxin and thioredoxin (TRX) [14, 15]. After bean pyralid larvae feeding for 48 h, many genes related to the ROS removal system were identified that were significantly up-regulated (Tables 2 and 3). For example, there were 23 POD, 5 PPO, 9 glutathione S-transferase (GST) and 3 TRX1 identified in Gantai-2-2. Additionally, there were 20 POD, 4 PPO and 2 GST identified in Wan82–178. When comparing Gantai-2-2 with Wan82–178 at 0 h feeding, 1 GST was down-regulated. When comparing Gantai-2-2 with Wan82–178 at 48 h feeding, 1 TRX1 was up-regulated.

JA, ethylene (ET), and other plant hormone signaling pathways can be activated after pest feeding, which in turn causes a rise in the defense gene expression levels, an accumulation of defensive compounds, and an increase in the release of volatiles, finally, the plants showed resistance to the pests [16–18]. Our results showed that the genes related to plant hormone had changed after bean pyralid feeding, including JA, ET and auxin (Tables 2 and 3). Eight DEGs related to JA biosynthesis were all up-regulated in Gantai-2-2, including 1 lipoxygenase (LOX), 3 linoleate 9S–LOX, 1 alpha-dioxygenase (α-DOX), 1 hydroperoxide dehydratase and 2 12-oxophytodienoic acid reductase (OPDA). Additionally, 6 DEGs related to JA biosynthesis were all up-regulated in Wan 82–178, including 1 LOX, 2 linoleate 9S–LOX, 1 α-DOX, 1 hydroperoxide alpha dehydratase and 1 OPDA. When comparing Gantai-2-2 with Wan82–178 at 0 h feeding, 1 linoleate 9S–LOX and 1 OPDA were down-regulated. When comparing Gantai-

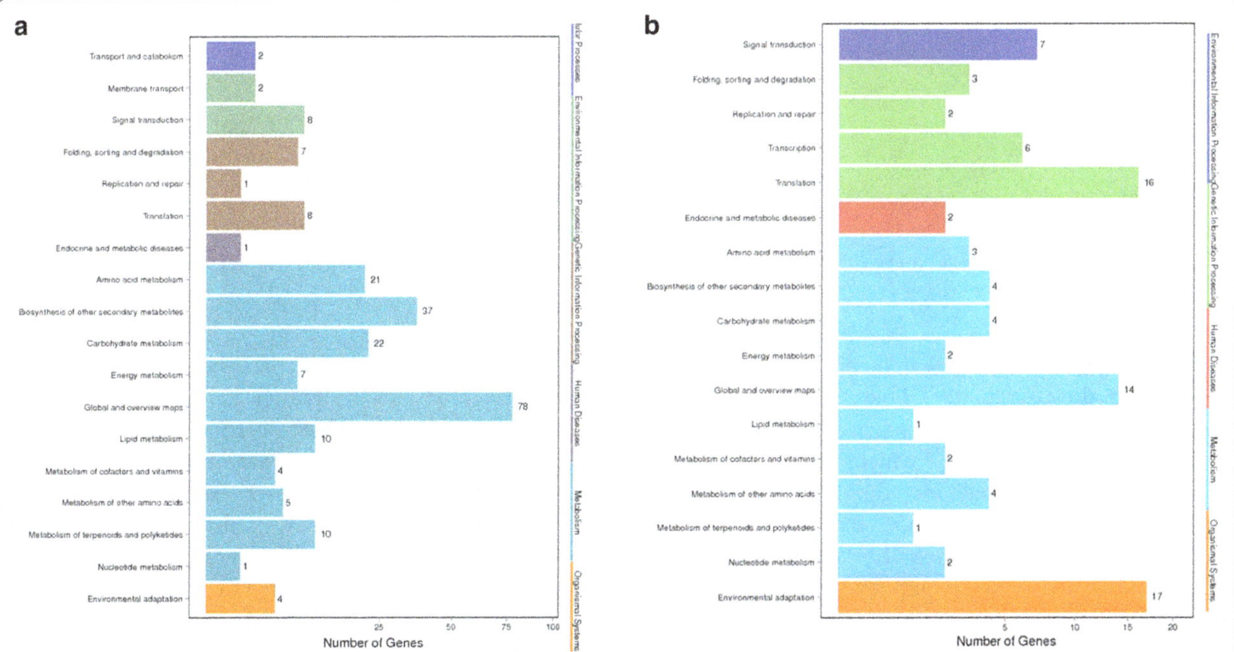

Fig. 7 Pathway classification of the DEGs. a Pathway classification of "bean pyralid-induced DEGs which appeared in both materials". **b** DEGs were identified in Gantai-2-2 compared to Wan 82–178 before and after bean pyralid feeding Note: The X axis represent the percent of genes (%), and the Y axis represent the metabolic process

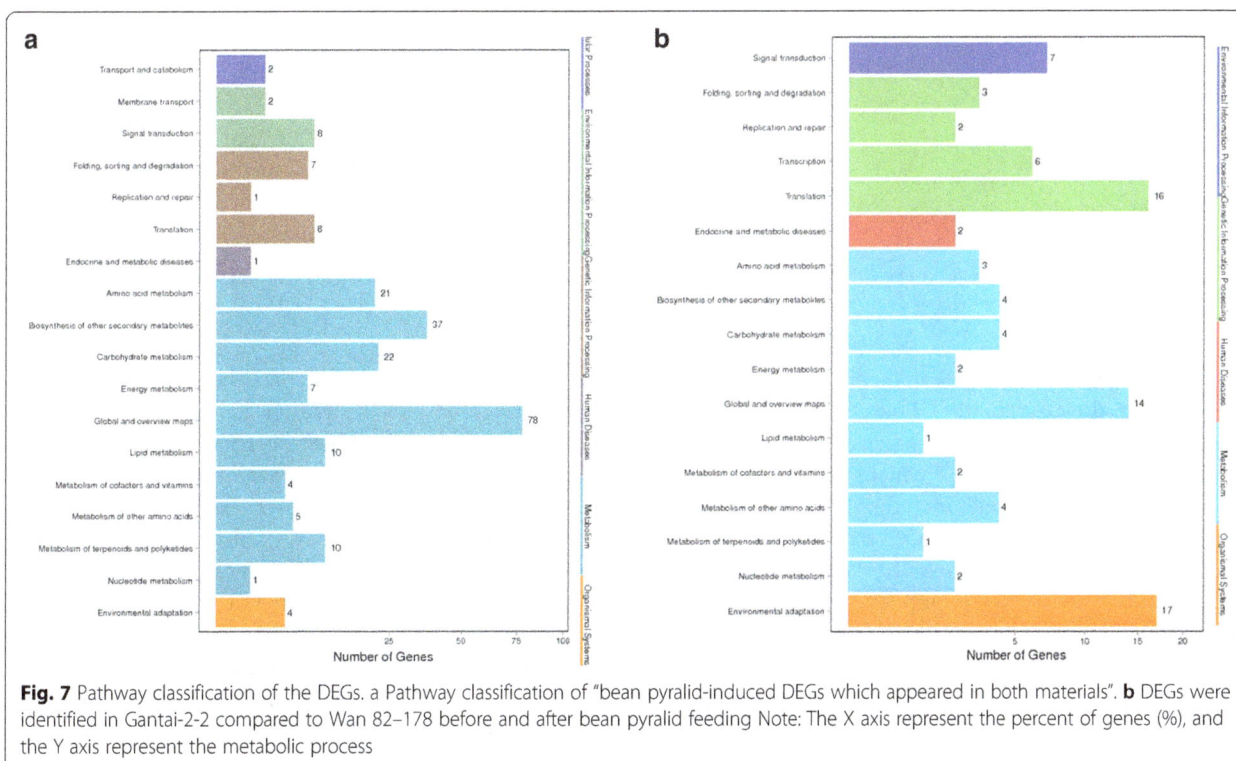

Table 2 Functional classification of DEGs

Functional category	Pathways	HRK48/HRK0		HSK48/HSK0		HRK0/HSK0		HRK48/HSK48	
		up	down	up	down	up	down	up	down
ROS removal	Peroxidases	23	0	20	0	0	0	0	0
	Polyphenol oxidase	5	0	4	0	0	0	0	0
	Glutathione S transferases	9	0	2	0	0	1	0	0
	Thioredoxin 1	3	0	0	0	0	0	1	0
Hormone metabolism	Jasmonate	8	0	6	0	0	2	0	2
	Ethylene	22	2	14	0	1	0	1	1
	Abscisic acid	1	0	2	0	0	2	0	1
	Auxin	12	0	6	0	1	0	1	0
	Brassinosteroid	1	0	0	0	0	0	0	0
	Cytokinin	0	2	1	0	0	0	0	0
	Gibberelin	3	0	1	0	0	0	0	0
	Salicylic acid	0	0	1	0	0	0	0	1
Signalling	Protein kinases	17	0	6	1	7	26	8	7
	Calcium	6	1	3	5	4	15	4	1
Stress	Biotic	17	2	8	0	1	3	2	1
	Biotic .signalling	1	0	0	0	0	0	0	0
	Biotic.PR-proteins	10	2	5	2	1	12	3	6
	Biotic. proteinase inhibitors	8	0	7	0	0	0	3	0
	Abiotic	2	0	1	0	0	0	0	0
	Abiotic.heat	6	0	4	2	0	6	0	2
	Abiotic.cold	1	0	0	1	0	1	2	0
	Abiotic.drought/salt	1	2	1	1	2	1	0	0
	Abiotic.touch/wounding	1	0	1	0	0	0	0	0
	Abiotic.unspecified	9	4	5	2	1	1	1	1
Secondary metabolism	Isoprenoid	6	0	6	0	0	1	0	1
	Phenylpropanoid	13	0	4	0	2	4	2	3
	Flavonoid	22	0	4	0	0	3	1	1
	Cytochrome P450	10	1	2	0	1	1	1	0
Cell wall	Simple phenol	4	0	1	0	0	0	0	0
	modification	7	3	2	2	0	0	3	0

HRK represented the highly resistant line Gantai-2-2; *HSK* represented the highly susceptible line Wan82–178; and the numbers 0 and 48 represented the processing times

2-2 with Wan82–178 at 48 h feeding, 1 LOX and 1 OPDA were down-regulated. There were 24 DEGs related to ET biosynthesis and signal transduction identified in Gantai-2-2, of which 22 DEGs were significantly up-regulated, including 7 aminocyclopropane carboxylate oxidase (ACC oxidase), 2 ethylene receptors, 3 ethylene responsive factor 1 (ERF1), 5 EREBP-like factor, and so on. Additionally, 14 DEGs related to ET biosynthesis and signal transduction identified in Wan 82–178 were all significantly up-regulated, including 5 ACC oxidase, 2 ethylene receptors, 2 EREBP-like factor, 3 ERF1, and so on. One ACC oxidase was found to be higher in Gantai-2-2 than in Wan82–178 at 0 h. In addition, 1 EREBP-like factor was found to be higher in Gantai-2-2 than in Wan82–178 at 48 h. There were 12 significantly up-regulated DEGs associated with auxin synthesis and signal transduction pathways in Gantai-2-2, including 4 IAA-amino acid hydrolase, 1 auxin responsive GH3 gene family, 3 SAUR family proteins, and so on. Six DEGs associated with auxin synthesis and signal transduction pathways identified in Wan82–178 were significantly up-regulated, including 1 IAA-amino acid hydrolase, 3 SAUR family protein, and so on. In addition, 1 SAUR family protein was found to be higher in Gantai-2-2 than in Wan82–178 at 48 h.

Table 3 Comparison of some DEGs of the Gantai-2-2 and Wan82–178 after bean pyralid larvae feeding

Gene ID	Gene Annotation	HRK48/ HRK0	HSK48/ HSK0	HRK0/ HSK0	HRK48/ HSK48
Peroxidases					
Glyma.20G169200.1	Peroxidases	4.34	3.79	–	–
Glyma.04G220600.1	Peroxidases	5.72	5.72	–	–
Glyma.10G222400.1	Peroxidases	4.36	4.78	–	–
Glyma.15G128800.1	Peroxidases	5.53	5.42	–	–
Glyma.09G277900.1	Peroxidases	6.46	5.68	–	–
Glyma.06G145300.1	Peroxidases	4.59	4.82	–	–
Glyma.09G277800.1	Peroxidases	5.42	4.19	–	–
Glyma.10G050800.1	Peroxidases	9.64	7.48	–	–
Glyma.06G275900.1	Peroxidases	9.64	5.10	–	–
Glyma.02G233800.1	Peroxidases	7.92	8.11	–	–
Glyma.09G022400.1	Peroxidases	4.81	6.44	–	–
Glyma.18G211000.1	Peroxidases	5.19	4.34	–	–
Glyma.08G179700.1	Peroxidases	5.46	4.91	–	–
Glyma.13G138300.1	Peroxidases	8.10	6.22	–	–
Glyma.18G211100.1	Peroxidases	6.68	5.39	–	–
Glyma.16G164400.1	Peroxidases	8.58	10.14	–	–
Glyma.12G129500.1	Peroxidases	4.64	5.70	–	–
Glyma.15G128700.1	Peroxidases	5.37	3.92	–	–
Glyma.08G179600.1	Peroxidases	3.47	3.76	–	–
Glyma.16G164200.1	Peroxidases	9.65	6.46	–	–
Glyma.15G052700.1	Peroxidases	6.73	–	–	–
Glyma.11G162100.1	Peroxidases	7.75	–	–	–
Glyma.09G022300.1	Peroxidases	5.92	–	–	–
Polyphenol oxidase					
Glyma.04G121700.1	Polyphenol oxidase	6.95	–	–	–
Glyma.07G193400.1	Polyphenol oxidase	7.54	–	–	–
Glyma.06G270400.1	Polyphenol oxidase	7.87	7.11	–	–
Glyma.15G071200.1	Polyphenol oxidase	9.17	8.94	–	–
Glyma.13G242300.1	Polyphenol oxidase	7.14	–	–	–
Glutathione S-transferase					
Glyma.11G198500.1	Glutathione S-transferase	6.40	6.15	–	–
Glyma.20G020300.4	Putative glutathione S-transferase	7.48	9.37	–	–
Glyma.13G129000.1	Glutathione S-transferase	3.63	–	–	–
Glyma.02G024800.1	Glutathione S-transferase	8.21	–	–	–
Glyma.10G192900.1	Glutathione S-transferase	6.52	–	–	–
Glyma.08G118700.1	Glutathione S-transferase	4.93	–	–	–
Glyma.02G024600.1	Glutathione S-transferase	9.35	–	–	–
Glyma.07G139800.1	Glutathione S-transferase	4.52	–	–	–
Glyma.07G139700.1	Glutathione S-transferase	5.19	–	−4.77	–
Thioredoxin					
Glyma.12G215000.1	Thioredoxin 1	5.87	–	–	–
Glyma.18G255200.3	Thioredoxin 1	8.15	–	–	–

Table 3 Comparison of some DEGs of the Gantai-2-2 and Wan82–178 after bean pyralid larvae feeding *(Continued)*

Gene ID	Gene Annotation	HRK48/ HRK0	HSK48/ HSK0	HRK0/ HSK0	HRK48/ HSK48
Glyma.06G266700.1	Thioredoxin 1	7.91	–	–	7.91
Hormone metabolism					
Glyma.07G034900.1	Linoleate 9S–lipoxygenase	5.23	6.54	–	–
Glyma.15G026400.2	Linoleate 9S–lipoxygenase	4.13	3.75	–	–
Glyma.07G034800.1	Linoleate 9S–lipoxygenase	4.73	–	–4.64	–
Glyma.13G030300.1	Lipoxygenase	5.90	3.07	–	–
Glyma.13G030300.2	Lipoxygenase	–	–	–	–8.93
Glyma.04G035000.1	Hydroperoxide dehydratase	5.99	5.74	–	–
Glyma.13G109800.1	12-Oxophytodienoic acid reductase	7.41	7.82	–	–
Glyma.15G223900.1	12-Oxophytodienoic acid reductase	9.26	–	–	–
Glyma.176209900.1	12-Oxophytodienoic acid reductase	–	–	–6.70	–10.45
Glyma.19G011700.1	Alpha-dioxygenas	4.43	3.15	–	–
Glyma.17G178300.2	Aminocyclopropanecarboxylate oxidase	7.02	8.71	–	–
Glyma.01G056100.1	Aminocyclopropanecarboxylate oxidase	7.73	6.08	–	–
Glyma.08G092800.1	Aminocyclopropanecarboxylate oxidase	8.18	5.71	–	–
Glyma.09G107100.1	Aminocyclopropanecarboxylate oxidase	4.64	4.79	–	–
Glyma.16G017500.1	Aminocyclopropanecarboxylate oxidase	6.79	5.11	–	–
Glyma.02G268200.1	Aminocyclopropanecarboxylate oxidase	9.18	–	–	–
Glyma.02G268000.4	Aminocyclopropanecarboxylate oxidase	9.49	–	–	–
Glyma.02G268000.3	Aminocyclopropanecarboxylate oxidase	–	–	3.66	–
Glyma.09G002600.15	Ethylene receptor	7.47	–	–	–
Glyma.09G002600.12	Ethylene receptor	–	9.06	–	–
Glyma.09G002600.14	Ethylene receptor	–	–	–	4.98
Glyma.03G251700.3	Ethylene receptor	–	–	–	–7.07
Glyma.10G007000.1	Ethylene-responsive transcription factor 1	6.46	–	–	–
Glyma.10G186800.1	Ethylene-responsive transcription factor 1	5.98	–	–	–
Glyma.20G070100.1	EREBP-like factor	3.48	3.42	–	–
Glyma.13G279200.1	IAA-amino acid hydrolase	3.38	3.03	–	–
Glyma.13G352400.2	IAA-amino acid hydrolase	8.65	–	–	–
Glyma.15G022300.1	IAA-amino acid hydrolase	3.83	–	–	–
Glyma.06G115100.1	IAA-amino acid hydrolase	3.21	–	–	–
Glyma.08G010400.1	SAUR family protein	9.27	5.23	–	–
Glyma.06G006500.1	SAUR family protein	8.60	9.21	–	–
Glyma.12G150500.1	SAUR family protein	6.19	–	–	–
Glyma.04G006600.1	SAUR family protein	–	10.13	–	–
Glyma.06G282000.1	SAUR family protein	–	–	–	5.33
Glyma.12G222400.1	Asparagine synthase (glutamine-hydrolysing)	3.87	3.85	–	–
Glyma.12G150500.1	Asparagine synthase (glutamine-hydrolysing)	3.87	7.34	–	–
Glyma.15G072400.1	Asparagine synthase (glutamine-hydrolysing)	3.74	–	–	–
Glyma.15G071300.3	Asparagine synthase (glutamine-hydrolysing)	9.47	–	–	–
Glyma.01G190600.1	Auxin responsive GH3 gene family	6.83	–	–	–

Table 3 Comparison of some DEGs of the Gantai-2-2 and Wan82–178 after bean pyralid larvae feeding *(Continued)*

Gene ID	Gene Annotation	HRK48/ HRK0	HSK48/ HSK0	HRK0/ HSK0	HRK48/ HSK48
Glyma.U019800.1	ARF	−7.57	–	–	–
Protein kinases					
Glyma.05G220200.4	Protein kinase	8.92	–	–	–
Glyma.14G116000.7	Protein kinase	7.77	–	–	–
Glyma.07G253900.1	Protein kinase A	7.99	–	–	–
Glyma.17G029200.1	Protein kinase A	–	8.43	–	–
Glyma.18G242700.2	Protein kinase A	–	–	−7.71	–
Glyma.17G029200.1	Protein kinase A	–	–	–	−8.43
Glyma.15G209300.1	LRR receptor-like serine/threonine-protein kinase FLS2	8.52	8.32	–	–
Glyma.08G128900.1	LRR receptor-like serine/threonine-protein kinase FLS2	5.16	–	–	–
Glyma.06G319700.1	LRR receptor-like serine/threonine-protein kinase FLS2	5.14	–	–	–
Glyma.19G145200.1	LRR receptor-like serine/threonine-protein kinase FLS2	7.35	–	–	–
Glyma.17G250800.2	LRR receptor-like serine/threonine-protein kinase FLS2	9.25	–	–	–
Glyma.08G079100.1	LRR receptor-like serine/threonine-protein kinase FLS2	–	6.36	–	–
Glyma.16G185100.1	LRR receptor-like serine/threonine-protein kinase FLS2	–	–	–	6.59
Glyma.01G004800.1	Serine/threonine-protein kinase PBS1(STK)	7.57	–	–	–
Glyma.09G272300.7	Serine/threonine-protein kinase PBS1(STK)	7.76	–	–	–
Glyma.18G217000.4	Serine/threonine-protein kinase PBS1(STK)	8.11	–	–	–
Glyma.09G063200.2	Serine/threonine-protein kinase PBS1(STK)	–	8.13	–	–
Glyma.13G216100.1	Serine/threonine-protein kinase PBS1(STK)	–	8.12	9.54	–
Glyma.06G081800.1	Serine/threonine-protein kinase PBS1(STK)	–	−7.67	–	–
Glyma.20G137300.1	Serine/threonine-protein kinase PBS1(STK)	–	–	9.15	–
Glyma.17G039800.2	Serine/threonine-protein kinase PBS1(STK)	–	–	8.75	–
Glyma.16G185100.1	Serine/threonine-protein kinase PBS1(STK)	–	–	5.92	–
Glyma.20G137300.1	Serine/threonine-protein kinase PBS1(STK)	–	–	–	9.76
Glyma.17G039800.2	Serine/threonine-protein kinase PBS1(STK)	–	–	–	8.60
Glyma.19G036600.2	Serine/threonine-protein kinase PBS1(STK)	–		–	7.78
Glyma.05G066700.1	Serine/threonine-protein kinase SRK2	7.86	–	–	–
Glyma.01G204200.4	Serine/threonine-protein kinase SRK2	8.39	–	–	–
Glyma.18G054100.1	Serine/threonine-protein kinase WNK1	4.26	3.80	–	–
Glyma.10G092400.1	Serine/threonine-protein kinase WNK1	–	–	7.69	–
Glyma.20G105300.5	Serine/threonine-protein kinase CTR1	8.33	–	–	–
Glyma.07G197200.2	Serine/threonine-protein kinase CTR1	5.89	–	–	–
Glyma.20G105300.1	Serine/threonine-protein kinase CTR1	7.87	–	–	–
Glyma.03G232400.2	Calmodulin	3.52	–	–	–
Glyma.05G237200.1	Calmodulin	–	−3.34	–	–
Glyma.09G182400.1	Calmodulin	–	−4.57	–	–
Glyma.05G028600.4	Calmodulin	–	–	8.98	–
Glyma.13G271800.4	Calmodulin	–	–	8.05	–
Glyma.08G127700.2	Calmodulin	–	–	7.71	–
Glyma.03G178200.2	Calmodulin	–	–	–	9.28
Glyma.08G044400.2	Calmodulin	–	–	−3.37	–
Glyma.07G093900.1	Calmodulin	–	–	−5.80	–

Table 3 Comparison of some DEGs of the Gantai-2-2 and Wan82–178 after bean pyralid larvae feeding *(Continued)*

Gene ID	Gene Annotation	HRK48/ HRK0	HSK48/ HSK0	HRK0/ HSK0	HRK48/ HSK48
Glyma.03G232500.2	Calmodulin	–	–	−4.66	–
Glyma.12G103600.1	Calmodulin	–	–	−4.82	–
Glyma.19G229400.1	Calmodulin	–	–	−4.67	–
Glyma.09G182400.1	Calmodulin	–	–	−5.32	–
Glyma.05G237200.1	Calmodulin	–	–	−4.32	–
Glyma.03G232400.2	Calmodulin	–	–	−4.67	–
Glyma.11G147500.4	Calcium/calmodulin-dependent protein kinase	–	7.67	–	–
Glyma.18G096500.3	Calcium/calmodulin-dependent protein kinase	–	–	−7.61	–
Glyma.14G068400.1	Calcium/calmodulin-dependent protein kinase	–	–	−4.06	–
Glyma.06G098900.2	Calcium/calmodulin-dependent protein kinase	–	–	–	7.96
Glyma.03G246800.1	Calcium-binding protein CML	4.35	–	–	–
Glyma.16G095700.5	Calcium-binding protein CML	8.09	–	–	–
Glyma.06G034700.1	Calcium-binding protein CML	–	−3.39	−4.34	–
Glyma.20G034200.1	Calcium-binding protein CML	–	−7.80	–	–
Glyma.08G265200.2	Calcium-binding protein CML	–	9.50	8.17	8.01
Glyma.02G207800.2	Calcium-binding protein CML	–	–	−8.41	
Glyma.02G186900.1	Calcium-binding protein CML	–	–	8.50	8.49
Glyma.16G214500.1	Calcium-binding protein CML	–	–	−4.78	−5.30
Glyma.18G260700.1	Calcium-binding protein CML	–	–	−3.50	–
Glyma.02G108700.1	Calcium-binding protein CML	–	–	−4.88	–
Glyma.16G059300.1	Calcium-binding protein CML	–	–	−6.04	–
Glyma.07G004300.2	Ca^{2+}-transporting ATPase	8.95	–	−8.45	–
Glyma.07G004300.4	Ca^{2+}-transporting ATPase	9.30	–	–	–
Glyma.15G167500.4	Ca^{2+}-transporting ATPase	8.15	–	–	–
Glyma.11G048300.2	Ca^{2+}-transporting ATPase	−7.52	–	–	–
Glyma.02G186100.2	Ca2 + −transporting ATPase	–	9.37	8.96	–
Glyma.19G136400.2	Ca2 + −transporting ATPase	–	−7.98	–	7.88
Glyma.05G108200.1	Extracellular signal-regulated kinase 1/2	9.60	5.28	–	–
Glyma.08G115300.1	Extracellular signal-regulated kinase 1/2	3.32	–	–	–
Biotic					
Glyma.15G206800.1	Chitinase	8.69	9.09	–	–
Glyma.12G156600.1	Chitinase	5.09	10.32	–	–
Glyma.19G245400.1	Chitinase	4.39	4.20	–	–
Glyma.17G076100.1	Chitinase	4.21	–	–	–
Glyma.02G042500.1	Chitinase	3.34	–	–	–
Glyma.11G124500.1	Chitinase	5.05	–	–	–
Glyma.16G173000.1	Chitinase	3.53	–	−3.07	–
Glyma.13G346700.1	Chitinase	4.58	–	–	–
Glyma.15G143600.1	Chitinase	−3.19	–	–	–
Glyma.15G062500.1	Pathogenesis-related protein 1	3.57	–	–	–
Glyma.15G062400.1	Pathogenesis-related protein 1	3.75	–	–	–
Glyma.13G094200.1	Pathogenesis-related protein 1	4.19	–	–	
biotic. Proteinase inhibitors					

Table 3 Comparison of some DEGs of the Gantai-2-2 and Wan82–178 after bean pyralid larvae feeding *(Continued)*

Gene ID	Gene Annotation	HRK48/ HRK0	HSK48/ HSK0	HRK0/ HSK0	HRK48/ HSK48
Glyma.12G234800.1		6.24	3.97	–	3.12
Glyma.08G341700.1		4.10	4.72	–	3.18
Glyma.08G341300.1		9.65	6.16	–	3.49
Glyma.09G163900.1		6.78	5.46	–	–
Glyma.08G235400.1		9.47	7.27	–	–
Glyma.08G341400.1		9.18	6.00	–	–
Glyma.16G212400.1		9.90	7.93	–	–
Glyma.09G163700.1		7.37	–	–	–
Glyma.02G156800.1	Lectin, mannose-binding 2	7.06	4.28	–	–
Isoprenoid					
Glyma.08G277000.1	1-Deoxy-D-xylulose-5-phosphate synthase	5.15	–	–	–
Glyma.01G134600.4	Homogenitisate phytyltransferase	5.30	2.40	–	–
Glyma.10G070200.1	Homogenitisate phytyltransferase	6.62	–	–	–
Glyma.02G188200.3	Prolycopene isomerase	7.49	–	−8.17	–
Glyma.12G197400.1	Isoprene synthase	9.73	9.14	–	–
Glyma.06G302200.1	Isoprene synthase	3.07	6.08	–	–
Glyma.13G326400.2	(E)-4-hydroxy-3-methylbut-2-enyl-diphosphate synthase	–	8.05	–	–
Glyma.14G004600.2	Acetyl-CoA C-acetyltransferase	–	9.16	–	−8.16
Glyma.15G121400.2	Farnesyl diphosphate synthase	–	7.72	–	–
	Phenylpropanoid				
Glyma.03G181700.1	Phenylalanine ammonia-lyase	4.14	–	–	–
Glyma.02G309300.1	Phenylalanine ammonia-lyase	4.83	–	–	–
Glyma.19G182300.1	Phenylalanine ammonia-lyase	3.10	–	–	–
Glyma.01G004200.4	Caffeoyl-CoA O-methyltransferase	9.08	–	−8.20	–
Glyma.05G147000.1	Caffeoyl-CoA O-methyltransferase	8.73	–	–	–
Glyma.01G187700.1	Caffeoyl-CoA O-methyltransferase	4.32	–	–	–
Glyma.09G281800.1	Caffeic acid 3-O-methyltransferase	5.28	6.62	–	−3.63
Glyma.09G281900.1	Caffeic acid 3-O-methyltransferase	8.60	4.37	–	–
Glyma.07G048900.1	Caffeic acid 3-O-methyltransferase	3.50	–	–	–
Glyma.01G021000.1	Cinnamyl-alcohol dehydrogenase	9.87	8.21	–	–
Glyma.13G255300.1	Cinnamyl-alcohol dehydrogenase	–	7.74	–	−7.74
Glyma.14G221200.1	Cinnamyl-alcohol dehydrogenase	–	–	−2.96	–
Glyma.04G039900.1	Shikimate O-hydroxycinnamoyl transferase	9.00	–	–	–
Glyma.02G283500.1	Shikimate O-hydroxycinnamoyl transferase	–	–	5.69	5.44
Glyma.08G220200.3	Shikimate O-hydroxycinnamoyl transferase	–	–	–	−4.63
Glyma.13G302500.1	Shikimate O-hydroxycinnamoyl transferase	–	–	−4.56	–
Glyma.04G040400.1	Shikimate O-hydroxycinnamoyl transferase	–	–	−4.76	–
Glyma.18G103500.1	Shikimate O-hydroxycinnamoyl transferase	–	–	5.88	–
Glyma.18G267800.1	Trans-resveratrol di-O-methyltransferase	4.35	–	–	7.24
Glyma.10G176500.1	Trans-resveratrol di-O-methyltransferase	7.73	–	–	–
Flavonoid					
Glyma.10G292200.1	Chalcone isomerase	6.19	3.67	–	–
Glyma.20G241700.1	Chalcone isomerase	3.07	–	–	–

Table 3 Comparison of some DEGs of the Gantai-2-2 and Wan82–178 after bean pyralid larvae feeding *(Continued)*

Gene ID	Gene Annotation	HRK48/ HRK0	HSK48/ HSK0	HRK0/ HSK0	HRK48/ HSK48
Glyma.20G241500.2	Chalcone isomerase	7.86	–	–	–
Glyma.06G143000.1	Chalcone isomerase	3.02	–	–	–
Glyma.08G110300.1	Chalcone synthase	8.63	–	–	–
Glyma.08G109500.1	Chalcone synthase	5.15	–	–	–
Glyma.09G075200.1	Chalcone synthase	4.29	–	–	9.44
Glyma.01G228700.1	Chalcone synthase	4.90	–	–	–
Glyma.11G011500.1	Chalcone synthase	3.38	–	–	–
Glyma.01G091400.1	Chalcone synthase	5.45	–	–	–
Glyma.08G110900.1	Chalcone synthase	11.16	–	–	–
Glyma.08G110500.1	Chalcone synthase	5.13	–	–	–
Glyma.09G269600.1	Bifunctional dihydroflavonol 4-reductase/flavanone 4-reductase	5.99	3.67	–	–
Glyma.18G220600.1	Bifunctional dihydroflavonol 4-reductase/flavanone 4-reductase	3.97	–	–	–
Glyma.09G269500.1	Bifunctional dihydroflavonol 4-reductase/flavanone 4-reductase	4.08	–	–	–
Glyma.11G070500.1	Leucoanthocyanidin reductase	5.91	4.47	–	–
Glyma.01G211800.1	Leucoanthocyanidin reductase	8.88	–	–	–
Glyma.01G172700.1	Leucoanthocyanidin reductase	4.78	–	–	–
Glyma.11G070200.2	Leucoanthocyanidin reductase	4.57	–	–	–
Glyma.16G103900.1	Leucoanthocyanidin reductase	3.50	–	–	–
Glyma.01G172900.3	Leucoanthocyanidin reductase	4.62	–	–	–
Glyma.01G172600.1	Leucoanthocyanidin reductase	6.37	–	–	–
Glyma.04G131100.1	Leucoanthocyanidin reductase	–	9.27	–	–
Cytochrome P450					
Glyma.11G062500.1	Cytochrome P450, family 71, subfamily D, polypeptide 9 (flavonoid 6-hydroxylase)	5.95	–	–	–
Glyma.11G062600.1	Cytochrome P450, family 71, subfamily D, polypeptide 9 (flavonoid 6-hydroxylase)	6.49	–	–	–
Glyma.18G080400.1	Cytochrome P450, family 71, subfamily D, polypeptide 9 (flavonoid 6-hydroxylase)	7.66	–	–	–
Glyma.20G008200.4	Cytochrome P450, family 71, subfamily D, polypeptide 9 (flavonoid 6-hydroxylase)	–	–	7.52	–
Glyma.07G083000.1	Cytochrome P450, family 76, subfamily M, polypeptide 7 (ent-cassa-12,15-diene 11-hydroxylase)	3.74	–	–	–
Glyma.11G197300.1	Cytochrome P450, family 79, subfamily A, polypeptide 2 (phenylalanine N-monooxygenase)	6.76	7.50	–	–
Glyma.03G030400.1	cytochrome P450, family 83, subfamily B, polypeptide 1	–	–	−5.19	–
Glyma.03G129200.1	cytochrome P450, family 86, subfamily A, polypeptide 1 (fatty acid omega-hydroxylase)	–	–	–	5.55
Glyma.08G125100.1	Cytochrome P450, family 90, subfamily B, polypeptide 1 (steroid 22-alpha-hydroxylase)	−7.49	–	–	–
Glyma.03G143700.1	Cytochrome P450, family 93, subfamily A, polypeptide 1 (3,9-dihydroxypterocarpan 6a–monooxygenase)	5.07	–	–	–
Glyma.19G144700.1	Cytochrome P450, family 93, subfamily A, polypeptide 1 (3,9-dihydroxypterocarpan 6a–monooxygenase)	4.42	–	–	–
Glyma.13G173500.1	Cytochrome P450, family 93, subfamily C (2-hydroxyisoflavanone synthase)	3.76	–	–	–
Glyma.17G227500.1	Cytochrome P450, family 97, subfamily A (beta-ring hydroxylase)	8.58	–	–	–
Glyma.03G226800.3	Cytochrome P450, family 734, subfamily A, polypeptide 1 (PHYB activation tagged suppressor 1)	6.24	9.44	–	–
Simple phenol					
Glyma.U027300.1	L-ascorbate oxidase	8.27	7.49	–	–
Glyma.01G108200.1	L-ascorbate oxidase	5.84	–	–	–
Glyma.07G142600.1	L-ascorbate oxidase	4.88	–	–	–

Table 3 Comparison of some DEGs of the Gantai-2-2 and Wan82–178 after bean pyralid larvae feeding *(Continued)*

Gene ID	Gene Annotation	HRK48/ HRK0	HSK48/ HSK0	HRK0/ HSK0	HRK48/ HSK48
Glyma.18G193400.1	L-ascorbate oxidase	5.53	–	–	
Transcription factors					
Glyma.16G054400.1	WRKY transcription factor 33/WRKY	5.32	–	–	
Glyma.15G186300.1	WRKY transcription factor 33/WRKY	8.95	–	–	–
Glyma.02G141000.5	WRKY transcription factor 22/WRKY	–	8.16		–
Glyma.12G221500.1	NAC	10.01	5.16	–	–
Glyma.12G149100.1	NAC	6.66	6.43	–	–
Glyma.08G173400.1	NAC	3.57	3.63	–	–
Glyma.16G151500.1	NAC	5.80	–	–	–
Glyma.09G032100.1	Myb proto-oncogene protein, plant/MYB	6.50	–	–	–
Glyma.11G142900.1	Myb proto-oncogene protein, plant/MYB	5.94	–	–	–
Glyma.16G006500.9	MYB	7.83	–	–	−7.57
Glyma.13G144600.3	Myb proto-oncogene protein, plant/MYB	–	7.91	–	–
Glyma.10G180800.1	myb proto-oncogene protein, plant/MYB	–	–	–	−4.06
Glyma.19G104200.1	EREBP-like factor/AP2-EREBP	8.49	–	–	–
Glyma.10G036600.1	Ethylene receptor/AP2-EREBP	4.31	5.95	–	–
Glyma.10G036700.1	Ethylene-responsive transcription factor 1/AP2-EREBP	4.63	7.78	–	–
Glyma.10G036600.1	EREBP-like factor/AP2-EREBP	4.31	5.95	–	–
Glyma.10G223200.1	EREBP-like factor/AP2-EREBP	8.24	–	–	–
Glyma.07G212400.1	EREBP-like factor/AP2-EREBP	−4.88	–	5.52	–
Glyma.16G164800.1	EREBP-like factor/AP2-EREBP	7.76	–	–	–
Glyma.10G061400.1	EREBP-like factor/AP2-EREBP	−8.71	–	–	–
Glyma.19G248900.2	Ethylene-responsive transcription factor 1/AP2-EREBP	–	8.36	–	–
Glyma.19G248900.1	Ethylene-responsive transcription factor 1/AP2-EREBP	–	4.98	–	–
Glyma.13G329700.2	AP2-like factor, euAP2 lineag/AP2-EREBP	–	–	−6.95	–
Glyma.18G159900.2	AP2-EREBP	–	–	–	−7.67

HRK represented the highly resistant line Gantai-2-2; *HSK* represented the highly susceptible line Wan82–178; and the numbers 0 and 48 represented the processing times

The results showed that many genes related to protein kinase could be induced by bean pyralid (Tables 2 and 3). After bean pyralid feeding for 48 h, 17 protein kinases were identified in Gantai-2-2, that were significantly up-regulated, including 2 protein kinase, 1 protein kinase A, 5 LRR receptor-like serine/threonine-protein kinase FLS2, 3 serine/threonine-protein kinase PBS1 (STK), 2 serine/threonine-protein kinase SRK2, 1 serine/threonine-protein kinase WNK1 and 3 serine/threonine-protein kinase CTR1. Additionally, 7 protein kinases were identified in Wan82–178, of which 6 DEGs were up-regulated, including 1 protein kinase A, 2 LRR receptor-like serine/threonine-protein kinase FLS2, 2 serine/threonine-protein kinase PBS1 (STK) and 1 serine/threonine-protein kinase WNK1. The results showed that many genes related to Ca^{2+} could be induced by bean pyralid (Table 3). After bean pyralid feeding for 48 h, 7 DEGs associated with Ca^{2+} signaling were identified in Gantai-2-2, of which 6 DEGs were up-regulated, including 1 calmodulin, 2 calcium-binding protein CML and 3 Ca^{2+}-transporting ATPase. Additionally, 8 DEGs associated with Ca^{2+} signaling were identified in Wan82–178, of which 3 DEGs were up-regulated, including 1 calcium/calmodulin-dependent protein kinase, 1 calcium-binding protein CML and 1 Ca^{2+}-transporting ATPase.

Some genes induced by abiotic stress may have been induced by insects as well, for example, the genes associated with harm and drought stresses are often induced by chewing insects [19–21]. Meanwhile, Our results showed that many genes related to biotic and abiotic stresses were induced by bean pyralid too. After bean pyralid feeding for 48 h, many genes related to biotic stress could be induced by bean pyralid (Tables 2 and 3). For example, 43 DEGs related to biotic stress were identified in Gantai-2-2, of which 39 DEGs were up-regulated, including 12 PR-proteins, 8 proteinase inhibitors, 9 chitinase, 1 lectin mannose-binding 2, and so on. Additionally, 22 DEGs were associated with biotic stress identified in Wan82–178, of which 20 DEGs were up-regulated, including 7 PR-proteins, 7 proteinase inhibitors, 3 chitinase, 1 lectin mannose-binding 2, and so on. In addition, 3 proteinase inhibitors were found to be higher in Gantai-2-2 than in Wan82–178 at 48 h. After bean pyralid feeding for 48 h, 23 DEGs associated with abiotic stress, such as pathogen infection, heat stress, cold stress and drought stress, were identified in Gantai-2-2, of which 17 DEGs were up-regulated. Additionally, 16 DEGs associated with abiotic stress were identified in Wan82–178, of which 9 DEGs were up-regulated (Tables 2 and 3). When comparing Gantai-2-2 with Wan82–178 at 0 h feeding, 3 DEGs associated with abiotic stress were up-regulated and 9 DEGs were down-regulated. When comparing Gantai-2-2 with Wan82–178 at 48 h feeding, 3 DEGs associated with abiotic stress were up-regulated and 3 DEGs were down-regulated.

It has widely been reported that secondary metabolism pathways, such as isoprenoid, phenylpropanoid, flavonoid, cytochrome P450 and simple phenol metabolism, were involved in plant resistance to insect feeding, or possibly functioned as direct defense compounds, antioxidants, signaling molecules or insect toxins [22–24]. Our results showed that many genes related to secondary metabolism pathways could be induced by bean pyralid (Tables 2 and 3). After bean pyralid feeding for 48 h, 6 DEGs related to isoprenoid biosynthesis pathway were identified in Gantai-2-2 that were all up-regulated, including 1 1-deoxy-D-xylulose-5-phosphate synthase (DXS), 2 homogentisate phytyltransferas (HPT), 1 prolycopene isomerase and 2 isoprene synthase. Additionally, 6 DEGs were identified in Wan82–178 that were all up-regulated, including 1 HPT, 2 isoprene synthase, 1 (E)-4-hydroxy-3-methylbut-2-enyl-diphosphate synthase, 1 acetyl-CoA C-acetyltransferase and 1 farnesyl diphosphate synthase. When comparing Gantai-2-2 with Wan82–178 at 0 h feeding, 1 prolycopene isomerase was down-regulated. When comparing Gantai-2-2 with Wan82–178 at 48 h feeding, 1 acetyl-CoA C-acetyltransferase was down-regulated. There were 13 DEGs associated with phenylpropanoid biosynthesis pathway were identified in Gantai-2-2 that were all up-regulated, including 3 phenylalanine ammonia-lyase (PAL), 3 caffeoyl-CoA 3-O-methyl transferase (CCOAOMT), 3 caffeic acid 3-O-methyltransferase (COMT), 1 cinnamyl-alcohol dehydrogenase (CAD), 2 trans-resveratrol di-O-methyltransferase and 1 shikimate O-hydroxycinnamoyl transferase. Additionally, 4 DEGs were identified in Wan82–178 were all up-regulated, including 2 COMT and 2 CAD. When comparing Gantai-2-2 with Wan82–178 at 0 h feeding, 1 CCOAOMT, 1 CAD and 2 shikimate O-hydroxycinnamoyl transferase were down-regulated, and 2 shikimate O-hydroxycinnamoyl transferase were up-regulated. When comparing Gantai-2-2 with Wan82–178 at 48 h feeding, 1 COMT, 1 CAD and 1 shikimate O-hydroxycinnamoyl transferase were down-regulated, and 2 shikimate O-hydroxycinnamoyl transferase was up-regulated. There were 22 DEGs associated with flavonoid biosynthesis pathway identified in Gantai-2-2 that were all up-regulated, including 4 chalcone isomerase (CHI), 8 chalcone synthase (CHS), 3 dihydroflavonol4-reductase (DFR) and 7 leucoanthocyanidin reductase (LAR). Additionally, 4 DEGs were all up-regulated identified in Wan82–178, including 1 CHI, 1 DFR and 2 LAR. When comparing Gantai-2-2 with Wan82–178 at 48 h feeding, 1 CHS was up-regulated. After bean pyralid feeding for 48 h, 26 DEGs related to cytochrome P450 (CPY) were identified in Gantai-2-2, of which 25 DEGs were up-regulated. Additionally, 8 DEGs were identified in Wan82–178, of which 7 DEGs were up-regulated. When comparing Gantai-2-2 with Wan82–178 at 0 h feeding, 1 CPY was up-regulated and 1 CPY was down-regulated. When comparing Gantai-2-2 with Wan82–178 at 48 h feeding, 1 CPY was up-regulated. After bean pyralid feeding for 48 h, 4 and 1 L-ascorbate oxidase (L-AO) were identified in Gantai-2-2 and Wan82–178, respectively, which were all up-regulated.

Bean pyralid-induced the transcription factor genes

Transcription factors, known as trans-acting factors, are a type of DNA binding protein that regulates the transcription of the target genes by combining with cis-acting elements in the gene promoter [25]. Research demonstrated that plant transcription factors, including NAC, MYB, WRKY, were involved in the plant defense responses [26]. Our results showed that 25 transcription factors were identified in Gantai-2-2, including 4 NAC, 7 AP2-EREBP, 3 MYB, 2 WRKY, 1 bHLH, 2 C3H, and so on, of which 21 were up-regulated and 4 were down-regulated, after bean pyralid larvae feeding for 48 h (Table 4). Fifteen transcription factors were identified in Wan82–178, including 3 NAC, 4 AP2-EREBP, 1 MYB, 1 WRKY, 2 bHLH, 3 C3H, and so on, which were all up-regulated, after bean pyralid larvae feeding for 48 h. When comparing Gantai-2-2 with Wan82–178 at 0 h feeding, 7 transcription factors were identified, of which 1 AP2-EREBP, 1 C2C2-YABBY and 1 FAR1 were up-regulated, and 1 AP2-EREBP, 2 MYB and 1 LIM were down-regulated. When comparing Gantai-2-2 with Wan82–178 at 48 h feeding, 5 transcription factors were identified, of which 1 FAR1 was up-regulated, and 1 AP2-EREBP, 2 MYB and 1 C3H were down-regulated (Table 4).

Analysis of DEGs by qRT-PCR

The quantitative real time-PCR (qRT-PCR) technology was used to verify the credibility of the RNA-Seq. Therefore, we used the qRT-PCR technology to verify 17 DEGs identified by RNA-Seq. It was found that the qRT-PCR expression patterns of 17 DEGs were all consistent with the RNA-Seq results (Fig. 8), which indicated that the RNA-Seq results were reliable in the present study.

Discussion

Influence of the genes related to ROS removal on bean pyralid response

As an electron acceptor of H_2O_2, POD oxidizes various materials in the process of secondary metabolites and can produce phenoxy and oxygen free radicals by interacting with some phenolic compounds, which may also directly interfere with the feeding of insects or reduce the nutrition level of leaves, thereby lowering the edibility of the plants [27]. Meanwhile, POD has been found to have significant insecticidal effects on Lepidoptera and Coleoptera [28]. Previous studies have shown that POD could be induced by insects in wheat [29], sorghum [30], cucumber [31] and rice [32, 33]. PPO is widely distributed in plants and can potentially catalyze the oxidation of polyphenols into a keto metal enzyme, it can directly oxidize acid into quinone using O_2 as an oxidation substrate, then produce amino acids and proteins that are difficult to digest and toxic to herbivores.

Table 4 Transcription factors (TF) related to the insect resistance identified in different alignment schemes

TF family	HRK48/HRK0		HSK48/HSK0		HRK0/HSK0		HRK48/HSK48	
	up	down	up	down	up	down	up	down
NAC	4	0	3	0	0	0	0	0
PLATZ	1	0	0	0	0	0	0	0
AP2-EREBP	5	2	4	0	1	1	0	1
MYB	3	0	1	0	0	2	0	2
bHLH	1	0	2	0	0	0	0	0
WRKY	2	0	1	0	0	0	0	0
C3H	2	0	3	0	0	0	0	1
C2H2	1	0	0	0	0	0	0	0
LIM	1	0	0	0	0	0	0	0
TUB	0	1	0	0	0	1	0	0
GRF	1	0	0	0	0	0	0	0
ARF	0	1	0	0	0	0	0	0
Tify	0	0	1	0	0	0	0	0
C2C2-YABBY	0	0	0	0	1	0	0	0
FAR1	0	0	0	0	1	0	1	0
Sum	21	4	15	0	3	4	1	4

HRK represented the highly resistant line Gantai-2-2; *HSK* represented the highly susceptible line Wan82–178; numbers 0 and 48 represented the processing time

Meanwhile, as an anti-nutritional protein, PPO can produce protein affinity, and therefore reduce the edibility of plants, which plays an important role in the defense of insect feeding [34, 35]. Previous research studies have shown that the activity of PPO increased after tomatoes were eaten by *Spodoptera exigua* or were treated with oral secretions of *Spodoptera exigua* [36, 37]. GST is encoded by a large and diverse family of genes, that plays important roles in the responses of plants to oxidative damage induced by various environmental conditions, especially in the resistance to resist the toxic effects of ROS on cells [38–40]. In oxygen stress reactions, TRX would pass restoring forces to the reductase, which could potentially remove the lipid peroxide or repair the oxidized protein, resulting in an alleviation of oxygen stress [41]. TRXh5 and TRXh8 in *Arabidopsis thaliana* were strongly induced under biotic or abiotic stress [42, 43]. After bean pyralid larvae feeding for 48 h, we found that POD, PPO and GST were all significantly up-regulated in Gantai-2-2 and Wan82–178. These results suggested that these genes were involved in the defense reactions of soybean to bean pyralid. TRX1 was up-regulated in Gantai-2-2, which suggested that TRX1 involved in defense responses to bean pyralid in the highly resistant material only.

Influence of plant hormones on bean pyralid response

JA signaling pathway is implicated in insect-induced responses in plant; it was involved in the signal transduction of insect defense, regulated the expression of plant downstream defense genes, and significantly induced responses of defense systems in plant, thereby effectively reducing pests [44]. After bean pyralid feeding, 3 key genes involved in JA biosynthetic pathway were identified in Gantai-2-2 and Wan82–178, namely, LOX, α-DOX and OPDA, which were significantly up-regulated. LOX is the key enzyme in the synthesis of the JA [45], and plays an important signal factor in plant induction defense pathways [46]. Adversity stresses, such as insects and diseases, could induce single or multiple LOX genes in plants [47, 48]. Additionally, α-DOX could catalyze the oxidation of fatty acids and produce 2-hydrogen peroxide fatty acids, and it is also an enzyme related to plant stress resistance in JA biosynthetic pathway [32]. For example, α-DOX can be induced by insects in rice [32, 49]. OPDA is the intermediate product of JA biosynthesis pathway with biological activities that can regulate plant defense genes. For example, OPDA has been found to induce defense genes in *Arabidopsis thaliana* which resist attacks from *Bradysia odoriphaga* [50]. As a plant endogenous hormone, ET regulates the defense responses of plants to pests and diseases [51, 52]. Our results found that 3 types of

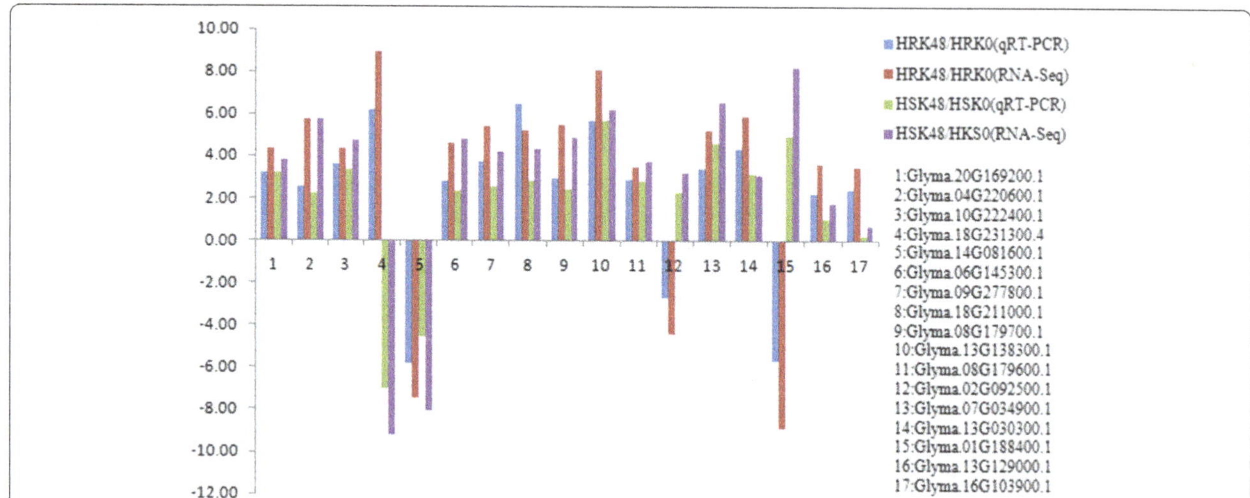

Fig. 8 DEGs confirmed by qRT-PCR using the same sample as that in RNA-Seq. X-axis represented gene name, the blue column represented qRT-PCR results in HRK48/HRK0, the red column represented RNA-Seq results in HRK48/HRK0, the green column represented qRT-PCR results in HSK48/HSK0, and the purple column represented RNA-Seq results in HSK48/HSK0; Y-axis represented the relative level of gene expression

genes related to ET biosynthesis and signal transduction pathway. For example, most of ACC oxidase, ethylene receptor and ERF1 genes were up-regulated after bean pyralid feeding. ACC oxidase is one of the key enzymes in ethylene biosynthesis pathway [52]. Ethylene receptor is an upstream component of ethylene signal transduction pathway, that plays an important role in plant growth and responses to adversity stresses [53]. ERF is a specific plant transcription factor that plays an important role in plant cell growth, hormone regulation, disease resistance and abiotic stresses [54–56]. Auxin can directly act on cell membranes or intracellular components, and affects some cellular responses. In addition, it can also indirectly regulate the expression of genes, and has a direct impact on plant stress [57, 58]. After bean pyralid feeding, some genes related to auxin synthesis and signal transduction pathways were identified, such as IAA-amino acid hydrolase, auxin responsive GH3 (Gretchen Hagen3) gene family and SAUR (small auxin up RNA) family protein. Early/primary auxin response genes were composed of three major gene families: Aux/IAA (auxin/indoleacetic acid), GH3 and SAUR transcription factor family [59]. GH3 gene family is a typical auxin-responsive gene family and can promote amino acids to be combined with IAA, JA and SA, then changed the concentration of biologically active forms within the cells. This aided in regulating plant growth, development and defense responses [60]. SAUR gene family is known to be the specific and largest family among the auxin response factors in plants and is related to environmental stimuli [61, 62]. These results indicated that JA, ET and auxin may be involved in elevating the basal resistance of soybean to herbivory.

Influence of the intracellular signal transduction pathway on bean pyralid response

Protein kinase is involved in the signal transduction pathway of plants and plays an important role in the defense responses of plants [63]. Protein kinases could be induced by bean pyralid, including protein kinase, protein kinase A, LRR receptor-like serine/threonine-protein kinase FLS2, serine/threonine-protein kinase PBS1 (STK), serine/threonine-protein kinase SRK, serine/threonine-protein kinase WNK1, and serine/threonine-protein kinase CTR1. LRR receptor-like kinase is the largest family of receptor protein kinase in plants; it plays an important role in the regulation of plant growth, development, biotic and abiotic stresses [64]. Serine/threonine-protein kinase (STK) is an important signal molecule; when plant suffers stimulation, such as pest feeding, salinity, drought stress, trauma, cytokines or hormones, STK is rapidly activated in the serine and threonine residue phosphorylation sites and is further activated in the downstream signal molecules through phosphorylation cascades, which activate specific signaling pathways. It is eventually transmitted by an outside signal to the nucleus, and activates or inhibits specific genes [64–66]. These results indicated that protein kinases played an important role in the defense against bean pyralid.

When the plants were stimulated by external environmental and received the signal, the receptor activated the calcium channels on the membrane through a series of phosphorylation reactions to cause calcium ions to be released from the calcium base into the cytoplasm. This response led to an increase in the concentration of Ca^{2+} in the cytoplasm, which then activated the plant defense response [67, 68]. Previous studies have shown that Ca^{2+} could be induced by *Spodoptera littoralis* in *Phaseolus lunatus* [69] and *Ginkgo biloba* [70]. After bean pyralid feeding, most of the DEGs related to Ca^{2+} signaling identified in Gantai-2-2 were up-regulated, but in Wan82–178 were down-regulated. It was suggested that Ca^{2+} signaling was involved in the defense responses to bean pyralid in the highly resistant material. After soybean was stimulated by bean pyralid, the concentrations of Ca^{2+} in the cytoplasm changed; therefore, calcium was sent to transfer the stimulation signal. However, Ca^{2+} signaling may have played a negative role in the regulation of defense responses in the highly susceptible material.

Influence of the genes related to biotic stress on bean pyralid response

When plants suffer from pest stress, a variety of defense proteins which inhibit the insect from producing digestive enzymes are produced to resist insect pests, including PR, proteinase inhibitors, chitinase and lectin [71, 72] and thereby destroy the insect's normal digestive absorption functions, then disrupt their nutrient uptake and utilization, which could ultimately lead to malnutrition and inhibited growth of the insects [73].

PR proteins are generated and accumulated after pathogen invasions or abiotic stress and are inducible components in the self defense mechanisms of plants [74–76]. Uknes et al. found that the contents of PR-1, PR-2 and PR-5 were increased, when the *Arabidopsis thaliana* were treated with TCV and INA, and its disease resistance was increased [77]. After bean pyralid feeding, most of the DEGs related to PR were induced. Our results speculated that the PR proteins might be involved in the defense responses of soybean to bean pyralid.

Proteinase inhibitors can inhibit the activity of chymotrypsin and trypsin in the digestive tract of the Lepidoptera and Diptera, to inactivate the related digestive enzyme in intestines, and thereby interfering with the insects' normal feeding and digestion. Therefore, the insects cannot attain enough nutrition, which in turn affects their growth and development, then leads to insect death [78, 79]. In other cases, it caused excessive production of proteases in insects, which resulted in a deficit of essential amino acids, then caused insect death [80]. Our results showed that the proteinase inhibitors could be induced by bean pyralid, so we assumed that the proteinase inhibitors were involved in the defense responses.

Chitinase could destroy the peritrophic membranes of insects [81]. When insect eats plants with sustained chitinase expression, its digestive tract becomes damaged and its epidermis cannot form normally [82]. Chitinase could be induced by aphid and spider mites in sorghum [83] and tomato [84], respectively. Our results showed that chitinase could be induced by bean pyralid as well. It was assumed that after the chitinase was absorbed into the insects' bodies, chitin was hydrolyzed, which inhibited the growth and development of insects and achieved the purpose of defense for the plants [85].

Lectin has at least one non-catalytic domain-specific reversible binding to monosaccharides or oligosaccharides, and after being absorbed into the gut of an insect, lectin induces local or systematic toxic effects, which bond to glycoproteins on the peritrophic membrane and damage the structure of the peritrophic membrane. This response causes insect growth arrest, antifeedant and even death [86, 87]. Previous results have shown that lectin displays insecticidal activities against a variety of insects. For example, after *Heliothis virescens* [88], *Myzus persicae* [89] and *Lacanobia oleracea* [90] fed on lectin-transgenic plants, their survival rates and fecundity were greatly reduced. Our results showed that the lectin, mannose-binding 2 could be induced by bean pyralid. It was assumed that the lectin that was released from damaged insect cells of the soybean combined with the chitin of the peritrophic membranes, sugar compounds of the digestive tract epithelial cells, and the glycosylation digestive enzymes, after bean pyralid feeding. This affected the normal absorption of nutritional materials. It also induced disease in the digestive tracts and promoted gastrointestinal bacterial reproduction, thereby causing insect growth inhibition or death, which achieved the defensive purpose of killing the pests [91].

Influence of the genes related to the secondary metabolism on bean pyralid response

Isoprenoid biosynthesis pathway is an important secondary metabolism route that widely exists in plants. It has the ability to synthetize isoprenoid compounds, and plays an important role in plant resistance reactions, in addition to participating in the growth and development of plants [92]. The genes related to isoprenoid biosynthesis pathway could be induced by bean pyralid, which speculated that isoprenoid biosynthesis pathway was involved in the defense reactions of soybean to bean pyralid.

Phenylpropanoid and its derivatives play an important role in plant development and stress responses. Environmental stresses can promote plant carbon synthesis, which changes into benzene propane synthesis, causing the accumulations of large amounts of substances related to plant stress resistance, such as lignin, flavone, flavonol and alkaloids. These substances have been found to improve the resistance of plants to various biotic and abiotic stresses [93]. The genes related to phenylpropanoid biosynthesis pathway, including PAL, CCOAOMT, COMT and CAD, could be induced by bean pyralid. PAL is not only the enzyme that catalyzes the first step of phenylpropanoid biosynthesis pathway but also the key and rate limiting enzyme for the pathway. Various metabolite substances that are catalyzed by PAL, such as lignin, flavonoids and phytoalexin, were related to plant insect resistance [94]. PAL can be induced by many types of biotic and abiotic stresses, including pathogen infections, insect feeding, mechanical injuries and exogenous plant hormones [95–97]. CCOAOMT is a type of important methyltransferase in plant lignin biosynthesis [98, 99]. COMT catalyzes the reaction that converts caffeic acid into ferulic acid, which is an important step in the biosynthesis of lignin monomers [100]. CAD is the last step in lignin synthesis metabolic pathway and plays a key role in lignin biosynthesis [101, 102]. PAL, CCOAOMT, COMT and CAD were related to lignin, which suggested that soybean might have enhanced its cell mechanical strength through the synthesis of lignin, which reinforced the cell walls, strengthened the cell to blocked insect feeding, and defended the insect pests by nutrient limitation [103, 104].

Flavonoid plays an important role in insect-induced, disease, and other stress responses in plants [105]. The genes related to flavonoid biosynthesis pathway, including CHI, CHS, DFR and LAR, were found to be induced by bean pyralid. CHS and CHI are two key enzymes in flavonoid biosynthesis pathway; CHS catalyzes the condensation reaction of 4-coumaric acid-CoA and acyl-CoA, which causes chalcone to be formed, and flavonoid is then formed under the action of CHI, a necessary enzyme that syntheses flavanone, flavone, flavonol and anthocyanin substances [106, 107]. The amount of flavonoid metabolites was directly affected by the expression of CHS and CHI, and the loss of expression ability or loss of enzyme function [108]. DFR and LAR are key enzymes in flavonoid synthesis, and during the process of plant flavonoid biosynthesis, anthocyanins, proanthocyanidins and other common synthesis precursors are leucoanthocyanins. DFR was catalyzed by flavanonols to generate leucoanthocyanidin, LAR converted leucoanthocyanidin into 2,3-trans-flavan-3-ols [109]. Previous studies have shown that over-expressions of DFR could improve the content and antioxidant capacities of flavonoid [110]. Our results suggested that flavonoid biosynthesis pathway played an important role in the defense against bean pyralid and that flavonoid accumulation is an important characteristic of responses to abiotic stress in plants.

Cytochrome P450 (CYP) is a type of pheme containing oxidoreductases. It can catalyze some substances with a defensive function, such as sterol, isoflavone, alkaloid, terpenoid, and so on [111, 112]. CYP plays an important role in the defense of organisms against pests and abiotic stress [113]. CYP genes can be induced by aphid in soybean [114]. Our results showed that some genes related to CYP can also be induced by bean pyralid. Previous studies have shown that cyanogentic glycoside which was catalyzed and synthesized by the CYP79A and CYP71E1 genes in sorghum, was harmful to pests [115]. Therefore, it was speculated that soybean would use the CYP family to mitigate the threat of insect infestation.

AO is a copper-containing oxidase family that is wide spread in plants. It is a type of simple phenol and can catalyze the ascorbic acid (AA) of protoplasts in vitro and can oxygenate AA to generate monodehydroascorbate (MDHA), which regulates the overall oxidation state of the AA library of plant protoplasts in vitro [116, 117]. When activity of AO increased, the ratio of oxidation and reduction in extracellular AA was also increased. There was a significant redox gradient on both sides of the plasma membrane, which modulated plant defense responses against insect attack and other stressed [118–120]. Our results showed that some genes related to L-AO can also be induced by bean pyralid. Therefore, L-AO was involved in the response reactions of soybean to bean pyralid.

Transcription factor on bean pyralid response

WRKY is one of the largest families of transcription factors in plants and plays an important role in the regulation of plant growth and development as well as biotic and abiotic stresses [121–123]. Previous studies have shown that WRKY could be induced by insect stress [124]. For example, one WRKY23 gene was induced by *Heterodera schachtii* in *Arabidopsis thaliana* [125] and 20 WRKY transcription factors were induced by cotton boll weevil in *Gossypium hirsutum* [126]. NAC is known to be a second major transcription factor in plants and is also a specific transcription factor. NAC can activate downstream genes when plants are exposed to biotic or abiotic stress, so it is involved in the stress responses of plants [127–129]. Some NAC genes were induced by insects in rape [130], sugarcane [131] and *Gossypium hirsutum* [126]. MYB is one of the central regulators of development, metabolism, and response to abiotic and biotic stresses [132]. One of the known mechanisms is that MYB transcription factor contribute to the accumulation of flavonoids to protect plants from radiation injury and enhance their resistance to cold and insect pests in plants [133]. MYB could be induced by small cabbage white caterpillars [134] and cotton boll weevil [126]. AP2/EREBP is a type of specific transcription factor which is bound with ethylene responsive elements. It is able to combine with the GCC box of the promoter region of some resistance related genes, and carry out the function of activating the expression of these genes [135]. AP2/EREBP can potentially regulate the molecular responses of plants to hormones, pathogens, low temperatures, drought and high salt, which could improve the tolerance of crops to stress [136–138]. For example, the over expression of *Tsi1* gene encoding EREBP/AP2 transcription factor increased the resistance to pathogens in tobacco, as well as the resistance ability to osmotic stress [138]. Our results showed that the transcription factor, such as WRKY, NAC, MYB and AP2/EREBP, could be induced by bean pyralid. It was confirmed that these transcription factors were involved in the defense reactions to bean pyralid.

Conclusions

To explore the defense mechanisms of soybean resistance to bean pyralid, the comparative transcriptome sequencing was completed between the leaves infested with bean pyralid larvae and no worm of soybean (Gantai-2-2 and Wan82–178) on the Illumina HiSeq™ 2000 platform. The results showed that there were 1064 and 680 DEGs were identified in the Gantai-2-2 and Wan82–178 after bean pyralid larvae feeding for 48 h, respectively, compared to 0 h. When comparing Gantai-2-2 with Wan82–178, 605 DEGs were identified at 0 h feeding, and 468 DEGs were identified at 48 h feeding. The DEGs were divided into three categories, including the DEGs with non-bean pyralid-induced genotype, bean pyralid-induced DEGs that appeared in both materials and bean pyralid-induced genotype DEGs. According to GO and KEGG functional and metabolic pathway analysis combined with the previously reported literatures, we concluded that the response to bean pyralid feeding might be related to the disturbed functions and metabolism pathways of some key DEGs, such as DEGs involved in the ROS removal system, mainly POD, PPO, GST and TRX; phyto-hormone signaling pathways, mainly JA, ET and auxin pathway; intracellular signal transduction pathways, including plant protein kinases and Ca^{2+} signaling; secondary metabolism, such as isoprenoid biosynthesis pathway, phenylpropanoid biosynthesis pathway, flavonoid biosynthesis pathway, CYP and L-AO; transcription factors, such as WRKY, NAC, MYB and AP2/EREBP. Meanwhile, bean pyralid activated a large number of genes related to biotic and abiotic stresses. These results will help to elucidate the molecular mechanism of response to bean pyralid feeding in soybean, and provide a valuable resource of soybean defense genes that will benefit other studies in this field. Future research will focus on the cloning and transgenic function validation of possible candidate genes associated with the anti-bean pyralid soybean.

Methods

Plant materials

The tested materials Gantai-2-2 (highly resistant line) [3] and Wan82–178 (highly susceptible line) [3] were planted inside insect net rooms in experimental fields at the Guangxi Academy of Agricultural Sciences. The plants were sown in three rows, with 10 strains per row. During the entire growth period of soybeans, pesticides and fertilizers were not used. When the plants grew to a level of 10 compound leaves, the seventh compound leaves on the left side were collected before the infestation. There was a total of five plants for each sample. Then, each of the samples was simultaneously artificially infested with 5 four-year-old bean pyralids on the right side of the seventh leaves, counted downward. There were two biological repetitions. In each sample, five leaves were mixed, and they were frozen with liquid nitrogen and stored at −80 °C for further use.

Total RNA extraction

The total RNA (5 μg) from the leaves (100 mg per sample) of Gantai-2-2 and Wan82–178 by using TRIzol kit (Invitrogen, Carlsbad, CA, USA) according to the manufacture's protocol. Briefly 1.3 ml of TRIzol kit was added into 100 mg of leaf sample, the sample was homogenized using power homogenizer and centrifuged at 12,000×g for 10 min at 4 °C. After the fatty layer was removed and discarded, the cleared supernatant was transferred into a new tube and mixed with 0.2 ml of chloroform. The sample tube was shaken for 15 s, followed by an incubation for 3–5 min at room temperature. Next, the sample was centrifuged 12,000×g for 10 min at 4 °C and the aqueous phase was moved into a new tube for RNA precipitation. For precipitating RNA from each sample, 10 μg of RNase-free glycogen was added to the aqueous phase as a carrier, followed by 0.5 ml of 100% isopropanol, then samples were placed at −20 °C for 1 h and centrifuged at 13,600×g for 20 min at 4 °C and discarded the supernatant. To wash the RNA pellet, we added 1.0 ml of 75% ethanol into the tube, vortexed the tube gently, centrifuged the tube 13,600×g for 3 min at 4 °C and discarded the wash. The RNA pellet was air-dried, suspended in 50 μl Nuclease-free water, incubated at 55–60 °C for 10 min. RNA concentration, purity and integrity were determined using a NanoDrop2000 (Thermo Fisher Scientific, Waltham, MA, USA) and an Agilent 2100 Bioanalyzer (Agilent, Santa Clara, CA, USA).

cDNA library construction and transcriptome deep sequencing

Equal amount of total RNA (5.0 μg) was used for cDNA library construction using TruSeq™ RNA Sample Preparation Kit v2 (Illumina) (Illumina, SanDiego, CA, USA) and the cDNA library was sequenced on an Illumina HiSeq™ 2000 platform (Hiseq2000 Truseq SBS Kit v3-HS (200 cycles), Illumina) following the manufacturers' protocols. Briefly, Dynal Oligo (dT) beads (Invitrogen) were obtained poly(A) mRNAs. Then mRNAs were chemically fragmented into ~200 nt fragments. First-strand cDNA was generated by using reverse transcriptase and random primers, the second strand cDNA synthesis using DNA Polymerase I (Invitrogen) and RNase H (Invitrogen) treatment. The cDNA fragments were end repaired by using End Repair Mix (Illumina) reagent, Finally, purified and enriched to create the final cDNA library. Eight cDNA libraries were sequenced by using pair-end (2 × 100 bp) sequencing technology with an Illumina HiSeq™ 2000 sequencer (The RNA-Seq test and its results were analyzed by the BGI Tech Solutions Co., Ltd. (BGI Tech) (BGI, Shenzhen, PR China).

Raw read filtering and mapping to the reference genome and gene sequences

SOAPfuse software was used to filter noise for the original sequencing reads [139]. Raw reads of eight libraries were performed by removing adapter sequences and low-quality reads, higher N rate sequences, and excessively short sequences. After passing the QC process of the alignment, the results were used for further analysis. The remaining high-quality reads were submitted for mapping analysis against the soybean reference genome (ftp://ftp.jgi-psf.org/pub/compgen/phytozome/v9.0/early_release/Gmax_275_Wm82.a2.v1/, version Glyma 2.0, 975 Mb), using BWA software and Bowtie software and allowing two base mismatches. We assembled the transcripts with reads using Cufflinks (http://cufflinks.cbcb.umd.edu/) [140].

DEG analysis

Genes and isoforms expression level are quantified by a software package: RSEMV1.2.12 (RNASeq by Expectation Maximization) [141]. RSEM computes maximum likelihood abundance estimates using the Expectation-Maximization (EM) algorithm for its statistical model, including the modeling of paired-end (PE) and variable-length reads, fragment length distributions, and quality scores, to determine which transcripts are isoforms of the same gene. The expression quantity of each gene (fragments per kilobase of exon model per million mapped fragments, FPKM) was used for the calculated expression level, the formula for which is as follows:

$$\mathrm{FPKM}\ (\mathrm{A}) = \frac{10^6 C}{N/10^3}$$

where FPKM (A) is the expression of gene A, C is the number of fragments that are uniquely aligned to gene A, N is the total number of fragments that are uniquely aligned to all genes, and L is the number of bases on gene A. The FPKM method can eliminate the influence of different gene lengths and sequencing discrepancies in the calculation of gene expression. Therefore, the calculated gene expression can be directly used for comparing the difference in gene expression among samples. According to the Noiseq method, an absolute value of |Log2FC (Fold Change)| ≥ 1 and diverge probability ≥0.8 were used as the threshold to screen DEGs. In the DESeq2 method, the absolute value of |Log2FC (Fold Change)| ≥ 1 and p-adj ≤ 0.001 were used as the threshold to judge the significance of the gene expression difference. For the EdgeR method, the absolute value of |Log2FC (Fold Change)| ≥ 1 and FDR (False Discovery Rate) ≤ 0.001 were used as the threshold to judge the significance of the gene expression difference.

Bioinformatics analysis

GO (GO, http://www.geneontology.org/) and functional enrichment analysis were conducted on all DEGs using TermFinder software (http://www.yeastgenome.org/help/analyze/go-term-finder). Then, all DEGs were mapped to a pathway in the KEGG database (http://www.genome.jp/kegg/pathway.html) using Blast_v2.2.26 software. A *p*-value was used as ≤0.05 was used as the threshold to judge the significance of the GO and KEGG pathway enrichment analyses.

The transcript version number of Wm82.a2.v1, which was obtained from the sequencing analysis was deleted from the website (http://www.soybaen.org/correspondence/) and only the names of the genes were retained. These names were converted into the differential gene transcription version number of Wm82.a2.v1. MapMan software (http://www.gabipd.de/projects/MapMan/) was used to assign the pathways and functional classifications to the DEGs. The latest Osa_MSU_v7 mapping file and pathways files were downloaded from the website.

Quantitative real time-PCR (qRT-PCR) analysis

qRT-PCR analysis was used to verify the RNA-Seq results. Primer Premier 5.0 (Premier Biosoft International, Palo Alto, CA) was used to design primers for qRT-PCR experiment (Additional file 13: Table S13). PrimeScript RT Reagent kits with gDNA Eraser (Takara, Dalian, China) were used to reverse transcriptase and synthesize the cDNA. The qRT-PCR reaction mixture (25 µl) contained 12.5 µl SybrGreen qPCR Master Mix (2 × concentration, Ruian Biotechnologies, Shanghai, China), 0.5 µl reverse and forward primers (10 µM), 2 µl cDNA and 9.5 µl ddH_2O. Then, the qRT-PCR reaction was performed on an ABI7500FAST Real-Time PCR System (Applied Biosystems, Foster City, CA, USA) as follows: 2 min at 95 °C, 40 cycles of heating at 95 °C for 10 s and annealing at 60 °C for 40 s. The *β-actin* gene was used as the internal control to calculate the relative expression. $2^{-\Delta\Delta cq}$ method was used to calculate the differential expression of a gene in two samples. Three technological and three biological reactions ($n = 3 \times 3$) were performed for every gene in each sample.

Additional files

Additional file 1: Table S1. Total number of DEGs between HRK48 and HRK0. (XLS 163 kb)

Additional file 2: Table S2. Total number of DEGs between HSK48 and HSK0. (XLS 111 kb)

Additional file 3: Table S3. Total number of DEGs between HRK0 and HSK0. (XLS 119 kb)

Additional file 4: Table S4. Total number of DEGs between HRK48 and HSK48. (XLS 101 kb)

Additional file 5: Table S5. GO analysis the DEGs of HRK0-VS-HRK48. (XLS 164 kb)

Additional file 6: Table S6. GO analysis the DEGs of HSK0-VS-HSK48. (XLS 141 kb)

Additional file 7: Table S7. GO analysis the DEGs of HSK0-VS-HRK0. (XLS 137 kb)

Additional file 8: Table S8. GO analysis the DEGs of HSK48-VS-HRK48. (XLS 119 kb)

Additional file 9: Table S9. Pathway analysis the DEGs of HRK0-VS-HRK48. (XLS 108 kb)

Additional file 10: Table S10. Pathway analysis the DEGs of HSK0-VS-HSK48. (XLS 92 kb)

Additional file 11: Table S11. Pathway analysis the DEGs of HSK0-VS-HRK0. (XLS 90 kb)

Additional file 12: Table S12. Pathway analysis the DEGs of HSK48-VS-HRK48. (XLS 90 kb)

Additional file 13: Table S13. DEGs confirmed by the qRT-PCR. (XLS 119 kb)

Abbreviations

AA: Ascorbic acid; ACC oxidase: Aminocyclopropane carboxylate oxidase; CAD: Cinnamyl-alcohol dehydrogenase; CCOAOMT: Caffeoyl-coA 3-O-methyl transferase; CHI: Chalcone isomerase; CHS: Chalcone synthase; COMT: Caffeic acid 3-O-methyltransferase; CYP: Cytochrome P450; DEGs: Differentially expressed genes; DFR: Dihydroflavonol 4-reductase; DXS: 1-Deoxy-D-xylulose-5-phosphate synthas; ERF1: Ethylene responsive factor 1; ET: Ethylene; GO: Gene Ontology; GST: Glutathione S-transferase; HPT: Homogenitisate phytyltransferas; JA: Jasmonic acid; KEGG: Kyoto Encyclopedia of Genes and Genomes; L-AO: L-ascorbate oxidase; LAR: Leucoanthocyanidin reductase; LOX: Lipoxygenase; MDHA: Monodehy-droaseorbate; OPDA: 12-oxophytodienoic acid reductase; PAL: Phenylalanine ammonia-lyase; POD: Peroxidase; PPO: Polyphenol oxidase; qRT-PCR: Quantitative real time-PCR; ROS: Reactive oxygen species; TRX: Thioredoxin; α-DOX: alpha-dioxygenase

Acknowledgements

Not applicable

Funding

This work was supported by the Natural Science Fund of Guangxi (2013GXNSFDA019009; 2016GXNSFAA380238), and the Development Foundation of Guangxi Academy of Agricultural Sciences (2016JM03). The funders had no role in study design, data collection and analysis, decision to publish, or preparation of the manuscript.

Authors' contributions

ZDS and WYZ conceived and designed the experiments. WYZ, ZYC, HZC, ZGL, SZY and XMT performed the experiments. WYZ analyzed the data. ZDS and WYZ contributed reagents/materials/analysis tools. WYZ and ZDS conceived the experiments and wrote the manuscript. All authors read and approved the final manuscript.

Competing interests

The authors declare that they have no competing interests.

References

1. Wlson RF. Chapter 1 soybean; market driven research needs. In: Stacey G, editor. Genetics and genomics of soybean. New York: Springer; 2008. p. 3–15.
2. Editorial committee of plate of Chinese diseases and insects on crop. Plate of Chinese diseases and insects on crop, fifth fascicule, diseases and insects on oil crop (first). Beijing: Agricultural press; 1982. p. 136–7.
3. Cui ZL, Gai JY, Ji DF, Ren ZJ. A study on leaf-feeding insect species on soybeans in Nanjing area. Soybean Sci. 1997;16(1):12–20.
4. Sun ZD, Yang SZ, Chen HZ, Li CY, Long LP. Identification of soybean resistance to bean pyralid (*Lamprosema indicate* Fabricicus) and oviposition preference of bean pyralid on soybean varieties. Chin J Oil Crop Sci. 2005; 27(4):69–71.
5. Sun ZD, Chen HZ, Wei DW. A study on leaf-feeding insect species on soybeans in Nanning. Guangxi Agric Sci. 2001;2:104–6.
6. Long LP, Yang SZ, Chen HZ, Qin JL, Li CY, Sun ZD. Effects of different genotypes of soybean varieties on the experimental population of bean pyralid (*Lamprosema indicata* Fabricius). Chin J oil. Crop Sci. 2004;26:67–70.
7. Xing GN, Tan LM, Liu ZXN, Yue H, Zhang HZ, Shi HF, et al. Morphological variation of pubescence on leaf blade and petiole and their correlation with resistance to bean pyralid (*Lamprosema indicate* Fabricius) in soybean landraces. Soybean Sci. 2012;31(5):691–6.
8. Xing GN, Zhao TJ, Gai JY. Inheritance of resistance to *Lamprosema indicata* Fabricius in soybean. Acta Agron Sin. 2008;34(1):8–16.
9. Li GJ, Cheng LG, Zhang GZ, He XH, Zhi HJ, Zhang YM. Mixed major-gene plus polygenes inheritanceanalysis for resistance in soybean to bean pyralid (*Lamprosema indicate* Fabricius). Soybean Sci. 2008;27(1):33–6. 41
10. Xing GN, Zhou B, Wang YF, Zhao TJ, DY Y, Chen SY, et al. Genetic components and major QTL confer resistance to bean pyralid (*Lamprosema indicate* Fabricius) under multiple environments in four RIL populations of soybean. Theor Appl Genet. 2012;125(5):859–75.
11. Zeng WY, Cai ZY, Zhang ZP, Chen HZ, Yang SZ, Tang XM, et al. Studies on the physiological and biochemical of bean pyralid (*Lamprosema indicata* Fabricius) resistance in different soybean varieties. J Southern Agric. 2015; 46(12):2112–6.
12. Schmutz J, Cannon SB, Schlueter J, Ma J, Mitros T, Nelson W, et al. Genome sequence of the palaeopolyploid soybean. Nature. 2010;463(7278):178–83.
13. Foyer CH, Shigeoka S. Understanding oxidative stress and antioxidant functions to enhance photosynthesis. Plant Physiol. 2011;155(1):93–100.
14. Slesak I, Libik M, Karpinska B, Karpinski S, Miszalski Z. The role of hydrogen peroxide in regulation of plant metabolism and cellular signaling in response to environmental stresses. Acta Biochim Pol. 2007;54(1):39–50.
15. Quan LJ, Zhang B, Shi WW, Li HY. Hydrogen peroxide in plants: a versatile molecule of the reactive oxygen species network. J Integr. Plant Biol. 2008; 50(1):2–18.
16. Ellis C, Turner JG. The Arabidopsis mutant cev1 has constitutively active jasmonate and ethylene signal pathways and enhanced resistance to pathogens. Plant Cell. 2001;13(5):1025–33.
17. Farmer EE, Ryan CA. Interplant communication: airborne methyl jasmonate induces synthesis of proteinase inhibitors in plant leaves. Proce Nati Acad Sci USA. 1990;87(19):713–6.
18. De Vos M, Van Oosten VR, Van Poecke RM, Van Pelt JA, Pozo MJ, Mueller MJ, et al. Signal signature and transcriptome changes of *Arabidopsis* during pathogen and insect attack. Mol Plant-Microbe Interact. 2005;18(9):923–37.

19. Brodbeck BV, Lii RFM, Andersen PC. Physiological and behavioral adaptations of three species of leafhoppers in response to the dilute nutrient content of xylem fluid. J Insect Physiol. 1993;39(1):73–81.
20. Reymond P, Weber H, Damond M, Fanner EE. Differential gene expression in response to mechanical wounding and insect feeding in *Arabidopsis*. Plant Cell. 2000;12(5):707–19.
21. Mozoruk J, Hunnicutt LE, Cave RD, Hunter WB, Bausher MG. Profiling transcriptional changes *Citrus sinensis* (L.) Osbeck challenged by herbivory from the xylem-feeding leafhopper *Homalodisca coagulate* (say) by cDNA macroarray analysis. Plant Sci. 2006;170(6):1068–80.
22. Schuler MA. The role of cytochrome P450 monooxygenases in plant-insect interactions. Plant Physiol. 1996;112(4):1411–9.
23. Treutter D. Significance of flavonoids in plant resistance: a review. Environ Chem Lett. 2006;4(3):147–57.
24. Korkina LG. Phenylpropanoids as naturally occurring antioxidants: from plant defense to human health. Cell Mol Biol. 2007;53(1):15–25.
25. Zhu JH, Verslues PE, Zheng XW, Lee BH, Zhan XQ, Manabe Y, et al. HOS10 encodes an R2R3-type MYB transcription factor essential for cold acclimation in plants. Proc Nati Acad Sci USA. 2010;102(28):9966–71.
26. Rushton PJ, Somssich IE. Transcriptional control of plant genes to responsive to pathogens. Curr Opin Plant Biol. 1998;1(4):311–5.
27. Felton GW, Donato K, Del Vecchio RJ, Duffey SS. Activation of plant foliar oxidases by insect feeding reduces nutritive quality of foliage for herbivores. J Chem Ecol. 1989;15(12):2667–94.
28. Estruch JJ, Carozzi NB, Desai N, Duck NB, Warren GW, Koziel MG. Transgenic plants: an emerging approach to pest control. Nat Biotechnol. 1997;15(2):137–41.
29. Leszczynski B. Changes in phenols content and metabolism in leaves of susceptible and resistant wheat cultivars infested by *Rhopalosiphum padi* (L.) (Hom.:Aphididae). J Appl Entomol. 1985;100(4):343–8.
30. Huang YH. Phloem feeding regulates the plant defense pathways responding to both aphid infestation and pathogen infection, Biotechnology and sustainable agriculture 2006 and beyond. New York: Springer; 2007. p. 215–9.
31. Zhang SZ, Zhang F, Hua BZ. Enhancement of phenylalanine ammonia lyase, polyphenoloxidase, and peroxidase in cucumber seedlings by *Bemisia tabaci* (Gennadius) (Hemiptera: Aleyrodidae) infestation. Agric Sci. 2008;7(1):82–7.
32. Wei Z, Hu W, Lin QS, Cheng XY, Tong MJ, Zhu LL, et al. Understanding rice plant resistance to the brown planthopper (*Nilaparvata lugens*): a proteomic approach. Proteomics. 2009;9(10):2798–808.
33. Du B, Wei Z, Wang ZQ, Wang XX, Peng XX, Du B, et al. Phloem-exudate proteome analysis of response to insect brown planthopper in rice. J Plant Physiol. 2015;183:13–22.
34. Royo J, Vancanneyt G, Pérez AG, Sanz C, Störmann K, Rosahl S, et al. Characterization of three potato lipoxygenases with distinct enzymatic activities and different organ-specific and wound-regulated expression patterns. J Biol Chem. 1996;271(35):21012–9.
35. Constabel CP, Ryan CA. A survey of wound- and methyl jasmonate-induced leaf polyphenol oxidase in crop plants. Phytochemistry. 1998;47(4):507–11.
36. Constabel CP, Bergey DR, Ryan CA. Systemin activates synthesis of wound-inducible tomato leaf polyphenol oxidase via the octadecanoid defense signaling pathway. Proc Nati Acad Sci USA. 1995;92(2):407–11.
37. Zebelo S, Piorkowski J, Disi J, Fadamiro H. Secretions from the ventral eversible gland of *Spodoptera exigua* caterpillars activate defense-related genes and induce emission of volatile organic compounds in tomato, *Solanum lycopersicum*. BMC Plant Biol. 2014;14(1):140.
38. Breusegem FV, Vranová E, Dat JF, Inzé D. The role of active oxygen species in plant signal transduction. Plant Sci. 2001;161(3):405–14.
39. Yu T, Li YS, Chen XF, Hu J, Chang X, Zhu YG. Transgenic tobacco plants overexpressing cotton glutathione S-transferase (GST) show enhanced resistance to methyl viologen. J Plant Physiol. 2003;160(11): 1305–11.
40. Gallé A, Csiszár J, Secenji M, Guóth A, Cseuz L, Tari I, et al. Glutathione transferase activity and expression patterns during grain filling in flag leaves of wheat genotypes differing in drought tolerance: response to water deficit. J Plant Physiol. 2009;166(17):1878–91.
41. Santos CVD, Rey P. Plant thioredoxins are key actors in the oxidative stress response. Trends Plant Sci. 2006;11(7):329–34.
42. Reichheld JP, Mestres-Ortega D, Laloi C, Meyer Y. The multigenic family of thioredoxin h in *Arabidopsis thaliana*: specific expression and stress response. Plant Physiol Biochem. 2002;40(6–8):685–90.
43. Laloi C, Mestresortega D, Marco Y, Meyer Y, Reichheld JP. The *Arabidopsis* cytosolic thioredoxin h5 gene induction by oxidative stress and its W-box-mediated response to pathogen elicitor. Plant Physiol. 2004;134(3):1006–16.
44. Zhou G, Wang X, Yan F, Wang X, Li R, Cheng J, et al. Genome-wide transcriptional changes and defence-related chemical profiling of rice in response to infestation by the rice striped stem borer Chilo Suppressalis. Physiol Plant. 2011;143(1):21–40.
45. Albeverio S, Khrennikov A. Biosynthesis and metabolism of jasmonates. J Plant Growth Regul. 2004;23(3):179–99.
46. Gardner HW. Biological roles and biochemistry of the lipoxygenase pathway. Hortscience. 1995;30(2):197–205.
47. Grechkin A. Recent developments in biochemistry of the plant lipoxygenase pathway. Prog Lipid Res. 1998;37(5):317–52.
48. Fidantsef AL, Stout MJ, Thaler JS, Duffey SS, Bostock RM. Signal interactions in pathogen and insect attack: expression of lipoxygenase, proteinase inhibitor II, and pathogenesis-related protein P4 in the tomato, *Lycopersicon esculentum*. Physiol Mol Plant Pathol. 1999;54(3):97–114.
49. Wang B, Hajano JUD, Ren Y, Lu C, Wang X. iTRAQ-based quantitative proteomics analysis of rice leaves infected by rice stripe virus reveals several proteins involved in symptom formation. Virol J. 2015;12(1):99–119.
50. Stintzi A, Weber H, Reymond P, Browse J, Farmer EE. Plant defense in the absence of jasmonic acid: the role of cyclopentenones. Proce Nati Acad Sci USA. 2001;98(22):12837–42.
51. Johnson PR, Ecker JR. The ethylene gas signal transduction pathway in plants: a molecular perspective. Annu Rev Genet. 1998;32(32):227–54.
52. Yang SF, Hoffman NE. Ethylene biosynthesis and its regulation in higher plants. Annu Rev Plant Physiol. 2003;35(1):155–89.
53. Niu YY, Chen M, Xu ZS, Li LC, Chen XP, Ma YZ. Characterization of ethylene receptors and their interactions with GmTPRA novel tetratricopeptide repeat protein (TPR) in soybean (*Glycine max* L.). JIA. 2013;12(4):571–81.
54. Nole-Wilson S, Krizek BA. DNA binding properties of the *Arabidopsis* floral development protein AINTEGMENGTA. Nucleic Acids Res. 2000; 28(21):4076–82.
55. Zhang Z, Huang R. Enhanced tolerance to freezing in tobacco and tomato overexpressing transcription factor TERF2/LeERF2 is modulated by ethylene biosynthesis. Plant Mol Biol. 2010;73(3):241–9.
56. Monroy AF, Dhindsa RS. Low temperature signal transduction: induction of cold acclimation specific genes of alfalfa by calcium at 25 degrees C. Plant Cell. 1995;7(3):321–31.
57. Guilfoyle TJ. Aux/IAA proteins and auxin signal transduction. Trends Plant Sci. 1998;3(6):205–7.
58. Guilfoyle TJ, Hagen G. Auxin response factors. Curr Opin Plant Biol. 2007; 10(5):453–60.
59. Hagen G, Guilfoyle T. Auxin-responsive gene expression: genes, promoters and regulatory factors. Plant Mol Biol. 2002;49(3):373–85.
60. Park JE, Park JY, Kim YS, Staswick PE, Jeon J, Yun J, et al. GH3-mediated auxin homeostasis links growth regulation with stress adaptation response in *Arabidopsis*. J Biol Chem. 2007;282(13):10036–46.
61. Kant S, Bi YM, Zhu T, Rothstein SJ. SAUR39, a small auxin-up RNA gene, acts as a negative regulator of auxin synthesis and transport in rice. Plant Physiol. 2009;151(2):691–701.
62. Kant S, Rothstein S. Auxin-responsive SAUR39 gene modulates auxin level in rice. Plant Signal Behav. 2009;4(12):1174–5.
63. León J, Rojo E, Sánchez-Serrano J. Wound signaling in plants. J Epx Bot. 2001;52(354):1–9.
64. Afzal AJ, Wood AJ, Lightfoot DA. Plant receptor-like serine threonine kinases: roles in signaling and plant defense. Mol Plant-Microbe Interact. 2008;21(5): 507–17.
65. Hasegawa PM, Bressan RA, Zhu JK, Bohnert HJ. Plant cellular and molecular responses to high salinity. Annu Rev Plant Physiol Plant Mol Biol. 2000;51:463–99.
66. Zipfel C. Pattern-recognition receptors in plant innate immunity. Curr Opin Immunol. 2008;20(1):10–6.

67. Chinnusamy V, Schumaker K, Zhu JK. Molecular genetic perspectives on cross-talk and specificity in abiotic stress signaling in plants. J Exp Bot. 2004;55(395):225–36.
68. Dey S, Ghose K, Basu D. Fusarium elicitor-dependent calcium influx and associated ros generation in tomato is independent of cell death. J Exp Bot. 2010;126(2):217–28.
69. Maffei ME, Mithöfer A, Arimura GI, Uchtenhagen H, Bossi S, Bertea CM, et al. Effects of feeding *Spodoptera littoralis* on lima bean leaves. III. Membrane depolarization and involvement of hydrogen peroxide. Plant Physiol. 2006; 140(3):1022–35.
70. Mohanta TK, Occhipinti A, Zebelo SA, Foti M, Fliegmann J, Bossi S, et al. Ginkgo biloba responds to herbivory by activating early signaling and direct defenses. PLoS One. 2012;7(3):e32822.
71. Karban R, Kuć J. Induced resistance against pathogens and herbivores: an overview, Induced plant defenses against pathogens and herbivores; 1999. p. 1–15.
72. Alagar M, Suresh S, Saravanakumar D, Samiyappan R. Feeding-induced changes in defence enzymes and proteins and their implications in host resistance to *Nilaparvata lugens*. J Appl Entomol. 2010;134(2):123–31.
73. Chen H, Wilkerson CG, Kuchar JA, Phinney BS, Howe GA. Jasmonate-inducible plant enzymes degrade essential amino acids in the herbivore midgut. Proc Nati Acad Sci USA. 2005;102(52):19237–42.
74. Stensjo K, Pettersson J, Bryngelsson T, Jonsson L. Aphid infestation induces PR-proteins differently in barley susceptible or resistant to the birdcherry-oat aphid (*Rhopalosiphum padi*). Physiol Plant. 2008;110(4):496–502.
75. Van Loon LC, Van Strien EA. The families of pathogenesis related proteins, their activities, and comparative analysis of PR-1 type proteins. Physiol Mol Plant Pathol. 1999;55(2):85–97.
76. Jwa NS, Agrawal GK, Rakwal R, Park CH, Agrawal VP. Molecular cloning and characterization of a novel jasmonate inducible pathogenesis-related class 10 protein gene, JIOsPR10, from rice (*Oryza sativa* L.) seedling leaves. Biochem Biophys Res Commun. 2001;286(5):973–83.
77. Uknes S, Winter AM, Delaney T, Vernooij B, Morse A, Friedrich L, et al. Biological induction of systemic acquired resistance in *Arabidopsis*. Mol Plant-Microbe Ineract. 1993;6(6):692–8.
78. Green TR, Ryan CA. Wound induced proteinase inhibitors in plant leaves: a possible defense mechanism against insects. Science. 1972;175(4023):776–7.
79. Tamhane VA, Chougule NP, Giri AP, Dixit AR, Sainani MN, Gupta VS. In vivo and in vitro effect of Capsicum Annum proteinase inhibitors on *Helicoverpa armigera* gut proteinases. Biochimica Et Biophvsica Acta. 2005;1722(2):156–67.
80. Gatehouse AMR, Hilder VA, Boulter D. Plant genetic manipulation for crop protection. Biotch Agric. 1992;22(6):1408–9(2).
81. Rao R, Fiandra L, Giordana B, De EM, Congiu T, Burlini N, et al. AcMNPV ChiA protein disrupts the peritrophic membrane and alters midgut physiology of Bombyx Mori larvae. Insect Biochem Mol Biol. 2004;34(11):1205–13.
82. Ding X, Gopalakrishnan B, Johnson LB, White FF, Wang XR, Morgan TD, et al. Insect resistance of transgenic tobacco expressing an insect chitinase gene. Transgenic Res. 1998;7(2):77–84.
83. Zhu-Salzman K, Salzman RA, Ahn JE, Koiwa H. Transcriptional regulation of sorghum defense determinants against a phloem-feeding aphid. Plant Physiol. 2004;134(1):420–31.
84. Kant MR, Ament K, Sabelis MW, Haring MA, Schuurink RC. Differential timing of spider mite-induced direct and indirect defenses in tomato plants. Plant Physiol. 2004;135(1):483–95.
85. Wang HB, Jin S, Tao Y. Insect resistance of the chitinase in *Vicia faba* leaves to *Aphis craccivove*. J Fudan University (Nat Sci). 1994;33(3):348–52.
86. Peumans WJ, Van Damme EJ. Lectins as plant defense proteins. Plant Physiol. 1995;109(2):347–52.
87. Komath SS, Kavitha M, Swamy MJ. Beyond carbohydrate binding: new direct ions in plant lectin research. Org Biomol Chem. 2006;4(6):973–88.
88. Boulter D, Edwards GA, Gatehouse AMR, Gatehouse JA, Hilder VA. Additive protective effects of incorporating two different higher plant insect resistence genes in transgenic tobacco plants. Corp Protect. 1990;9:351–4.
89. Hilder VA, Powell KS, Gatehouse AMR, Gatehouse JA, Gatehouse LN, Shi Y, et al. Expression of snowdrop lectin in transgenic tobacco plants results in added protection against aphids. Transgenic Res. 1995;4(1):18–25.
90. Van Damme EJM. Handbook of plant lectins: properties and biomedical applications. Chester: John Willey and Sons Ltd; 1998. p. 452.
91. Chrispeels MJ, Raikhel NV. Lectins, lectin genes, and their role in plant defense. Plant Cell. 1991;3(1):1–9.
92. Lange BM, Rujan T, Martin W, Croteau R. Isoprenoid biosynthesis: the evolution of two ancient and distinct pathways across genomes. Proc Nati Acad Sci USA. 2000;97(24):13172–7.
93. Herrmann KM. The shikimate pathway: early steps in the biosynthesis of aromatic compounds. Plant Cell. 1995;7(3):907–19.
94. Chaman ME, Copaja SV, Argandoña VH. Relationships between salicylic acid content, phenylalanine ammonia-lyase (PAL) activity, and resistance of barley to aphid infestation. J Agric Food Chem. 2003;51(8):2227–31.
95. Ritter H, Schulz GE. Structural basis for the entrance into the phenylpropanoid metabolism catalyzed by phenylalanine ammonia-lyase. Plant Cell. 2004;16(12):3426–36.
96. Li JB, Fang LP, Zhang YN, Yang WJ, Guo Q, Li L, et al. The relationship between the resistance of ctton against cotton aphid, aphis gossypii, and the activity of phenylalanine ammonia-lyase. Chin Bull Entomol. 2008;45(3):422–5.
97. Hartley SE, Firn RD. Phenolic biosynthesis, leaf damage, and insect herbivory in birch (*Betula Pendula*). J Chem Ecol. 1989;15(1):275–83.
98. Kühnl T, Koch U, Heller RW, Wellmann E. Elicitor induced S-adenosyl-l-methionine: caffeoyl-CoA 3-O-methyltransferase from carrot cell suspension cultures. Plant Sci. 1989;60(1):21–5.
99. Pakusch AE, Kneusel RE, Matern U. S-adenosyl-l-methionine: trans-caffeoyl-coenzyme a 3-O-methyltransferase from elicitor-treated parsley cell suspension culture. Arch Biochem Biophys. 1989;271(2):488–94.
100. Gowri G, Bugos RC, Campbell WH, Maxwell CA, Dixon RA. Stress responses in alfalfa (*Medicago sativa* L.): X. Molecular cloning and expression of S-adenosyl-l-methionine: caffeic acid 3-O-methyltransferase, a key enzyme of lignin biosynthesis. Plant Physiol. 1991;97(1):7–14.
101. Wyrambik D, Grisebach H. Purification and properties of isoenzymes of cinnamyl-alcohol dehydrogenase from soybean-cell-suspension cultures. Eur J Biochem. 1975;59(1):9–15.
102. Mansell RL, Gross GG, Stöckigt J, Franke H, Zenk MH. Purification and properties of cinnamyl alcohol dehydrogenase from higher plants involved in lignin biosynthesis. Phytochemistry. 1974;13(11):2427–35.
103. Henzell RF, Hall WT. Substituted phenols as repellents for male Costelytra zealandica (*Goleoptera: Scarabaeidae*) in mating flights. New Zealand J Zool. 1974;1(4):509–13.
104. Schroeder FC, del Campo ML, Grant JB, Weibel DB, Smedley SR, Bolton KL, et al. Pinoresinol: a lignol of plant origin serving for defense in a caterpillar. Proc Nati Acad Sci USA. 2006;103(42):15497–501.
105. Winkel-Shirley B. Biosynthesis of flavonoids and effects of stress. Curr Opin Plant Biol. 2002;5(3):218–23.
106. Koes RE, Quattrocchio F, Mol JNM. The flavonoid biosynthetic pathway in plants: function and evolution. BioEssays. 1994;16(2):123–32.
107. Martin CR. Structure, function, and regulation of the chalcone synthase. Int Rev Cytol. 1993;147:233–84.
108. Muir SR, Collins GJ, Robinson S, Hughes S, Bovy A, Ric De Vos CH, et al. Overexpression of petunia chalcone isomerase in tomato results in fruit containing increased levels of flavonols. Nat Biotechnol. 2001;19(5):470–4.
109. Johnson ET, Ryu S, Yi H, Shin B, Cheong H, Choi G. Alteration of a single amino acid changes the substrate specificity of dihydroflavonol 4-reductase. Plant J. 2001;25(3):325–33.
110. Kumar V, Yadav SK. Overexpression of CsDFR and CsANR enhanced flavonoids accumulation and antioxidant potential of roots in tobacco. Plant Roots. 2013;7:65–76.
111. Schuler MA, Daniele WR. Functional genomics of P450s. Annu Rev Plant Biol. 2003;54(1):627–9.
112. Harvey PJ, Campanella BF, Castro PM, Harms H, Lichtfouse E, Schäffner AR, et al. Phytoremediation of polyaromatic hydrocarbons, anilines and phenols. Environ Sci Pollut Res Int. 2002;9(1):29–47.
113. Morant M, Bak S, Møller BL, Werck-Reichhart D. Plant cytochromes P450: tools for pharmacology, plant protection and phytoremediation. Curr Opin Biotechnol. 2003;14(2):151–62.
114. Bansal R, Mian M, Mittapalli O, Miche AP. RNA-Seq reveals a xenobiotic stress response in the soybean aphid, *Aphis glycines*, when fed aphid-resistant soybean. BMC Genomics. 2014;15:972.
115. Kahn RA, Bak S, Svendsen J, Halkier BA, Møller BL. Isolation and reconstitution of cytochrome P450ox and in vitro reconstitution of the entire biosynthetic pathway of the cyanogenic glucoside dhurrin from sorghum. Plant Physiol. 1997;115(4):1661–70.
116. Sanmartin M, Pateraki L, Chatzopoulou F, Kanellis AK. Differential expression of the ascorbate oxidase multigene family during fruit development and in response to stress. Planta. 2007;225(4):873–85.

117. Gaspard S, Monzani E, Casella L, Gullotti M, Maritano S, Marchesini A. Inhibition of aseorbate oxidase by phenolic compounds. Enzymatic and spectroscopic studies. Biochemistry. 1997;36(16):4852–9.
118. Sanmartin M, Drogoudi PD, Lyons T, Pateraki I, Barnes J, Kanellis AK. Over-expression of ascorbate oxidase in the apoplast of transgenic tobacco results in altered ascorbate and glutathione redox states and increased sensitivity to ozone. Planta. 2003;216(6):918–28.
119. Barbehenn RV, Jaros A, Yip L, Tran L, Kanellis AK, Constabel CP. Evaluating ascorbate oxidase as a plant defense against leaf-chewing insects using transgenic poplar. J Chem Ecol. 2008;34(10):1331–40.
120. Fotopoulos V, Sanmartin M, Kanellis A. Effect of ascorbate oxidase over-expression on ascorbate recycling gene expression in response to agents imposing oxidative stress. J Exp Bot. 2006;57(14):3933–43.
121. Eulgem T, Rushton PJ, Schmelzer E, Hahlbrock K, Somssich IE. Early nuclear events in plant defence signalling: rapid gene activation by WRKY transcription factors. EMBO J. 1999;18(17):4689–99.
122. Eulgem T, Rushton PJ, Robatzek S, Somssich IE. The WRKY superfamily of plant transcription factors. Trends Plant Sci. 2000;5(5):199–206.
123. Rushton PJ, Somssich IE, Ringler P, Shen QJ. WRKY transcription factors. Trends Plant Sci. 2010;15(5):247–58.
124. Zhou X, Jiang Y, Yu D. WRKY22 transcription factor mediates dark-induced leaf senescence in Arabidopsis. Mol Cell. 2011;31(4):303–13.
125. Grunewald W, Karimi M, Wieczorek K, Van de Cappelle E, Chnitzki E, Grundler F, et al. A role for AtWRXY23 in feeding site establishment of plant-parasitic nematodes. Plant Physiol. 2008;148:358–68.
126. Artico S, Ribeiro-Alves M, Oliveira-Neto OB, Leonardo Macedo LPD, Silveira S, et al. Transcriptome analysis of *Gossypium hirsutum* flower buds infested by cotton boll weevil (*Anthonomus grandis*) larvae. BMC Genomics. 2014; 15(854):1–24.
127. Fang Y, You J, Xie K, Xie W, Xiong L. Systematic sequence analysis and identification of tissue-specific or stress-responsive genes of NAC transcription factor family in rice. Mol Gen Genomics. 2008;280(6):547–63.
128. Nakashima K, Takasaki H, Mizoi J, Shinozaki K, Yamaguchi-Shinozaki K. NAC transcription factors in plant abiotic stress responses. Biochim Biophys Acta. 2012;1819(2):97–103.
129. Wang Z, Dane F. NAC (NAM/ATAF/CUC) transcription factors in different stresses and their signaling pathway. Acta Physiol Plant. 2013;35(5): 1397–408.
130 Hegedus D, Yu M, Baldwin D, Gruber M, Sharpe A, Parkin I, et al. Molecular characterization of *brassicanapus* NAC domain transcriptional activators induced in response to biotic and abiotic stresses. Plant Mol Biol. 2003;53(3): 383–97.
131. Nogueira FTS, Schlogl PS, Camargo SR, Fernandez JH, de Jr Rosa VE, Pompermayer P, et al. *SsNAC23*, a member of the NAC domain protein family, is associated with cold, herbivory and water stress in sugarcane. Plant Sci. 2005;169(1):93–106.
132. Kaur H, Heinzel N, Schöttner M, Baldwin IT, Gális I. R2R3-NaMYB8 regulates the accumulation of phenylpropanoid-polyamine conjugates, which are essential for local and systemic defense against insect herbivores in *Nicotiana attenuata*. Plant Physiol. 2010;152(3):1731–47.
133. Toledo-Ortiz G, Huq E, Quail PH. The *Arabidopsis* basic/helix-loop-helix transcription factor family. Plant Cell. 2003;15(8):1749–70.
134. Van Verk MC, Gatz C, Linthorst HJM. Transcriptional regulation of plant defense responses. J Regul Econ. 2009;20(2):191–211.
135. Fujimoto SY, Ohta M, Usui A, Shinshi H, Ohme-Takagi M. Arabidopsis ethylene-responsive element binding factors act as transcriptional activators or repressors of GCC box-mediated gene expression. Plant Cell. 2000;12(3):393–404.
136. Leubner-Metzger G, Petruzzelli L, Waldvogel R, Vögeli-Lange R, Meins FJR. Ethylene responsive element binding protein (EREBP) expression and the transcriptional regulation of class I beta-1, 3-glucanase during tobacco seed germination. Plant Mol Biol. 1998;38(5):785–95.
137. Liu Q, Kasuga M, Sakuma Y, Miura S, Yamaguchi-Shinozaki K, Shinozaki K. Two transcription factors, DREB1 and DREB2, with an EREBP/AP2 DNA-binding domain separate two cellular signal transduction pathways in drought- and low-temperature-responsive gene expression in *Arabidopsis*. Plant Cell. 1998;10(8):1391–406.
138. Park JM, Park CJ, Lee SB, Ham BK, Shin R, Paek KH. Overexpression of the tobacco Tsi1 gene encoding an EREBP/AP2-type transcription factor enhances resistance against pathogen attack and osmotic stress in tobacco. Plant Cell. 2001;13(5):1035–46.
139. Jia W, Qiu K, He M, Song P, Zhou Q, Zhou F, et al. SOAPfuse: an algorithm for identifying fusion transcripts from paired-end RNA-Seq data. Genome Biol. 2013;14(2):R12.
140. Trapnell C, Roberts A, Goff L, Pertea G, Kim D, Kelley DR, et al. Differential gene and transcript expression analysis of RNA-Seq experiments with TopHat and cufflinks. Nat Protoc. 2012;7(3):562–78.
141. Li B, Dewey CN. RSEM: accurate transcript quantification from RNA-Seq data with or without a reference genome. BMC bionformatics. 2011;12(1):323.

3

Genomic prediction accuracies in space and time for height and wood density of Douglas-fir using exome capture as the genotyping platform

Frances R. Thistlethwaite[1], Blaise Ratcliffe[1], Jaroslav Klápště[1,2,3], Ilga Porth[4], Charles Chen[5], Michael U. Stoehr[6] and Yousry A. El-Kassaby[1*]

Abstract

Background: Genomic selection (GS) can offer unprecedented gains, in terms of cost efficiency and generation turnover, to forest tree selective breeding; especially for late expressing and low heritability traits. Here, we used: 1) exome capture as a genotyping platform for 1372 Douglas-fir trees representing 37 full-sib families growing on three sites in British Columbia, Canada and 2) height growth and wood density (EBVs), and deregressed estimated breeding values (DEBVs) as phenotypes. Representing models with (EBVs) and without (DEBVs) pedigree structure. Ridge regression best linear unbiased predictor (RR-BLUP) and generalized ridge regression (GRR) were used to assess their predictive accuracies over space (within site, cross-sites, multi-site, and multi-site to single site) and time (age-age/ trait-trait).

Results: The RR-BLUP and GRR models produced similar predictive accuracies across the studied traits. Within-site GS prediction accuracies with models trained on EBVs were high (RR-BLUP: 0.79–0.91 and GRR: 0.80–0.91), and were generally similar to the multi-site (RR-BLUP: 0.83–0.91, GRR: 0.83–0.91) and multi-site to single-site predictive accuracies (RR-BLUP: 0. 79–0.92, GRR: 0.79–0.92). Cross-site predictions were surprisingly high, with predictive accuracies within a similar range (RR-BLUP: 0.79–0.92, GRR: 0.78–0.91). Height at 12 years was deemed the earliest acceptable age at which accurate predictions can be made concerning future height (age-age) and wood density (trait-trait). Using DEBVs reduced the accuracies of all cross-validation procedures dramatically, indicating that the models were tracking pedigree (family means), rather than marker-QTL LD.

Conclusions: While GS models' prediction accuracies were high, the main driving force was the pedigree tracking rather than LD. It is likely that many more markers are needed to increase the chance of capturing the LD between causal genes and markers.

Keywords: Douglas-fir, Genomic selection, Exome capture, Full-sib families, Genotype x environment interaction, Predictive model

* Correspondence: y.el-kassaby@ubc.ca
[1]Department of Forest and Conservation Sciences, Faculty of Forestry, The University of British Columbia, 2424 Main Mall, Vancouver, BC V6T 1Z4, Canada
Full list of author information is available at the end of the article

Background

Novel advancements in genomics technologies and statistical genetics, have paved the way for an increasingly prosperous environment for breeding industries. Notably in the dairy sector, the traditional phenotype-dependent selection has been successfully replaced by genotype-dependent selection (aka genomic selection, GS) which reduces the time required for evaluating genetic traits [1]. Tree selective breeding (aka tree improvement) challenges are somewhat similar to those of the livestock industry; namely, long generation times and low heritability and late expressing traits. GS, if successful, can offer unprecedented gains in forestry through reducing trait evaluation time, allowing faster breeding generations turn-over, and hence, an increased genetic gain per unit time can be reached. Additionally, the implementation of GS to forestry would offer a certain resilience to spontaneous market or environmental, and/or extraneous influences (e.g., disease/pest resistance, climate change); as breeding programmes adapt accordingly in a shorter time frame [2].

Selective breeding has traditionally focused on phenotypic selection, and more recently the linkage disequilibrium (LD) based indirect method of marker-assisted selection (MAS) [3]. MAS is suited for major-gene traits (such as resistance to white-pine blister rust (*Cronartium ribicola*)) and proven to be less effective for predicting complex quantitative traits that closely reflect Fisher's infinitesimal model [4]. MAS models could not effectively describe a complex trait, since small effect loci are not readily discovered and considered 'not significant', thus leaving a substantial proportion of genetic control unaccounted for (i.e., missing heritability). Meuwissen et al. [5] proposed the method of 'genomic selection' to alleviate this problem by simultaneously considering all marker effects, and in doing so, all genetic contributions are captured regardless of size (significance). GS enables complex quantitative trait selection using genomic marker data alone and the method does not require a *priori* knowledge regarding the specific genetic architecture of the trait in question. Instead, markers throughout the entire genome (or in this case exome) are incorporated into the estimate. The resulting genomic estimated breeding values (GEBVs) for each individual derived from the GS models provide a basis, upon which selection decisions are made. The effect of this is a paradigm shift, in which the model unit of these breeding analyses shifts from being the ***line of decent*** to the ***allele***. This means that the phenotypic values of individuals are determined from genotypic data, enabling early selection of traits, leading to a significantly shorter breeding cycle and higher selection differential, particularly for the "difficult to assess" attributes.

The feasibility of applying GS to forest trees was initially assessed through deterministic simulations and the results indicated that GS has the potential to radically improve the efficiency of tree selective breeding [6]. Grattapaglia and Resende [6] also recommended further experimentation and proof of concept investigations. Since then, several forest tree species "proof-of-concept investigations" have been conducted with encouraging results [7–18].

Neves et al. [19] recognized that one of the major barriers to the application of genomic technologies to tree selective breeding was the large size of the genome of many forest species. This is particularly true of conifers, and although conifers have a large genome [20–22], their transcriptomes are comparable with other plants such as Arabidopsis whose genome is more than 100 times smaller [23]. Therefore, as an alternative to the prohibitive costs in both monetary terms and time, and the complexity of sequencing the whole genome, Neves et al. [19] focused on the coding region. This they refer to as "sequence capture", and proposed that it would enable more efficient genetic variant identification in conifers; as it had previously been done in human and maize genomic studies [24–26]. Although sources of variation may not be exclusively found in the exome, the reduced cost and time compared to that of whole genome sequencing, as well as the ability to still capture a significant proportion of variants and rare variants, makes this method desirable as it is harder to find functional variants in non-coding regions [27, 28]. Exome capture has been recognized as an effective method for capturing rare variants in the field of medicine [29], and for increasing knowledge of unmapped large genomes [19, 28]. In addition, Suren et al. [20] have shown it to be a cost effective method for reducing complexity in large genomes, such as those of conifers.

The present GS study was based on the exomic information collected from 1372, 38-year-old coastal Douglas-fir (*Pseudotsuga menziesii* Mirb. (Franco)) trees. The samples represented 37 full-sib families with replications over three sites in coastal BC. The study objectives were: 1) to compare two GS methods; ridge regression best linear unbiased predictor (RR-BLUP) and generalised ridge regression (GRR), and 2) to test the GS prediction accuracy for within-, cross-, pooled multi-site, and time- time (age-age /trait-trait) between age 12 and 38 years. Two phases to this analysis were carried out, firstly the two GS models were trained on estimated breeding values (EBVs). This represents an analysis in which model predictions are based on pedigree (both historical and contemporary) and marker-QTL LD information. Secondly the models were trained on deregressed breeding values (DEBVs). In this analysis the pedigree information (parental average) is removed, resulting in model predictions based on marker-QTL LD and co-segregation. Results of ABLUP cross-validations are provided as a reference for comparison.

Results

To assess the studied attributes' variation, boxplots were produced showing the variance of the estimated (EBVs) and the deregressed (DEBVs) breeding values (Fig. 1). It is interesting to note that the deregression process maintained the within

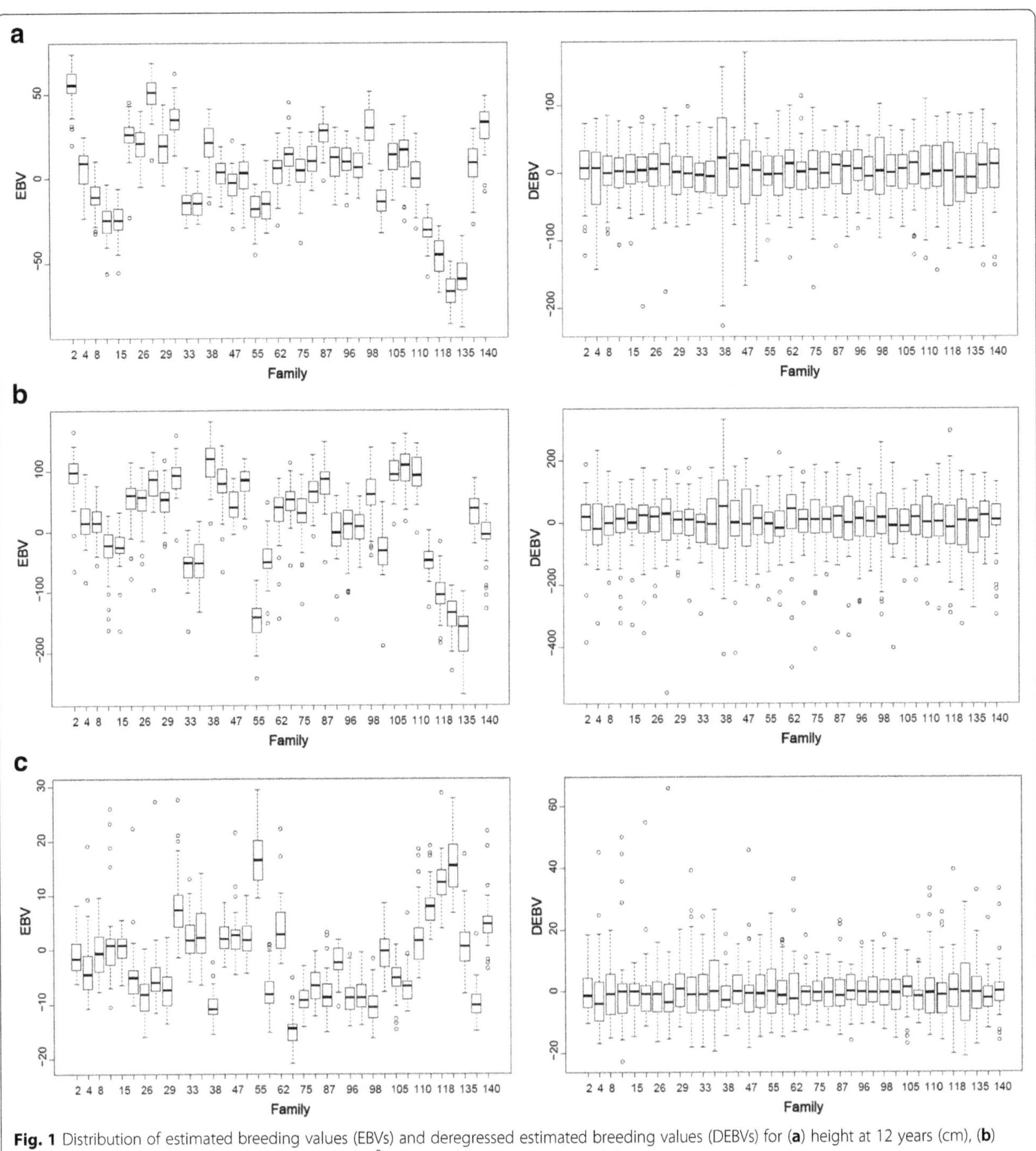

Fig. 1 Distribution of estimated breeding values (EBVs) and deregressed estimated breeding values (DEBVs) for (**a**) height at 12 years (cm), (**b**) height at 35 years (cm), and (**c**) wood density (g/cm^3) calculated from resistance to drilling

family variation for the three studied attributes (HT12, HT35, and WD_{res}); however, it virtually eliminated among family variation resulting in similar family means (Fig. 1).

Traits' heritabilities and EBV accuracy

Pedigree-based relationship matrix (ABLUP) heights heritability estimates varied among sites ranging between 0.13 (Lost Creek) and 0.24 (Adam), and 0.05 (Adam) and 0.23 (Fleet) for age 12 and 35 years, respectively (Fig. 2). The multi-site height heritability estimates were similar to the average of the single-site estimates at age 12 (0.17 vs. 0.19); and was slightly higher than the average single site estimate by age 35 (0.17 vs. 0.14) (Fig. 2). Pedigree-based relationship wood density heritability estimates generally were higher than those obtained for height and substantially varied among sites (range: 0.22 (Lost Creek) and 0.45 (Fleet)), with higher multi- than single-site average estimates (0.43 vs. 0.37) (Fig. 2). The average theoretical accuracies for the EBVs

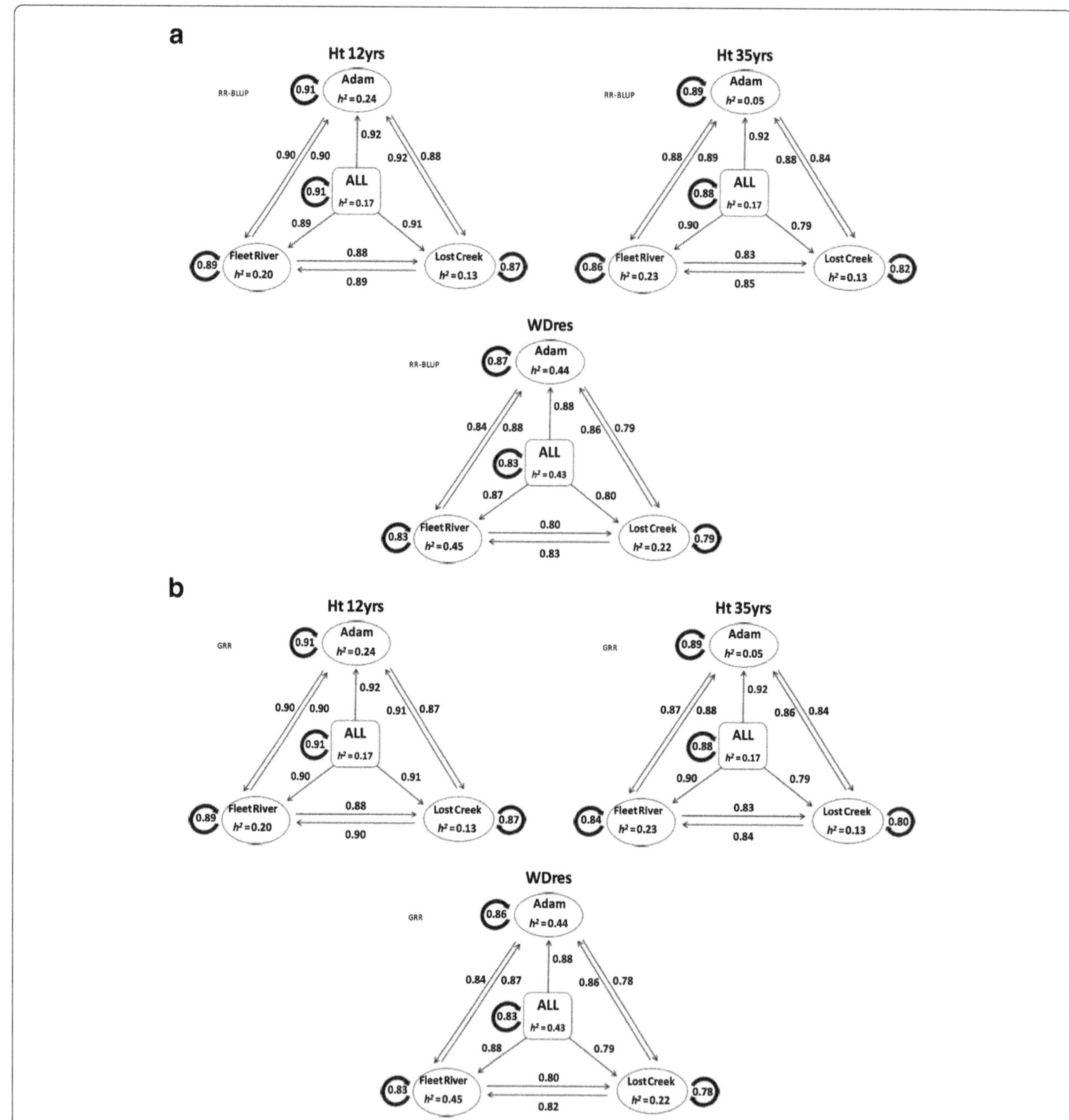

Fig. 2 Heritabilities and GS prediction accuracies for models trained on EBVs and predicting GEBVs for each of the traits. Showing the results of within-site, cross-site, combined-sites, and multi-site to single-site cross-validation (Top (**a**), using RR-BLUP, Bottom (**b**), using GRR). The direction of the arrows depicts what information (site) is used to train the model (shaft end), and which is being predicted (head). Traits key: HT 35 yrs. = height at 35 years (cm); HT 12 yrs. = height at 12 years (cm); WD_{res} = wood density. Sites are Adam, Fleet River, Lost Creek, and multi/combined-site (ALL)

were: 0.61, 0.68, and 0.76 for HT12, HT35, and WD_{res} respectively.

Cross-validation across space and time

Within-site prediction accuracy

Within-site prediction accuracies, determined based on the correlations between EBVs and GEBVs for the two genomic selection (RR-BLUP and GRR) models, generally produced very similar results (correlations and standard errors) across EBV traits and sites (Table 1). For all EBV traits, Adam produced the highest prediction accuracies, with Fleet River second, and Lost Creek always producing the lowest accuracies. The two models, RR-BLUP and GRR, produced almost identical results in analysed EBV traits. The only differences occurring in the prediction of GEBV for HT35, in which the GRR method produced slightly lower

Table 1 Within single-site prediction accuracies and their standard errors for ABLUP and genomic selection (ridge regression (RR-BLUP) and generalized ridge regression (GRR)) models for EBVs and GEBVs of heights (HT12 and HT35) and wood density (WD_{res})

Trait	Model	Site		
		Adam	Fleet River	Lost Creek
HT12	ABLUP	0.81 ± 0.002	0.77 ± 0.002	0.88 ± 0.001
	RR-BLUP	0.91 ± 0.010	0.89 ± 0.012	0.87 ± 0.013
	GRR	0.91 ± 0.010	0.89 ± 0.012	0.87 ± 0.012
HT35	ABLUP	0.85 ± 0.002	0.83 ± 0.002	0.88 ± 0.002
	RR-BLUP	0.89 ± 0.007	0.86 ± 0.020	0.82 ± 0.010
	GRR	0.89 ± 0.008	0.84 ± 0.020	0.80 ± 0.013
WD_{res}	ABLUP	0.94 ± 0.001	0.96 ± 0.001	0.94 ± 0.001
	RR-BLUP	0.87 ± 0.007	0.83 ± 0.026	0.79 ± 0.015
	GRR	0.86 ± 0.008	0.83 ± 0.026	0.84 ± 0.016

prediction accuracies than RR-BLUP (0.84 vs. 0.86, and 0.80 vs. 0.82) for Fleet River and Lost Creek, respectively. The same was found in the analysis of WD_{res} for Adam (GRR = 0.86, RR-BLUP = 0.87). In contrast the WD_{res}, GRR results for Lost Creek (0.84) were slightly more accurate than in the RR-BLUP model (0.79). It is worthy to mention that the overall predictive accuracy of the studied genomic selection models was generally high across sites and EBV traits (RR-BLUP: average = 0.86 and range of 0.79–0.91 and GRR: average = 0.86 and range of 0.80–0.91) (Table 1). Generally, all prediction accuracies' standard error estimates were small reflecting good model fit.

Using the deregressed breeding values to train the two GS models, we obtained predictive accuracy results approximating 0.0 for WD_{res} for within-site cross-validation (RR-BLUP: Adam = −0.10 ± 0.055, Fleet River = −0.04 ± 0.046, and Lost Creek = −0.06 ± 0.049; GRR: Adam = −0.05 ± 0.074, Fleet River = −0.03 ± 0.045, and Lost Creek = −0.04 ± 0.046). The other models for HT12 and HT35 failed to converge.

The ABLUP within-site cross-validation, provided results of a similar nature as the GS models trained on EBVs. The average prediction accuracy for ABLUP within-site was 0.87 (both RR-BLUP and GRR had averages of 0.86), with a range of 0.77–0.96. WDres was predicted with the highest accuracy of all three traits, and surpassed the accuracy of the GS models: 0.94 ± 0.0009, 0.96 ± 0.0009, and 0.94 ± 0.001, for Adam, Fleet River, and Lost Creek, respectively (Fig. 3).

Cross-sites prediction accuracy

The average predictive accuracy of the RR-BLUP and GRR genomic selection models was very similar for cross- and within-site analyses. However, some trends in the predictive ability of the sites did occur (Fig. 2a and b). The sites Adam and Fleet River always produced the same or higher prediction accuracies for Lost Creek than the within site estimate for Lost Creek. In addition, Lost Creek cross-site prediction accuracies for Adam and Fleet River were also always higher than for itself. Fleet River always produced higher prediction accuracies for Adam than itself. This was true for all traits using EBVs. This may be an indication that whilst there may be some GxE occurring as expected, it may not be significant enough to employ single-site testing and breeding. The accuracy of cross-site predictions varied and ranged from 0.78 (GRR for WD_{res}: Adam - Lost Creek) to 0.92 (RR-BLUP for HT12: Lost Creek - Adam,), and both selection models provided comparable results (Fig. 2a and b). Overall, across sites prediction accuracy decreased from HT12 to HT35 to WD_{res} with averages of 0.90, 0.86, and 0.83, respectively (all from RR-BLUP).

When the studied attributes' deregressed breeding values (DEBVs) were used to train the two GS models, surprising outcomes were obtained and the resulting predictive accuracies were extremely low approximating 0.0 for WD_{res}, while the remaining models for HT12 and HT35 failed to converge.

The results of the ABLUP cross-sites validation provided evidence of stronger GxE interaction than was predicted by the GS models (Fig. 3). This particular trend can occur as site-specific GxE interactions tend to over-estimate within-site prediction accuracy due to the sharing of a common environment, whilst seemingly inhibiting the efficacy of cross-site analyses. The average cross-sites prediction accuracies for ABLUP were as follows: 0.68, 0.70, and 0.79 for HT12, HT35, and WD_{res}, respectively (Fig. 3). A drop from the accuracies obtained by the GS models trained on EBV data, and a complete reversal of the order of those accuracies.

Within multi-site (combined) prediction accuracy

The RR-BLUP and GRR models gave almost identical results, with HT12 having the highest combined-site prediction accuracy, followed by HT35 and lastly WD_{res} (0.91, 0.88, and 0.83, respectively for both RR-BLUP and GRR) (Fig. 2a and b, Table 2). In general, the combined site prediction accuracies were higher than the average of the within-site accuracies, except for WD_{res} which produced the same accuracy for both (RR-BLUP: 0.91 vs. 0.89, 0.88 vs. 0.86, 0.83 vs. 0.83; for HT12, HT35, and WD_{res}, respectively) (Fig. 2a). Finally, it is also interesting to note the exceedingly small standard error values associated with all within multi-site prediction accuracies, highlighting the model(s) fit (Table 2).

Again, using the deregressed breeding values (DEBVs) to train the two GS models, we obtained results approximating 0.0 for all within multi-site cross-validation analyses for all traits considered (HT12, HT35, and WD_{res}) (Table 2).

The multi-site cross-validation accuracies for the ABLUP model were: 0.88 ± 0.002, 0.86 ± 0.003, and 0.84 ± 0.003, for HT12, HT35, and WD_{res}, respectively (Fig. 3). Similar in magnitude and sequence to the GS models trained with EBVs.

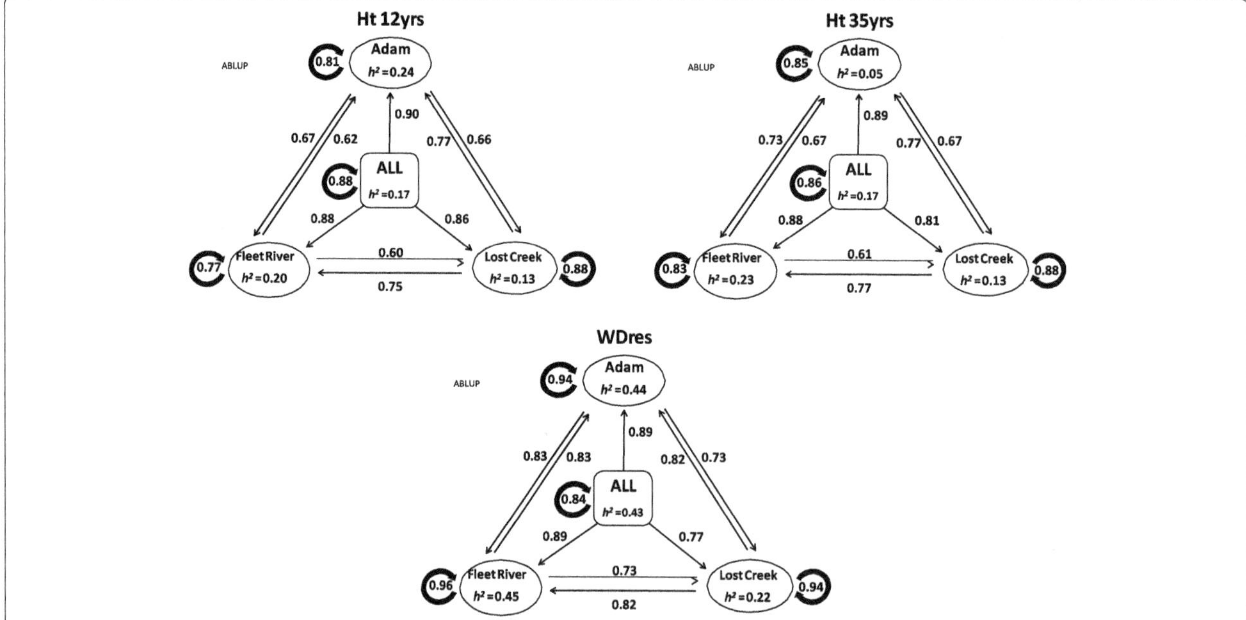

Fig. 3 Heritabilities and prediction accuracies of the ABLUP model for each of the traits. Showing the results of within-site, cross-site, combined-sites, and multi-site to single-site cross-validation. The direction of the arrows depicts what information (site) is used to train the model (shaft end), and which is being predicted (head). Traits key: HT 35 yrs. = height at 35 years (cm); HT 12 yrs. = height at 12 years (cm); WD_{res} = wood density (g/cm^3) calculated from resistance drilling to drilling. Sites are Adam, Fleet River, Lost Creek, and multi/combined-site (ALL)

Multi-site to single-site prediction accuracy

On average the multi- to single-site predictability for each trait was slightly higher than the within multi-site average estimates (for RR-BLUP, HT12: 0.91 vs. 0.89, HT35: 0.87 vs. 0.86, and WD_{res}: 0.85 vs. 0.83) (Fig. 2a). In most cases the multi- to single-site accuracy predictions were the same or higher, than the corresponding single-site predictions. Except for HT35 at Lost Creek for both RR-BLUP and GRR which gave slightly lower prediction accuracies for multi- to single-site than within site (Lost Creek) analyses (RR-BLUP: 0.79 vs. 0.82; GRR: 0.79 vs. 0.80) (Fig. 2a and b). In most cases Adam was the most predictable site, and on average, Lost Creek was the least predictable.

Again, using the deregressed breeding values (DEBVs) to train the two GS models, we obtained results approximating 0.0 for all multi- to single-site cross-validation analyses.

The multi- to single-site predictions for the ABLUP model closely followed the pattern of predictions from the GS models trained on EBV data. In each case (for HT12, HT35, and WD_{res}) Adam was the most predictable site (joint first with Fleet River for WD_{res}), Fleet River the second, and Lost Creek the least predictable site (Fig. 3). The average multi- to single-site predictability for the traits were again the same or slightly higher than the within multi-site predictions for ABLUP (HT12; 0.88 vs. 0.88, HT35: 0.86 vs. 0.86, and WD_{res}: 0.85 vs. 0.84) (Fig. 3).

Table 2 Within multi-site genomic selection prediction accuracies and their standard errors for ridge regression (RR-BLUP) and generalized ridge regression (GRR) models), estimating GEBVs and GEDBVs for heights (HT12 and HT35) and wood density (WD_{res}) (g/cm3) calculated from resistance drilling to drilling

Trait	GS Model	GEBVs	GEDBVs
HT12	RR-BLUP	0.91 ± 0.004	−0.09 ± 0.019
	GRR	0.91 ± 0.003	−0.04 ± 0.017
HT35	RR-BLUP	0.88 ± 0.006	−0.02 ± 0.021
	GRR	0.88 ± 0.006	−0.01 ± 0.029
WD_{res}	RR-BLUP	0.83 ± 0.009	0.00 ± 0.032
	GRR	0.83 ± 0.010	−0.01 ± 0.025

Two sites predicting one site accuracy

The RR-BLUP and GRR models for this analysis produced similar results overall (Fig. 4a and b). Generally, the highest prediction accuracies were obtained for HT12 (average = 0.90), followed by HT35 (average = 0.88), and lastly WD_{res} (average = 0.83) for both RR-BLUP and GRR. There were only minor differences between the two models (Fig. 4a and b). Similar to the multi- to single-site (above), the prediction accuracy of two sites to one site indicated that, in all cases, Adam was the most predictable site using this two to one analysis, despite the discrepancy in heritabilities, and on average Lost Creek was the least predictable in most cases.

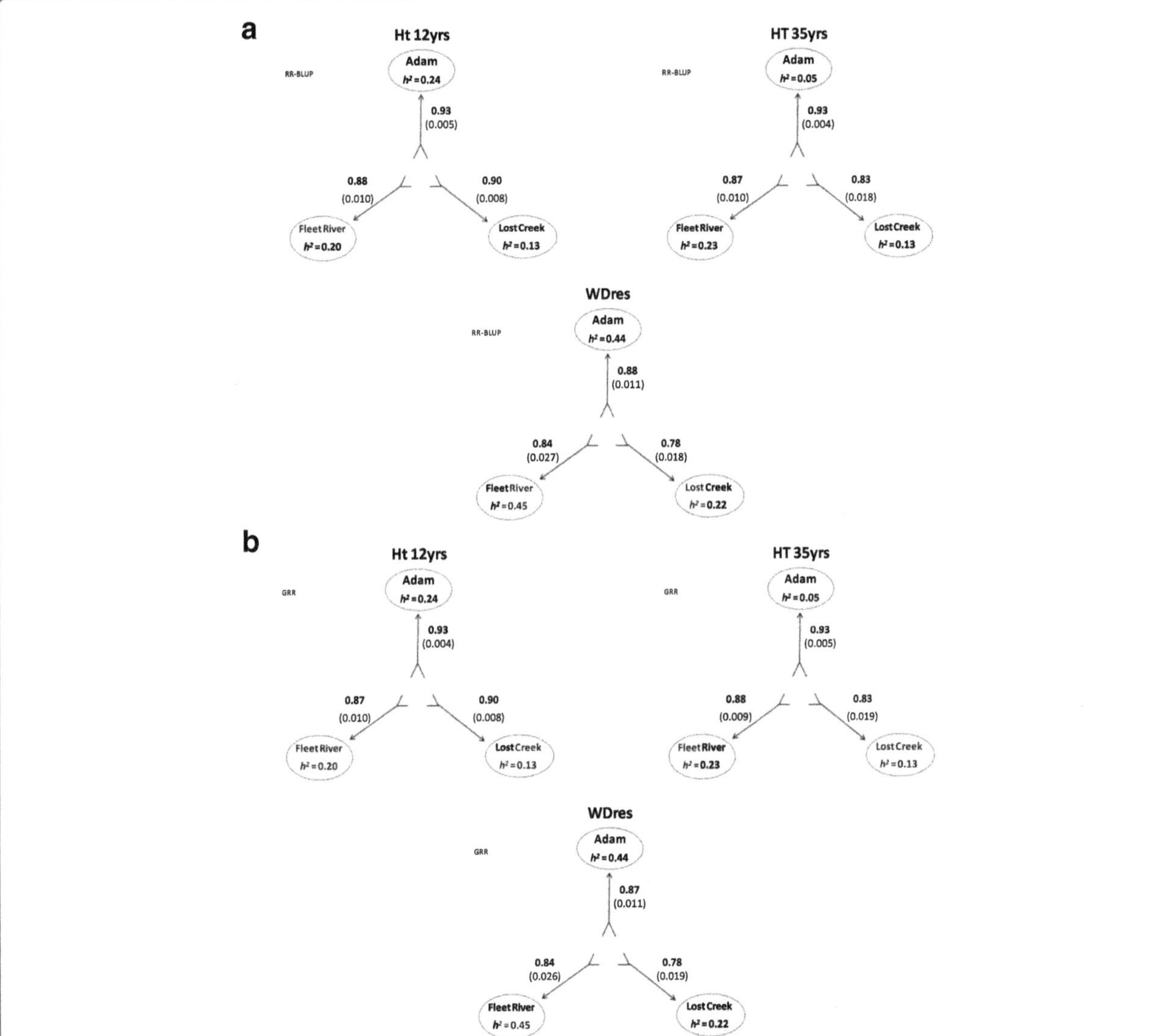

Fig. 4 Heritabilities and GS prediction accuracies for models trained on EBVs and predicting GEBVs for each of the traits. Showing the results of two sites predicting one site cross-validation (Top (**a**) using RR-BLUP, Bottom (**b**), using GRR). The direction of the arrows depicts what information (site) is used to train the model (shaft end), and which is being predicted (head). Traits key: HT 35 yrs. = height at 35 years (cm); HT 12 yrs. = height at 12 years (cm); WD_{res} = wood density (g/cm^3) calculated from resistance to drilling. Sites are Adam, Fleet River, Lost Creek

The deregressed breeding values produced similar results to those obtained from within, cross- and multi-sites GS models, with two sites to one site cross-validation analyses prediction accuracy approximating 0.0 for HT12, HT35, and WD_{res}.

The ABLUP cross-validation for two sites predicting one site resembled the results of the GS model trained on EBVs. Prediction accuracies for HT12 and HT35 were lower than the GS models using EBVs (HT12 average = 0.81 vs. 0.90, and HT35 average = 0.81 vs. 0.88) (Fig. 5). However, the average predictability for WD_{res} remained the same for ABLUP as the GS models using EBVs (0.83) (Fig. 5). In general Adam was the most predictable site for all three traits (HT12, HT35 and WD_{res}), followed by Fleet River and lastly Lost Creek in this ABLUP analysis (Fig. 5). This is the same site predictability trend displayed by the GS models trained on EBV data.

Time- time prediction accuracy (age- age/ trait-trait correlation)

In order to test the theory that the target time for forward selection can be reduced, prediction models were assessed on their accuracy when trained on younger trees (12 years: Ht_{12}) followed by validation on the same trees for height at age 35 (Ht_{35}) and wood density at age 38 (WD_{res38}) (i.e., correlations between GEBVs at age 12 and EBVs at ages 35 (HT_{12}-HT_{35}) and 38 ($trait_{HT12}$-$trait_{WDres38}$). The GEBVs values for Ht_{12} have significant positive correlations with

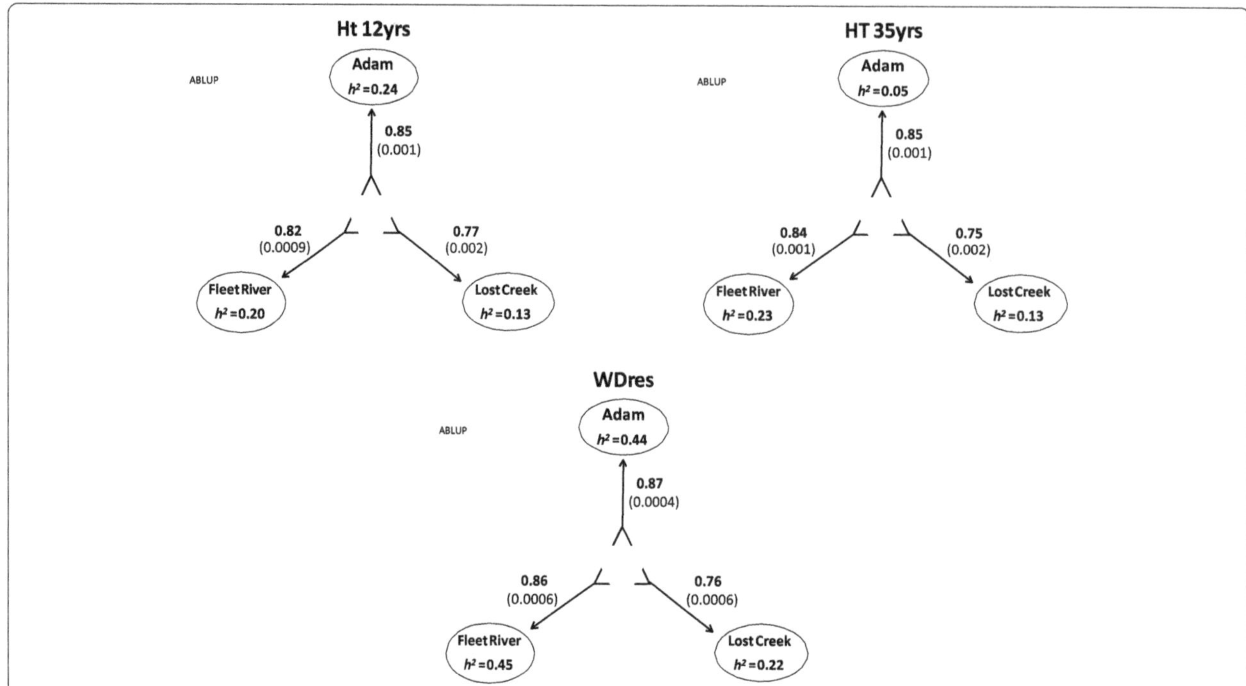

Fig. 5 Heritabilities and prediction accuracies of the ABLUP model for each of the traits. Showing the results of two sites predicting one site cross-validation. The direction of the arrows depicts what information (site) is used to train the model (shaft end), and which is being predicted (head). Traits key: HT 35 yrs. = height at 35 years (cm); HT 12 yrs. = height at 12 years (cm); WDres = wood density (g/cm3) calculated from resistance to drilling. Sites are Adam, Fleet River, and Lost Creek

EBVs for HT_{35} for both RR-BLUP (0.71 ± 0.0004) and GRR (0.71 ± 0.0004) (Table 3). Trait- trait correlation between HT_{12} and WD_{res} (recorded at 38 years: $trait_{HT12}$-$trait_{WDres38}$) produced significant negative correlations (RR-BLUP: -0.46 ± 0.0005; GRR: -0.46 ± 0.0006) (Table 3). These are the most useful correlations to make as they reflect the direction in which selections will be made.

Using the deregressed breeding values to train the two GS models, we obtained results approximating 0.0 for all time-time cross-validation analyses (Table 3).

Time-time and trait-trait cross-validation of ABLUP resembles closely that of the GS models trained on EBV data. EBVs for HT_{12} have a medium to strong positive correlation with EBVs at HT_{35} (0.74 ± 0.001), and a significant negative correlation with EBVs of wood density WD_{res} at age 38 ($trait_{HT12}$-$trait_{WDres38}$: −0.48 ± 0.002).

Discussion

Exome capture

Exome capture is a target enrichment method for sequencing the protein coding regions in a genome. This makes analysis much more efficient. Through targeting this region, the focus is immediately resolved to those areas that are likely to contain sources of variation for the phenotype. The reduced cost and time compared to that of whole genome sequencing, as well as the ability to still capture a significant proportion of variants makes exome capture desirable as it is harder to find functional variants in noncoding regions [20, 27, 28]. Added to which, forest trees already tend to have large and complex genome sizes [20–22]. Exome capture has been recognized as an effective method for increasing knowledge of unmapped large genomes [19, 28].

Table 3 Genomic selection prediction accuracies for time- time correlations for HT12 for RR-BLUP and GRR models (standard errors)

GS Model	EBVs		DEBVs	
	WD_{res}	HT35	WD_{res}	HT35
RR-BLUP	−0.46 ± 0.0005	0.71 ± 0.0004	0.02 ± 0.0038	−0.03 ± 0.0039
GRR	−0.46 ± 0.0006	0.71 ± 0.0004	0.002 ± 0.0040	−0.004 ± 0.0047

Prediction accuracies based on correlations between GEBVs at age 12 and EBVs at age 35 (HT) and 38 (WD); and correlations between GEDBVs at age 12 and DEBVs at age 35 (HT) and 38 (WD). Traits key: HT35 = height at 35 years (cm); WDres = wood density (g/cm3) calculated from resistance drilling to drilling; HT12 = height at 12 years (cm)

Whilst exome capture is expected to produce some missing marker information, the method surpasses whole genome shotgun sequencing or genotyping methods using genome complexity reduction, in depth of coverage. Given a restricted budget (with no option for re- sequencing) the depth achieved using exome sequencing is expected to be much greater than genotyping-by-sequencing [30], with the added benefit of obtaining more reliable calls. After the sequencing process, those markers with a large proportion of missing information can be filtered out, however Rutkoski et al. [31] admit that this step is unnecessary since it affects the GS prediction accuracy little. In order to maintain power in the subsequent analyses, marker imputation should be used to infer missing data in those records not filtered out, conserving the majority of the original SNPs. This was carried out in the current study, although the ratio of missing data was very low and therefore the impact of this procedure is expected to be minimal. Given that currently there is no genetic map available for Douglas-fir, imputation methods in this case are restricted to those which support unordered data. There have been various imputation methods proposed for unordered markers however their efficacy and suitability is yet to be fully determined in genomic selection [31]. Though some methods do show some promising results when compared to mean imputation, notably random forest regression produced greater GS prediction accuracy than its counterparts in both Rutkoski et al. [31] and Poland et al. [32] studies.

Although we managed to capture significant family effects using this type of SNP data, we found little evidence that the GS models were able to capture marker-QTL LD using this genotypic data set (see below for discussion). It is highly likely that substantially more SNPs will be required to capture significant effects for these traits (i.e., LD). In addition to this, our genotyping efforts were focused on a restricted portion of the genome, the exome, which is limited to the available 40 K probes used in the present study. In humans, the exome constitutes a mere 1% of the total genome [33], thus considering the unique genome size and complexity of conifers [34] we expect that the population of SNPs used in this study represents a very small fraction of the Douglas-fir genome. Additionally, in this case the revealed exome which represents functional genes that, by default, are under selection and thus are conserved. By focusing the present study on this region we were not able to capture the variation among families, as this was not represented in the exome. In order to seize this variation in intergenic regions, an alternative, whole genome approach must be used for genotyping, for example genotyping-by-sequencing as it uncovers unordered SNPs across the entire genome. These intergenic regions are not under the same selection pressure as the exome, and may contain important regulatory sequences which correspond to the control of the traits we investigated. Another approach used by Fuentes-Utrilla et al. [16], was to use restriction-site associated DNA sequencing (RADseq) technology. They found that in a species with no whole genome assembly (in this case Sitka spruce (*Picea sitchensis* (Bong.) Carr)), a SNP panel could be constructed from randomly located markers generated from RADseq. However, it must be noted that their GS analysis was performed strictly within a single family.

Heritability

The use of ABLUP instead of GBLUP is likely falsely inflating the heritability estimate. Though the impact of heritability on the predictive accuracy seems to be low in this study. Our results show that even with modest heritability, predictive accuracies can be high. As shown by the combined site analyses, the heritability for HT12, HT35, and wood density were modest (0.17, 0.17, and 0.43, respectively), yet the prediction accuracies were 0.91 (± 0.004), 0.88 (± 0.006), and 0.83 (± 0.009), respectively (RR-BLUP) (Table 2). The large sample size and low effective population size of the present study (N_e = 21) likely helped in negating the effect of low trait heritability [5]. M rtens et al. [35] provided evidence that increased relatedness between training and validation populations leads to higher prediction accuracy in their study on yeast. Furthermore, the use of correlation between pedigree-based and marker-based breeding values, approximates correlation between unknown true breeding value and genomic breeding value. The accuracy can go above heritability in this case because both values are representing only genetic effects, this is in line with results obtained by Gamal El-Dien et al. [12].

Genomic selection

The desired outcome of genomic selection is to produce unbiased marker effect estimates [5], and to avoid the Beavis effect [36] which hinders MAS causing marker effects to be overestimated [37]. Instead of selecting markers based on a significance threshold, GS estimates all marker effects simultaneously causing a different problem; there are more predictor effects (p, markers in this case) to be estimated than there are observations (n, samples) [3]. Least squares cannot be used to estimate all the effects at once since there are not enough degrees of freedom. In addition, multi-collinearity of markers would cause any model to be over fitted [3].

To address this issue (large p, small n), various statistical models have been proposed. They generally fall into the following categories: shrinkage models, variable selection models, and kernel methods. Shrinkage models (e.g., ridge regression BLUP [RR-BLUP], Whittaker et al. [38]) fit all marker effects which are all shrunk to the same degree. With RR-BLUP, it is assumed that the trait in question more closely resembles Fisher's infinitesimal model (many loci with small effects), and marker effects are samples from a normal distribution (with equal variance). Variable selection (e.g., generalized ridge-regression [GRR], Shen et al. [39]) however, reduces the number of markers used, and in doing so assumes that the trait is controlled by fewer, strong effect loci [3]. Kernel methods convert the predictor variables to distances in effect creating a matrix similar to an additive genetic relationship matrix. The kernel matrix quantifies the distance between individuals but also smoothing parameters are added [3]. This method is flexible and can incorporate complex relationships between markers, for this reason it is useful in cases where non-additive effects are suspected to occur [40]. Indeed, Douglas-fir height and wood density have proven to have non-additive genetic variance component [41]; however, its unpredictability has driven the species' advanced generation breeding and selection methods to be additive genetic gain-dependent [41–43]. Since different traits have differing genetic architecture, there does not exist one model that is necessarily the best for all traits or populations [3].

In the present study, we assessed the prediction accuracies of two GS models (RR-BLUP and GRR) in two phases; first when trained on EBVs, and second when trained on DEBVs. In the first instance, using EBVs and by virtue of retaining family means, our data contained both contemporary and historical pedigree information as well as marker-QTL LD information. Without further adjustment, all this information was parsed into the GS models and resulted in high prediction accuracies (for both model types). Subsequent to this, in phase two, we deregressed the EBVs, removing the parent average effect in order to disassociate the pedigree information from the marker-QTL LD. This resulting in DEBVs that contained the marker-QTL LD information only without the between family genetic variance. Using these DEBVs to train the GS models, we obtained prediction accuracies which for all practical purposes were 0.0. The success of the first phase can be attributed to the large within and among family genetic variances. This pedigree-driven approach dominated the analysis and produced the observed high prediction accuracies, which were comparable to the ABLUP accuracies. When this influence of pedigree was removed, the generated models were no longer able to provide useful predictions. A similar effect was seen in interior spruce (*Picea glauca* x *Picea engelmannii*) where cross-validation using family folding resulted in decreased prediction accuracy, because between family variance was dominating the analysis (Gamal El-Dien et al. 2017, unpublished). In addition to this Fuentes-Utrilla et al. [16], when studying GS in Sitka spruce, found that within family predictions cannot be extrapolated to between families. Furthermore, when examining marker transferability between families in white spruce, Beaulieu et al. [44] also found within family predictions to be more precise than between family predictions. Similarly, to Gamal El-Dien et al. (2017, unpublished) they also found that when a family had no representation in the training group, the accuracies obtained for that family were very small, occasionally negative, and often not statistically significant from zero. Much like using the DEBVs here, which try to predict between family variance, but have family means stripped away. Fuentes-Utrilla et al. [16] concluded that species with large effective population sizes (notably conifers), have a reduced ability to make predictions across families. With this in mind, GS may best be employed to produce GEBVs for within large full-sib families in conifers as it captures the Mendelian sampling term.

This discrepancy in the results from the two GS phases, is indicative that we simply were not able to capture the marker-QTL LD with the available SNPs. Whilst other studies have had some success in this area, it is important to note that these particular investigations have focused on tree species with much smaller genomes, for example Resende et al. [7]. Resende et al. [17] make this point in their 2017 study of *Eucalyptus*, that although other studies may have failed to produce significant predictions between unrelated populations, this may be due to low marker density. This is likely inhibiting the ability to capture short-range LD, the result being that prediction models rely almost entirely on relatedness.

In their study, Resende et al. [7] used 7680 DArT markers on *Eucalyptus* which is estimated to have a genome size of 609 Mbp [7], equivalent to 12.61 markers/Mbp. In contrast, here we used 69,951 SNP markers on the Douglas-fir genome which is estimated at 18,700 Mbp, giving 3.74 markers/Mbp. To obtain a similar coverage as Resende et al. [7] would require approximately 235,800 markers, many more than we currently have. With such a large genome, it is likely that many more SNPs will be required (≈235,800+), with greater genomic coverage in order to capture this, so far elusive, LD. In a more recent study, also using *Eucalyptus*, Müller et al. [18] managed to capture some short-range historical LD using 5000–10,000 SNPs. Yet they too concluded that the genomic prediction in this case was largely driven by relatedness.

The similar predictions given by the RR-BLUP and GRR methods, and the similarly high ABLUP accuracies, was again the result on heavy reliance on between family variance, and thus we have gained no new information as to the genetic control or architecture of the traits in question. Although similar findings have been noted in more successful GS studies. For example, Resende et al. [8] when comparing GS methods, found there to be little difference between the predictive abilities of shrinkage and variable selection methods (4 in total) even considering they have different *a priori* assumptions. They used a *Pinus taeda* training population with 951 individuals and 4853 SNPs in the analyses. Prediction accuracies for

17 traits (including growth, disease resistance and development) ranged from 0.17 to 0.51. Only one trait (fusiform rust resistance) showed any significant difference between the models. Higher prediction accuracies were obtained using variable selection methods for this trait. This reflects the genetic architecture of the trait which is controlled by few, large effect loci [8].

Single and multi-site cross-validation

The combined site GS analysis produced higher prediction accuracies than the single site analyses on average (Fig. 2a and b). The combined site training population having more individuals than the single site populations. This is in agreement with what the literature states should happen; Grattapaglia [2] noted that increasing the training population size increases accuracy up to a point (around 1000 individuals). In addition to increased sample size for predictive accuracy improvement, the multi-site approach incorporated the present GxE in the model, resulting in further improvement.

The prediction accuracies are high for HT12, HT35, and WD_{res} GEBVs (0.87–0.92, 0.79–0.92, and 0.78–0.88, respectively) (Fig. 2a and b). They are moderately higher than those in previous studies including other forest tree species [7, 8, 10, 11, 15]. Largely due to the inclusion of the pedigree structure. In this case the GS methods are not giving much advantage over ABLUP (ABLUP multi-site cross-validation accuracies: 0.88 ± 0.002, 0.86 ± 0.003, and 0.84 ± 0.003, for HT12, HT35, and WD_{res} respectively), thus both are predicting only family means. The prediction accuracies for both the GS models and the ABLUP model are much higher than theoretical accuracy of the EBVs (0.61, 0.68, and 0.76; for HT12, HT35, and WD_{res}, respectively), which indicates that both the EBVs and GEBVs are converging on the family means, and are far from the true breeding values.

The high prediction accuracies for the GEBVs may also partially be the result of using a relatively large training population, known to correlate with accuracy. In addition, there is an interacting effect of the relatively low effective population size (N_e = 21), and both these data characteristics increase the accuracy of predictions [3]. Although in this case the low effective population/family size also meant that the Mendelian sampling term could not be defined. Indeed, Gamal El-Dien et al. [12] using an interior spruce population with N_e = 93.8 (estimated assuming the OP families are unrelated and contributed equally to the experiment) had lower prediction accuracies than obtained here. But these in turn were vastly higher than results from a study with 214 open-pollinated white spruce (*Picea glauca*) families in Quebec with N_e = 622.5 [9]. Another component responsible for the increased accuracy of predictions, was the training of the models on EBVs rather than raw phenotypes [12].

Prediction accuracies fell dramatically for models trained on DEBVs, as marker-QTL LD could not be recovered using the available SNP data set, indicating that the SNP markers effectively tracked the pedigree.

Cross-site validation

Similar prediction accuracies were observed in the cross-site compared to the single-site and combined-site GS analyses (with models trained on EBVs). Genotype x Environment interaction is an important consideration of forest tree selective breeding, more so than in animal breeding where individuals are considered to share a common environment [45–48]. Prediction accuracy for HT12 across sites was relatively high. An unexpected result given the fact that the heritability was only 0.17 (combined sites), and there is expected to be a competition effect at early stand development (i.e., strong environmental component). This is possibly an indication of the partially controlled experimental environment (not natural stands) in which the trees were growing. Given these experimental conditions, it is also conceivable that GxE effects are minimal. Which is reflected in the cross-site predictions (Fig. 2a and b), which are of a similar magnitude to the within-site prediction accuracies for all attributes.

Time- time and trait- trait correlations

It is thought that forward selection in Douglas-fir should be carried out at a minimum of 17 years [49], when accurate predictions of phenotype at time of harvest can be measured. We tested the correlation of height at 12 years and height at 35 years (HT_{12}-HT_{35}), and wood density (38 years) ($trait_{HT12}$-$trait_{WDres38}$) and positive (0.71 for height) and negative (−0.46 for wood density) correlations were detected by the GS models trained on EBV data. Although they did not offer accuracy above that of ABLUP, indicative of a strong reliance on pedigree information rather than marker-QTL LD. The results give an indication of how useful early selection could be. Correlations for height are moderately strong and low-moderate for wood density. Marker-trait associations are known to vary according to the tree age, limiting any correlation. This would certainly hamper efforts to create a highly correlated age-age model in trees [9, 50]. At the moment providing such predictions at an age any younger than 12 years would not be recommended (note that height at age 12 was the earliest measurement available in the present study). Since larger age differences have been shown to produce less accurate models [51]. The discrepancy in prediction accuracy between the time-time correlations and the cross-validations suggest that there is some inconsistency between EBVs and marker effects at these two ages (12 and 35 years). Although the discrepancy is relatively small, and so meaningful results may still be obtained through age-age and trait-trait correlations.

This is, in addition to the varied environmental conditions the trees endure over their long lifespans, lessening time-time correlations. To this effect and as White et al. [52] suggested, training models will need to be updated with mid-rotation phenotypes in order to accurately predict mature trait values.

Genomic selection and forest tree breeding

Several genomic selection "proof-of-concept" studies have been conducted on few coniferous tree species (e.g., loblolly pine, maritime pine, Sitka spruce, white spruce, and white-Engelmann spruce hybrid), all concluded that GS has the potential to increase the genetic gain through speeding breeding generation turn-over and increasing the selection differential. It should be stated that, with the exception of the maritime pine study [14], none of the derived GS predictive models have been validated on independent "validation" populations. The success of GS is dependent on the linkage disequilibrium between the markers used (i.e., SNP panels) and causal genes underpinning the traits of interest, and the degree of relatedness between the training and validation populations. Therefore, caution is required during GS implementation as LD changes after every round of breeding (i.e., recombination); the fact that it does rapidly decay called for using dense marker coverage to overcome this caveat. Still we have found that by only using sequence capture data, we were unable to successfully resolve this marker-QTL LD. We only had success in capturing among family effects (i.e., pedigree). Even with a SNP panel designed appropriately based on an informative SNP library, and large enough to handle a conifer genome, there are additional hurdles. LD only survives over relatively short distances in conifers compared to livestock species due to their relatively large N_e [53]. This led Fuentes-Utrilla et al. [16] to conclude that GS may only be useful in tree populations with reduced N_e, for example seed orchards, or lines/ subgroups which have been produced through selective breeding. Though as they demonstrate with their analysis, it is possible to generate very large full-sib families in trees, by controlled crossing. In this type of population, LD extends over longer distances than in open-pollinated populations. As a result, they suggest that GS could be employed to make selections within families.

Based on comparable studies: 1) a greater number of markers, and 2) wider coverage throughout the whole genome or dense unordered SNP genotyping platform (e.g., GBS), would be needed to capture this LD and additional intergenic variation [17]. Or indeed a shift in the level at which GS is applied, i.e. to within families rather than across families [16]. Whilst GS still has the potential to deliver unprecedented gains, it does not seem likely that was achieved in the present study as the prediction driving force was the pedigree rather than LD. Despite the low N_e, the small family size prevented the accurate assessment of the Mendelian sampling term in this population. Therefore, the EBVs were heavily shrunk toward the family mean and all within family deviations were not estimated precisely.

Finally, it should be emphasized that any gains captured through GS require unique tree improvement delivery methods as the traditional seed orchards' production mode requires time for reaching sexual maturity, even under intensive management such as top grafting or hormonal applications, and its sexual-production effectively breaks the LD between selected traits and markers.

Conclusions

The results suggest that the population of SNP markers used, along with their low coverage across the Douglas-fir genome was not successful for capturing the LD with the causal genes underpinning the studied attributes. In this case the impressive results of the investigated genomic selection models relied heavily on relatedness rather than the LD. Alternative marker generation methods such as whole genome sequencing or other dense unordered SNP genotyping methods such as GBS are needed, as is a larger SNP array. Exome capture provide enough markers to successfully capture/track the pedigree (contemporary and historical) and thus it is useful for genetic variance decomposition of conifer traits, thus providing better genetic parameter estimates. Since we were only able to resolve the between family effects, we gained no new information regarding the genetic architecture of the traits. Whilst low N_e may help boost prediction accuracy in similar studies, there may be a lower limit to this. Beyond which Mendelian sampling is not captured. However, using a single, large full-sib family causes LD to extend over further distances compared to open-pollinated trees, therefore it is possible to make within family selections using this type of population as Fuentes-Utrilla et al. [16] have demonstrated.

Methods

Experimental population

Predictive models were trained on a replicated 38-year-old pedigreed coastal Douglas-fir (*Pseudotsuga menziesii* Mirb. (Franco)) progeny testing population. The trial was established by the Ministry of Forests, Lands and Natural Resource Operations of British Columbia, Canada in 1975 and consists of 165 full-sib families (54 parents), of which 37 families were selected for sampling from three test sites (**Adams** (Lat. 50 24′42″ N, Long. 126 09′ 37″W, Elev. 576 mas), **Fleet River** (Lat. 48 39′25″ N, Long. 128 05′ 05″ W, Elev. 561 mas), and **Lost Creek** (Lat. 49 22′15″ N, Long. 122 14′07″ W, Elev. 424 mas)) with a total of 1372 trees (N ≈ 500 per site) and effective population size (N_e) of 21.

Tissue sampling, DNA extraction and genotyping

Cambial tissue was collected using a hammer and hollow punch tool (approx. 2 cm diameter) to remove two small circular pieces of bark/ cambium and developing tissue from each tree. The cambium disks were separated from the bark layer and immediately placed in a 2 ml collection tube with 1 ml storage buffer (10 mM EDTA pH 8.0, 10 mM Na_2SO_3) and kept at 4 °C until DNA extraction. DNA extraction followed the modified procedure developed by Ivanova et al. [54] (R. Whetten, unpublished, North Carolina State University, personal communications). Genotyping was done using the exome capture method in a commercial facility (RAPiD Genomics©, Florida, US). A total of 40 K probes were designed using the available Douglas-fir transcriptome assembly [55]. From the Douglas-fir reference transcriptome, a total of 325,372 non-overlapping 120-bp probes were initially designed. After filtering out redundant and organelle matching probes, this number was reduced to 117,135 probes. Of the remaining probes, we selected 7464 that contained 17,096 SNPs previously reported [55]. A further 32,536 probes were selected bringing the total to 40 K. Selection was performed by randomly sampling the remaining probes restricting the selection to a maximum of two probes per transcript. These 40 K (7464 + 32,536) probes cover a total of 21,187 transcripts. The raw sequenced reads were demultiplexed in each individual barcodes. Low quality bases with less than 20 quality score in the 3′ end were trimmed out followed by a low quality filter that removed reads with more than 10% of the read with less than 20 quality score. The filtered reads were aligned against the reference transcriptome using Mosaik v2.2.3 [56] with the following parameters -mmp 0.05 -m all -a all -hs 15. SNP markers were identified at a population level using Freebayes [57] without considering indels, multi-nucleotide polymorphisms and complex events. This analysis resulted in approximately 550,000 SNPs. These SNPs were filtered to identify the highest quality sites. These included only biallelic SNPs with less than 40% missing data. Further filtering was applied so that data was located on contigs with a mean read depth less than or equal to 60. Additional filters included minor allele frequency (MAF) >5%, Hardy-Weinberg disequilibrium cut-off <−0.05, and maximum site depth < 60. This process resulted in a total of 1372 samples with 69,551 SNPs for use in the study, and mean imputation was used for the missing data. (for more details on exome capture see, Neves et al. [19]).

Phenotyping

Early- (1988) and mid- (2011) rotation growth trait measurements of the studied trees had been assessed for height (HT: in meters) and mid-rotation (age 38 years, 2014) wood density (WD_{res}) was assessed indirectly using the average resistance measurements recorded with a Resistograph® (Instrumenta Mechanik Labor, Germany). Resistograph readings were subsequently converted to wood density indices (g/cm^3) by scaling them by the DBH measurements, as performed by El-Kassaby et al. [58].

Estimated breeding values (EBVs) and deregressed breeding values (DEBVs)

EBVs were fitted in ASReml 3.0 [59], using the following mixed linear model for a single site:

$$y = X\boldsymbol{\beta} + Z\boldsymbol{a} + e \tag{1}$$

where $\boldsymbol{y}$ is the phenotypic trait measurement, $\boldsymbol{\beta}$ is a fixed effect vector (overall mean), $\boldsymbol{a}$ is a random effects vector (additive genetic) which are normally distributed ($\sim N(0, \boldsymbol{A}\sigma_a^2)$) where $\boldsymbol{A}$ is average numerator relationship matrix and σ_a^2 is additive genetic variance, $\boldsymbol{e}$ is the random residual effects which are normally distributed ($\sim N(0, \boldsymbol{I}\sigma_e^2)$) where $\boldsymbol{I}$ is identity matrix and σ_e^2 is residual variance. $\boldsymbol{X}$ and $\boldsymbol{Z}$ are incidence matrices relating the fixed and random effects to the observations. Since GxE plays an important role in forestry [60] the combined site EBVs were estimated with terms accounting for site, replication, and family structure. Thus minimising biases in BV calculation caused by environmental variations between and within sites, and full-sib genetic effects.

Narrow-sense heritability was calculated as $h^2 = \sigma_a^2 / (\sigma_a^2 + \sigma_e^2)$, where σ_a^2 and σ_e^2 are the variances of additive genetic and residual effects, respectively. Combined site heritability estimates included the GxE model term in the denominator. The breeding values ($\hat{a}$) are fitted using BLUP as follows:

$$a = AZ' \sigma_a^2 \; V^{-1} \left(y - X\hat{\beta}\right) \tag{2}$$

where $\boldsymbol{V}$ is the variance- covariance matrix of $\boldsymbol{y}$ obtained by:

$$V = ZAZ' \sigma_a^2 + I\,\sigma_e^2 \tag{3}$$

Breeding values were deregressed using the deregression procedure of Garrick et al. [61]. This adjusts the BV data to account for family means, resulting in DEBVs that contain information regarding individuals only, without parental BV influence.

The theoretical accuracy of the EBVs (r) was calculated following the procedure of Dutkowski et al. [62].

$$\hat{r} = \sqrt{1 - \frac{SE_i^2}{(1+F_i)\hat{\sigma}_a^2}} \tag{4}$$

where SE_i is the standard error of breeding value, and F_i is the inbreeding coefficient of the ith individual.

Cross-validation across time and space

The two GS models (RR-BLUP and GRR) were compared to assess their application to various traits, and in addition they were cross-validated in order to assess their prediction accuracy across environments/ spatial divisions and time. The intention was to give an indication to which extent (if any) GS increases gain per unit time over traditional methods.

The validation processes used was a replicated randomly assigned 10-fold cross validation. Models were trained on 9/10 of these folds, with the remaining 1/10 fold used as a validation set. Prediction accuracy was determined as the mean of the replications of the Pearson product-moment correlation between EBVs of the validation set and their predicted GEBVs. Or in the case where DEBVs were used to train the models, the correlation between DEBVs and predicted GEDBVs (genomic estimated deregressed breeding values). The following combinations were used: 1) within-site, using information from a single site to estimate the GEBVs within that same site, 2) Cross-site/ between sites in all combinations, with information from single sites used to predict other sites, 3) Combined sites, pooled information from all the sites, 4) Multi-site to single site, the pooled information from all three sites used in the estimation of single sites only, 5) Two sites to predict one site, pooled information from two sites used in the estimation of the remaining site only, and 6) Time-time (age-age)/ trait-trait, using information from individual trees to obtain correlations between GEBVs at age 12 and EBVs at ages 35 for height and 38 for wood density. In addition, the same 10-fold cross-validation process was used to assess the predictive accuracy of the ABLUP model for all the spatial cross-validation analyses for all the attributes (HT12, HT35, and WD_{res}) in the study. In this case predictions made were based on relatedness as given by the pedigree.

Genomic selection analysis

Two GS methods were compared: ridge regression (RR-BLUP) and generalized ridge regression (GRR) [3]. The performance of each of the methods was assessed according to their predictive accuracy determined by the correlation between GEBVs and EBVs, or DEBVs and GEDBVs.

Genomic estimated breeding values (GEBVs) and genomic estimated deregressed breeding values (GEDBVs)

RR-BLUP

Ridge regression (RR-BLUP: Whittaker et al. [38]) was proposed for use as a selection tool based on marker information. The model was fitted using the R package 'RRBLUP' [63], the GEBV (or GEDBV if using deregressed BVs) is obtained by the sum of p marker effects:

$$g(x_i) = \sum\nolimits_{k=1}^{p} x_{ik}\beta_k \tag{5}$$

where x_{ik} is the score (genotype) of SNP k in individual i, β_k is the marker effect of k. This method assumes that the marker effects are normally distributed (mean = 0) and so the BLUP solutions for marker effects are derived from solving mixed linear model equations of Henderson [64] so that λ is optimised in the following Eq:

$$\beta = \left(Z'Z + \lambda I\right)^{-1} Z'y \tag{6}$$

where $\boldsymbol{Z}$ is an incidence matrix relating markers to individuals, $\boldsymbol{I}$ is an identity matrix, and $\boldsymbol{y}$ is the vector of EBVs fitted in ASReml. The shrinkage parameter (λ) can be written as $\lambda = \sigma_e^2/\sigma_\beta^2$ (residual variance / common marker effect variance). Since marker effects are assumed to be identically distributed, all effects are shrunk equally towards zero. This method is equivalent to using lines (as opposed to markers) as random effects in a mixed model analysis, where the covariance is modelled by a kinship matrix calculated from the marker data (a genomic relationship matrix [$\boldsymbol{G}$ matrix]) (this is sometimes referred to as GBLUP).

GRR

Generalized ridge regression (GRR) is a two-step variable selection method, the first step obtains estimates in the same way that RR-BLUP does using linear mixed model analysis to solve for optimum λ, and in the second step the BLUP for β is subjected to an alternative shrinkage parameter which is marker specific, using GRR to solve a heterogeneous error model which replaces $I\lambda$ in (5) with *diag(*λ*)*:

$$\hat{\beta} = \left(Z'Z + \operatorname{diag}(\lambda)\right)^{-1} Z'y \tag{7}$$

In this case λ is a vector of p shrinkage parameters. For the kth element: $\lambda_k = \widehat{\sigma_e^2}/\widehat{\sigma_{\beta k}^2}$, is the parameter, where $\widehat{\sigma_{\beta k}^2}$ is the variance of marker effect k ($\widehat{\sigma_{\beta k}^2} = \hat{\beta}_k^2/(1-h_{kk})$). β is from step 1 (the BLUP marker effect) and h_{kk} is the influence of the dependant variable on the fitted value for observation k. In other words, h_{kk} represents the diagonal element ($n + k$) of the influence matrix $\boldsymbol{H} = \boldsymbol{T(T'T)^{-1} T'}$ and:

$$T = \begin{pmatrix} X & Z \\ 0 & \operatorname{diag}(\lambda) \end{pmatrix} \tag{8}$$

Abbreviations

ABLUP: Pedigree-based Best Linear Unbiased Predictor; DEBV: Deregressed estimated breeding value; EBV: Estimated breeding value; GBLUP: Genomics-based Best Linear Unbiased Predictor; GRR: Generalized Ridge Regression; GS: Genomic selection; GxE: Genotype-environment interaction; HT12: Height at age 12 years; LD: Linkage disequilibrium; MAS: Marker-assisted selection; QTL: Quantitative trait loci; RR-BLUP: Ridge Regression Best Linear Unbiased Predictor; WD_{res}: Wood density estimated using a Resistograph

Acknowledgements
We thank British Columbia Ministry of Forests, Lands and Natural Resource Operations, Victoria, BC for data and trials access, M.F. Resende Jr. and L.G. Neves of RAPiD Genomics, Florida, USA for genotyping.

Funding
This work was supported by Genome British Columbia (User Partnership Program (UPP-001) to YAK and MUS, NSERC Discovery Grant to YAK, TimberWest Forest Corp., Western Forest Products Inc. We declare that the funding agencies did not participate in the design of the study and collection, analysis, and interpretation of data and in writing the manuscript.

Authors' contributions
YAK and MUS conceived and designed the experiment, FRT, BR, IP collected the data, FRT performed the data analyses and wrote the manuscript, BR, JK, IP, CC, MUS, YAK advised on the data analyses and edited the manuscript. All authors read and approved the final manuscript.

Competing interests
The authors declare that they have no competing interests.

Author details
[1]Department of Forest and Conservation Sciences, Faculty of Forestry, The University of British Columbia, 2424 Main Mall, Vancouver, BC V6T 1Z4, Canada. [2]Scion (New Zealand Forest Research Institute Ltd.), 49 Sala Street, Whakarewarewa, Rotorua 3046, New Zealand. [3]Department of Genetics and Physiology of Forest Trees, Faculty of Forestry and Wood Sciences, Czech University of Life Sciences Prague, Kamycka 129, 165 21 Praha 6, Czech Republic. [4]Département des sciences du bois et de la forêt, Université Laval, QC, Québec G1V 0A6, Canada. [5]Department of Biochemistry and Molecular Biology, Oklahoma State University, Stillwater, OK 74078-3035, USA. [6]British Columbia Ministry of Forests, Lands and Natural Resource Operations, Victoria, BC V8W 9C2, Canada.

References
1. Hayes B, Bowman P, Chamberlain A, Goddard M. Genomic selection in dairy cattle: progress and challenges. J Dairy Sci. 2009;92:433–43.
2. Grattapaglia D. Breeding forest trees by genomic selection: current progress and the way forward. In: Tuberosa R, et al., editors. Genomics of plant genetic resources. Netherlands: Springer; 2014. p. 651–82.
3. Lorenz AJ, Chao S, Asoro FG, Heffner EL, Hayashi T, Iwata H, Smith KP, Sorrells MK, Jannink JL. Genomic selection in plant breeding: knowledge and prospects. Adv Agron. 2011;110:77.
4. Fisher RA. The correlation between relatives on the supposition of Mendelian inheritance. Trans R Soc Edinburgh. 1918;52:399–433.
5. Meuwissen T, Hayes B, Goddard M. Prediction of total genetic value using genome-wide dense marker maps. Genetics. 2001, 1819;157
6. Grattapaglia D, Resende MD. Genomic selection in forest tree breeding. Tree Genet Genomes. 2011;7:241–55.
7. Resende MD, Resende MF, Sansaloni CP, Petroli CD, Missiaggia AA, Aguiar AM, Abad JM, Takahashi EK, Rosado AM, Faria DA, Pappas GJ Jr, Kilian A, Grattapaglia D. Genomic selection for growth and wood quality in eucalyptus: capturing the missing heritability and accelerating breeding for complex traits in forest trees. New Phytol. 2012;194:116–28.
8. Resende MF, Muñoz P, Resende MD, Garrick DJ, Fernando RL, Davis JM, Jokela EJ, Martin TA, Peter GF, Kirst M. Accuracy of genomic selection methods in a standard data set of loblolly pine (*Pinus taeda* L.). Genetics. 2012;190:1503–10.
9. Beaulieu J, Doerksen T, Clément S, Mackay J, Bousquet J. Accuracy of genomic selection models in a large population of open-pollinated families in white spruce. Heredity. 2014;113:343–52.
10. Beaulieu J, Doerksen TK, MacKay J, Rainville A, Bousquet J. Genomic selection accuracies within and between environments and small breeding groups in white spruce. BMC Genomics. 2014;15:1048.
11. Resende MFR, Muñoz P, Acosta JJ, Peter GF, Davis JM, Grattapaglia D, Resende MD, Kirst M. Accelerating the domestication of trees using genomic selection: accuracy of prediction models across ages and environments. New Phytol. 2012c;193:617–24.
12. Gamal El-Dien O, Ratcliffe B, Klápště J, Chen C, Porth I, El-Kassaby YA. Prediction accuracies for growth and wood attributes of interior spruce in space using genotyping-by-sequencing. BMC Genomics. 2015;16:370.
13. Ratcliffe B, El-Dien OG, Klápště J, Porth I, Chen C, Jaquish B, El-Kassaby YA. Comparison of genomic selection models across time in interior spruce (*Picea engelmannii* × *glauca*) using unordered SNP imputation methods. Heredity. 2015;115:547–55.
14. Bartholome J, van Heerwaarden J, Isik F, Boury C, Vidal M, Polmion C, Bouffier L. Performance of genomic prediction within and across generations in maritime pine. BMC Genomics. 2016;17:604.
15. Isik F, Bartholome J, Farjat A, Chancerel E, Raffin A, Sanchez L, Plomion C, Bouffier L. Genomic selection in maritime pine. Plant Sci. 2016;242:108–19.
16. Fuentes-Utrilla P., Goswami C., Cottrell J.E, Pong-Wong R., Law A., A'Hara S. W., Lee S.J., Woolliams J.A., QTL analysis and genomic selection using RADseq derived markers in Sitka spruce: the potential utility of within family data. Tree Genet Genomes 2017: 13: 33.
17. Resende RT, Resende MDV, Silva FF, Azevedo CF, Takahashi EK, Silva-Junior OB, Grattapaglia D. Assessing the expected response to genomic selection of individuals and families in Eucalyptus breeding with an additive-dominant model. SSS. 2017:1–11.
18. Müller BSF, Neves LG, de Almeida Filho JE, Resende MFR Jr, Muñoz PR, dos Santos PET, Filho EP, Kirst M, Grattapaglia D. Genomic prediction in contrast to a genome-wide association study in explaining heritable variation of complex growth traits in breeding populations of Eucalyptus. BMC Genomics. 2017;18:524.
19. Neves LG, Davis JM, Barbazuk WB, Kirst M. Whole-exome targeted sequencing of the uncharacterized pine genome. Plant J. 2013;75:146–56.
20. Suren H, Hodgins KA, Yeaman S, Nurkowski KA, Smets P, Rieseberg LH, Aitken SN, Holliday JA. Exome capture from the spruce and pine giga-genomes. Mol. Ecol Res. 2016;16:1136–46.
21. Neale DB, McGuire PE, Wheeler NC, Stevens KA, Crepeau MW, Cardeno C, Zimin AV, Puiu D, Pertea GM, Sezen UU, Casola C, Koralewski TE, Paul R, Gonzalez-Ibeas D, Zaman S, Cronn R, Yandell M, Holt C, Langley CH, Yorke JA, Salzberg SL, Wegrzyn JL. The Douglas-fir genome sequence reveals specialization of the photosynthetic apparatus in Pinaceae. G3. 2017; https://doi.org/10.1534/g3.117.300078.
22. Neale DB, Kremer A. Forest tree genomics: growing resources and applications. Nat Rev Genet. 2011;12:111–22.
23. Nystedt B, Street NR, Wetterbom A, Zuccolo A, Lin Y, Scofield DG, Vezzi F, Delhomme N, Giacomello S, Alexeyenko A, Vicedomini R, Sahlin K, Sherwood E, Elfstrand M, Gramzow L, Holmberg K, Hällman J, Keech O, Klasson L, Koriabine M, Kucukoglu M, Käller M, Luthman J, Lysholm F, Niittylä T, Olson Å, Rilakovic N, Ritland C, Rosselló JA, Sena J, Svensson T, Talavera-López C, Theißen G, Tuominen H, Vanneste K, Wu Z, Zhang B, Zerbe P, Arvestad L, Bhalerao R, Bohlmann J, Bousquet J, Gil RG, Hvidsten TR, de Jong P, MacKay J, Morgante M, Ritland K, Sundberg B, Lee Thompson S, Van de Peer Y, Andersson B, Nilsson O, Ingvarsson PK, Lundeberg J, Jansson S. The Norway spruce genome sequence and conifer genome evolution. Nature. 2013;497:579–84.
24. Gnirke A, Melnikov A, Maguire J, Rogov P, LeProust EM, Brockman W, Fennell T, Giannoukos G, Fisher S, Russ C, Gabriel S, Jaffe DB, Lander ES, Nusbaum C. Solution hybrid selection with ultra-long oligonucleotides for massively parallel targeted sequencing. Nat Biotechnol. 2009;27:182–9.
25. Fu Y, Springer NM, Gerhardt DJ, Ying K, Yeh CT, Wu W, Swanson-Wagner R, D'Ascenzo M, Millard T, Freeberg L, Aoyama N, Kitzman J, Burgess D, Richmond T, Albert TJ, Barbazuk WB, Jeddeloh JA, Schnable PS. Repeat subtraction-mediated sequence capture from a complex genome. Plant J. 2010;62:898–909.

26. Walsh T, Shahin H, Elkan-Miller T, Lee MK, Thornton AM, Roeb W, Abu Rayyan A, Loulus S, Avraham KB, King MC, Kanaan M. Whole exome sequencing and homozygosity mapping identify mutation in the cell polarity protein GPSM2 as the cause of nonsyndromic hearing loss DFNB82. Am J Hum Genet. 2010;87:90–4.
27. Kiezun A, Garimella K, Do R, Stitziel NO, Neale BM, McLaren PJ, Gupta N, Sklar P, Sullivan PF, Moran JL, Hultman CM, Lichtenstein P, Magnusson P, Lehner T, Shugart YY, Price AL, de Bakker PIW, Purcell SM, Sunyaev SR. Exome sequencing and the genetic basis of complex traits. Nat Genet. 2012;44:623–30.
28. Mertes F, El Sharawy A, Sauer S, van Helvoort JMLM, van der Zaag PJ, Franke A, Nilsson M, Lehrach H, Brookes AJ. Targeted enrichment of genomic DNA regions for next-generation sequencing. Brief Funct Genomics. 2011;10:374–86.
29. Choi M, Scholl UI, Ji W, Liu T, Tikhonova IR, Zumbo P, Nayir A, Bakkaloğlu A, Ozen S, Sanjad S, Nelson-Williams C, Farhi A, Mane S, Lifton RP. Genetic diagnosis by whole exome capture and massively parallel DNA sequencing. Proc Natl Acad Sci U S A. 2009;106:19096–101.
30. Bodi K, Perera AG, Adams PS, Bintzler D, Dewar K, Grove DS, Kieleczawa J, Lyons RH, Neubert TA, Noll AC, Singh S, Steen R, Zianni M. Comparison of commercially available target enrichment methods for next-generation sequencing. J Biomol Tech. 2013;24:73–86.
31. Rutkoski JE, Poland J, Jannink J-L, Sorrells ME. Imputation of unordered markers and the impact on genomic selection accuracy. G3: Genes| Genomes| Genetics. 2013;3:427–39.
32. Poland J, Endelman J, Dawson J, Rutkoski J, Wu S, Manes Y, Dreisigacker S, Crossa J, Sánchez-Villeda H, Sorrells M, Jannink J. Genomic selection in wheat breeding using genotyping-by-sequencing. Plant Genome. 2012;5:103–13.
33. Ng SB, Turner EH, Robertson PD, Flygare SD, Bigham AW, Lee C, Shaffer T, Wong M, Bhattacharjee A, Eichler EE, Bamshad M, Nickerson DA, Shendure J. Targeted capture and massively parallel sequencing of 12 human exomes. Nature. 2009;461:272–6.
34. De La Torre AR, Birol I, Bousquet J, Ingvarsson PK, Jansson S, Jones SJM, Keeling CI, MacKay J, Nilsson O, Ritland K, Street N, Yanchuk A, Zerbe P, Bohlmann J. Insights into conifer Giga-genomes. Plant Physiol. 2014;166(4):1724–32.
35. Märtens K, Hallin J, Warringer J, Liti G, Parts L. Predicting quantitative traits from genome and phenome with near perfect accuracy. Nat Commun. 2016;7:11512.
36. Xu S. Theoretical basis of the Beavis effect. Genetics. 2003;165:2259–68.
37. Lande R, Thompson R. Efficiency of marker-assisted selection in the improvement of quantitative traits. Genetics. 1990;124:743–56.
38. Whittaker JC, Thompson R, Denham MC. Marker-assisted selection using ridge regression. Genet Res. 2000;75:249–52.
39. Shen X, Alam M, Fikse F, Rönnegård LA. Novel generalized ridge regression method for quantitative genetics. Genetics. 2013;193:255–1268.
40. Gianola D, van Kaam JB. Reproducing kernel Hilbert spaces regression methods for genomic assisted prediction of quantitative traits. Genetics. 2008;178:2289–303.
41. Yanchuk AD. General and specific combining ability from disconnected partial diallels of coastal Douglas-fir. Silvae Genet. 1996;45:37–45.
42. El-Kassaby YA, Park Y-S. Genetic variation and correlation in growth, biomass, and phenology pf Douglas-fir diallel progeny at different spacings. Silvae Genet. 1993;42:289–97.
43. Krakowski J, Park Y-S, El-Kassaby YA. Early testing of Douglas-fir: wood density and ring width. For Genet. 2005;12:99–105.
44. Beaulieu J, Doerksen T, Boyle B, Clément S, Deslauriers M, Beauseigle S, Blais S, Poulin P-L, Lenz P, Caron S, Rigault P, Bicho P, Bousquet J, Mackay J. Association genetics of wood physical traits in the conifer white spruce and relationships with gene expression. Genetics. 2011;188:197–214.
45. Burdon RD. Genetic correlation as a concept for studying genotype-environment interaction in forest tree breeding. Silvae Genet. 1977;26:168–75.
46. Owino F. Genotype x environment interaction and genotypic stability in loblolly pine. Silvae Genet. 1977;26:21–6.
47. Matheson AC, Raymond CA. The impact of genotype x environment interaction on Australian Pinus Radiata breeding programs. Aust. For Res. 1984;14:11–25.
48. Matheson AC, Cotterill PP. Utility of genotype x environment interactions. For Ecol Manag. 1990;30:159–74.
49. Magnussen S, Yanchuk AD. Selection age and risk: finding the compromise. Silvae Genet. 1993;42:25–40.
50. Lerceteau E, Szmidt AE, Andersson B. Detection of quantitative trait loci in Pinus Sylvestris L. across years. Euphytica. 2001;121:117–22.
51. Ratcliffe B, Gamal el-Dien O, Klápště J, Porth I, Chen C, Jaquish B, el-Kassaby YA. A comparison of genomic selection models across time in interior spruce (Picea engelmannii x glauca) using unordered SNP imputation methods. Heredity. 2015;115:547–55.
52. White TL, Adams WT, Neale DB. Forest genetics. Cabi. 2007;
53. Neale DB, Savolainen O. Association genetics of complex traits in conifers. Trends Plant Sci. 2004;9:325–30.
54. Ivanova NV, Fazekas AJ, Hebert PDN. Semi-automated, membrane-based protocol for DNA isolation from plants. Plant Mol Biol Rep. 2008;26:186–98.
55. Howe GT, Yu J, Knaus B, Cronn R, Kolpak S, Dolan P, Lorenz WW, Dean JFDASNP. Resource for Douglas-fir: de novo transcriptome assembly and SNP detection and validation. BMC Genomics. 2013;14:137.
56. Lee WP, Stromberg MP, Ward A, Stewart C, Garrison EP, Marth GTMOSAIK. A hash-based algorithm for accurate NextGeneration sequencing short-read mapping. PLoS One. 2014;9(3):e90581.
57. Garrison E, Marth G. Haplotype-based variant detection from short-read sequencing. arXiv preprint arXiv:1207.3907 [q-bio.GN]. 2012;
58. El-Kassaby YA, Mansfield S, Isik F, Stoehr M. In situ wood quality assessment in Douglas-fir. Tree Genet Genomes. 2011;7:553–61.
59. Gilmour A.R., Gogel B., Cullis B., Thompson R.;ASReml user guide release 3.0. 2009
60. Cappa EP, Stoehr MU, Xie C-Y, Yanchuk AD. Identification and joint modeling of competition effects and environmental heterogeneity in three Douglas-fir (Pseudotsuga Menziesii Var. Menziesii) trials. Tree Genet Genomes. 2016;12:102.
61. Garrick DJ, Taylor JF, Fernando RL. Deregressing estimated breeding values and weighting information for genomic regression analyses. Genet Sel Evol. 2009:41–55.
62. Dutkowski GW, eSilva JC, Gilmour AR, Lopez GA. Spatial analysis methods for forest genetic trials. Can J For Res. 2002;32:2201–14.
63. Endelman JB. Ridge regression and other kernels for genomic selection with R package rrBLUP. Plant Genome. 2011;4:250–5.
64. Henderson C. A simple method for computing the inverse of a numerator relationship matrix used in prediction of breeding values. Biometrics. 1976; 32:69–83.

4

Genome-wide characterization and expression profiling of *PDI* family gene reveals function as abiotic and biotic stress tolerance in Chinese cabbage (*Brassica rapa* ssp. *pekinensis*)

Md. Abdul Kayum[1], Jong-In Park[1], Ujjal Kumar Nath[1], Gopal Saha[1], Manosh Kumar Biswas[1], Hoy-Taek Kim[2] and Ill-Sup Nou[1*]

Abstract

Background: Protein disulfide isomerase (PDI) and PDI-like proteins contain thioredoxin domains that catalyze protein disulfide bond, inhibit aggregation of misfolded proteins, and function in isomerization during protein folding in endoplasmic reticulum and responses during abiotic stresses.Chinese cabbage is widely recognized as an economically important, nutritious vegetable, but its yield is severely hampered by various biotic and abiotic stresses. Because of, it is prime need to identify those genes whose are responsible for biotic and abiotic stress tolerance. PDI family genes are among of them.

Results: We have identified 32 *PDI* genes from the Br135K microarray dataset, NCBI and BRAD database, and in silico characterized their sequences. Expression profiling of those genes was performed using cDNA of plant samples imposed to abiotic stresses; cold, salt, drought and ABA (Abscisic Acid) and biotic stress; *Fusarium oxysporum* f. sp. *conglutinans* infection. The Chinese cabbage *PDI* genes were clustered in eleven groups in phylogeny. Among them, 15 *PDI* genes were ubiquitously expressed in various organs, while 24 *PDI* genes were up-regulated under salt and drought stress. By contrast, cold and ABA stress responsive gene number were ten and nine, respectively. In case of *F. oxysporum* f. sp. *conglutinans* infection 14 *BrPDI* genes were highly up-regulated. Interestingly, *BrPDI1–1* gene was identified as putative candidate against abiotic (salt and drought) and biotic stresses, *BrPDI5–2* gene for ABA stress, and *BrPDI1–4, 6–1* and *9–2* were putative candidate genes for both cold and chilling injury stresses.

Conclusions: Our findings help to elucidate the involvement of *PDI* genes in stress responses, and they lay the foundation for functional genomics in future studies and molecular breeding of *Brassica rapa* crops. The stress-responsive *PDI* genes could be potential resources for molecular breeding of *Brassica* crops resistant to biotic and abiotic stresses.

Keywords: Pdi, Chinese cabbage, Gene expression, Gene evolution, Biotic and abiotic stresses

* Correspondence: nis@sunchon.ac.kr
[1]Department of Horticulture, Sunchon National University, 255 Jungang-ro, Suncheon, Jeonnam 57922, Republic of Korea
Full list of author information is available at the end of the article

Background

Protein disulfide isomerases (PDIs) are enzymes found primarily in the endoplasmic reticulum (ER) in eukaryotes play a vital role in protein folding. Proteins, immediately after biosynthesis must be folded into correct three-dimensional shape for proper functioning. Misfolded proteins are often nonfunctional by producing aggregates and interferes cellular functions. PDIs can bind into misfolded or unfolded proteins preventing to produce such aggregates [1]. Venetianer and Straub, [2] firstly identified PDIs as protein-folding catalysts and demonstrated their catastrophic consequences of a defective protein folding process. Plants utilize multiple mechanisms to fold proteins correctly for proper functioning. PDIs are one of the mechanisms, which work through formation and breakage of connections between cysteine residues by producing disulfide bonds, which was described by Anfinsen [3]. These bonds stabilize the folded protein and provide correct structure to perform its particular function.

PDIs contain four thioredoxin (TRX)-like domains, two of which contain a catalytic site for disulfide bond formation. The reduced (dithiol) forms of PDIs catalyze the reduction of impaired thiol residues of a particular substrate, acting as an isomerase [4]. In human, PDIs usually contain four TRX-like domains (a, b, b' and a'), a linker (x) and a C-terminal extension domain (c) [5]. The 'a' and 'a' domains are TRX domains, containing an active Cys-Gly-His-Cys motif joined to α-helices and β-strands (β-α-β-α-β-α-β-β-α) that is essential for polypeptide redox and isomerization [6]. Although the '*b*' and '*b*' domains share some structural identity with TRX domain, they do not possess specific active motif site [6]. The α-helices are particularly important for DNA binding motifs, including helix-turn-helix motifs, leucine zipper motifs and zinc finger motifs. The helix-turn-helix (HTH) motif is a major structural motif capable of DNA binding. This motif is composed of two α-helices joined by a short strand of amino acids found in many regulatory proteins for gene expression [7]. The C-terminal region contains a C-domain rich in acidic residues typical as calcium binding proteins and ends with an ER retention signal composed of four amino acids, generally KDEL/GKNF/VASS [8].

PDI genes have been identified in many higher plants, such as 21 *PDI* genes in *Arabidopsis thaliana*, 12 in *Oryza sativa*, 12 in *Zea mays*, 10 in *Vitis vinifera*, 9 in *Triticum aestivum* and so on [9]. In eukaryotes, PDI family proteins are divided into eleven groups based on phylogenetic analysis. Proteins in groups' I–V contain two thioredoxin-like active domains, whereas those in groups VI–XI contain a single thioredoxin-like active domain. PDI proteins in higher plants are involved in signal transduction pathways and in transcriptional complexes that regulate the responses of genes to environmental stimuli. PDI proteins ERP57, PDIp, P5, ERP72, PDIR, and PDI-D act as redox catalysts and isomerases and exhibit differential functions, such as peptide binding, cell adhesion, and chaperone activities [10]. These proteins play key role as storage protein and plasma membrane maturation [9] *GmPDIL-3a* and *GmPDIL-3b* are highly expressed during seed maturation suggesting that they are involved in folding or accumulation of storage proteins in soybean [11]. *PDIL2–1* of *A. thaliana* is directly involved in ovule structure and embryo sac development and determining proper direction of pollen tube growth [12]. The PDI-like protein RB60 plays important roles providing redox potential to regulate photosynthesis [13]. Most wheat *PDI* and *PDIL* genes are expressed during endosperm development, indicating their association with storage protein biosynthesis and deposition, which is directly related to gluten quality [14]. *BdPDIL1–1* is significantly up-regulated under abiotic (drought, salt, ABA and H_2O_2) stress, suggesting their involvement in multiple stress responses [15]. All maize *PDI* genes are highly responsive to cold, salt, dehydration, and ABA stress [16]. During drought, heat and cold stress, *TaPDI1*, *TaPDI2* and *TaPDI3* are highly up-regulated in roots and other tissues [17]. There are many family genes involved in tolerance to abiotic and biotic stresses in plants; however we have selected PDI because of its abundance of ubiquitous sulfydryl oxidoreductase protein, which is an important cellular protein with multiple biological functions of all eukaryotic cells. This sulfydryl oxidoreductase protein of PDI displays versatile redox behavior, which also interact with other proteins and assumed its potential role against various diseases and abiotic stresses [18]. In addition, PDI is an important redox proteins regulating reactive oxygen species (ROS) production in the cells and alter redox status of cells to activate defense mechanism [19]. In this study, we make effort to identify the members of *PDI* gene family in *Brassica rapa* through genome wide exploration by using different bioinformatic tools. In addition, the putative candidates of *PDI* genes responsive to abiotic and biotic stresses are predicted through expression profiling using stress induced Chinese cabbage materials.

Results

Identification of *PDI* genes in *B. rapa*

A total of 32 *BrPDI* genes were identified using SWISSPROT of the *B. rapa* genomic database (http://brassicadb.org/brad/searchAll.php) (BRAD; [20]); using key word "PDI", NCBI, and annotations of microarray Br135K dataset from cold-treated *B. rapa* (Chiifu & Kenshin), removing any duplicates. A BLAST search was performed using *Arabidopsis* PDI sequences as the query, and picked the encoded protein and genomic sequences of *PDI* genes from BRAD database for 32 *BrPDI* genes. A high degree of similarity of these 32 BrPDI protein sequences was also picked for other plant species. Isoelectric points, molecular weights and residual size (ranged from 144 to 596 aa) of putative 32 BrPDI proteins are presented in Table 1.

Phylogenetic analysis and domain location

A phylogenetic tree was constructed using 76 full-length protein sequences of *PDI* and *PDI-like* genes, which included 32 from Chinese cabbage, 11 from *Brachypodium distachyon,* 12 from maize, and 21 from *Arabidopsis* to investigate evolutionary relationship. Eleven phylogenetic groups were denoted among the considered *PDI* genes from different plant species (Fig. 1) with distinct 4 clades. Clade 1 consisted phylogenetic groups I, II, III and VII, whose members contain two active thioredoxin domains, whereas members of group VII encoded proteins with a single N-terminal active domain (Fig. 2). Clade 2 contained phylogenetic group IV, V and VI, among them members of group IV and V possessing two active thioredoxin domains in tandem at their N-termini. Group members of VI may have lost one of the two active thioredoxin domains due to shared structural

Table 1 In silico analysis of PDI genes identified in Chinese cabbage with their *Arabidopsis* orthologs

Gene name	Gene ID	Chromosome			Stand	Sub-genome	Iso electric point (Pi)	Molecular weight (Mw)	Protein length	Orthologous gene
		No.	Start	End						
BrPDI1–1	Bra016405	A08	17,525,813	17,528,427	-	MF1	4.66	55.78	501	AT1G21750
BrPDI1–2	Bra012293	A07	11,001,072	11,003,432	-	MF2	4.92	55.04	496	AT1G21750
BrPDI1–3	Bra017948	A06	8,725,538	8,728,086	+	LF	5.21	55.52	498	AT1G21750
BrPDI1–4	Bra008311	A02	13,630,848	13,633,652	+	MF1	4.79	56.24	511	AT1G77510
BrPDI1–5	Bra015665	A07	24,746,925	24,749,554	+	LF	4.85	55.8	509	AT1G77510
BrPDI2–1	Bra007120	A09	28,945,749	28,948,773	-	LF	4.70	64.28	579	AT3G54960
BrPDI2–2	Bra002464	A10	9,445,124	9,447,998	-	LF	4.59	66.16	596	AT5G60640
BrPDI2–3	Bra020239	A02	4,767,927	4,771,274	+	MF2	4.58	65.06	588	AT5G60640
BrPDI3–1	Bra014319	A08	1,670,650	1,673,502	-	MF1	4.74	46.36	413	AT1G52260
BrPDI3–2	Bra018958	A06	1,024,377	1,027,518	-	LF	4.98	59.04	529	AT1G52260
BrPDI4–1	Bra000454	A03	11,214,808	11,217,122	+	MF2	5.80	39.48	362	AT2G47470
BrPDI4–2	Bra004455	A05	202,365	204,674	-	LF	6.05	39.45	361	AT2G47470
BrPDI5–1	Bra015375	A10	1,739,233	1,741,963	+	LF	6.43	46.64	432	AT1G04980
BrPDI5–2	Bra005546	A05	6,147,811	6,150,674	+	LF	5.71	47.85	443	AT2G32920
BrPDI6–1	Bra018672	A06	2,704,568	2,705,424	-	LF	4.93	16.43	144	AT1G07960
BrPDI7–1	Bra034408	A05	13,683,334	13,685,270	-	MF2	5.28	49.08	435	AT1G35620
BrPDI8–1	Bra001793	A03	18,521,034	18,525,290	+	MF2	6.85	54.29	484	AT3G20560
BrPDI8–2	Bra035770	A05	17,795,565	17,799,462	-	LF	6.96	54.31	483	AT3G20560
BrPDI8–3	Bra019071	A03	26,497,077	26,500,750	+	MF1	6.82	53.75	478	AT4G27080
BrPDI8–4	Bra010413	A08	13,621,192	13,624,765	-	MF2	6.72	53.89	480	AT4G27080
BrPDI8–5	Bra030465	A05	11,654,854	11,657,768	-	MF2	7.73	54.20	483	AT1G50950
BrPDI8–6	Bra018881	A06	1,468,257	1,472,162	-	LF	6.26	53.57	477	AT1G50950
BrPDI9–1	Bra026786	A09	35,442,828	35,445,717	+	MF2	6.85	60.45	532	AT1G15020
BrPDI9–2	Bra014330	A08	1,583,063	1,585,667	+	MF1	6.65	58.34	523	AT2G01270
BrPDI10–1	Bra001092	A03	14,721,770	14,723,230	+	MF2	6.75	33.86	303	AT3G03860
BrPDI10–2	Bra031969	A05	24,598,071	24,599,646	-	MF1	7.09	34.22	302	AT3G03860
BrPDI10–3	Bra036758	A08	7,293,480	7,294,725	+	MF1	8.57	33.54	298	AT3G03860
BrPDI10–4	Bra036429	A01	26,156,526	26,158,281	_	LF	8.91	36.75	326	AT1G34780
BrPDI11–1	Bra019406	A03	24,393,210	24,394,793	_	MF1	6.02	51.69	468	AT4G21990
BrPDI11–2	Bra013579	A01	6,445,389	6,446,955	_	LF	6.39	51.58	468	AT4G21990
BrPDI11–3	Bra034466	Scaffold000096	53,632	55,924	_	LF	6.74	53.67	479	AT1G62180
BrPDI11–4	Bra029505	A09	17,850,361	17,852,000	+	LF	6.74	53.57	479	AT4G04610

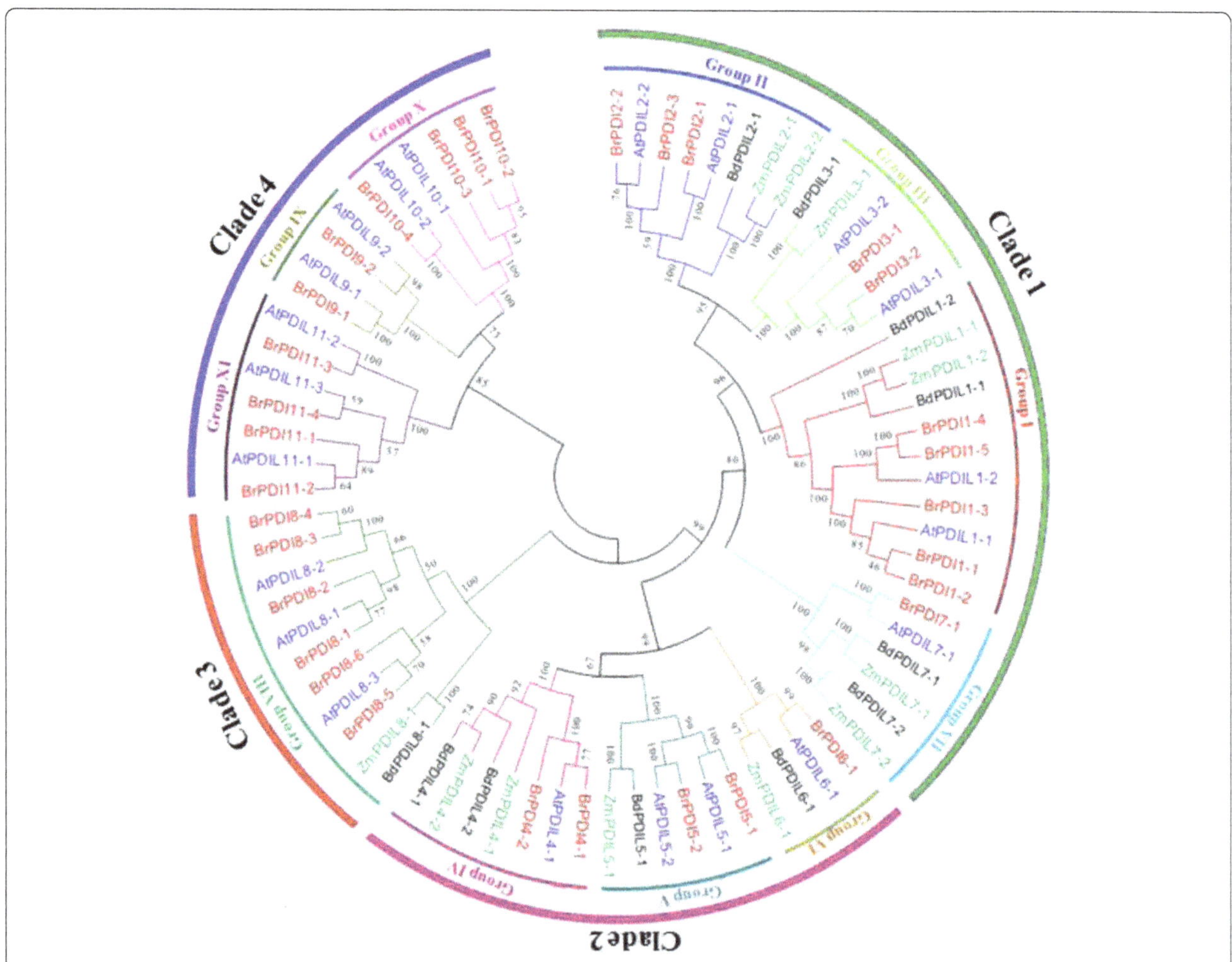

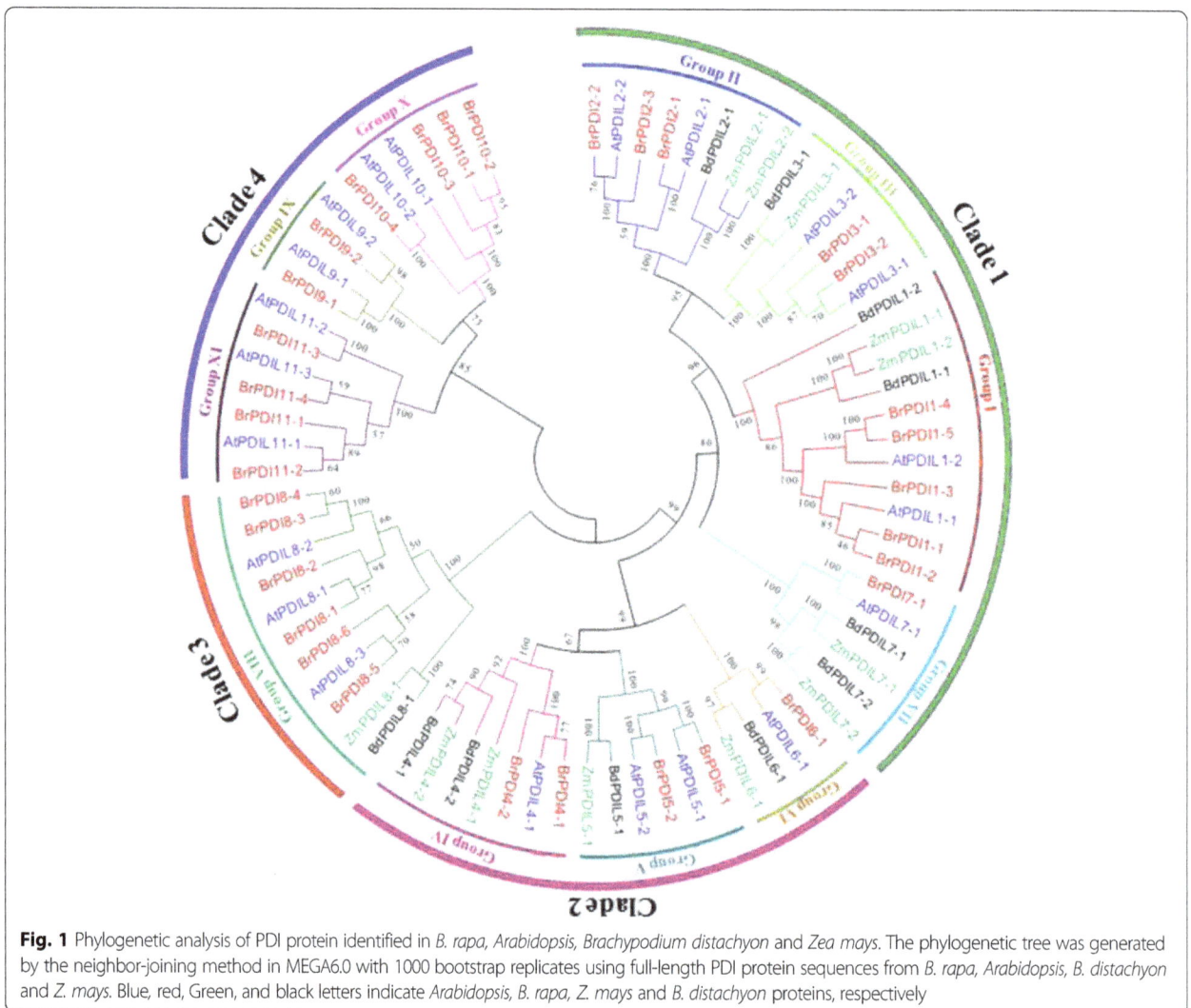

Fig. 1 Phylogenetic analysis of PDI protein identified in *B. rapa, Arabidopsis, Brachypodium distachyon* and *Zea mays*. The phylogenetic tree was generated by the neighbor-joining method in MEGA6.0 with 1000 bootstrap replicates using full-length PDI protein sequences from *B. rapa, Arabidopsis, B. distachyon* and *Z. mays*. Blue, red, Green, and black letters indicate *Arabidopsis, B. rapa, Z. mays* and *B. distachyon* proteins, respectively

features of a common progenitor. Clade 3 contained phylogenetic groups VIII, which included genes encoding proteins with a single active thioredoxin domain. The members of groups IX, X and XI formed clade 4; encoded proteins with an active thioredoxin domain. Phylogenetic and domain analyses revealed that these *PDI* family genes are divergent in plant. Most BrPDI proteins are contained an N-terminal signal peptide and a C-terminal KDEL signal responsible for translocation and ER retention, respectively. In the domain structures, 'a' and 'a' are homologous to thioredoxin and contain a -CXXC- active site for isomerase and redox activities. By contrast, the 'b' and 'b' domains have no homology to thioredoxin and lack of -CXXC- catalytic site. However, the secondary structures of four (a, a', b, b') domains are similar to thioredoxin rather than the active catalytic site. The thioredoxin-like domain comprises $\beta\alpha\beta\alpha\beta\alpha\beta\beta\alpha$ motifs forming four layers of β-sheets that are sandwiched three layers of α-helixes (Fig. 3).

We performed multiple sequence alignment of 'a'- and 'a'-type domain sequences of BrPDI proteins to analyze the features of them. These domain sequences contained five β-sheets, four α-helices and the -CGHC- motif of the active site. They also contained conserved arginine, glutamic acid, proline, and lysine residues (Fig. 3). All BrPDI members 'a'-type domain with conserved arginine residue, critical for their catalytic function, except BrPDI3–1 and BrPDI3–2 which contained phenylalanine or leucine residues instead of arginine residue. In addition, BrPDI9–1 contained tryptophan and BrPDI10–1, BrPDI10–2 and BrPDI10–3 contained lysine in the place of arginine residues. BrPDI2–3 accomplished with RGHC non-characteristic active sites; but BrPDI3–1 and BrPDI3–2 with CARS. Whereas, BrPDI8–1, BrPDI8–2, BrPDI8–5, and BrPDI8–6 comprised with CYWS; and BrPDI10–1, BrPDI10–2, BrPDI10–3, and BrPDI10–4 contained CPFS non-characteristic active sites (Fig. 2) which may affect their redox potential and lose their function.

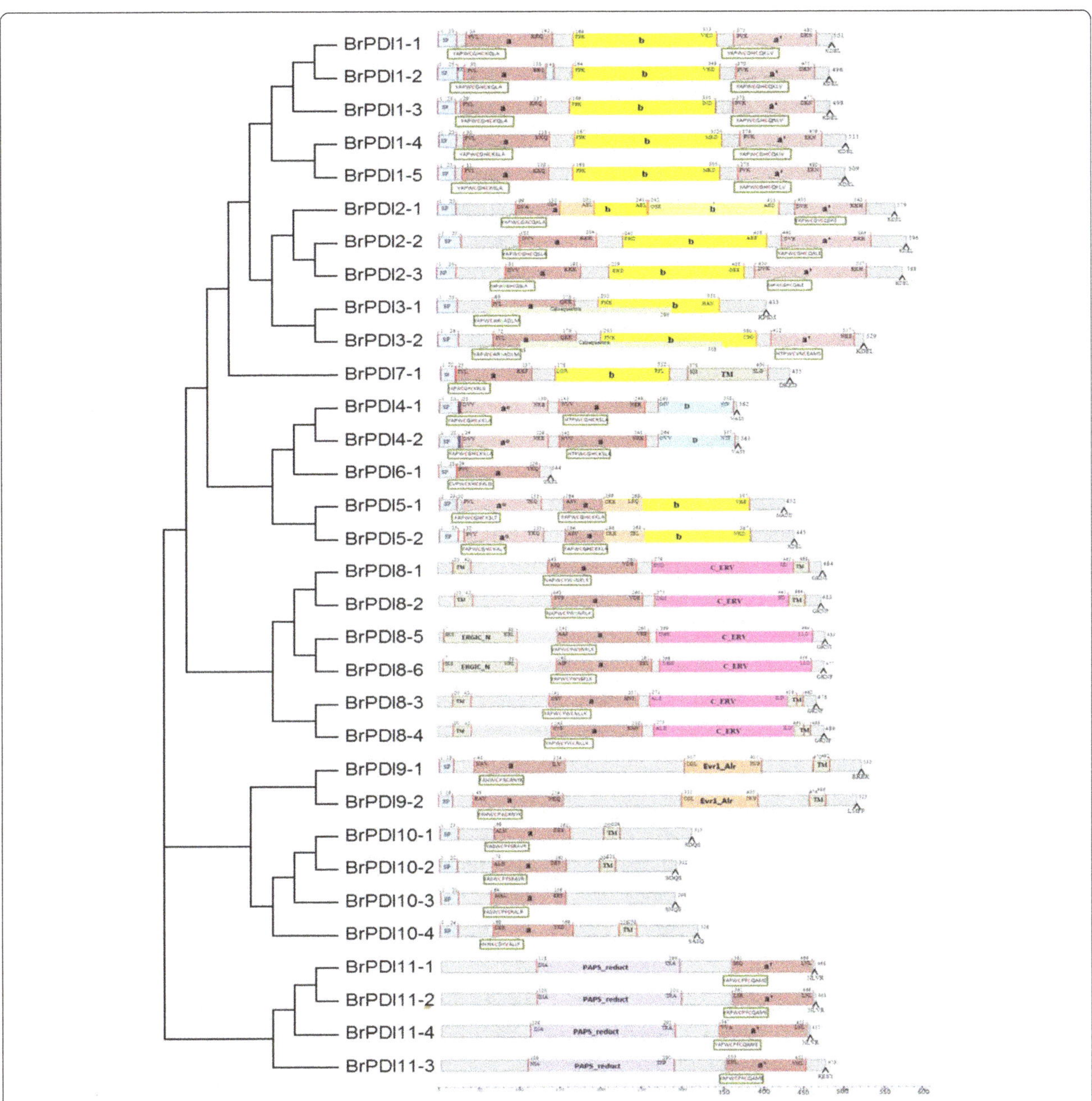

Fig. 2 Domain structure of the deduced amino acid sequences of *B. rapa PDI* genes. The putative signal peptides (SP), the a and b type domains, the N-terminal calcium binding domain calsequestrin, the D domains (Erp29), the transmembrane domains (TM) and the C_ERV (COPII-coatedERV) domain are shown. The thioredoxin-like catalytic domains with two active sites (shown in detail in the boxes) are also shown. Numbers above indicate domain boundaries (aa), and numbers on the right indicate ORF (aa). Domains a° (light gray) and a' (gray) are homologous to TRX and contain the catalytic CxxC motif (red). Domains b (yellow) and b' (light yellow) also exhibit a TRX fold, but they do not share high sequence similarity with each other or with domains a or a'. The C-terminal extension (red) contains a (K/H)DEL retention signal for the ER

Chromosomal distribution of *BrPDI* genes

Thirty two *BrPDI* genes are located on different chromosomes not less than one on each except chromosome 04. The highest number (six) of *BrPDI* genes was identified on chromosome A05 (19.35%), while chromosome A01, A02, A07, and A10 contained two *BrPDI* genes (6.45%) for each, and chromosome A04 has no *BrPDI* genes (0%). Only one gene (*BrPDI11–3*) is located in scaffold (Fig. 4a and b). None of the *BrPDI* gene clusters was detected in Chinese cabbage. The sequence alignment of BrPDI proteins showed higher similarity within groups. BrPDI proteins showed ≥68% similarity within groups, except group10 and group 2. By contrast, similarity between groups was ≤65%. Seventeen pairs of BrPDI proteins showed 80% similarity indicating duplications are predominant for those genes (Table 2).

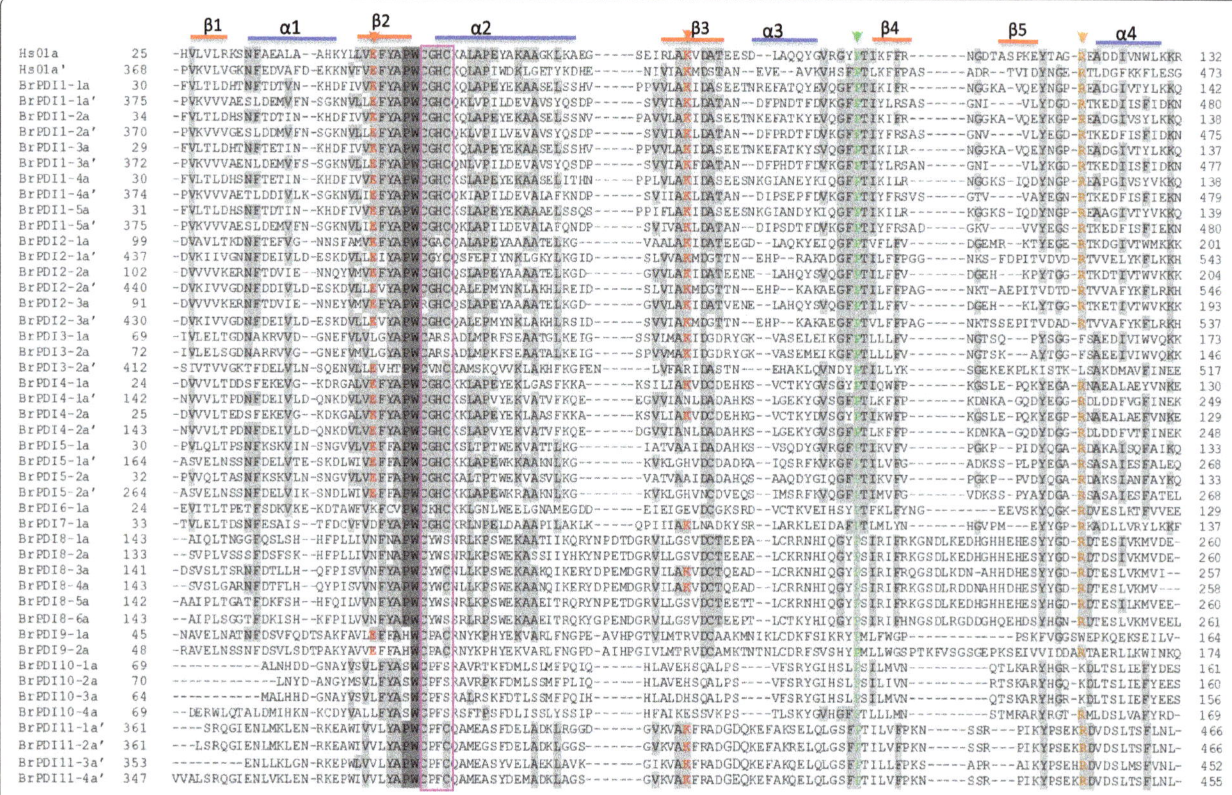

Fig. 3 Multiple sequence alignment of the a-type domains of *B. rapa* PDI proteins and a typical human PDI. Residues highlighted in gray and light gray share 100% and >50% identity, respectively. Elements of the secondary structure are specified by blue colour bars (α-helices) and red colour bars (β-sheets). Red arrows indicate the two buried charged residues in the vicinity of the active site, orange arrow indicates the conserved arginine (R) and green arrow indicates the *cis* prolines (P) near each active site. Active-site residues within a domain are pink colour boxed

Three fractionated subgenomes, like least fractionated (LF), medium fractionated (MF1), and most fractionated (MF2) subgenomes are found in *B. rapa* genome. Notably, 32 *BrPDI* genes are distributed onto ten chromosomes with fractionation into three subgenomes, of which are 15 LF (46.87%), 8 MF1 (25%), and MF2 9 (28.13%; Fig. 4c and Table 1). In addition, 32 *BrPDI* genes are distributed in different block during evolution. Among them, 7 genes (22.58%) belong to AK8 block, followed by 19.45% of *BrPDI* genes into AK1, AK3 blocks; while only 6.45% of *BrPDI* genes were allocated to AK2, AK4, AK5 and AK6 (Fig. 4d). We have also found a total of 17 segmental duplicated *PDI* genes pairs in *B. rapa* genome (Fig. 5 and Table 3). Furthermore, the substitution rate of non-synonymous (Ka) and synonymous (Ks) ratios (Table 3) were calculated to assess the selection pressures among duplicated *BrPDI* gene pairs. In these analysis, Ka/Ks ratios were identified as lower than 1, 1 and above 1 indicating negative or purifying selection, neutral selection and positive selection, respectively. Nine out of seventeen *BrPDI* duplicated gene pairs, had Ka/Ks value below 1 (purifying selection) and rest of them had Ka/Ks value above 1(positive selection). Moreover, the estimated divergence time of *BrPDI* genes showed that duplication event started at 29.74 million years ago (MYA) and continued up to 1.62 MYA (Table 3).

Motif composition and intron–exon analysis

The intron–exon structures were almost consistent among the groups of *BrPDI* genes. Group I contained nine to ten exons; whereas, members of group V contained nine exons (Additional file 1: Figure S1). All members of group II possessed 10 to 11 exons, while members of group IV and IX contained 11 exons, and the member of group VIII composed with the highest number (15) of exons. Member of groups X and XI possessed 3–4 exons. Finally, member of groups VI and VII contained four and five exons, respectively. Conserved motifs among BrPDI proteins were investigated using the MEME motif search tool. Motifs 1 and 2 contained-CXXC- catalytic sites, which are necessary for isomerase and redox activity. Motifs 3 was present in all groups, but motifs 2 was absent in groups VI, VII, VIII, IX and X. Motif 1 and 5 were absent in group IV and XI, respectively. Relatively less conserved motif 6 was found only in groups I, II, III

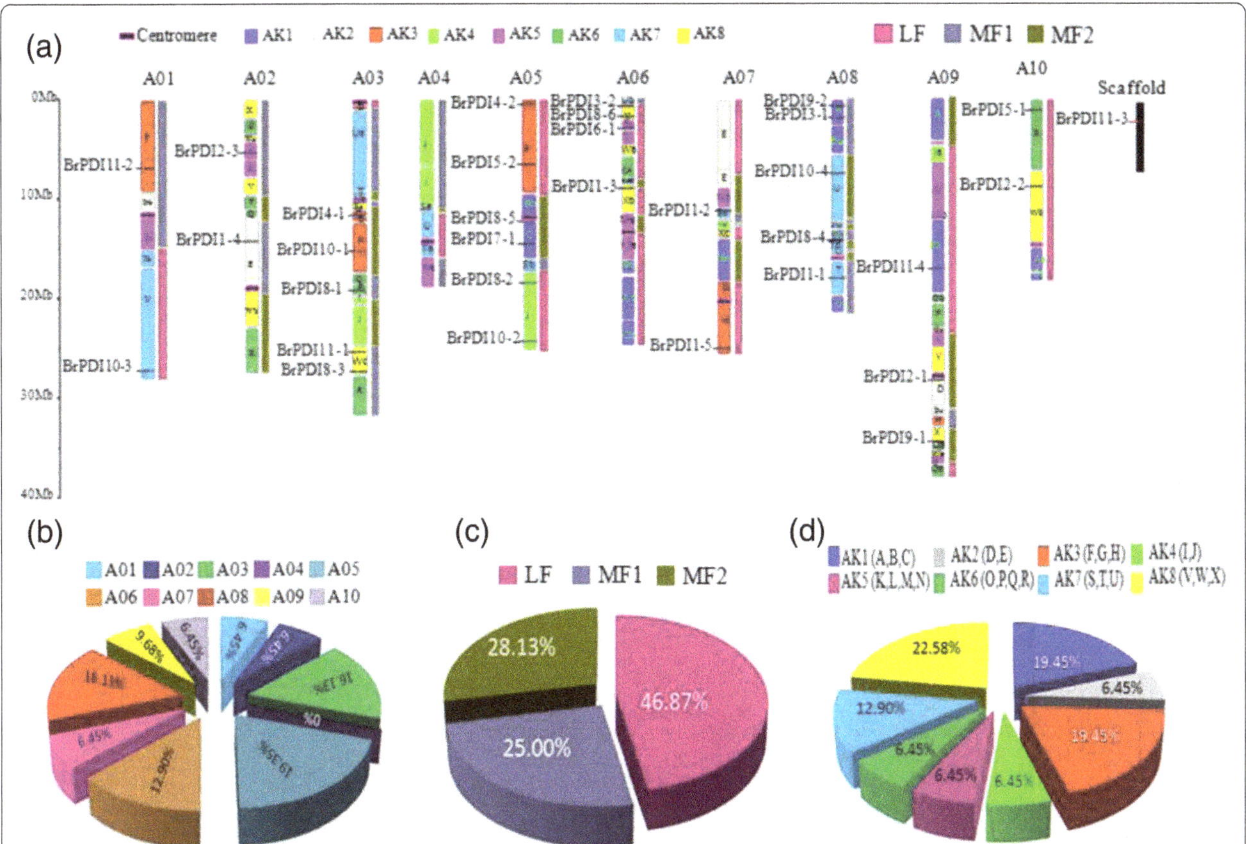

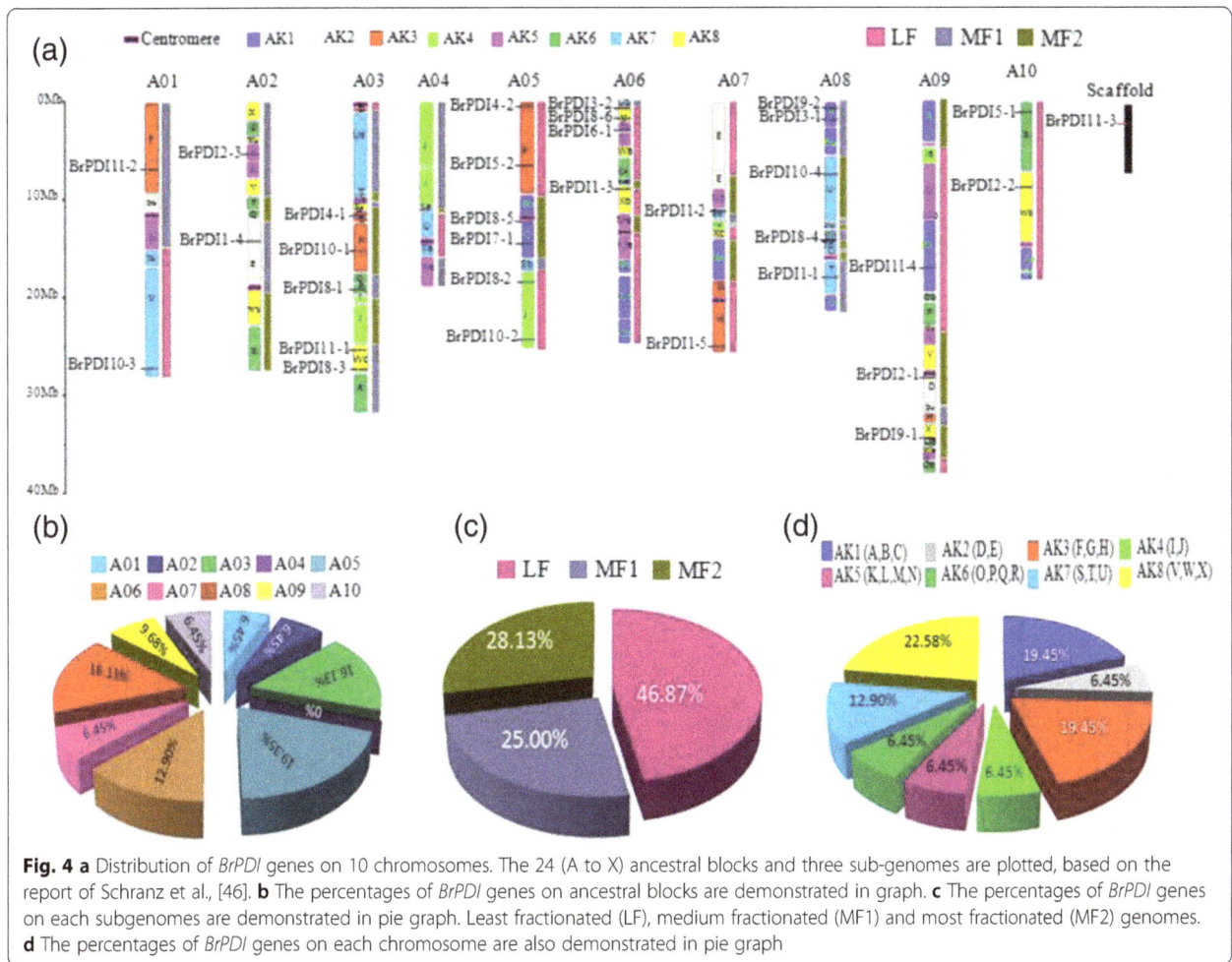

Fig. 4 a Distribution of *BrPDI* genes on 10 chromosomes. The 24 (A to X) ancestral blocks and three sub-genomes are plotted, based on the report of Schranz et al., [46]. **b** The percentages of *BrPDI* genes on ancestral blocks are demonstrated in graph. **c** The percentages of *BrPDI* genes on each subgenomes are demonstrated in pie graph. Least fractionated (LF), medium fractionated (MF1) and most fractionated (MF2) genomes. **d** The percentages of *BrPDI* genes on each chromosome are also demonstrated in pie graph

and VII. Motifs 4, 7, 8 and 9 were unique to group VIII and motif 10 was present in group VIII and XI only (Additional file 1: Figure S2).

Microsyntenic analysis

A microsynteny map was constructed to find out orthologous gene pairs of *PDI* genes among *B. rapa*, *B.distachyon* and *A. thaliana* for exploration of evolutionary history and relationships among the genomes (Fig. 5). We found 22 orthologous gene pairs between *B. rapa* and *A. thaliana*; whereas, 13 orthologous gene pairs were identified between *B. rapa* and *B. distachyon* (Fig. 5). We found 17 duplicated PDI gene pairs in *B. rapa* genome. All of the segmental duplications are denoted with an orange line in fig. 5, no tandem duplications were found in *PDI* genes in *B. rapa*.

Microarray data analysis

We used our previously published microarray data set to analyze the expression patterns of 32 *BrPDI* genes using two contrasting inbred lines 'Chiifu' and 'Kenshin', treated with cold and freezing stress (4 °C, 0 °C, −2 °C and −4 °C) for 2 h [20]. All of the *BrPDI* genes were differentially expressed in response to cold or freezing stress in two lines (Fig. 6). In Chiifu, most of the *BrPDI* genes were highly expressed upon exposure to cold and freezing temperatures. While, *BrPDI2–1, 4–1, 8–2, 8–6, 9–1* and *10–1* genes showed higher expression during cold stresses in 'Kenshin' compared to 'Chiifu'.

Promoter analysis of *BrPDI* genes

PDI genes of Chinese cabbage might contain biotic and abiotic stress-responsive *cis*-acting elements in their promoter regions. Most of the *BrPDI* genes bear cold-responsive LTR *cis*-acting element, drought inducible MBS *cis*-element, and salt- or environmental stresses *cis*-acting element (WBOX and TC-rich repeats; Additional file 2: Table S1). The genes *BrPDI2–1, BrPDI2–2, BrPDI2–3, BrPDI3–2, BrPDI6–1, BrPDI7–1, BrPDI8–1, BrPDI8–2, BrPDI8–4, BrPDI10–1, BrPDI10–2, BrPDI10–3, BrPDI10–4, BrPDI11–1, BrPDI11–2, BrPDI11–3,* and *BrPDI11–4* have no ABA-responsive CE3 *cis*-acting element in their promoter regions. Majority of *BrPDI* genes contained more than one disease

Table 2 For each pair wise alignment, the similarity (relative to the maximum similarity) and the number of identical amino acids (in %) is given

Protein	BrPDI 1–2	BrPDI 1–3	BrPDI 1–4	BrPDI 1–5	BrPDI 2–1	BrPDI 2–2	BrPDI 2–3	BrPDI 3–1	BrPDI 3–2	BrPDI 4–1	BrPDI 4–2	BrPDI 5–1	BrPDI 5–2	BrPDI 6–1	BrPDI 7–1	BrPDI 8–1	BrPDI 8–2	BrPDI 8–3	BrPDI 8–4	BrPDI 8–5	BrPDI 8–6	BrPDI9–1	BrPDI 9–2	BrPDI 10–1	BrPDI 10–2	BrPDI 10–3	BrPDI 10–4	BrPDI 11–1	BrPDI 11–2	BrPDI 11–3	BrPDI 11–4
BrPDI1–1	**90**	**84**	73	76	30	33	32	19	21	24	26	19	18	25	20	9	9	12	10	32	32	26	18	20	23	24	28	27	25	28	30
BrPDI1–2		**83**	74	77	32	33	32	18	18	25	26	20	19	25	20	8	12	12	10	32	32	25	19	20	24	23	29	29	29	32	34
BrPDI1–3			72	74	28	31	32	19	19	24	26	19	19	24	21	9	9	10	10	27	32	26	22	21	24	24	29	27	25	27	29
BrPDI1–4				**82**	28	32	33	19	17	28	26	21	20	21	16	12	12	12	11	33	31	28	20	20	20	24	27	23	23	24	25
BrPDI1–5					29	29	31	17	21	24	25	21	19	25	19	11	12	12	13	34	32	29	20	20	24	24	27	31	25	24	27
BrPDI2–1						55	54	28	29	25	25	18	21	27	19	11	15	11	14	27	28	29	23	22	23	22	24	36	35	38	36
BrPDI2–2							**86**	30	30	27	28	20	18	22	17	11	13	12	13	33	31	25	23	23	20	21	26	34	33	37	41
BrPDI2–3								29	28	27	28	19	18	27	20	12	11	14	14	31	30	24	22	23	17	20	25	37	32	37	34
BrPDI3–1									**80**	18	14	15	15	19	16	13	11	11	11	26	25	25	21	25	26	29	25	35	32	34	30
BrPDI3–2										18	17	16	16	23	18	9	8	11	9	26	26	24	12	29	29	31	29	31	33	40	30
BrPDI4–1											**93**	28	23	31	18	13	13	14	15	31	35	27	15	27	26	27	31	29	29	28	29
BrPDI4–2												29	25	34	15	13	13	16	16	33	37	32	14	27	25	27	31	30	28	28	28
BrPDI5–1													**81**	27	15	11	13	15	13	29	32	29	16	25	26	27	27	34	33	32	29
BrPDI5–2														23	13	13	14	13	10	28	27	27	14	27	28	29	25	30	30	34	30
BrPDI6–1															26	27	27	28	29	31	34	27	21	27	32	22	23	25	25	29	32
BrPDI7–1																12	11	11	11	21	23	23	16	26	19	20	25	29	27	29	26
BrPDI8–1																	**86**	70	70	74	69	31	18	28	25	23	23	25	28	23	30
BrPDI8–2																		68	69	74	70	31	16	27	23	27	41	28	24	24	26
BrPDI8–3																			**89**	68	65	32	11	65	22	22	41	24	23	23	27
BrPDI8–4																				67	63	32	18	65	23	25	41	26	26	26	25
BrPDI8–5																					**82**	31	14	24	25	26	47	23	27	23	26
BrPDI8–6																						35	16	25	28	25	23	23	27	28	27
BrPDI9–1																							69	21	33	27	30	22	24	22	35
BrPDI9–2																								32	23	24	27	21	22	18	31
BrPDI10–1																									**83**	**83**	34	29	40	44	27
BrPDI10–2																										**80**	35	27	27	41	28
BrPDI10–3																											36	29	29	23	25
BrPDI10–4																												26	30	30	27
BrPDI11–1																													**92**	75	**84**
BrPDI11–2																														75	**83**
BrPDI11–3																															75

Duplications are given bold

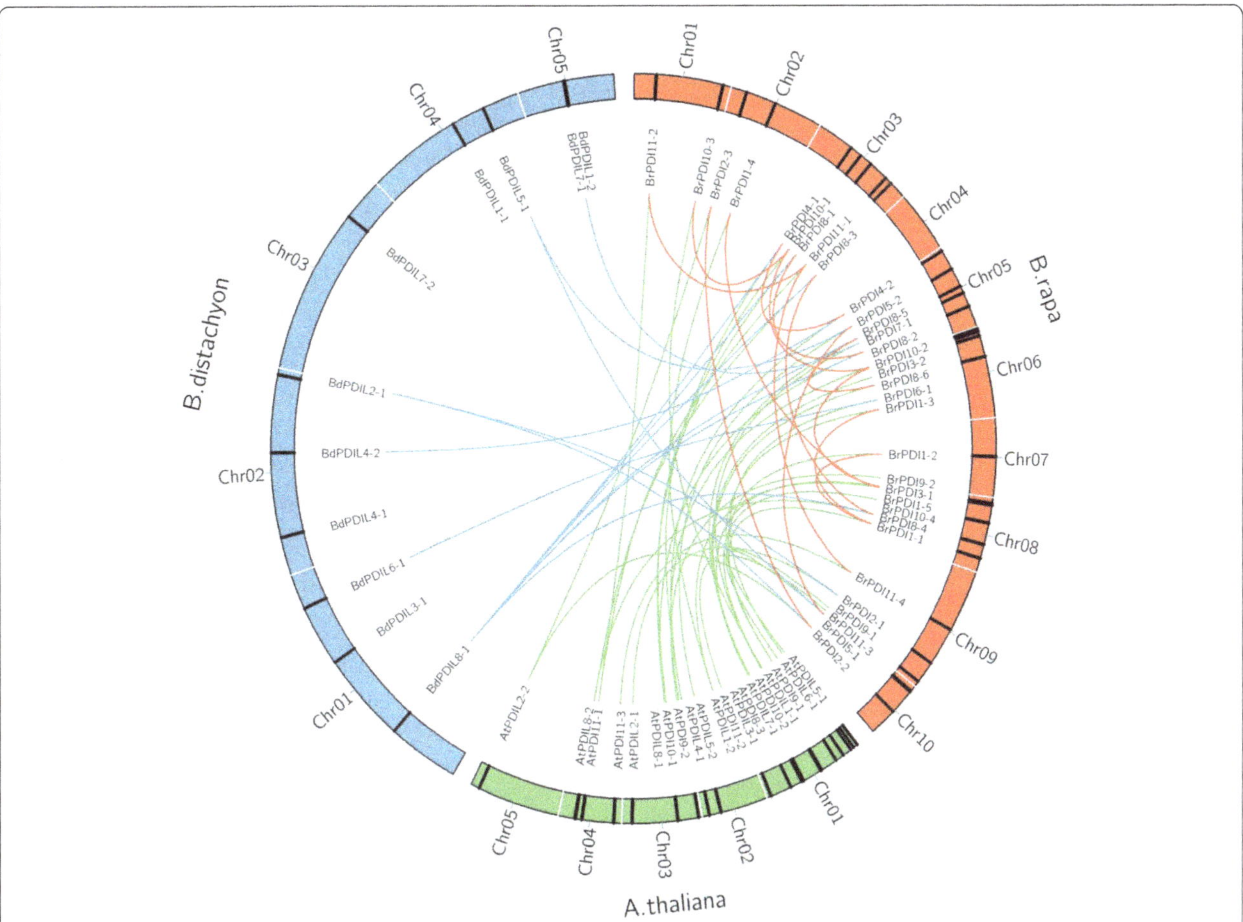

Fig. 5 Microsynteny analysis of *PDI* genes among *B. rapa*, *Brachypodium distachyon* and *A. thaliana*. The chromosomes from the three species are indicated in different colors, orange, green and blue colors represent *B. rapa*, *A. thaliana* and *B. distachyon* chromosome respectively. Orange lines depiction of duplicated *BrPDI* genes on 10 *B. rapa* chromosomes

resistance or defense responsive *cis*-acting element (Box-W1, WBOX, and TC-rich repeats) except *BrPDI8–5, BrPDI10–3, BrPDI11–3* and *BrPDI11–4.*

Organ-specific gene expression

We have used gene-specific primers of 32 *BrPDIs* for RT-PCR to check the expression patterns of *BrPDI*s in different organs (root, stem, leaf, flower buds, sepals, petals, stamens, and pistils) in Chinese cabbage line SUN-3061. Semi-quantitative RT-PCR analysis revealed that among the 32 genes, 15 were ubiquitously expressed in all organs (Fig. 7). Three genes (*BrPDI 1–2, 9–1* and *11–4*) were slightly expressed in all of the tested organs. *BrPDI 7–1* was highly expressed in floral parts but slightly expressed in roots, stems, leaves and flower buds. On the other hands, *BrPDI 1–2, 1–3, 1–4, 2–3* and *8–1* were totally absent in roots, stems and leaves but only expressed in flower buds and floral parts. In addition, *BrPDI 2–1* was weakly expressed in roots, leaves and pistil; while *BrPDI3–1* was expressed slightly in roots, stems, and leaves. *BrPDI8–5* was expressed in leaves, flower buds, and pistils; the *BrPDI 8–6* was expressed in stamens only. *BrPDI10–4* was appeared with very low signal in leaves, sepals and pistils. *BrPDI 11–1* and *BrPDI1 11–2* were highly expressed in vegetative organs but weekly expressed in reproductive organs (Fig. 7).

Analysis of abiotic stress-responsive gene expression

Gene-specific primers of *BrPDI* genes were used to obtain expression profiles in response to various abiotic stresses. Expression patterns of *BrPDI* genes in response to cold was obtained by analyzing qPCR data from cold treated two contrasting Chinese cabbage lines, 'Chiifu' and 'Kenshin'. In addition, we have calculated the expression patterns of different *BrPDI* genes during salt, drought and ABA stresses using qPCR data of Chinese cabbage line 'kenshin'. Most of the *BrPDI* genes were significantly down regulated over the time course of cold stress, while six genes of 'Chiifu' (*BrPDI 1–3, BrPDI 3–1, BrPDI 6–1, BrPDI 9–2, BrPDI 10–2* and *BrPDI 11–1*) were significantly up-regulated on the advancement of

Table 3 Estimated Ka/Ks ratios of the duplicated *BrPDI* genes with their divergence time in Chinese cabbage

Duplicated gene pairs			Ks	Ka	Ka/Ks	Duplication type	Types of selection	Time (mya)
BrPDI1–1	vs.	BrPDI1–2	0.0569	0.1138	2.0000	Segmental	Positive	1.89
BrPDI1–1	vs.	BrPDI1–3	0.1064	0.0726	0.6823	Segmental	Purifying	3.54
BrPDI1–2	vs.	BrPDI1–3	0.1113	0.0715	0.6424	Segmental	Purifying	3.71
BrPDI1–4	vs.	BrPDI1–5	0.2421	0.0349	1.4415	Segmental	Positive	8.07
BrPDI2–2	vs.	BrPDI2–3	0.0486	0.1554	3.1975	Segmental	Positive	1.62
BrPDI3–1	vs.	BrPDI3–2	0.1203	0.0698	0.5802	Segmental	Purifying	4.01
BrPDI4–1	vs.	BrPDI4–2	0.1093	0.1515	1.3860	Segmental	Positive	3.64
BrPDI5–1	vs.	BrPDI5–2	0.1010	0.4750	4.7029	Segmental	Positive	3.37
BrPDI8–1	vs.	BrPDI8–2	0.1773	0.0527	0.2972	Segmental	Purifying	5.91
BrPDI8–3	vs.	BrPDI8–4	0.1179	0.0343	0.2909	Segmental	Purifying	3.93
BrPDI8–5	vs.	BrPDI8–6	0.8922	0.3621	0.4058	Segmental	Purifying	29.74
BrPDI10–1	vs.	BrPDI10–2	0.2892	0.1337	0.4623	Segmental	Purifying	9.64
BrPDI10–1	vs.	BrPDI10–3	0.4183	0.2768	0.6613	Segmental	Purifying	13.94
BrPDI10–2	vs.	BrPDI10–3	0.7166	0.2026	0.2827	Segmental	Purifying	23.89
BrPDI11–1	vs.	BrPDI11–2	0.0895	0.1179	1.3173	Segmental	Positive	2.98
BrPDI11–1	vs.	BrPDI11–4	0.0969	0.2040	2.1052	Segmental	Positive	3.23
BrPDI11–2	vs.	BrPDI11–4	0.0985	0.3570	3.6243	Segmental	Positive	3.28

Ks the number of synonymous substitutions per synonymous site, *Ka* the number of nonsynonymous substitutions per nonsynonymous site, *MYA* million years ago

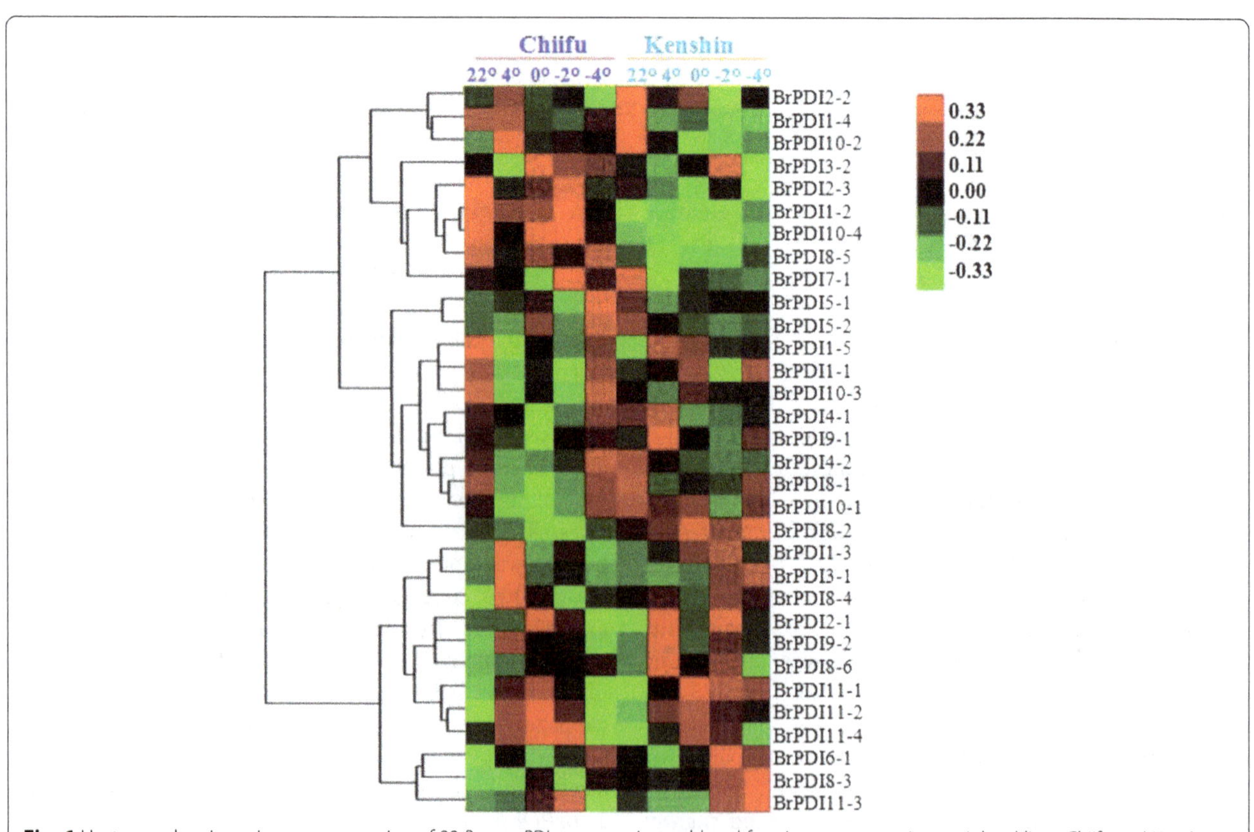

Fig. 6 Heat map showing microarray expression of 32 *B. rapa PDI* genes against cold and freezing treatments in two inbred lines Chiifu and Kenshin

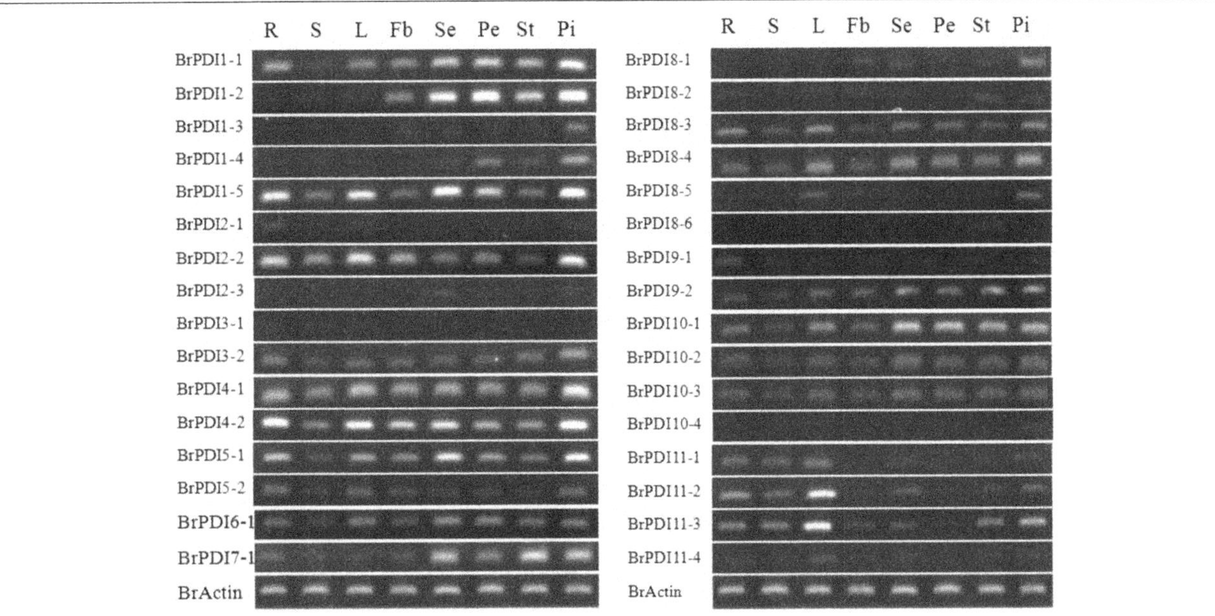

Fig. 7 Expression profiles of *BrPDI* genes in various tissues, as determined by RT-PCR analysis. Eight amplified bands (from left to right) per gene represent amplified products from R- roots; S- stems, L- leaves; Fb- flower buds; Se- sepals; Pe- petals; St- stamens and Pi- pistils

cold stress durations (Fig. 8a). By contarst, most of the *BrPDI* genes in 'Kenshin' were significantly down-regulated throughout the stress period, however, only four genes (*BrPDI 1–1, BrPDI 4–1, BrPDI 4–2,* and *BrPDI 5–2*) showed high expression in 'Kenshin' compared to 'Chiifu' (Fig. 8a).

High transcript levels were detected in 24 *BrPDI* genes over time courses and these genes were up-regulated in response to salt stress. *BrPDI 1–1* gene was exhibited approximately 42- fold higher expression at 24 h, whereas *BrPDI 4–2, BrPDI 5–1* and *BrPDI 5–2* were exhibited 15-, 13-, and 19-fold higher expression, respectively than control at 12 h time point (Fig. 8b). In addition, *BrPDI 1–3, BrPDI 6–1, BrPDI 7–1, BrPDI 8–4, BrPDI 9–1, BrPDI 10–1, BrPDI 10–2,* and *BrPDI 10–3* were significantly up-regulated up to 12 h time point (Fig. 8b). Besides, *BrPDI 1–4, BrPDI 1–5, BrPDI 2–1, BrPDI 2–3, BrPDI 3–2, BrPDI 4–1* and *BrPDI 8–3* were up-regulated approximately 5-, 8-, 7-, 11-, 11-, 7-, and 5-fold, respectively at 48 h.

In drought stress, most of the *BrPDI* genes were up-regulated at 12 h and 24 h; then gradually down-regulated with the advancement of time courses (Fig. 8c). Eight genes (*BrPDI 1–1, BrPDI 3–1, BrPDI 3–2, BrPDI 4–2, BrPDI 5–2, BrPDI 8–1, BrPDI 10–1,* and *BrPDI 10–2)* were significantly up-regulated at 12 h. Five genes (*BrPDI 1–2, BrPDI 1–5, BrPDI 2–1, BrPDI 2–2,* and *BrPDI 5–1*) were significantly up-regulated at 24 h drought stress, thereafter being down-regulated (Fig. 8c). *BrPDI 1–1, BrPDI 1–5, BrPDI5–1* and *BrPDI 10–2* exhibited about 27-, 15-, and 14- fold higher expression, respectively compared to control. Most of the *BrPDI* genes had low level of expression in response to ABA treatment (Fig. 8d). *BrPDI 1–1, BrPDI 1–3, BrPDI 1–5, BrPDI 3–1, BrPDI 4–2, BrPDI 5–1,* and *BrPDI 5–2* were significantly up-regulated at 6 h and 12 h time point. However, those *BrPDI* genes were down-regulated at the beginning of ABA stress, but start up-regulation at 6 h and continued up to 12 h time point, thereafter gradually down-regulated over the remaining time courses. Four genes (*BrPDI 11–1, BrPDI 11–2, BrPDI 11–3 and BrPDI 11–4*) were down-regulated over the time courses. *BrPDI 1–1* gene was 10-fold up-regulated at 12 h; while *BrPDI 5–2* gene showed 13-fold up-regulation at 6 h time points compared to control (Fig. 8d).

Expression profiling of the *BrPDI* genes during chilling injury treatment

Chilling injury experiment with cold tolerance 'Chiifu' and cold sensitive 'Kenshin' was conducted to get the deep insight of the expression of the target cold tolerance gene as an alternative of functional analysis of the predicted genes. We exposed the 'Chiifu' and 'Kenshin' seedlings in cold injury treatment at 0 °C for 72 h until complete chilling injury of the susceptible (Kenshin). First remarkable chilling injury was recognized in 'Kenshin' seedlings at 24 h time point and gradually progress with the enhancement of time courses, 'Kenshin' seedlings were completely injured at 72 h time course. Whereas, no chilling injury was observed in 'Chiifu' seedlings up to 72 h time point, but plant growth become stunted (Fig. 9). In 'Chiifu',

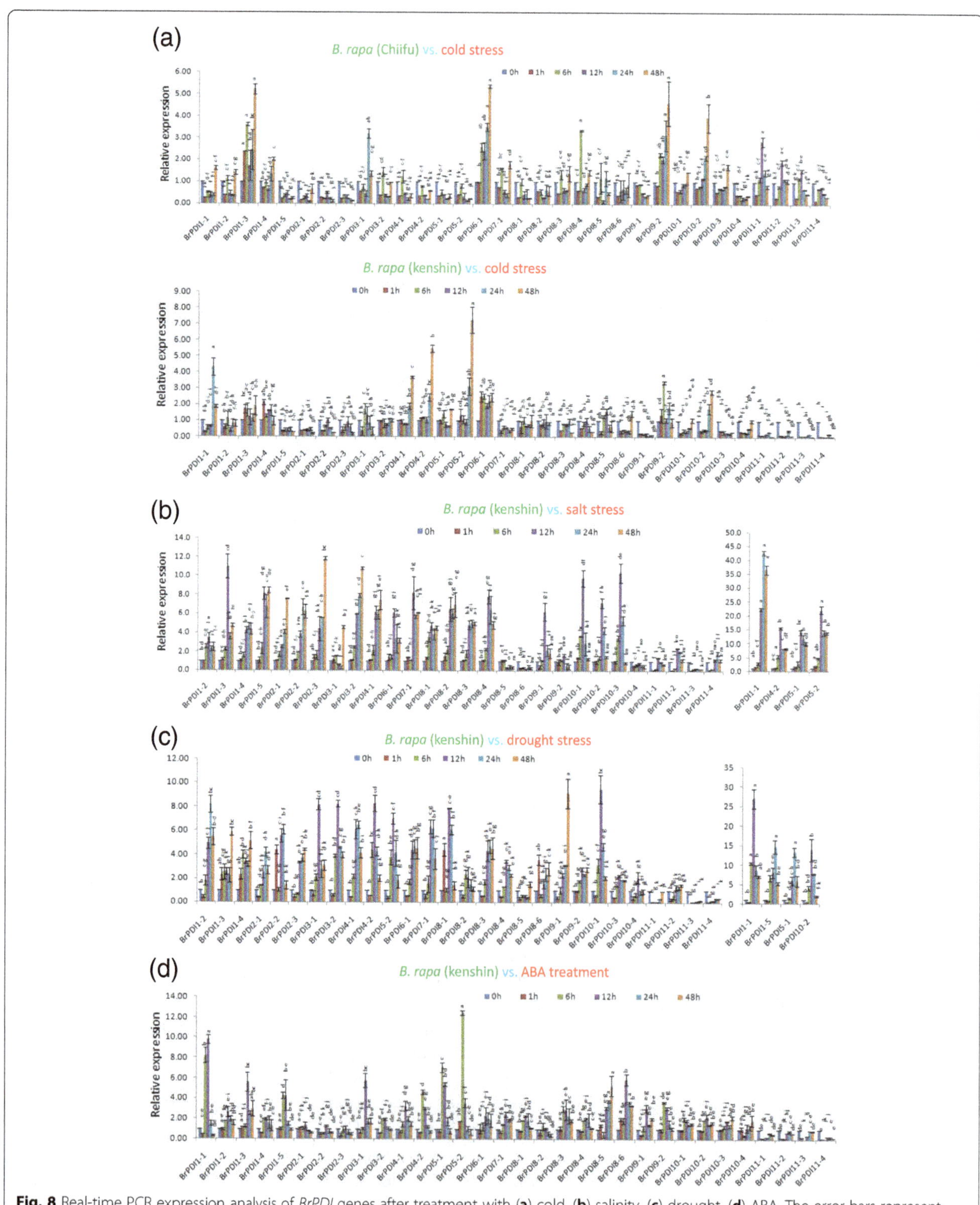

Fig. 8 Real-time PCR expression analysis of *BrPDI* genes after treatment with (**a**) cold, (**b**) salinity, (**c**) drought, (**d**) ABA. The error bars represent the standard error of the means of three replications

eight genes showed higher transcript level. *BrPDI1–4* and *BrPDI6–1* genes showed significant up-regulation about 12- and 9-fold higher expressions, respectively at 72 h time point in 'Chiifu' compared to 'Kenshin' (Fig. 10). Moreover, *BrPDI9–2* and *BrPDI11–1* showed about 6- and 8-fold higher expressions at 48 h and 24 h time point, respectively. In addition, *BrPDI10–2* gene exhibited higher expression in both line at 72 h

Fig. 9 Comparative chilling injury symptoms in cold tolerant 'Chiifu' and 'Kenshin' plant at different time points (0 h, 24 h, 48 h, and 72 h) against chilling treatment (0 ° C). Chilling injury first appeared at 24 h in 'Kenshin' thereby gradually progress with the advancement of time points. The 'kenshin' plants are completely chill injured at 72 h time point but no chilling injury symptoms appeared in 'Chiifu' plants at any time points

time point. In 'kenshin', four genes (*BrPDI4–2, BrPDI5–1, BrPDI5–2,* and *BrPDI7–1*) showed higher expression against chilling injury treatment (0 °C).

Analysis of biotic stress-related gene expression

To elucidate the expression patterns of *BrPDI* genes in response to *Fusarium oxysporum* f. sp. *conglutinans* infection, we have collected leaf samples from infected and mock-treated Chinese cabbage line 'Chiifu' for performing qPCR. This fungus causes wilt and root rot diseases in *Brassica* crops. The *BrPDI* genes exhibited distinct expression patterns in response to *Fusarium* infection. Out of 32 genes, 14 showed high expression at 6 h time point (Fig. 11). The expression levels of *BrPDI*

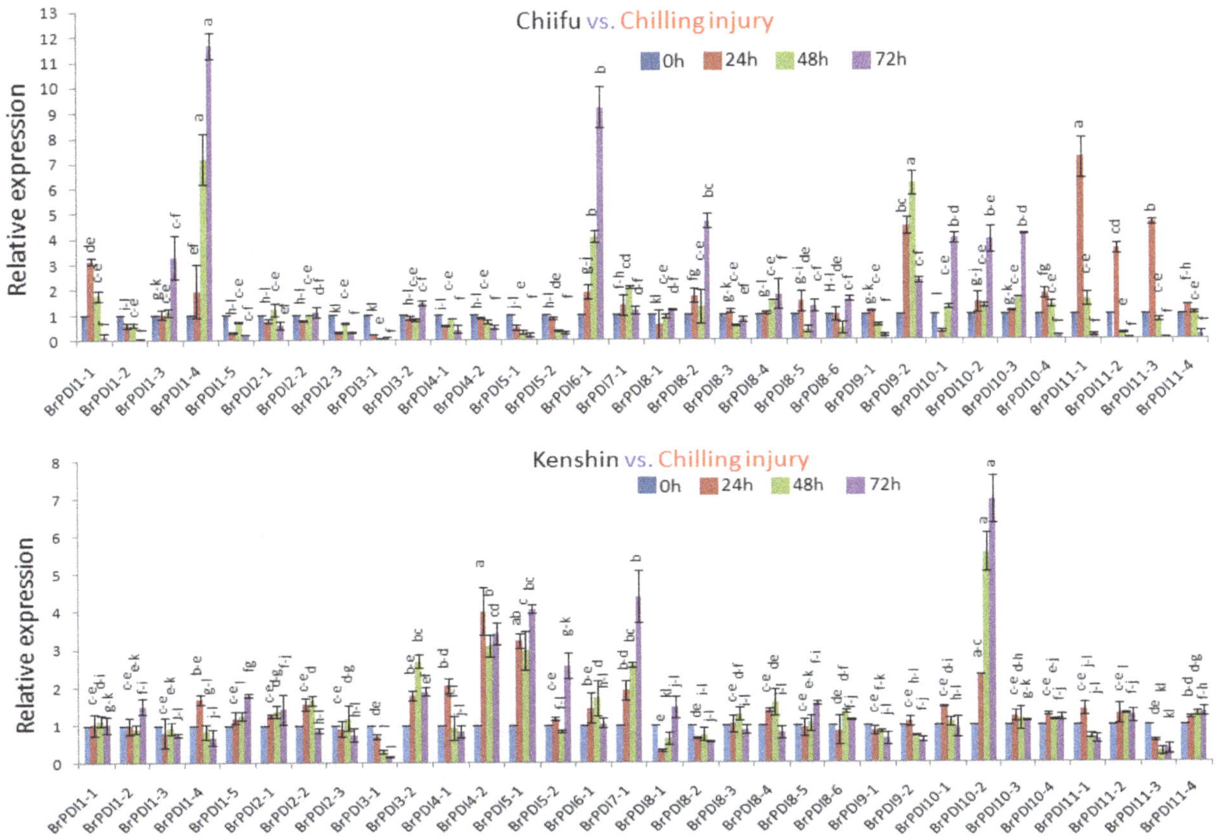

Fig. 10 Real-time PCR expression analysis of *BrPDI* genes after treatment with Chilling injury. The error bars represent the standard error of the means of three replications

1–1, BrPDI 2–3, and *BrPDI 5–1* were approximately 27-, 10-, and 11-fold higher, respectively compared to mock-treated plants. Fifteen genes (*BrPDI 6–1, BrPDI 7–1, BrPDI 8–1, BrPDI 8–2, BrPDI 8–3, BrPDI 8–4, BrPDI 8–5, BrPDI 8–6, BrPDI 9–1, BrPDI 10–1, BrPDI 10–2, BrPDI 10–4, BrPDI 11–1, BrPDI 11–2,* and *BrPDI 11–4*) had very little or insignificant expression compared to mock-treated (uninfected) plants. However, *BrPDI10–3* and *BrPDI11–3* showed no response against *F. oxysporum* infection.

Discussion

In genome-wide exploration study of Chinese cabbage *PDI* genes; we have identified 32 *PDI* genes identical to 21 *PDI* genes of *Arabidopsis.* The Chinese cabbage genome has gone under two duplication events since its divergence from *Arabidopsis* [21]. The evolutionary relationship between *Arabidopsis* and Chinese cabbage is also supported by our results. We analyzed 32 *BrPDI* genes expression pattern based on the microarray data set in two contrasting Chinese cabbage inbred lines, 'Chiifu' and 'Kenshin' exposed to cold and freezing stress (4 °C, 0 °C, −2 °C, and −4 °C) [20]. From an evolutionary point of view, gene duplications, tandem and segmental duplications can increase the gene numbers of a particular gene family [22]. Distribution of *BrPDI* genes in the *B. rapa* genome are also affected by segmental duplication, tandem duplication, polyploidization etc. during evolution [23, 24]. In addition, genome triplication events of *B. rapa* may also be played important role in the expansion of *PDI* gene family. We found 17 pairs of segmental duplicated *PDI* genes in Chinese cabbage genome (Fig. 5 and Table 3), this result indicates that the expansion of *PDI* gene family in *B. rapa* genome depicted on segmental gene duplication.

We also investigated evolutionary history of *BrPDI* gene family and calculated the Ka (the number of synonymous substitutions per synonymous site), Ks (the number of nonsynonymous substitutions per nonsynonymous site) and Ka/Ks ratios of segmental duplicated gene pairs. Nine gene pairs had Ka/Ks < 1 (purifying selection) and eight gene pairs had Ka/Ks >1(positive selection; Table 3) indicating purifying selection and positive selection together accelerated evolution and functional divergence of *BrPDI* genes in *B. rapa* genome. Based on the synonymous substitution rate, we calculated divergence time of the *BrPDI* genes and found that duplication started on 29.74 MYA and stop at 1.62 MYA (Table 3) which indicate *PDI* genes divergence take place after the genome triplication events in *B. rapa.* Genome triplication event of *B. rapa* has occurred from the ancestor of *A.*

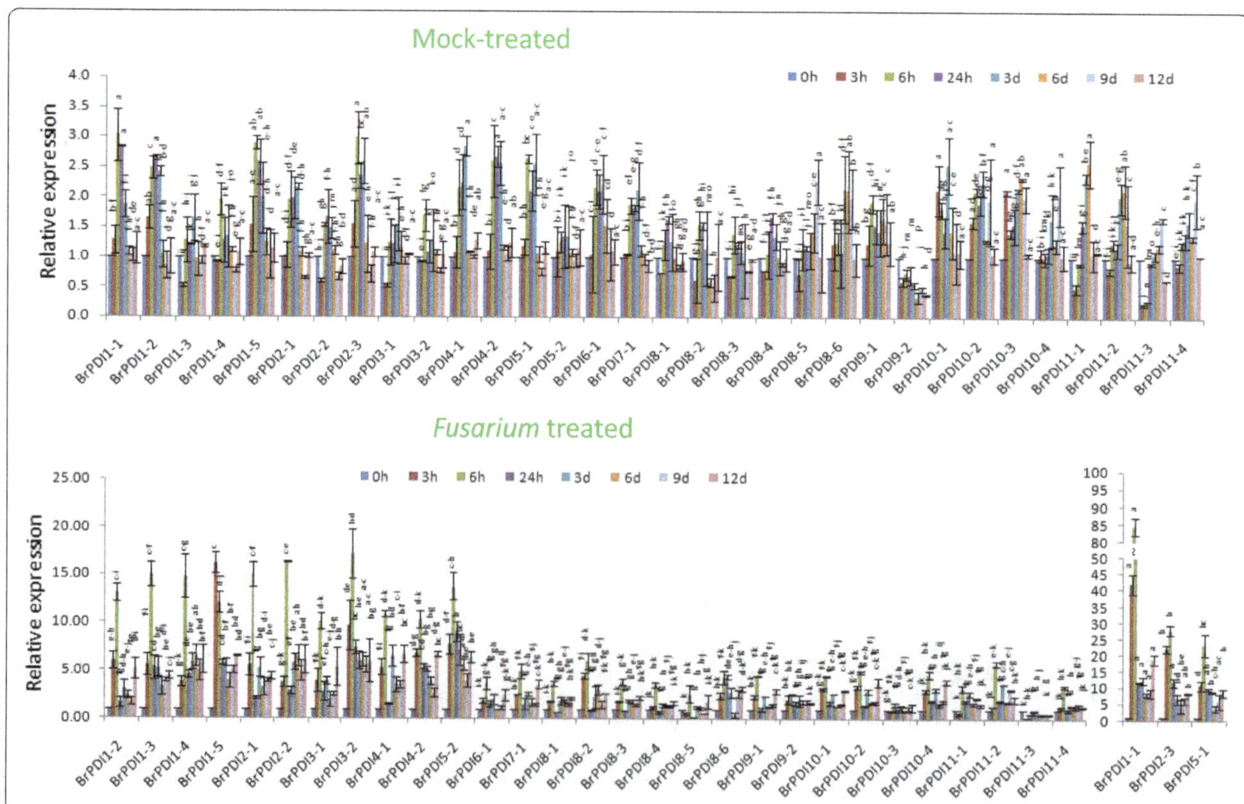

Fig. 11 Real-time PCR expression analysis of *BrPDI* genes after treatment with *Fusarium oxysporum* f.sp. *conglutinans*. The error bars represent the standard error of the means of three replications

thaliana between five to nine million years ago (MYA; [25]). Besides this, the exons and introns distribution of *BrPDI* genes are also consistent with the number of exons and introns of *PDI* genes in *Arabidopsis*, which indicate their close evolutionary relationship. In addition with, by drawing microsyntenic map, we conclude that *BrPDI* genes are strongly related to those of *B. distachyon* and *A. thaliana.*

The Chinese cabbage *PDI* genes appeared as functionally differentiated for short time, and keeps as functional genes in the genome and maintaining complex functions in organ development. The features and functions of *PDI* family genes have been extensively studied in *Triticum aestivum* [14] and *Brachypodium distachyon* [15]. Our identified and characterized *PDI* genes in Chinese cabbage help to increase the understanding of role of this family gene. Most of the *BrPDI* genes were predominantly expressed in all organs suggested their functions in regulation of plant growth and development. Based on our organ-specific expression analysis, all *BrPDI* genes are expressed at least in one organ at variable level. *BrPDI 2–1* and *BrPDI 9–1* were highly expressed in root, which is consistent with the previous findings [14, 15]. In addition to their involvement in plant growth and development, *PDI* genes are also responsive to adverse environmental conditions [15]. We have found *BrPDI 1–2, BrPDI 1–3, BrPDI1–4, BrPDI 2–3 BrPDI 8–1* and *BrPDI 8–6* were expressed in flower buds and floral parts, suggesting their possible role in flower development.

The expression patterns of 32 *BrPDI* genes were analyzed using a whole-genome microarray dataset of two inbred lines of *B. rapa*, 'Chiifu' and 'Kenshin' imposed to various temperature treatment [26]. Thereof, we have selected *BrPDI* family genes for analyzing their differential expression patterns compared to control by constructing heat-map (Fig. 6). Our results suggest that *BrPDI* genes play a vital role against abiotic stress responses in Chinese cabbage. In addition, our main objectives were to identify putative candidate *PDI* genes related to stress response during plant growth and development. It is established fact that environmental stresses like, cold, salt, and drought are severely affected crop production and make threaten in food security worldwide. Therefore, application of artificial stress treatment helps to know, how plants adapt to stresses via molecular, morphological, and physiological mechanisms. There is intense interest in identifying stress-responsive genes and elucidate them to develop stress-tolerant crop cultivars. Relative expression of PDI family genes were estimated from the sample of the plants subjected to abiotic stresses (cold, salt and drought), hormone (ABA), chilling injury, and biotic (*F. oxysporum* f.sp.*conglutinans*) stress treatment over different time periods (Fig. 8a–d, 9~11) using qPCR data. In plants, PDIs are localized to the ER and other cellular compartments, such as nucleus and chloroplasts [27, 28]. Proteins in the ER of plants can lead to unfolding or misfolding due to complex physiological processes during different environmental stresses [29]. Plants possess multiple mechanisms for ensuring correct folding of proteins. PDI was the first reported catalyst of protein folding (2). In the current study, *BrPDI1–3, BrPDI3–1, BrPDI6–1, BrPDI9–2, BrPDI10–2,* and *BrPDI11–1* were highly expressed in cold-tolerant 'Chiifu' than in cold-susceptible 'Kenshin' during cold stress treatment. *PDI* genes that are highly expressed under cold stress may encode proteins catalyze for breakage of disulfide bonds in misfolded proteins, restoring their proper functioning through correction of folding patterns. Lu & David [30] reported six *Arabidopsis* genes; those were up-regulated due to chemicals such as dithiothreitol (DTT), β-mercaptoethanol and tunicamycin induction. In current study, we have found *BrPDI1–1, BrPDI4–1, BrPDI4–2,* and *BrPDI5–2* were highly expressed under cold stress in cold-susceptible 'Kenshin' compared to cold-tolerant 'Chiifu', suggesting that these four genes might be related to the cold-susceptibility in 'Kenshin'. However, *BrPDI2–1, BrPDI2–2, BrPDI5–2, BrPDI8–1, BrPDI8–6, BrPDI9–1, BrPDI10–4,* and *BrPDI11–4* were not expressed or even down-regulated in 'Chiifu' under cold treatment confirmed that these genes contain putative cold-inducible LTR *cis*-acting elements. Notably, these genes also contain a heat-inducible HSE *cis*-acting element. Hence, both cold- and heat-inducible *cis*-acting elements are presented in the promoter regions of the same genes suggesting that the latter *cis*-acting element may down-regulate these genes under cold stress. Two contrasting cold tolerant Chinese cabbage lines were used in this study to predict the putative cold susceptible and cold tolerant *BrPDI* gene(s) of this crop. Four *BrPDI* genes (*BrPDI1–4, BrPDI6–1, BrPDI9–2,* and *BrPDI11–1*) exhibited higher expression in 'Chiifu' compared to 'Kenshin' at the level of chilling injury. Among these four genes *BrPDI6–1, BrPDI9–2,* and *BrPDI11–1* genes were common in the higher expression in cold stress and chilling injury level treatments revealed their active involvement to overcome the cold and chilling injury at different time points of the cold tolerant line 'Chiifu'. Therefore, from the in depth expression data, it is clearly evident that *BrPDI6–1,* and *BrPDI9–2* genes are the putative candidate in Chinese cabbage for overcoming cold and even chilling injury temperature (0 ° C).

With few exceptions, *BrPDI* genes are highly expressed in Chinese cabbage under salt and drought stress, among them *BrPDI1–1* to *BrPDI5–2* genes encode the proteins which containing two active sites and a C-terminal ER retention signal composed of four amino acids (KDEL/ VASS/ MADD), might be involved in enhancing protein-folding in the ER. *BrPDI1–1, BrPDI1–3, BrPDI1–5, BrPDI3–1, BrPDI4–2, BrPDI5–1, BrPDI5–2, BrPDI8–5*, and *BrPDI8–6* were highly expressed under ABA stress compared to control. Mittler et al. [31] and Sakamoto et al. [32] observed rapid accumulation of reactive oxygen species (ROS) in plant cells subjected to environmental stresses such as salt, drought, and ABA. Indeed, during environmental stress (e.g., drought, salt, ABA, UV, and heat exposure) ROS levels can increase dramatically, which may significantly damage cell structure. PDI proteins play a significant role in thioredoxin-based redox pathway, which comprises part of the anti-oxidative defense system [33]. Abiotic stress usually leads to protein unfolding, misfolding, and aggregation; which represent a common threat to living cell [34]. Efficient protein repair systems and/or protein folding stability helps plant to survive in adverse environmental conditions [35]. Most *BrPDI* genes were differentially expressed in response to salt, drought, and ABA treatment, which is consistent with the findings of Zhu et al. [15]. Those genes are highly expressed in response to salt, drought, and ABA might function in the repair of misfolded proteins, thereby playing a role in plant resistance to the stresses. The up-regulated genes contain the respective stress-responsive *cis*-elements in their promoter regions (Additional file 2: Table S1).

PDI genes have been previously shown to be contributed against powdery mildew resistance in common wheat [36]. In this study, we have analyzed expression profiles of these genes in response to *Fusarium oxysporum* f.sp. *conglutinans* infection. Moreover, 14 genes were up-regulated compare to mock treated one due to infection of fungal pathogen at 6 h time point, whose genes containing a disease resistance Box-W1 *cis*-acting element (Fig. 11, Additional file 2: Table S1). By contrast, *BrPDI 6–1~ BrPDI 11–4* do not contain the Box-W1 element, those exhibited almost no expression due to fungal infection. Thus, the highly expressed *BrPDI* genes might play a vital role in response against fungal pathogen. Therefore, this might be the putative candidate genes for developing *Fusarium oxysporum* f.sp. *conglutinans* resistant Chinese cabbage cultivars through marker assisted back crossing (MAB) or gene technology. Besides, the up-regulated *BrPDI* genes might play differential roles in signal transduction pathways and/or cooperate with other genes to form networks for defending plants against adverse environmental conditions.

Conclusion

In conclusion, this is the first report of genome-wide characterization of PDIs in Chinese cabbage. We identified 32 *BrPDI* genes in the Chinese cabbage genome and characterized them based on motif distribution, protein structure, classification, number of introns -exons, sequence similarities, and expression patterns in response to abiotic (cold, salt, and drought) stresses, ABA treatment and biotic stress (*F. oxysporum* f.sp. *conglutinans* inoculation). We have also recognized *BrPDI* genes may play roles in biotic and abiotic stress responses. Several *BrPDI* genes (*BrPDI1–1, BrPDI1–4, BrPDI1–5, BrPDI3–1, BrPDI3–2, BrPDI4–1, BrPDI4–2, BrPDI5–1, BrPDI5–2, BrPDI6–1, BrPDI7–1, BrPDI8–2, BrPDI9–1, BrPDI9–2*, and *BrPDI11–1*) might function in responses to multiple stresses. The identified stress-induced *BrPDI* genes help to elucidate the complex regulatory network underlying stress resistance mechanisms. In addition, these genes represent a useful resource to molecular breeders for marker assisted back crossing (MAB) and/ or engineering transgenic plants in future research with increased resistance to biotic and abiotic stresses.

Methods

Identification and sequence analysis of *BrPDI* genes

Chinese cabbage PDI family members were identified using the SWISSPROT tool of the *B. rapa* database (http://brassicadb.org/brad/searchAll.php; 20) and NCBI using the keyword "PDI". We also searched these genes from a microarray Br135K dataset of two contrasting Chinese cabbage inbred lines, Chiifu and Kenshin exposed to low-temperature stress. The *Arabidopsis* PDI sequences were used as the query to perform a BLAST search setting a cutoff e-value of $<10^{-10}$. The CDS (coding DNA sequences) and protein sequences of the identified PDIs were obtained from the *B. rapa* genomic database. The PDI protein sequences were further analyzed for the presence of a thioredoxin domain using the web tool "SMART" (http://smart.embl-heidelberg.de/ [37]) and NCBI (https://www.ncbi.nlm.nih.gov/cdd). Additionally, the primary structures of the genes (protein length, molecular weight and isoelectric point) were analyzed using ExPasy (http://web.expasy.org/compute_pi/). ORFs were identified using ORF finder at NCBI (https://www.ncbi.nlm.nih.gov/orffinder/). Sequence alignment of PDI proteins were carried out using CLUSTAL Omega (http://www.ebi.ac.uk/Tools/msa/clustalo/). Sub-genome fractionation and positional information for all idetified *BrPDI* genes on 10 chromosomes of *B. rapa* were retrieved from the *Brassica* database, and draft maps of the locations of the *BrPDI* genes were constructed using Map Chart version 2.30 (http://www.wageningenur.nl/en/show/Mapchart.htm). The Gene Structure Display Server (GSDS) web tool (http://gsds.cbi.pku.edu.cn/) was used to determine the number of introns and exons by aligning CDS and genomic sequences of the *PDI* genes [38]. Putative *cis*-acting regulatory elements in the *BrPDI* genes were predicted in regions of approximately1500-bp upstream of the translation initiation site [ATG] using the PlantCARE (http://bioinformatics.psb.ugent.be/webtools/plantcare/html/).

Phylogenetic and conserved motif analysis of Chinese cabbage *PDI* genes

The predicted BrPDI protein sequences were aligned with those of *Arabidopsis*, *Brachypodium distachyon and Zea mays* and Chinese cabbage using Clustal Omega, and a phylogenetic trees was constructed based on the condensed alignment in MEGA6.06 using the Neighbor-Joining (NJ) algorithm [39], with the parameters set at 1000 replications for bootstrap values and complete deletion mode to analyze tree topology and reliability. The genes which were used in phylogenetic analysis, name and accession number provided in Additional file 2: Table S2. Motif analysis of proteins was performed using MEME (Expectation Maximization for Motif Elicitation v4.10.1) [40] with the following parameters: (1) number of motifs = 10, (2) Motif width ≥ 6 and ≤50.

Chromosome localization and gene duplications

The physical locations of *BrPDI* genes were collected from *B. rapa* genomic database and the positions of the *BrPDI* genes were drafted to ten *B. rapa* chromosomes by map chart program. Duplicated *BrPDI* genes were identified using BLAST searched against themselves, notably identity and query coverage were >80% of those particular genes [41]. Tandem duplicated genes were marked as an array of two or more homologous genes within a distance of 100 kb. The synonymous rate (*Ks*), nonsynonymous rate (*Ka*), and evolutionary constraint (*Ka*/*Ks*) were calculated between the duplicated PDI gene pairs (Table 3), using the method of Nei & Gojobori [42] by Mega 6.0 software. The value of Ka/Ks ratio like >1, <1 and =1 are depicted for positive selection, purifying selection and neutral selection, respectively [43]. The divergence time was calculated using the formula T = *Ks*/2r Mya (Millions of years) where, *Ks* being the synonymous substitutions per site and r is considered 1.5 × 10–8 substitutions rate per site per year for dicot plants [44]. We reconstructed 24 conserved chromosomal blocks (labelled A–X) of *B. rapa* genome. Colour coding of the blocks was assigned based on their positions in a proposed ancestral karyotype (AK1–8) following the procedure stated by Cheng et al. [45] and Schranz et al. [46].

Microarray and microsynteny of PDI gene family

Two contrasting inbred Chinese cabbage (*B. rapa* ssp. *pekinensis*) lines, 'Chiifu' and 'Kenshin' in respect to cold tolerance and cold susceptible, were used for microarray expression analysis Kayum et al. [47]. A heat map was generated based on transcript abundance value of *PDI* genes using Cluster 3.0 (http://bonsai.hgc.jp/~mdehoon/software/cluster/software.htm). The microsyntenic relationship of PDI genes among *B. rapa*, *Brachypodium distachyon* and *A. thaliana* were detected using Blast against whole genome of such crop species. Chromosomal positions of *PDI* genes were collected from respective databases and the relationship among the three crop species were plotted using circos software (http://circos.ca/) [48].

Plant growth conditions, treatments and sampling

Surface sterilize seeds of two Chinese cabbage inbred lines 'Chiifu' and 'Kenshin' were grown in incubation room maintaining 24 °C temperature with 14/10 h (light/dark) condition at the Department of Horticulture, Sunchon National University, Korea. Abiotic stress treatments were imposed to four-week-old seedlings according to the methods described by Ahmed et al. [49], three biological replicates were maintained for abiotic stress treatments. Two contrasting Chinese cabbage inbred lines, cold tolerance 'Chiifu' and cold susceptible 'Kenshin' were used for abiotic stress cold and chilling injury stress experiment. Whereas, only 'Kenshin' was used to predict the *BrPDI* genes against other stress treatments (salt, drought, and ABA) due to unavailability of any contrasting genotypes for other abiotic stress treatments (salt, drought, and ABA) in our hand. By contrast, we assumed 'Kenshin' would be suitable for molecular characterizing of *BrPDI* genes for abiotic stress responsiveness rather than cold, because these type of genotypes are widely grown in tropics and sub-tropics. Fresh roots and leaves (third and fourth leaves) were excised from stress-treated plants over time points (0 h, 1 h, 4 h, 12 h, 24 h, and 48 h). Samples were immediately frozen in liquid nitrogen and stored at −80 °C for RNA extraction. For chilling injury experiment, the Chinese cabbage inbred line 'Chiifu' and 'Kenshin' plants were grown in culture room under a 16 h light photoperiod maintaining 24 °C. The three week old seedlings were transferred to incubator (TOGA clean system; model: TOGA UGSR01) maintained 0 °C temperature and keep the seedling until clear remarkable cold injury symptoms appeared. Cold treated plants leaves were excised at 0 h, 24 h, 48 h and 72 h time points with the progression of cold injury in cold susceptible line 'Kenshin' (Fig. 9), thereby leaves were immediately frozen in liquid nitrogen and stored at −80 °C for RNA extraction. Total five biological replications were maintained for chilling injury treatment. However, inbred line 'Chiifu' was used for organ-specific expression profiling of *BrPDI* genes and also infection with *F. oxysporum* f. sp. *conglutinans* for biotic stress as described by Ahmed et al. [50]. *Fusarium*- and mock-infected leaves (4th and 5th leaves) were collected at 0 h, 3 h, 6 h, 24 h, 3d, 6d, 9d, 12d after inoculation and immediately frozen in liquid nitrogen and stored at −80 °C for RNA extraction.

RNA extraction and cDNA synthesis

Total RNA was extracted from the samples (roots and leaves) using an RNeasy mini kit (Qiagen, USA) following the manufacturer's instructions. DNA contamination was removed using RNase-free DNase (Promega, USA) following manufacturer's protocol. The extracted RNA was quantified by UV spectrophotometry at A260 using a NanoDropND-1000 and NanoDrop v3.7 software (Nano Drop Technologies, USA). Complementary DNA (cDNA) was synthesised from total RNA using a First-Strand cDNA synthesis kit (Invitrogen, Japan) following the manufacturer's instructions.

Qualitative and quantitative PCR expression analysis

Qualitative expression analysis was conducted using one-step Emerald Amp GT PCR Master Mix (Takara, Japan) by RT-PCR. Gene-specific primers for *BrPDIs* were used for RT-PCR and the *BrActin* gene of Chinese cabbage was used as an internal control. RT-PCR was performed using 1 μL of 50 ng template cDNA using master mix contained 10 pmol of each primer (forward and reverse), 9 μL sterile water and 8 μL Emerald mix in a total volume of 20 μL. The PCR conditions were initial denaturation 94 °C for 5 min followed by 30 cycles of denaturation at 94 °C for 30s, annealing at 58 °C for 30s and extension at 72 °C for 45 s, with a final extension at 72 °C for 5 min. The PCR products were visualized on 1.2% agarose gel.

Quantitative Real-time PCR (qRT-PCR) was conducted using 10 μL reaction volume consisted of 5 μL 2× Quanti speed SYBR mix, 1 μL (10 pmol) each of forward (F) and reverse (R) gene-specific primers (Additional file 2: Table S3), 1 μL template cDNA (50 ng) and 2 μL ultra-pure water (ddH_2O). The conditions for real-time PCR was initial denaturation at 95 °C for 5 min, followed by 40 cycles of denaturation at 95 °C for 10 s, annealing at 58 °C for 10 s and extension at 72 °C for 15 s. The qRT-PCR reactions were normalized using the Chinese cabbage *Actin* gene as a reference for all comparisons [51]. Fluorescence was measured at last step of each cycle, and three replicates were used for each sample. Amplification detection and data were processed using the Light cycler® 96 SW 1.1 software and the cq value was calculated using the $2^{-\Delta\Delta}C_T$ method to determine the relative expression. The relative expression data was statistically analyzed (Tukey HSD test) and lettering was done using Minitab 17 software (*https://www.minitab. com/products/minitab/*).

Additional files

Additional file 1: Figure S1. The genomic structures of *BrPDI* genes. Solid green boxes and red lines indicate exons and introns, respectively. The bottom scale indicates length of exons and introns. **Figure S2.** Schematic representation of motif compositions in the BrPDI protein sequences. Different motifs logo, numbered 1–10, motif are displayed in different colored boxes. The names of all members are displayed on the left-hand side. (PPT 932 kb)

Additional file 2: Table S1. Putative *cis*-elements, more than 6 bp, were identified in 32 *BrPDI* genes in Chinese cabbage. **Table S2.** A total of 76 *PDI* and *PDIL* genes name and accession numbers from 4 species used for constructing phylogenetic tree, including 32 from *Brassica rapa,(Br)* 21 from *Arabidopsis thaliana (At)*,11 from *Brachypodium distachyon (Bd)* and 12 from *Zea mays (Zm)*. **Table S3.** Primers for RT-PCR andreal-time PCR analysis of 32 *BrPDI* genes (DOC 472 kb)

Abbreviations

ABA: Abscisic acid; AK: Karyotype; At: *Arabidopsis thaliana*; Bd: *Brachypodium distachyon*; Br: *Brassica rapa*; BRAD: Brassica database; C: Chiifu; CDS: Coding DNA sequence; Cl: Cluster; dai: Days after infection; Er: Endoplasmic reticulum; *F.oxysporum*: *Fusarium oxysporum*; f.sp.: Fungal species; HTH: Helix-turn-helix; K: Kenshin; Ka: Nonsynonymous substitutions per nonsynonymous site; kb: Kilo basepair; Ks: Synonymous substitutions per synonymous site; LF: Least fractionated; MF1: Medium fractionated; MF2: Most fractionated; MYA: Million year; PD: Protein disulfide isomerase; qPCR: Quantitative polymerase chain reaction; RT-PCR: Reverse transcription polymerase chain reaction; TRX: Thioredoxin; Zm: *Zea mays*

Acknowledgments

Special thanks to Professor YoonKang Hur, Department of Biology, College of Biological Sciences and Biotechnology, Chungnam National University, Daejeon, Korea Republic for providing seeds of *B. rapa* inbred lines 'Chiifu' and 'Kenshin'.

Funding

This research was supported by Golden Seed Project (Center for Horticultural Seed Development, No. 213007-05-01-SB510) Ministry of Agriculture, Food and Rural Affairs (MAFRA), Ministry of Oceans and Fisheries (MOF), Rural Development Administration (RDA) and Korea Forest Service (KFS), South Korea. The funding bodies were not involved in the design of the study or in any aspect of the collection, analysis and interpretation of the data or paper writing.

Authors' contributions

The work presented here was carried out in collaboration among all authors. MAK carried out the computational analysis, plant culture and sample preparation, performed RT-PCR and real-time PCR, analyzed the data and drafted the manuscript. JIP and UKN collected primary data regarding genes and UKN revised the manuscript and made necessary corrections for giving its final shape. HTK and GS designed the stress experiments and cultured the plants and gave stress treatments to the inbred lines 'Chiifu' and 'Kenshin'. MKB and JIP perform in silico analysis. ISN designed and participated in all the experiments and assisted in improving the technical sites of the project. All authors have read and approved the final manuscript.

Competing interests

The authors declare that they have no competing interests.

Author details

[1]Department of Horticulture, Sunchon National University, 255 Jungang-ro, Suncheon, Jeonnam 57922, Republic of Korea. [2]University-Industry Cooperation Foundation, Sunchon National University, 255 Jungang-ro, Suncheon, Jeonnam 57922, Republic of Korea.

References

1. Hayano T, Hirose M, Kikuchi M. Protein disulfide isomerase mutant lacking its isomerase activity accelerates folding in the cell. FEBS Lett. 1995;377:505–11.
2. Venetianer P, Straub FB. The enzymatic reactivation of reduced ribonuclease. Biochim Biophys Acta. 1963;67:166–8.
3. Anfinsen C. Principles that govern the folding of protein chains. Science. 1973;181:223–30.
4. Hatahet F, Ruddock LW. Substrate recognition by the protein disulfide isomerases. FEBS J. 2007;274(20):5223–34. https://doi.org/10.1111/j.1742-4658.2007.06058.x. 17892489
5. Gruber CW, Cemažar M, Heras B, Martin JL, Craik DJ. Protein disulfide isomerase: the structure of oxidative folding. Trends Biochem Sci. 2006;31: 455–64.
6. Kemmink J, Darby NJ, Dijkstra K, Nilges M, Creighton TE. The folding catalyst protein disulfide isomerase is constructed of active and inactive thioredoxin modules. Curr Biol. 1997;7:239–45.
7. Kevin S. Helix-Turn-Helix, zinc-finger, and leucine-zipper motifs for eukaryotic transcriptional regulatory proteins. Trends Biochem Sci. 1989; 14(4):137–40.
8. Denecke J, DeRycke R, Botterman J. Plant and mammalian sorting signals for protein retention in the endoplasmic reticulum contain a conserved epitope. EMBO J. 1992;11:2345–55.
9. Houston NL, Fan CZ, Xiang QY, Schulze JM, Jung R, Boston RS. Phylogenetic analyses identify 10 classes of the protein disulfide isomerase family in plants, including single-domain protein disulfide isomerase-related proteins. Plant Physiol. 2005;137:762–78.
10. Ferrari DM, Soling HD. The protein disulphide-isomerase family: unravelling a string of folds. J Biol Chem. 1999;339:1–10.
11. Iwasaki K, Kamauchi S, Wadahama H, Ishimoto M, Kawada T, Urade R. Molecular cloning and characterization of soybean protein disulfide isomerase family proteins with nonclassic active center motifs. FEBS J. 2009; 276:4130–41.
12. Wang HZ, Leonor C, Boavida MR, McCormick S. Truncation of a protein disulfide isomerase, PDIL2-1, delays embryo sac maturation and disrupts pollen tube guidance in Arabidopsis Thaliana. Plant Cell. 2008;20:3300–11.
13. Trebitsh T, Levitan A, Sofer A, Danon A. Translation of chloroplast psbA mRNA is modulated in the light by counter acting oxidizing and reducing activities. Mol Cell Biol. 2000;20:1116–23.
14. d'Aloisio E, Paolacci AR, Dhanapal AP, Tanzarella OA, Porceddu E, Ciaffi M. The protein disulfide isomerase gene family in bread wheat (T. Aestivum L.). BMC Plant Biol. 2010;10:101.
15. Zhu C, Luo N, He M, Chen G, Zhu J, Yin G, Li X, Hu Y, Li J, Yan Y. Molecular characterization and expression profiling of the protein disulfide Isomerase gene family in Brachypodiumdistachyon L. PLoS One. 2014;9(4):e94704. https://doi.org/10.1371/journal. pone. 0094704.
16. Liu YH, Wang XT, Shi YS, Huang YQ, Song YC, Wang TY, Li Y. Expression and characterization of protein disulfide isomerases in maize (Zea Mays L.). Chinese J Biochem Mol Biol. 2009;25:229–34.
17. Han HC, Khurana N, Tyagi AK, Khurana JP, Khurana P. Identification and characterization of high temperature stress responsive genes in bread wheat (Triticumaestivum L.) and their regulation at various stages of development. Plant Mol Biol. 2011;75:35–51.
18. Hatahet F, Ruddock LW. Protein disulfide isomerase: a critical evaluation of its function in disulfide bond formation. Antioxid Redox Signal. 2009;11: 2807–50.
19. Stolf BS, Smyrnias I, Lopes LR, Vendramin A, Goto H, Laurindo FR, Shah AM, Santos CX. Protein disulfide isomerase and host-pathogen interaction. Sci World J. 2011;11:1749–61.
20. Cheng F, Liu S, Wu J, Fang L, Sun S, Liu B, Wang X. BRAD, the genetics and genomics database for *Brassica* plants. BMC Plant Biol. 2011;11(1):136. https://doi.org/10.1186/1471-2229-11-136.
21. Song XM, Huang ZN, Duan WK, Ren J, Liu TK, Li Y, et al. Genome-wide analysis of the HLH transcription factor family in Chinese cabbage (Brassica Rapa Ssp. Pekinensis). Mol Genet Genomics. 2014;289(1):77–91.
22. Bancroft I. Duplicate and diverge: the evolution of plant genome microstructure. Trends Genet Tig. 2001;17(2):89–93.
23. Lynch M, Conery JS. The evolutionary fate and consequences of duplicate genes. Science. 2000;290:1151–5.
24. Paterson AH, Bowers JE, Chapman BA. Ancient polyploidization predating divergence of the cereals, and its consequences for comparative genomics. Proc Natl Acad Sci. 2004;101:9903–8.
25. Wang Y, Tang H, Debarry JD, Tan X, Li J, Wang X, Lee TH, Jin H, Marler B, Guo H, Kissinger JC, Paterson AH. MCScanX: a toolkit for detection and evolutionary analysis of gene synteny and collinearity. Nucleic Acids Res. 2012;40(7):e49. https://doi.org/10.1093/nar/gkr1293.
26. Jung HJ, Dong X, Park JI, Thamilarasan SK, Lee SS, Kim YK, Lim YP, Nou IS, Hur Y. Genome-wide transcriptome analysis of two contrasting Brassica Rapa doubled haploid lines under cold-stresses using Br135K oligomeric chip. PLoS One. 2014;9(8):e106069.
27. Kim J, Mayfield PS. Protein disulfide Isomerase as a regulator of chloroplast translational activation. Science. 1997;278(5345):195421957.
28. Trebitsh T, Meiri E, Ostersetzer O, Adam Z, Danon A. The protein disulfide isomerase2like RB60 is partitioned between stroma and thylakoids Chlamydomonas Reinhardtii chloroplasts. J Biol Chem. 2001;276(3): 456424569.
29. Liu JX, Howell SH. Endoplasmic reticulum protein quality control and its relationship to environmental stress responses in plants. Plant Cell. 2010;22: 2930–42.
30. Lu DP, David AC. Endoplasmic reticulum stress activates the expression of a sub-group of protein disulfide isomerase genes and AtbZIP60 modulates the response in Arabidopsis Thaliana. Mol Genet Genomics. 2008;280:199–210.
31. Mittler R, Vanderauwera S, Gollery M, Van Breusegem F. The reactive oxygen gene network of plants. Trends Plant Sci. 2004;9:490–8.
32. Sakamoto H, Matsuda O, Iba K. ITN1, a novel gene encoding an ankyrinrepeat protein that affects the ABA-mediated production of reactive oxygen species and is involved in salt-stress tolerance in Arabidopsis Thaliana. Plant J. 2008;56:411–22.
33. Gilbert HF. Protein disulfide isomerase and assisted protein folding. J Biol Chem. 1997;272:29399–402.
34. Wang W, Vinocur B, Shoseyov O, Altman A. Role of plant heat-shock proteins and molecular chaperones in the abiotic stress response. Trends Plant Sci. 2004;9(5):244–52.
35. Ortbauer M. Abiotic Stress Adaptation: Protein Folding Stability and Dynamics. INTECH 2013; https://doi.org/10.5772/53129.
36. Faheem M, Li Y, Arshad M, Jiangyue C, Jia Z, et al. A disulphide isomerase gene (*PDI-V*) from *Haynaldia villosa* contributes to powdery mildew resistance in common wheat. Sci Rep. 2016;6:24227. https://doi.org/10.1038/srep24227.
37. Letunic I, Doerks T, Bork P. SMART 6: recent updates and new developments. Nucleic Acids Res. 2009;37:D229–32.
38. Guo AY, Zhu QH, Chen X, Luo JC. GSDS: a gene structure display server. Yi Chuan. 2007;29:1023e1026.
39. Tamura K, Dudley J, Nei M, Kumar S. MEGA4: molecular evolutionary genetics analysis (MEGA) software version 4.0. Mol Biol Evol. 2007;24:1596–9.
40. Bailey TL, Williams N, Misleh C, Li WW. MEME: discovering and analyzing DNA and protein sequence motifs. Nucleic Acids Res. 2006;34:W369–73. https://doi.org/10.1093/nar/gkl198.
41. Kong X, Lv W, Jiang S, Zhang D, Cai G, Pan J, Li D. Genomewide identification and expression analysis of calcium-dependent protein kinase in maize. BMC Genomics. 2013;14:433.
42. Nei M, Gojobori T. Simple methods for estimating the numbers of synonymous and nonsynonymous nucleotide substitutions. Mol Biol Evol. 1986;3:418–26.
43. Tang J, Wang F, Hou XL, Wang Z, Huang ZN. Genome-wide fractionation and identification of WRKY transcription factors in Chinese cabbage (Brassica Rapa ssp.pekinensis) reveals collinearity and their expression patterns under Abiotic and biotic stresses. Plant Mol Bio Rep. 2014;32(4): 781–95.

44. Koch MA, Haubold B, Mitchell-Olds T. Comparative evolutionary analysis of chalcone synthase and alcohol dehydrogenase loci in *Arabidopsis*, *Arabis*, and related genera *(Brassicaceae)*. Mol Biol Evol. 2000;17(10):1483–98.
45. Cheng F, Mandáková T, Wu J, Xie Q, Lysak MA, Wang X. Deciphering the diploid ancestral genome of the mesohexaploid *Brassica rapa*. Plant Cell Online. 2013;25:1541–54.
46. Schranz ME, Lysak MA, Mitchell-Olds T. The ABC's of comparative genomics in the Brassicaceae: building blocks of crucifer genomes. Trends Plant sci. 2006;11:535–42.
47. Kayum MA, Jung HJ, Park JI, Ahmed NU, Saha G, Yang TJ, Nou IS. Identification and expression analysis of *WRKY* family genes under biotic and abotic stresses in *Brassica rapa*. Mol Genet Genomics. 2014;290(1):79–95. https://doi.org/10.1007/s00438-014-0898-1.
48. Krzywinski M, Schein J, Birol İ, Connors J, Gascoyne R, Horsman D, Jones SJ, Marra MA. Circos: an information aesthetic for comparative genomics. Genome Res. 2009;9(9):1639–45.
49. Ahmed NU, Park JI, Jung HJ, Seo M, Kumar TS, Lee I, Nou IS. Identification and characterization of stress resistance related genes of *Brassica rapa*. Biotechnol Lett. 2012;34:979–87.
50. Ahmed NU, Park JI, Jung HJ, Kang KK, Lim YP, Hur Y, Nou IS. Molecular characterization of thaumatin family genes related to stresses in *Brassica rapa*. Sci Hortic. 2013;152:26–34.
51. Guo P, Baum M, Grando S, Ceccarelli S, Bai G, Li R, Korff MV, Varshney RK, Graner A, Valkoun J. Differentially expressed genes between drought-tolerant and drought sensitive barley genotypes in response to drought stress during the reproductive stage. J Exp Bot. 2009;60:3531–44.

5

Transcriptome sequencing and analysis of zinc-uptake-related genes in *Trichophyton mentagrophytes*

Xinke Zhang[1†], Pengxiu Dai[1†], Yongping Gao[1], Xiaowen Gong[1], Hao Cui[1], Yipeng Jin[2] and Yihua Zhang[1*]

Abstract

Background: *Trichophyton mentagrophytes* is an important zoonotic dermatophytic (ringworm) pathogen; causing severe skin infection in humans and other animals worldwide. Fortunately, commonly used fungal skin disease prevention and treatment measures are relatively simple. However, *T. mentagrophytes* is primarily studied at the epidemiology and drug efficacy research levels, yet current study has been unable to meet the needs of clinical medicine.

Zinc is a crucial trace element for the growth and reproduction of fungi and other microorganisms. The metal ions coordinate within a variety of proteins to form zinc finger proteins, which perform many vital biological functions. Zinc transport regulatory networks have not been resolved in *T. mentagrophytes*. The *T. mentagrophytes* transcriptome will allow us to discover new genes, particularly those genes involved in zinc uptake.

Result: We found *T. mentagrophytes* growth to be restricted by zinc deficiency; natural *T. mentagrophytes* growth requires zinc ions. *T. Mentagrophytes* must acquire zinc ions for growth and development.

The transcriptome of *T. mentagrophytes* was sequenced by using Illumina HiSeq™ 2000 technology and the de novo assembly of the transcriptome was performed by using the Trinity method, and functional annotation was analyzed. We got 10,751 unigenes. The growth of *T. mentagrophytes* is severely inhibited and there were many genes showing significant up regulation and down regulation respectively in *T. mentagrophytes* when zinc deficiency. Zinc deficiency can affect the expression of multiple genes of *T. mentagrophytes*. The effect of the zinc deficiency could be recovered in the normal medium. And we finally found the zinc-responsive activating factor (*ZafA*) and speculated that 4 unigenes are zinc transporters. We knocked *ZafA* gene by ATMT transformation in *T. mentagrophytes*, the result showed that *ZafA* gene is very important for the growth and the generation of conidia in *T. mentagrophytes*. The expression of 4 zinc transporter genes is potentially regulated by the zinc-responsive activating factor. The data of this study is also sufficient to be used as a support to study *T. mentagrophytes*.

Conclusion: We reported the first large transcriptome study carried out in *T. mentagrophytes* where we have compared physiological and transcriptional responses to zinc deficiency, and analyzed the expression of genes involved in zinc uptake. The study also produced high-resolution digital profiles of global genes expression relating to *T. mentagrophytes* growth.

Keywords: *Trichophyton mentagrophytes*, Transcriptome sequencing, Functional annotation, Zinc uptake, Zinc-responsive activating factor

* Correspondence: zyh19620207@163.com
[†]Equal contributors
[1]The College of Veterinary Medicine of the Northwest Agriculture and Forestry University, No. 3 Taicheng Road, Yangling, Shaanxi, People's Republic of China
Full list of author information is available at the end of the article

Background

Trichophyton mentagrophytes is an important zoonotic dermatophytic (ringworm) pathogen that can cause severe skin infections in humans and other animals, seriously threatening to human health and animal husbandry [1, 2].

Particular nutrients play an extremely important role during *T. mentagrophytes* invasion; these elements include zinc, iron, nitrogen, and selenium [3]. Among these, zinc is a crucial trace element, which often coordinates within a variety of proteins to form zinc finger proteins, which perform many vital biological functions. Although zinc is essential for fungi, it can also be toxic. When the intracellular zinc level rises to some critical level, the zinc ions can affect other important physiological processes [4]. Therefore, fungi have successfully evolved zinc transporter systems to maintain a homeostatic balance of zinc ions for survival and virulence.

Zinc transporter system expression in the model fungus *Saccharomyces cerevisiae* is primarily regulated by the C_2H_2-type zinc finger transcription factor *Zap1* at the transcriptional level [5, 6]. Studies have shown that various fungi can secrete functionally similar zinc finger transcriptional factors. For example in the fungi, *Aspergillus fumigatus, Candida albicans*, and *Cryptococcus gattii*, mutations in similar zinc transport mechanism genes stop growth and development, and can even cause loss of virulence [7–11]. *T. mentagrophytes* expresses various zinc finger proteins; of prime importance for growth and virulence is the exocrine zinc finger protein. Perhaps most representative are the zinc metalloproteinases, which can digest and absorb nutrients, and invade the body cuticle [12]. Previous research has shown that metalloproteinase gene mutations can affect *T. mentagrophytes* virulence at different levels [13]. At the same time, only if zinc metalloproteinase combines zinc element will it be able to exhibit biological activity. We speculated that *T. mentagrophytes* has a C_2H_2-type zinc finger transcription factor that can serve as an upstream regulator in the absorption of zinc.

The *T. mentagrophytes* zinc transport regulation network has not been determined. Our study sequenced the *T. mentagrophytes* transcriptome using Illumina HiSeq™ 2000 technology, a de novo assembly of the transcriptome was performed using the Trinity method, and we performed functional annotation analysis. This allowed us to produce high-resolution digital profiles of global gene expression relating to *T. mentagrophytes* growth. The *T. mentagrophytes* transcriptome will be characterized further, and zinc-uptake-related gene families will be systematically explored as discovered, in future work.

Methods

Fungal culture and RNA extraction

The *T. mentagrophytes* wild-type strain ATCC 28185 (a gift from Ruoyu Li, Peking University First Hospital, China) was maintained at 28 °C on solid Sabouraud dextrose medium (SDA) for 14 days. Five mL of sterile saline was used to wash off spores so as to collect fungus liquid. The fungus liquid concentration was adjusted to 10^8 CFU/mL by cell count plate. SDA with 1 mM EDTA was supplemented to generate zinc deficient SDA, named SDA-Zn (zinc ions have been chelated). The 150-μL fungus liquid was inoculated to SDA (sufficient zinc ions, grouped into Norm) and SDA-Zn with 200, 400, 600, and 1000 μM of zinc sulfate (grouped into Zn200, Zn400, Zn600, and Zn1000) respectively. Culture conditions were 28 °C for 14 days.

Total RNA was extracted using TRIzol® reagent (Invitrogen, USA) following the manufacturer's protocol, and DNase I (Takara, Japan) was used to remove genomic DNA. Integrity and size distributions were checked using an Agilent 2100 (Agilent, USA) with an RNA integrity number (RIN: 8.0) and GE Image Quant 350 (GE Healthcare, USA).

cDNA library construction and Illumina sequencing

The extracted RNA samples (Norm, Zn400, and Zn1000) were used for cDNA synthesis. Poly(A) mRNA was enriched by Oligo (dT) beads (Qiagen, German). Next, the enriched mRNA was fragmented and reverse transcribed into first-strand cDNAs with random hexamers. Use DNA polymerase I (Thermo Fisher Scientific, USA), RNase H, dNTP, and buffer to synthesize second-strand cDNA. Then using QiaQuick PCR extraction kit (Qiagen, German) to purify the cDNA fragments, and the cDNA fragments were ligated to Illumina sequencing adapters. The ligation products were size selected by agarose gel electrophoresis, PCR amplified, and sequenced using Illumina HiSeq™ 2000 by Gene De novo Biotechnology Co (Guangzhou, China). Sequence data were deposited at the NCBI Short Read Archive database (https://trace.ncbi.nlm.nih.gov/Traces/sra/sra.cgi?cmd=search_obj&m=&s=&term=SRR5097230&go=Search) under the accession numbers SRR5097135, SRR5097227, SRR5097226, SRR5097228 (Norm), SRR5097229, SRR5097230, SRR5097231, SRR5097232 (Zn10 00), and SRR5895930 (Zn400).

De novo assembly and functional annotation

High quality, clean reads for the assembly library were generated by filtering according to the following rules: reads containing adapters, more than 10% of unknown nucleotides (N) and more than 50% low quality (Q-value ≤10) bases were removed. The quality-filtered reads obtained were then de novo assembled into contigs by the

Trinity Program [14]. Trinity is a modular method and software package, which combines three components: *Inchworm, Chrysalis,* and *Butterfly*. Initially, *Inchworm* assembles reads, resulting in a collection of linear contigs. Next, *Chrysalis* clusters related contigs, and then builds de Bruijn graphs for each cluster of related contigs. Finally, *Butterfly* analyzes the paths and outputs one linear sequence and ultimately generates unigenes. We used the BLASTx program (https://blast.ncbi.nlm.nih.gov/Blast.cgi) with an E-value threshold of 1×10^{-5} to obtain protein functional annotations, by aligning our unigenes to protein sequences from NCBI Nr (non-redundant protein database, https://blast.ncbi.nlm.nih.gov/Blast.cgi), Swiss-Prot (annotated protein sequence database, http://www.expasy.org/), KEGG (Kyoto encyclopedia of genes and genomes, http://www.genome.jp/kegg/), and COG (clusters of orthologous groups of protein, https://www.ncbi.nlm.nih.gov/COG/). The Blast2GO program [15] was used to obtain gene ontology (GO) annotation of our unigenes from Nr annotation, and then WEGO software [16] was used to perform GO functional classifications. KEGG is a major public pathway-related database [17] with which one is able to analyze gene products within the context of metabolic and cellular processes.

Identification of Differentially Expressed Genes (DEGs)

The RPKM (reads per kb per million reads) was used to calculate and normalize the number of unique-match reads. The formula follows: RPKM = (1000,000 × C)/(N × L/1000), with RPKM set as the expression of unigene A, C as the number of reads that are uniquely mapped to unigene A, N as the total number of reads that are uniquely mapped to all unigenes, and L as the length (base number) of unigene A. The RPKM measure can provide normalized values of gene expression to enable transcript comparisons between Norm, Zn400, and Zn1000. We used the edgeR package (https://bioconductor.org/packages/release/bioc/html/edgeR.html) to identify differentially expressed genes across samples. We specified $|\log2FC| > 1$ with the false discovery rate (FDR) < 0.05, as the thresholds necessary to determine significant differences in gene expression between Norm, Zn400, and Zn1000. Differentially expressed genes (DEGs) were then subjected to GO functional and KEGG pathway enrichment analyses. GO enrichment analysis provides all GO terms that are significantly enriched in DEGs compared with the genome background, and filter DEGs corresponding to biological function. Initially all DEGs are mapped to GO terms in the Gene Ontology database (http://geneontology.org/), gene numbers are calculated for every term, and significantly enriched GO terms in DEGs compared with the genome background are defined by a hypergeometric test. The *P*-value formula is:

$$P = 1 - \sum_{i=0}^{m-1} \frac{\binom{M}{i}\binom{N-M}{n-i}}{\binom{N}{n}}$$

Here N is the number of all genes with GO annotation; M is the number of all genes that are annotated to the certain GO terms; n is the number of DEGs in N; m is the number of DEGs in M. The calculated *p*-value then goes through FDR correction, taking FDR ≤ 0.05 as a threshold. GO terms meeting this condition are defined as significantly enriched GO terms in DEGs. Our analysis successfully recognized the putative biological functions of our DEGs.

Genes usually interact with each other to play roles in particular biological functions. Pathway-based analysis helps to further determine genes' biological functions. KEGG is the major public pathway-related database [17]. Pathway enrichment analysis identifies significantly enriched metabolic pathways or signal transduction pathways in DEGs. The significance formula is the same as that in GO analysis. Here N is the number of all genes with KEGG annotation, M is the number of all genes annotated to specific pathways, n is the number of DEGs in N, and m is the number of DEGs in M. The calculated p-value then goes through FDR Correction, taking FDR ≤ 0.05 as a threshold. Pathways meeting this condition are defined as significantly enriched pathways in those DEGs.

Validation of differential expression using qRT-PCR

Total RNA was extracted as described above, and cDNA was generated from total RNA. Primers for quantitative real time PCR (qRT-PCR) were designed using Primer Premier 6.0 software (Premier, Canada), and synthesized by Gene De novo Biotechnology Co (Guangzhou, China). All primers are shown in Additional file 1. The 18S gene was used as an internal control. qRT-PCR was performed on a Step One Plus™ Real-Time PCR System (Thermo Fisher Scientific, USA). Each 20-μL reaction mixture contained 10 μL of Maxima SYBR Green/ROX qPCR Master Mix (2X) (Thermo Scientific, USA), 0.3 μL of each primer (10 μM), 0.8 μL of cDNA, and 8.6 μL of nuclease-free water. The qRT-PCR run protocol was as follows: 95 °C, 10 min; followed by 40 cycles of 95 °C, 15 s; 60 °C, 30 s; and 72 °C, 15 s in 96-well optical reaction plates. Three biological replicates with three technical replicates for each value determined the Ct values. Expression levels of the tested reference genes were determined by Ct values and calculated by2^-△△Ct.

Construction of transformation vectors and ATMT transformation

The binary vector pDHt/*ZafA::hph* used for site-directed mutagenesis was constructed by reorganizing the hygromycin B resistance gene (*hph*) of plasmid pAN7–1, the left and right flanking sequences of the *ZafA* gene of *T.mentagrophytes* simultaneously into *XhoI/HindIII* digested plasmid *pDHt/SK* (a gift from Dr. K. J. Kwon-Chung). The *ZafA* gene was knocked by Agrobactirium tumfacience mediated-transformant (ATMT) in *T. mentagrophytes*. The two pairs of primers (*ZafA*-F: CCAGA CTGAAGGTGCTAAG, *ZafA*-R: CCTGTTAGTATCG TCGTGTT; *hph*-F: TACATCCATACTCCATCCTTC, *hph-R*: CGGCATCTACTCTATTCCTT) designed by *ZafA* gene fragment disrupted and *hph* gene fragment were used to verified *ZafA* gene mutant strain, and its amplification length is 400 and 1200 bp, respectively.

Results

Effect of zinc deficiency on the growth of *T. mentagrophytes*

The *T. mentagrophytes* cells of five groups were maintained at 28 °C for 14 days. We found *T. mentagrophytes* can grow well on SDA in Norm, presenting a white colony with fluffy, fine mycelium on its surface. The colony morphology of the Zn1000 group was not significantly different from that of the Norm group, but the growth rate was slower. In contrast, with the decreasing zinc ion concentration in its medium, the growth of Zn400 and Zn600 *T. mentagrophytes* was severely inhibited, with pale-yellow mucus-covered colonies in which mycelium could not be seen. The growth of Zn200 *T. mentagrophytes* was most seriously inhibited, with the colony appearing to be folded over (Fig. 1a).

Upon microscopy Norm and Zn1000 mycelium can be seen to grow well, with numerous round microconidia in grape-like clusters. A small number of microconidia and mycelium can be observed in Zn600, and even fewer in Zn400. However, we could not detect microconidia in Zn200 at all, and its mycelium were particularly weak (Fig. 1b).

T. mentagrophytes from Zn200, Zn400 and Zn600 were then inoculated into normal medium and Zn1000 medium, and the *T. mentagrophytes* growth traits returned to normal (Fig. 1c).

These results collectively indicate that when *T. mentagrophytes* cells grow in zinc-deficient conditions growth

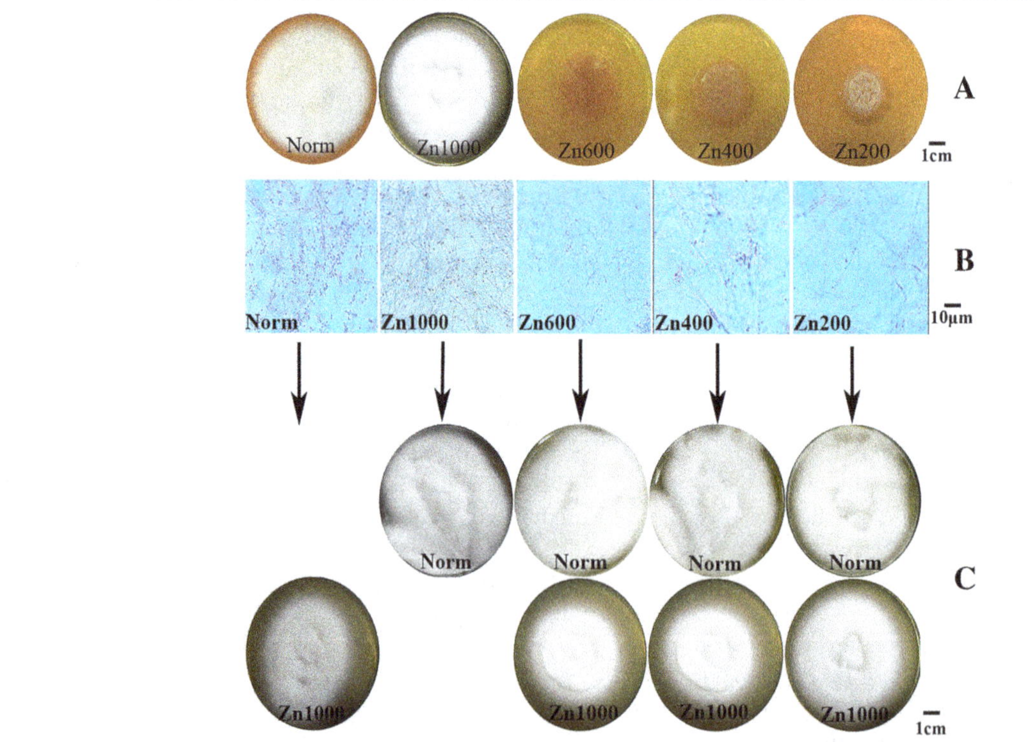

Fig. 1 a The growth situation of *T. mentagrophytes* in 5 groups. SDA with 1 mM EDTA was supplemented to generate zinc deficient SDA, named SDA-Zn (zinc ions have been chelated)**.** The 150-μL fungus liquid was inoculated to SDA (sufficient zinc ions, grouped into Norm) and SDA-Zn with 200, 400, 600, and 1000 μM of zinc sulfate (grouped into Zn200, Zn400, Zn600, and Zn1000) respectively. *T. mentagrophytes* can grow well on SDA in Norm and Zn1000, the growth of Zn200、Zn400 and Zn600 *T. mentagrophyte* was inhibited, especially in Zn200. **b** The each group was stained by lacto phenol cotton blue. Norm and Zn1000 mycelium grow well and with numerous round microconidia. A small number of microconidia and mycelium can be observed in Zn600, and even fewer in Zn400. We could not detect favorable microconidia and the mycelium was particularly weak in Zn200. **c** *T. mentagrophytes* from Zn200, Zn400 and Zn600 were inoculated into normal medium and Zn1000 medium, the *T. mentagrophytes* growth traits returned to normal, *T. mentagrophytes* can grow well on SDA in Norm and Zn1000

status is adversely affected, showing that zinc is very important for the growth of *T. mentagrophytes*.

De novo assembly and sequence annotation

A total of 36,793,459 raw reads and 36,227,708 quality filtered (clean) reads were obtained from the Norm library. We obtained 30,028,768 raw reads and 29,576,580 quality filtered (clean) reads from Zn1000. And in Zn400, we obtained 47,306,862 raw reads and 46,394,758 quality filtered (clean) reads. The saturation curves shown depict the detected number of genes that tend to be saturated (Additional file 2). The Q20 percentages (percentage of sequences with sequencing error rates) of the three libraries, Norm, Zn1000, and Zn400, were 97.69, 97.72, and 94.58% respectively, and the GC content ranged from 51.52 to 52.83%. All clean reads were pooled together and then de novo assembled by Trinity. The assembly produced a substantial number of contigs and 10,751 unigenes.

These unigenes were annotated using the Nr, Swiss-Prot, KEGG, and KOG databases. A final number of 9593, 6113, 3765, and 5172 unigenes had matches in the Nr, Swiss-Prot, KEGG, and KOG databases, respectively (Additional file 3). Up to 89.23% of all machine annotated unigenes showed similarity to known proteins in the Nr database. Additionally, the unigenes were searched against the Nr database using BLASTx, and homologous sequences and species identification were ascertained. The five highest number of homologous sequences corresponding to particular species follows: 24.86% *Trichophyton equinum* CBS 127.97, 23.98% *Trichophyton tonsurans* CBS 112818, 10.40% *Trichophyton interdigitale* H6, 6.60% *Trichophyton rubrum* CBS 118892, and 4.35% *Microsporum gypseum* CBS 118893.

As shown in Fig. 2a, 5172 unigenes (48.11% of total) were classified into 25 functional KOG classifications, based on sequence similarity. The predominant term was "general function prediction only," for which 1648 unigenes (31.86%) were qualified. "Posttranslational modification, protein turnover, chaperone" (1193 unigenes), "signal transduction mechanism" (1048 unigenes), and "RNA processing and modification" (721 unigenes) were other major categories selected, and only eight, 22, 53, and 54 unigenes matched the terms "cell motility,"

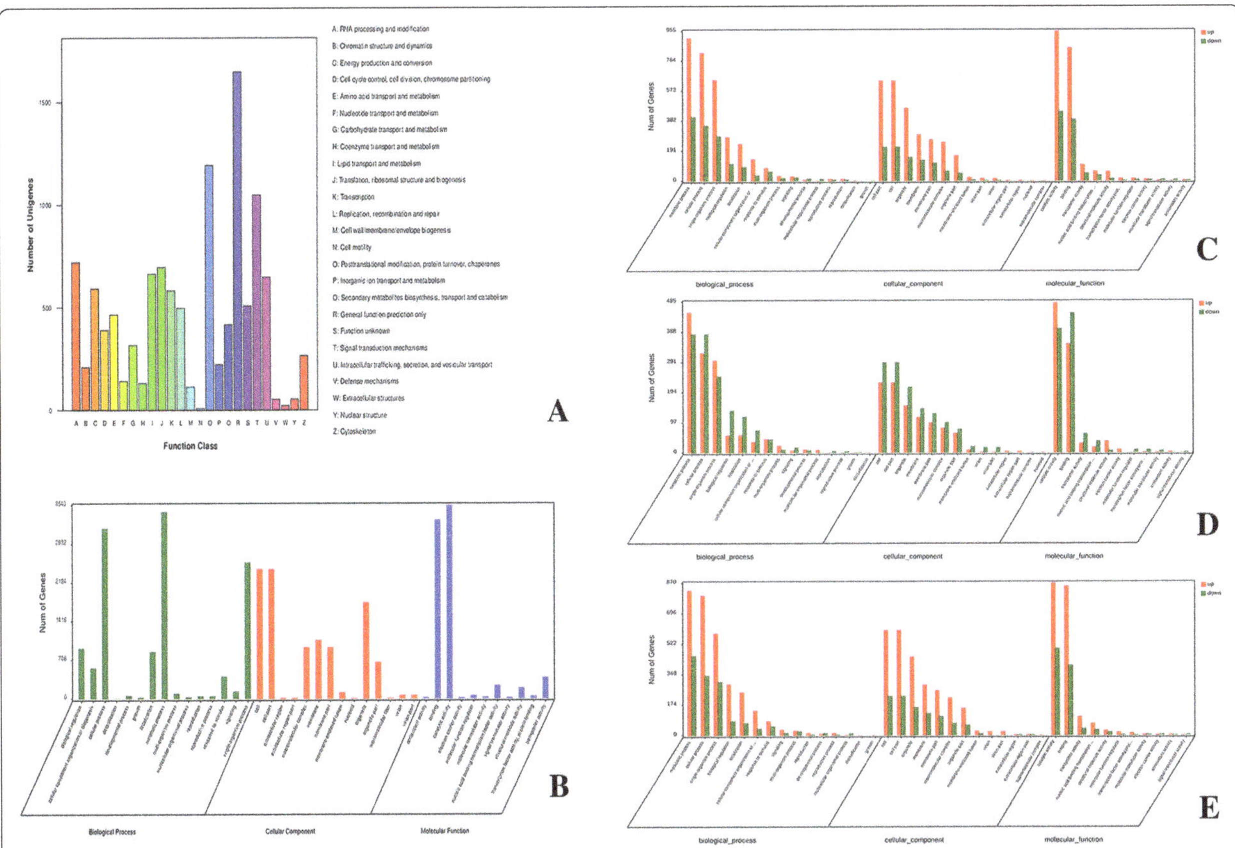

Fig. 2 a The KOG classification of unigenes, 5172 unigenes (48.11% of total) were classified into 25 functional KOG classifications. **b** The GO functional classification of unigenes. A total of 6053 machine annotated unigenes were grouped into 40 functional group categories using GO assignment. **c**, **d** and **e** The GO functional classification of DEGs. These DEGs were classified into three main categories including cellular component, biological process and molecular function

"extracellular structure," "defense mechanism," and "nuclear structure," respectively.

A total of 6053 machine annotated unigenes were grouped into 40 functional group categories using GO assignment (Fig. 2b). Among these categories, 15 are involved in "biological process," 11 in "molecular function," and 14 in "cellular component." "Metabolic process" (3422 unigenes) and "cellular process" (3121 unigenes) are dominant among these; "detoxification (four unigenes) and growth" (21 unigenes) are the scarcest in the "biological process" category. Within the "molecular function" category, a high percentage of genes are associated with "catalytic activity" (3538 unigenes) and "binding" (3272 unigenes). Minimal "molecular function" GO assignments included "signal transducer activity" (21 unigenes) and "electron carrier activity" (22 unigenes). Within the "cellular components category," "cell" (2374 unigenes) and "cell part" (2374 unigenes) are predominant; "nucleoid" (six unigenes) and "supramolecular fiber" (six unigenes) had the fewest matches.

A total of 2292 unigenes were annotated to 114 pathways in this study; the pathways of all our unigenes are shown in Additional file 4. "Metabolic pathway" represented the largest group (1916 unigenes), with most being involved in the "biosynthesis of amino acids" (138 unigenes), "carbon metabolism" (118 unigenes), "purine metabolism" (117 unigenes), and "oxidative phosphorylation" (113 unigenes). Secondary pathways included "genetic information processing" (1103 unigenes), which included "ribosome" (151 unigenes), "RNA transport" (99 unigenes), "spliceosome" (94 unigenes), "protein processing in endoplasmic reticulum" (93 unigenes), and "endocytosis" (92 unigenes). These probable pathways provide a valuable resource for investigating specific metabolic processes and gene functions in *T. mentagrophytes.*

Overview of differentially expressed genes

FDR and log2FC were both used to filter our DEGs, with the filter conditions of FDR < 0.05 and |log2FC| > 1. We compared Norm (as a control) to Zn400 and Zn1000. Results showed 2314 and 2127 genes are significantly up- and down-regulated, respectively, in Zn400. In Zn1000, 1395 and 1446 genes displayed significant up-regulation and down-regulation, respectively. A comparison was also performed using Zn1000 as the control, against Zn400. This result showed significant up- and down-regulation: 2268 and 2040 genes were regulated in Zn400 as compared with Zn1000, respectively. All DEGs are shown in Additional file 5, and a group diffuse analysis 'volcano plot' is shown in Additional file 6.

The Zn400 DEGs were subjected to GO-term analysis (Fig. 2c); these DEGs partitioned into three major categories: "cellular component" (1165), "biological process" (1880), and "molecular function" (2024). These major categories sorted into several subcategories (based on Pvalue <0.05 and Qvalue <0.05): "electron carrier activity" within the "molecular function" category, and "generation of precursor metabolites and energy," "oxidation-reduction process," "electron transport chain," "purine nucleoside biosynthetic process," "purine ribonucleoside biosynthetic process," "multi-organism process," "monovalent inorganic cation transport," and "proton transport" within the "biological process" category.

The Zn1000 DEGs were also subjected to GO-term analysis (Fig. 2d), These DEGs mainly tagged "preribosome," "oxidation-reduction process," "oxidoreductase activity," and "acting on other nitrogenous compounds as donor" subcategories (based on Pvalue <0.05 and Qvalue <0.05). DEGs in the comparison of Zn1000 with Zn400 mainly tagged "ion binding" and "cation binding" subcategories (based on Pvalue <0.05 and Qvalue <0.05) (Fig. 2e).

All Zn400 DEGs were subjected to pathway enrichment analysis. Up to 19.34% of the DEGs could be annotated, and 112 pathways were obtained (Additional file 5). Many pathways were significantly enriched (Pvalue <0.05, Qvalue <0.05) including oxidative phosphorylation; valine, leucine, and isoleucine biosynthesis; biosynthesis of amino acids; carbon metabolism; pantothenate and CoA biosynthesis; and the citrate cycle (TCA cycle). In Zn1000, 19.39% of the DEGs could be annotated, and 109 pathways were obtained. Oxidative phosphorylation is significantly enriched (Pvalue <0.05, Qvalue <0.05) (Additional file 7).

Identifying *T. mentagrophytes* zinc-uptake-related genes

Zinc uptake system expression in the model fungus *S. cerevisiae* is primarily regulated at the transcriptional level by the C_2H_2-type zinc finger transcription factor *Zap1* [5]. Subsequent studies have shown other fungi, for example *A. fumigatus*, can secrete functionally similar zinc finger transcriptional factor *ZafA* proteins [18]. A BLASTx [19] sequencing similarity search was performed using the *A. fumigatus ZafA* and *S. cerevisiae Zap1* protein sequences against our unigenes. We found a total of 100 unigenes similar to *A. fumigatus ZafA*, and 53 unigenes to *S. cerevisiae Zap 1*, respectively. A total of 36 of these unigenes have similarities with both *ZafA* and *Zap1*, but only 28 have a zinc finger structure (Additional file 8), according to functional annotation matches. Of these 28 unigenes with annotated zinc fingers, the sequence with the highest similarity to both *ZafA* and *Zap1* was Unigene0008014, which was a DEG in the transcriptome sequencing comparison of Norm versus Zn400, but was not a DEG in Norm versus Zn1000. Regardless, the predicted protein encoded by this gene is the most similar to the ZafA protein of *A. fumigatus* [18], unambiguously fits the zinc finger consensus [20], and has putative zinc-binding domains [21] (Fig. 3). Furthermore, Unigene0008014 qualified for "zinc-responsiveness transcriptional activator" (*Trichophyton equinum* CBS 127.97) and "zinc-responsive transcriptional regulator Zap1" (*S. cerevisiae* strain ATCC 204508/S288c) Nr and Swiss-Prot annotations in our analyses, respectively. We hypothesize that Unigene0008014 plays an important role in regulating zinc ion uptake in *T. mentagrophytes,* and be named a zinc-responsive activating factor. The zinc-responsive activating factor genomic sequence can be amplified using three different pairs of primers in *T. mentagrophytes* (primers shown in Additional file 9). The nucleotide sequence of our putative zinc-responsive activating factor has been submitted to NCBI GenBank under the accession KY420911.

Four of our unigenes, Unigene0002709, Unigene0002593, Unigene0004712, and Unigene0005637, are likely zinc transporters. Nr annotations for these transcripts, "zinc/iron transporter" (*Trichophyton tonsurans CBS* 112818), "plasma membrane zinc ion transporter" (*Trichophyton equinum CBS* 127.97), "membrane zinc transporter" (*Trichophyton tonsurans CBS* 112818), and "ZIP family zinc transporter" (*Trichophyton tonsurans CBS* 112818), respectively, support this hypothesis. Moreover, Swiss-Prot annotations, "zinc-regulated transporter" (*S. cerevisiae* strain ATCC 204508/S288c), "RNA polymerase II transcription factor B subunit" (*Candida glabrata* strain ATCC 2001/CBS 138/JCM 3761/NBRC 0622/NRRL Y-65), "zinc-regulated transporter" (*S. cerevisiae* strain ATCC 204508/S288c), "zinc-regulated transporter" (*S. cerevisiae* strain ATCC 204508/S288c), respectively, corroborate the assertion. Additionally, these four unigenes have high BLASTx [19] similarity with *ZrfA*, *ZrfB*, *ZrfC*, and *Aspf2* of *A. fumigatus* based on E-value.

MADQFDERSFCWECYWPNLDADNNTVLSSGCVDASCWDQADFSH
QSFDLCGTGTECCDNDQCSPGCATICDGFVDCDKSTACSEPQCNDM
NCNNQATACFDMNCVTEGEQSLRSLDQGHFFNQSIPLNWGCLDLQ
NTAAAAATSAATTSAVNDVSSLQHCNLDNDKLRSHFHFHHVGSINS
LDINSASKGLDNGIIDPKHSRPVISTPCSNHAGNVWPMASAAGECN
NNLNNHNNTHHPYLNHLHQRGNLDKAFDIAQWSAQMQRHRHAPS
FHSQTSSAGFIHDVMTNNSTVSTRSTPGLSLESSPMTALTSREVSIQS
MDYSGDELHICKWVANGDSGLACGITFPDAKSLQEHLVQEHANQT
EGAKGTGYYCCWEGCHRPHEPFSQKSKLQGHYLTHSNYKNFQCSI
CGKSFARQATLERHERSHRGDKPYKCKICGKAFTDSSELKTHRRIH
TGEKPFKCKHPGCTFETGDSSNMSSHKLTHGERRHKCPFPGCSKSF
TRPDQLKRHAKTAHKDDLFSLQRMASPVTPSLPVI

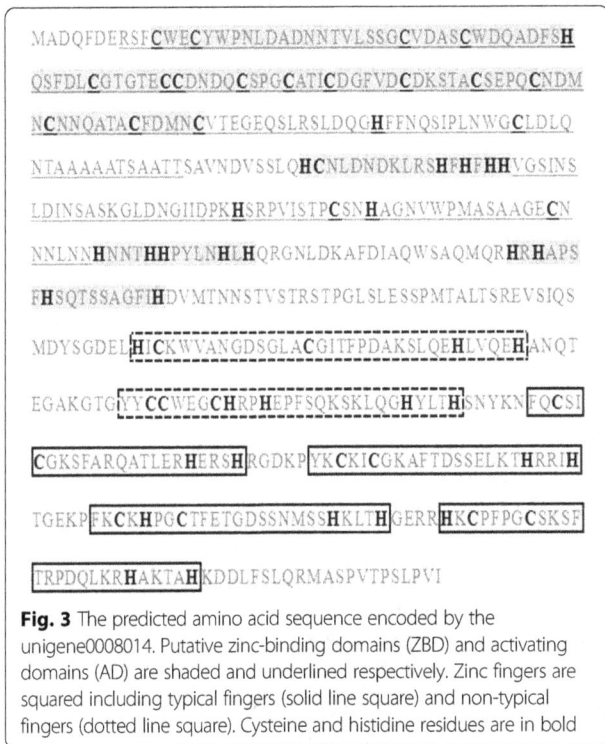

Fig. 3 The predicted amino acid sequence encoded by the unigene0008014. Putative zinc-binding domains (ZBD) and activating domains (AD) are shaded and underlined respectively. Zinc fingers are squared including typical fingers (solid line square) and non-typical fingers (dotted line square). Cysteine and histidine residues are in bold

To validate changes in gene expression patterns we used qRT-PCR against six unigenes: Unigene0008014, Unigene0002709, Unigene0002593, Unigene0002886, Unigene0005062, and Unigene0005193. Unigene0002709, Unigene0008014, Unigene0002886, and Unigene0005062 exhibited differential expression levels, identical to those obtained by sequencing in Zn400. Unigene0002593 and Unigene0005193 exhibited differential expression levels, identical to those obtained by sequencing in Zn1000 (Fig. 4). A statistical analysis was performed on the qRT-PCR differential expression analyses and is shown in Additional file 10.

Data from this study sufficiently support our investigation into genes related to zinc ion uptake regulation in *T. mentagrophytes.*

The change of phenotype in *ZafA* gene mutant strain

To observe the changes of phenotype and the growth ability after *ZafA* gene deleted in *T. mentagrophytes.* The *ZafA* gene was knocked by ATMT transformation in *T. mentagrophytes.* The fragment of *hph* gene could be amplified, and the fragment of *ZafA* gene could not be amplified in *ZafA* gene mutant strain (Fig. 5), this means that the *ZafA* gene is completely removed. The *T. mentagrophytes* wild-type strain and *ZafA* gene mutant strain were maintained at 28 °C on SDA-Zn medium with 800, 1000, 1200, 1400 and 1600 μM zinc sulfate for 16 days. The changes of phenotypic and growth ability are shown in Fig. 6. The wild-type strain can begin growing normally in third day, and with increasing of zinc ion concentration, the growth rate is accelerated. But the *ZafA* gene mutant strain can begin growing in eighth day, and the growth rate and state are much lower than the wild-type strain in same situation. Under the microscope, there was no significant difference in the quality and quantity of mycelium between wild-type strain and *ZafA* gene mutant strain, but the number of conidia of *ZafA* gene mutant strain was obviously less than wild-type strain in same culture situation. The result showed that the deletion of *Zafa* gene can negatively affect the growth and the number of conidia of *T. mentagrophytes.*

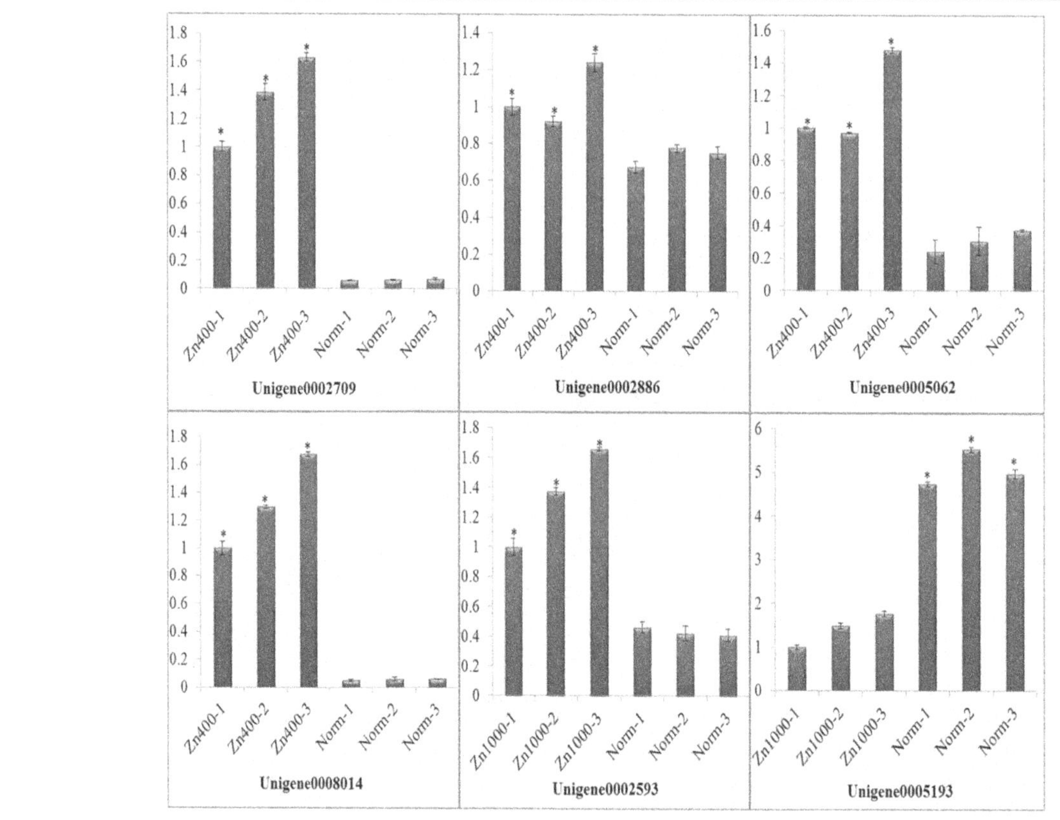

Fig. 4 The result of qRT-PCR in 6 unigenes. Unigene0002709, Unigene0008014, Unigene0002886, and Unigene0005062 exhibited differential expression levels, in Zn400. Unigene0002593 and Unigene0005193 exhibited differential expression levels in Zn1000

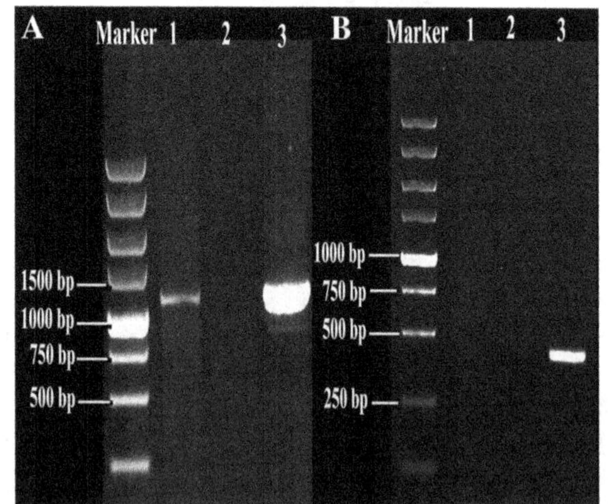

Fig. 5 PCR analysis of transformants. **a** Amplification of *hph* (1200 bp) using the primers *hph*-F and *hph-R*. *Lane* 1, DNA sample from the *T. mentagrophytes ZafA* gene mutant strain; lane 2, DNA sample from the wild-type *T. mentagrophytes* strain 28,185, lane 3, the transformation vector *pDHt/ZafA:: hph*. **b** Amplification of the *ZafA* gene fragment disrupted (400 bp) using the primers *ZafA*-F and *ZafA-R*. *Lane* 1, DNA sample from the *T. mentagrophytes ZafA* gene mutant strain; lane 2, the transformation vector *pDHt/ZafA:: hph*; lane 3, DNA sample from the wild-type *T. mentagrophytes* strain 28,185

Discussion

Illumina sequencing and sequence annotation in *T. mentagrophytes*

The infection of skin, nail, hair, and fur caused by dermatophytes is called dermatophytosis. As early as 1839, scientists confirmed that dermatophytes can cause human disease. Furthermore, at least 10–20% of the world's population may be infected with dermatophytes [22]. Dermatophytes comprise three genera: *Microsporum*, *Trichophyton*, and *Epidermophyton*. An important member of *Trichophyton*, *T. mentagrophytes* can cause severe skin infections in humans and other animals, and has a wide distribution around the world [23]. Therefore, *T. mentagrophytes* warrants investigation. Our study aimed to generate a large amount of cDNA sequence data to facilitate more detailed transcriptomics studies in *T. mentagrophytes*, and, in particular, to identify genes related to the regulation of zinc ion uptake in that organism. RNA-seq is a powerful tool that can provide a global overview of genes expression at the transcriptome level [24]; however, it has not been extensively applied to fungi. We believe RNA-seq will prove to be a powerful method for fungi study. The availability of our *T. mentagrophytes* transcriptome data will meet the initial

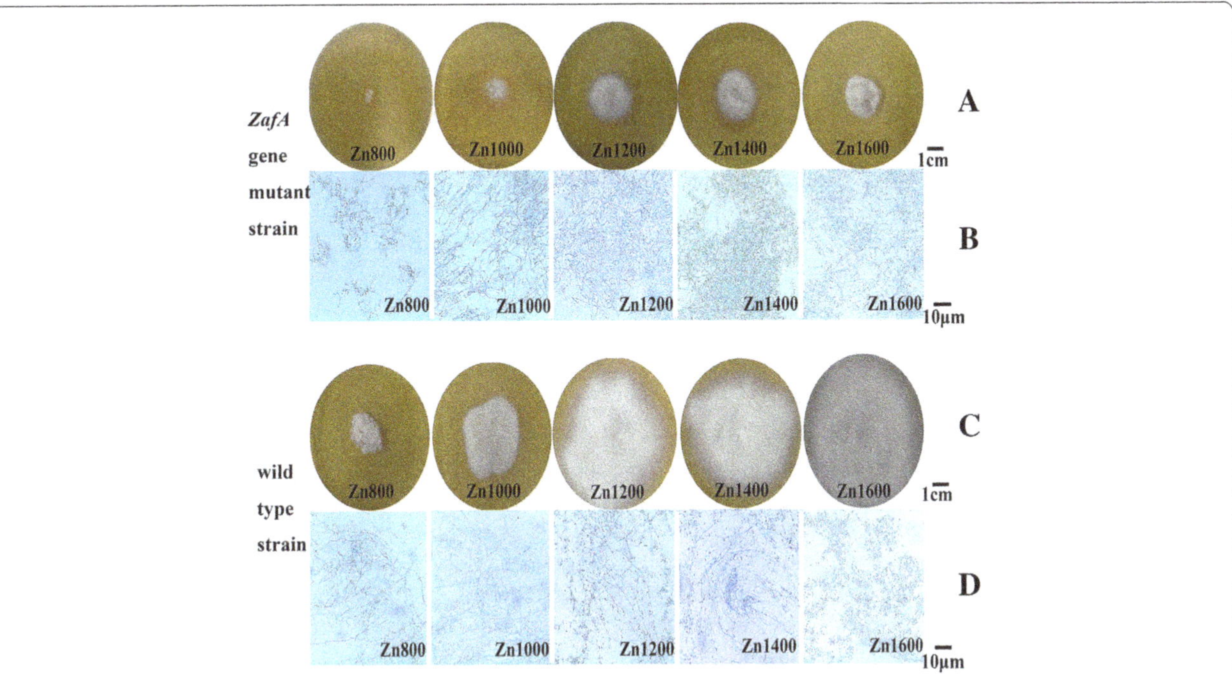

Fig. 6 The growth situation of *T. mentagrophytes* wild-type strain and *ZafA* gene mutant strain on SDA-Zn medium with 800, 1000, 1200, 1400 and 1600 μM zinc sulfate. **a** The growth situation of *ZafA* gene mutant strain, the growth of *T. mentagrophyte* is inhibited, especially in Zn800 and Zn1000. **b** Under the microscope, there is good quantity of mycelium in *ZafA* gene mutant strain, but the number of conidia is reduced. **c** The wild-type strain grows normally, and with increasing of zinc ion concentration, the growth rate is accelerated. **d** Under the microscope, there are good quantity of mycelium and a mass of conidia in wild-type strain

information needs for functional studies of this species and its relatives.

We chose to sequence the transcriptomes of Norm, Zn400, and Zn1000, based on our growth experiment results. Zn1000 and Norm samples grew similarly in our trials; there was a significant difference between Zn400 and Norm, microconidia and mycelium morphology are also significantly different between the two. The RNA-seq method was then performed on these samples using Illumina sequencing, which generated a total of 10,751 unigenes, of which more than 89% were annotated by our analyses. These data will provide a valuable resource for the study of *T. mentagrophytes*.

Zn-deficiency induced changes in *T. mentagrophytes* growth and gene expression

Fungi rely on zinc for growth; the zinc ions serve as a cofactor in numerous proteins, including important transcription factors [25]. Zinc chelation can reduce fungal growth in both rich and defined media [26]. Zinc chelation occurs during infection, and is an important strategy evolved in immune cells to hamper pathogen growth [27]. Zn1000 cells grew with no obvious morphological differences compared with Norm in our study. However, with sequential zinc ion concentration decreases in Zn200, Zn400, and Zn600, microconidia and mycelium morphology showed *T. mentagrophytes* growth to be increasingly restricted by zinc deficiency. Furthermore, when *T. mentagrophytes* growth was inhibited in Zn200, Zn400, and Zn600 zinc deficient samples, which were then inoculated into normal medium or Zn1000 zinc deficiency medium, *T. mentagrophytes* cells restore to normal growth patterns due to the increase in zinc concentration. This suggests that the growth inhibition we observed in *T. mentagrophytes* is primarily caused by a lack of zinc ions, rather than other metal ions, and it reinforces the hypothesis that *T. mentagrophytes* natural growth requires zinc ions. Zinc ion acquisition is, therefore, crucial for the growth and development of *T. Mentagrophytes*.

The *S. cerevisiae* yeast cell employs several different strategies to cope with stress caused by zinc deprivation [28]. Zinc deprivation induced by TPEN also induces a variety of changes in the gene expression in *C. gattii* cells [9]. Studies have shown that the growth and gene expression, in particular, high-levels of zinc transporter system expression, within a variety of fungi, including *A. fumigatus* [18], *C. albicans* [10], and *C. gattii* [9], are affected by zinc deficiency. Similarly, our RNA-seq data differential expression analyses revealed 4441 DEGs (2314 up-regulated and 2127 down-regulated) in Zn400, versus 2841 DEGs (1395 up-regulated and 1446 down-regulated) in Zn1000. Because of the higher zinc ion concentration, Zn1000 has fewer DEGs than Zn400.

This further indicates that the change of gene expression in *T. mentagrophytes* is predominantly caused by a lack of zinc ions, versus other metal ions. Zinc deficiency definitely affects gene expression in *T. mentagrophytes.* We speculate that *T. mentagrophytes* regulates the expression of many genes under conditions of zinc deficiency.

A genome-wide, functional analysis revealed that almost 400 different gene products are necessary for proper growth in zinc-limiting conditions, using a *S. cerevisiae* mutant library [29]. Of these proteins, most are associated with oxidative stress response, endoplasmic reticulum function, peroxisome biogenesis, or zinc uptake. Furthermore, as revealed by transcriptomic and functional analyzes, also in *S. cerevisiae,* low zinc conditions lead to alterations in lipid synthesis, sulfate metabolism, and oxidative stress tolerance [30, 31]. Our KEGG and GO DEG analyses showed that *T. mentagrophytes* has the same response as *S. cerevisiae* under zinc deficient environments. Most of our DEGs partitioned into either oxidation-reduction process, electron carrier activity, purine ribonucleoside biosynthetic process, monovalent inorganic cation transport, or proton transport, in the GO analysis. Furthermore, oxidative phosphorylation; valine, leucine and isoleucine biosynthesis; biosynthesis of amino acids; carbon metabolism; and the citrate cycle (TCA cycle) were significantly enriched in our pathway enrichment analysis. These results suggest that zinc deprivation can affect *T. mentagrophytes* development, growth, and gene expression. We speculate *T. mentagrophytes* can change particular metabolic pathways to resist zinc deficiency. However, this stressed cellular state cannot last forever, thus *T. Mentagrophytes* growth and development are eventually negatively impacted, to the point of cellular death under extended zinc deficiency.

Zinc-uptake-related genes in *T. mentagrophytes*

Fungi cells must acquire zinc ions for proper life cycle development, even as saprophytes, or during infective processes [32]. Zinc transport mechanisms were initially characterized in fungi with *S. cerevisiae Zap1, Zrt1,* and *Zrt2* genes [33]. *Zap1* can activate the transcription of *Zrt1* and *Zrt2* by binding to the zinc-response element in its promoter region; binding affinity is controlled by zinc level [5]. Zinc uptake and homeostasis is also important in the physiology and virulence of *A. fumigatus, C. albicans,* and *C. gattii.* The main function of the *A. fumigatus ZafA* protein is to regulate zinc uptake; it is a requisite for the growth of *A. fumigatus* in zinc-limited conditions [18]. Furthermore, *ZafA* mutants do not survive and have no pathogenicity in the mice lung, supporting the essential role of *A. fumigatus ZafA* in growth and virulence [18]. In addition, *ZafA* can induce *Zrf* and *Aspf2* gene expression under zinc deficient conditions, and its expression is also influenced by zinc concentration [8, 34]. *Csr1/Zap1* is considered a homolog of *S. cerevisiae Zap1* in *C. albicans.* Mutants lacking *Csr1/Zap1* alleles show growth deficiencies under zinc deficient conditions, and cannot form germ tubes or hyphae, demonstrating that *Csr1/Zap1* contributes to zinc uptake and homeostasis, as well as morphological transitions, in *C. albicans* [10]. Another pathogenic *Candida* species, *C. dubliniensis* also possesses the *S. cerevisiae Zap1* homolog *Csr1.* Mutants show growth defects under zinc-limited conditions. However, unlike *C. albicans,* these *Csr1* mutants are able to form germ tubes and undergo morphological transition, although the mutants do exhibit reduced virulence [35]. Similar to other fungi, the *S. cerevisiae* zinc finger transcription factor *Zap1* homolog was identified in *C. gattii.* The *Zap1* homolog mutant showed impaired growth under zinc-limited conditions compared with wild-type. Furthermore, the *Zap1* mutant displayed attenuated virulence in a murine cryptococcosis model. This suggests that *Zap1* plays critical roles in zinc uptake and virulence in *C. gattii* [9].

In the present study we found 28 unigenes with similarity to both *ZafA* of *A. fumigatus* and *Zap1* of *S. cerevisiae* according to functional annotation matches and BLASTx. We hypothesize that Unigene0008014 is functionally similar to both *ZafA* of *A. fumigatus* and *Zap1* of *S. cerevisiae,* based on sequence similarity, functional annotation, and the predicted protein product. The significant up-regulation of Unigene0008014 expression levels, in particular for Zn400, as detected by qRT-PCR, with such a dramatic difference in our transcriptome sequencing comparison for Zn400, support this hypothesis. We also knocked *ZafA* gene by ATMT transformation in *T. mentagrophytes.* By observing the changes of phenotype and the growth ability after *ZafA* gene deleted in *T. mentagrophytes.* We found that *ZafA* gene is very important for the growth and the generation of conidia in *T. mentagrophytes.* We think that Unigene0008014 can regulate zinc uptake at the transcriptional level as a C_2H_2-type zinc finger transcription factor in *T. mentagrophytes,* and name it a zinc-responsive activating factor.

The other 27 *ZafA*/*Zap1* putative homologs do not appear to be zinc-responsive activating factors, but do contain zinc finger structures, and many show significant expression changes under zinc deficiency, suggesting that these 27 unigenes have important metabolic functions. Therefore, we consider these 27 unigenes to possibly function as a regulator of zinc uptake; however, this speculation requires further research.

We also found four *T. mentagrophytes* unigenes, Unigene0002709, Unigene0002593, Unigene0004712, and Unigene0005637, that may be zinc transporters. These four sequences are similar to the zinc transporters of *A. fumigatus* and *S. cerevisiae* based on E-value, and our gene function annotation analyses show these four unigenes to be zinc transporters. Furthermore, Unigene0002709 and Unigene0002593 qRT-PCR results showed significant up-regulation. Unigene0005637 was not a DEG, and Unigene0004712 was significantly down-regulated. We think that these results may be due to the pH value of the medium partly inhibiting Unigene0005637 and Unigene0004712 expression levels. A similar effect has been observed in *A. fumigatus* [7]. Indeed, different pH values can affect the expression of various zinc transporters [8]. Thus, we think that the zinc transporter system in *T. mentagrophytes* comprises five unigenes: Unigene0008014, Unigene0002709, Unigene0002593, Unigene0004712, and Unigene0005637. However, the possibility that these five unigenes are just a part of a larger zinc transporter system exists, and requires further analysis.

Conclusion

In this study we report the first large transcriptome study carried out in *T. mentagrophytes* where we have compared physiological and transcriptional responses to zinc deficiency. A total of 10,751 unigenes were obtained and more than 89% of them were annotated. This provided more adequate resources to study *T. mentagrophytes*. Evidence from physiological observations, transcriptome and qRT-PCR analysis indicated that zinc deficiency could induce arrested development, and numerous genes expression changes in *T. mentagrophytes*. Importantly, we found the zinc-responsive activating factor (unigene0008014) and we speculated that 4 unigenes (unigene0002709, unigene000
2593, unigene0004712, unigene0005637) are zinc transporters. The expression of these 4 zinc transporter genes is potentially regulated by the zinc-responsive activating factor. And we knocked *ZafA* gene in *T. mentagrophytes*, the result showed that *ZafA* gene is very important for the growth and the generation of conidia in *T. mentagrophytes*.

Additional files

Additional file 1: The primers of qRT-PCR. (DOCX 15 kb)

Additional file 2: The saturation curves of RNA-seq. (TIFF 2098 kb)

Additional file 3: The Nr, Swiss-Prot, KEGG and KOG annotation of unigenes. (XLSX 3000 kb)

Additional file 4: The pathways of unigenes. (DOCX 29 kb)

Additional file 5: All DEGs. (XLSX 1503 kb)

Additional file 6: Group diffuse analysis 'volcano plot'. (TIFF 1807 kb)

Additional file 7: The pathways of DEGS. (XLSX 27 kb)

Additional file 8: The unigenes that have zinc finger structure. (DOCX 16 kb)

Additional file 9: Three pairs of primers that can amplify zinc-responsive. (DOCX 15 kb)

Additional file 10: The qRT-PCR differential expression analyses. (DOCX 17 kb)

Abbreviations

A. fumigatus: *Aspergillus fumigatus*; ATMT: Agrobacterium tumefaciens-mediated transformation; *C. albicans*: *Candida albicans*; *C. gattii*: *Cryptococcus gattii*; COG: Clusters of orthologous groups of protein; DEGs: Differentially Expressed Genes; EDTA: Ethylene Diamine Tetraacetic Acid; GO: Gene ontology; *hph*: Hygromycin B resistance gene; KEGG: Kyoto encyclopedia of genes and genomes; Nr: Non-redundant protein database; qRT-PCR: Quantitative real time PCR; *S. cerevisiae*: *Saccharomyces cerevisiae*; SDA: Sabouraud dextrose medium; SDA-Zn: Zinc deficient SDA; Swiss-Prot: Annotated protein sequence database; *T. mentagrophytes*: *Trichophyton mentagrophytes*; *ZafA*: Zinc-responsive activating factor

Acknowledgements

We thank the Gene De novo Biotechnology at Guangzhou for its assistance in related bioinformatics analysis.
We thank Steven M. Thompson, from Liwen Bianji, Edanz Editing China, for editing the English text of a draft of this manuscript.

Funding

This work was supported by the Youth Science Fund Project of National Natural Science Foundation of China (project number: 31,402,262) in the design of the study and collection, analysis of data and in writing the manuscript, the basic scientific research business special funds of the Northwest Agriculture and Forestry University (project number: 2014YB017 and 2,452,015,312) in the collection, analysis, and interpretation of data.

Authors' contributions

ZYH, ZXK, DPX and GYP conceived the experiments and performed the experiment. ZXK and DPX carried out the data analysis and wrote the paper. GYP, GXW, CH and JYP helped to carry out the data analysis and participated in the drafted manuscript. All authors read and approved the final manuscript.

Competing interests

The authors declare that they have no competing interests.

Author details

[1]The College of Veterinary Medicine of the Northwest Agriculture and Forestry University, No. 3 Taicheng Road, Yangling, Shaanxi, People's Republic of China. [2]Clinical Department, College of Veterinary Medicine, China Agricultural University, Beijing, People's Republic of China.

References

1. Knudtson WU, et al. *Trichophyton mentagrophytes* dermatophytosis in wild fox 1. J Wildl Dis. 1980;16(4):465–8.
2. Weitzman I, Summerbell RC. The dermatophytes. Clin Microbiol Rev. 1995; 8(2):240–59.
3. Youssef N, et al. Antibiotic production by dermatophyte fungi. Microbiology. 1978;105(1):105–11.
4. Pagani MA, et al. Disruption of iron homeostasis in *Saccharomyces cerevisiae* by high zinc levels: a genome-wide study. Mol Microbiol. 2007;65(2):521–37.
5. Zhao H, Eide DJ. *Zap1p*, a metalloregulatory protein involved in zinc-responsive transcriptional regulation in *Saccharomyces cerevisiae*. Mol Cell Biol. 1997;17(9):5044–52.

6. Eide DJ. The molecular biology of metal ion transport in *Saccharomyces cerevisiae*. Annu Rev Nutr. 1998;18(1):441–69.
7. Vicentefranqueira R, et al. The *zrfA* and *zrfB* genes of *Aspergillus fumigatus* encode the zinc transporter proteins of a zinc uptake system induced in an acid, zinc-depleted environment. Eukaryot Cell. 2005;4(5):837–48.
8. Amich J, et al. *Aspergillus fumigatus* survival in alkaline and extreme zinc-limiting environments relies on the induction of a zinc homeostasis system encoded by the *zrfC* and *aspf2* genes. Eukaryot Cell. 2010;9(3):424–37.
9. de Oliveira Schneider R, et al. *Zap1* regulates zinc homeostasis and modulates virulence in *Cryptococcus gattii*. PLoS One. 2012;7(8):e43773.
10. Kim M-J, et al. Roles of zinc-responsive transcription factor *Csr1* in filamentous growth of the pathogenic yeast *Candida albicans*. J Microbiol Biotechnol. 2008;18(2):242–7.
11. Rutherford JC, Bird AJ. Metal-responsive transcription factors that regulate iron, zinc, and copper homeostasis in eukaryotic cells. Eukaryot Cell. 2004; 3(1):1–13.
12. Brouta F, et al. Humoral and cellular immune response to a *Microsporum canis* recombinant keratinolytic metalloprotease (r-MEP3) in experimentally infected guinea pigs. Med Mycol. 2003;41(6):495–501.
13. Zhang X, et al. Metalloprotease genes of *Trichophyton mentagrophytes* are important for pathogenicity. Med Mycol. 2014;52(1):36–45.
14. Grabherr MG, et al. Full-length transcriptome assembly from RNA-Seq data without a reference genome. Nat Biotechnol. 2011;29(7):644–52.
15. Conesa A, et al. Blast2GO: a universal tool for annotation, visualization and analysis in functional genomics research. Bioinformatics. 2005;21(18):3674–6.
16. Ye J, et al. WEGO: a web tool for plotting GO annotations. Nucleic Acids Res. 2006;34(suppl 2):W293–7.
17. Kanehisa M, et al. KEGG for linking genomes to life and the environment. Nucleic Acids Res. 2008;36(suppl 1):D480–4.
18. Moreno MÁ, et al. The regulation of zinc homeostasis by the *ZafA* transcriptional activator is essential for *Aspergillus fumigatus* virulence. Mol Microbiol. 2007;64(5):1182–97.
19. Altschul SF, et al. Gapped BLAST and PSI-BLAST: a new generation of protein database search programs. Nucleic Acids Res. 1997;25(17):3389–402.
20. Iuchi S. Three classes of C2H2 zinc finger proteins. Cell Mol Life Sci. 2001; 58(4):625–35.
21. Auld DS. Zinc coordination sphere in biochemical zinc sites[J]. Biometals. 2001;14(3-4):271–313.
22. Emmons CW, Binford CH, Utz JP, et al. Medical mycology[M]. London: Henry Kimpton Publishers; 1977.
23. Samanta, I., Veterinary mycology. 2015: Springer.
24. Wang W, et al. Transcriptomic and physiological analysis of common duckweed *Lemna minor* responses to NH 4+ toxicity. BMC Plant Biol. 2016;16(1):1.
25. Colvin RA, et al. Cytosolic zinc buffering and muffling: their role in intracellular zinc homeostasis. Metallomics. 2010;2(5):306–17.
26. Lulloff SJ, Hahn BL, Sohnle PG. Fungal susceptibility to zinc deprivation. J Lab Clin Med. 2004;144(4):208–14.
27. Corbin BD, et al. Metal chelation and inhibition of bacterial growth in tissue abscesses. Science. 2008;319(5865):962–5.
28. Staats CC, et al. Fungal zinc metabolism and its connections to virulence. In: Metal economy in host-microbe interactions; 2015.
29. North M, et al. Genome-wide functional profiling identifies genes and processes important for zinc-limited growth of *Saccharomyces cerevisiae*. PLoS Genet. 2012;8(6):e1002699.
30. Iwanyshyn WM, Han G-S, Carman GM. Regulation of phospholipid synthesis in *Saccharomyces cerevisiae* by zinc. J Biol Chem. 2004;279(21):21976–83.
31. Carman GM, Han G-S. Regulation of phospholipid synthesis in *Saccharomyces cerevisiae* by zinc depletion. Biochim Biophys Acta. 2007; 1771(3):322–30.
32. Ballou ER, Wilson D. The roles of zinc and copper sensing in fungal pathogenesis. Curr Opin Microbiol. 2016;32:128–34.
33. Eide DJ. Zinc transporters and the cellular trafficking of zinc. Biochim Biophys Acta. 2006;1763(7):711–22.
34. Amich J, Leal F, Calera JA. Repression of the acid *ZrfA/ZrfB* zinc-uptake system of *Aspergillus fumigatus* mediated by *PacC* under neutral, zinc-limiting conditions. Int Microbiol. 2010;12(1):39–47.
35. Böttcher B, et al. *Csr1/Zap1* maintains zinc homeostasis and influences virulence in *Candida dubliniensis* but is not coupled to morphogenesis. Eukaryot Cell. 2015;14(7):661–70.

6

Proteomic insight into fruit set of cucumber (*Cucumis sativus* L.) suggests the cues of hormone-independent parthenocarpy

Ji Li[†], Jian Xu[†], Qin-Wei Guo, Zhe Wu, Ting Zhang, Kai-Jing Zhang, Chun-yan Cheng, Pin-yu Zhu, Qun-Feng Lou and Jin-Feng Chen[*]

Abstract

Background: Parthenocarpy is an excellent agronomic trait that enables crops to set fruit in the absence of pollination and fertilization, and therefore to produce seedless fruit. Although parthenocarpy is widely recognized as a hormone-dependent process, hormone-insensitive parthenocarpy can also be observed in cucumber; however, its mechanism is poorly understood. To improve the global understanding of parthenocarpy and address the hormone-insensitive parthenocarpy shown in cucumber, we conducted a physiological and proteomic analysis of differently developed fruits.

Results: Physiological analysis indicated that the natural hormone-insensitive parthenocarpy of 'EC1' has broad hormone-inhibitor resistance, and the endogenous hormones in the natural parthenocarpy (NP) fruits were stable and relatively lower than those of the non-parthenocarpic cultivar '8419 s-1.' Based on the iTRAQ technique, 683 fruit developmental proteins were identified from NP, cytokinin-induced parthenocarpic (CP), pollinated and unpollinated fruits. Gene Ontology (GO) analysis showed that proteins detected from both set and aborted fruits were involved in similar biological processes, such as cell growth, the cell cycle, cell death and communication. Kyoto Encyclopedia of Genes and Genomes (KEGG) analysis revealed that 'protein synthesis' was the major biological process that differed between fruit set and fruit abortion. Clustering analysis revealed that different protein expression patterns were involved in CP and NP fruits. Forty-one parthenocarpy-specialized DEPs (differentially expressed proteins) were screened and divided into two distinctive groups: NP-specialized proteins and CP-specialized proteins. Furthermore, qRT-PCR and western blot analysis indicated that NP-specialized proteins showed hormone- or hormone-inhibitor insensitive expression patterns in both ovaries and seedlings.

Conclusions: In this study, the global molecular regulation of fruit development in cucumber was revealed at the protein level. Physiological and proteomic comparisons indicated the presence of hormone-independent parthenocarpy and suppression of fruit abortion in cucumber. The proteomic analysis suggested that hormone-independent parthenocarpy is regulated by hormone-insensitive proteins such as the NP-specialized proteins. Moreover, the regulation of fruit abortion suppression may be closely related to protein synthesis pathways.

Keywords: Cucumber (*Cucumis sativus* L.), Parthenocarpy, Proteome, iTRAQ, Hormone dependent/independent

* Correspondence: jfchen@njau.edu.cn
[†]Equal contributors
State Key Laboratory of Crop Genetics and Germplasm Enhancement, Nanjing Agricultural University, Nanjing 210095, China

Background

Parthenocarpy is considered the most cost-effective solution for improving the fruit set rate when pollination or fertilization is suppressed by sub-optimum growth conditions, such as low temperature, weak light intensity or facility environments, and ensuring yields of vegetable and fruit crops that are self-sterile or gynoecious. Moreover, seedless fruits produced by parthenocarpy have a better texture, appearance and shelf life [1] and avoid yield loss caused by seed development [2–4].

Parthenocarpy is a widely recognized hormone-dependent biological process. Independent evidence indicates that auxins play a vital role in parthenocarpic fruit set. The genes linking the auxin signal transduction pathway to fruit set have been identified in recent decades. The involvement of IAA9, a member of the tomato Aux/IAA gene family of transcriptional regulators, was confirmed in tomato fruit set. Auxin dose-response assays showed that the down-regulation of IAA9 led to auxin hypersensitivity and resulted in parthenocarpy [5]. Another auxin signaling component involved in fruit set is ARF8, which was identified as a candidate gene for two parthenocarpy QTLs in tomato [6]. Based on the findings of Goetz et al. [7], a model was proposed for the mechanism of parthenocarpic induction. According to this model, ARF8 forms an inhibitory complex together with an AUX/IAA protein, possibly IAA9, to repress the transcription of the auxin response genes and consequently induce parthenocarpy. Wittwer et al. [8] showed that a second class of hormone, gibberellins (GAs), could also stimulate parthenocarpic fruit set. The only known gibberellin signaling component shown to be involved in fruit set is *DELLA*. The reduction in *SlDELLA* mRNA levels induces the formation of parthenocarpic tomato fruit [9]. Null and loss-of-function recessive mutations in the *DELLA* genes of *Arabidopsis* provoke a constitutive GA-response phenotype, including parthenocarpy [10]. Besides Auxin and GAs, Cytokinin (CK) is also involved in parthenocarpy, which accumulates to high levels in ovaries during fruit set [11–15]. Recent studies suggested that CKs may induce parthenocarpy partially through modulation of IAA and GA metabolisms [16–18]. Ethylene and abscisic acid (ABA) also play important roles in the regulation of fruit set and development. Ethylene is likely involved in the fruit set program by functioning coordinately with auxin [19–21]. ABA may acts as an antagonist of GA or auxin to induce and maintain the dormant state of ovaries, likely by repressing their transition to fruit [20]. These studies demonstrate the complicated and confusing relationships among hormone responses during fruit set. However, the key integrating molecular players remain largely undiscovered, and a global understanding of the mechanisms underlying parthenocarpy has yet to be attained.

Genetic studies have suggested that a majority of natural parthenocarpic properties in crops are quantitative traits regulated by both genetic and environmental factors [6, 22–26]. Photoperiod, temperature, light intensity and nutritional conditions have considerable influences on parthenocarpy [27–29]. A hypothesis proposed in the 1930s suggested that plant developmental responses to environmental stimuli were due to the spatiotemporal variations in phytohormone synthesis and transport [30, 31]. Studies have suggested that short-daylight conditions could enhance parthenocarpy by increasing the activity of auxin, while high temperatures suppressed the parthenocarpy rate by inhibiting the synthesis of auxin and gibberellin in the ovary of cucumber [32, 33]. Kim et al. [34] also found that ovaries had twice the auxin content at 15 °C than that at 25 °C, resulting in a higher rate of parthenocarpy in cucumber. Anyhow, the agricultural application of parthenocarpy was limited by its environmental sensitivity. In practice, the excessive application of exogenous hormones was often used to overwhelm the environmental effects on hormone synthesis, thereby inducing environmentally stable parthenocarpy.

Cucumber (*Cucumis sativus* L.) is emerging as a model for plants in the *Cucurbitaceae* family because of its small and fully sequenced genome ($2n = 2x = 14$, 367 Mb genome) [35]. The mechanism for sex determination, vascular system development, and typical pepo fruit are also well documented in cucumber. The rich parthenocarpic germplasm of cucumber offers an opportunity to investigate the coordination and communication of hormone signals and genes during parthenocarpic fruit set. Genetic studies of parthenocarpy in cucumber started in 1930s. Early studies suggested that parthenocarpy in cucumber is controlled by single genes [36–40]. While most recent studies confirmed that inheritance of parthenocarpy in cucumber is consistent with characteristics of quantitative traits [25, 26, 41–43]. Although cucumber is rich in parthenocarpic germplasm resources e.g. main branch specialized/lateral branch specialized parthenocarpy, temperature/photoperiods sensitive parthenocarpy and parthenocarpy with accelerated ovary expansion before anthesis, the agricultural application of parthenocarpic cucumber was limited by their environmental sensitivity [43]. A serial of studies indicated that stable parthenocarpy of cucumber can be induced by auxin or auxin transport inhibitors [34, 44–48]. Meanwhile artificially increasing of endogenous auxin in the ovary by introducing the DefH9-iaaM auxin-synthesizing gene into cucumber might also stimulate parthenocarpy [49]. Besides auxin, application of other hormones such as cytokinins, gibberenllins and brassinosteroids (BRs) could also promote parthenocarpy in cucumber [50, 51]. However, it reported that auxin and GAs had less potential to induce parthenocarpic fruit growth than CKs in cucumber [34, 51, 52]. Therefore, in practice application of exogenous cytokinin, particularly CPPU (N-(2-chloro-4-pyridyl)-N′-phenyl urea, a type of synthetic cytokinin), to induce parthenocarpy is widely used in cucumber production.

In previous studies, an excellent parthenocarpic cucumber cultivar, 'EC1,' was found, which showed environmentally stable parthenocarpy under different culture conditions [25, 43]. Previous transcriptome studies have demonstrated that the natural parthenocarpy (NP) of 'EC1' has many different aspects compared with cytokinin-induced parthenocarpy (CP) at the mRNA level [53]. However, mRNA levels are not always in accordance with protein activity. To improve the global understanding of parthenocarpy and address the environmental stability of parthenocarpy in cucumber, we conducted a physiological analysis and an iTRAQ (isobaric tags for relative and absolute quantitation)-based proteomic analysis in the natural parthenocarpic fruits of 'EC1' and cytokinin-induced parthenocarpic fruits of '8419 s-1' (a non-parthenocarpic variety).

Results

Physiological comparison of the parthenocarpic and non-parthenocarpic cucumber cultivars

Experiments were conducted to investigate the physiological differences between parthenocarpic cultivar 'EC1' and non-parthenocarpic cultivar '8419 s-1' during natural/cytokinin-induced parthenocarpy, pollinated fruit set and unpollinated fruit abortion. The detail information of the two cultivars was described in Material and Methods section. The longitudinal and radial growth of the natural parthenocarpic fruits of 'EC1' and CPPU -induced parthenocarpy of '8419 s-1,' pollinated and unpollinated fruits of '8419 s-1' were measured (Fig. 1a). Our results showed that the length and diameter of the parthenocarpic and pollinated fruits linearly increased from 0 to 6 dpa (days post-anthesis), and the natural and CPPU-induced parthenocarpic fruits showed similar growth curves, wherein the fruit size was generally larger than the pollinated fruits. In contrast, the growth of the unpollinated fruits of '8419 s-1' was blocked, and the length and diameter of the abortive fruits also decreased slightly.

Kim et al. [34] suggested that genetic factor for parthenocarpy in cucumber may be associated with high content of IAA in the ovaries at anthesis. In this study, endogenous auxins, cytokinins, and gibberellins were analyzed in the cucumber fruits noted above (Fig. 1b). The induction of both naturally occurring and hormone induced parthenocarpy is attributed to the presence of sufficient phytohormones in the ovaries [54–58]. However, parthenocarpic fruits of 'EC1' had relatively low and stable hormone levels compared with the fruits of '8419 s-1.' Moreover, the auxin and gibberellin

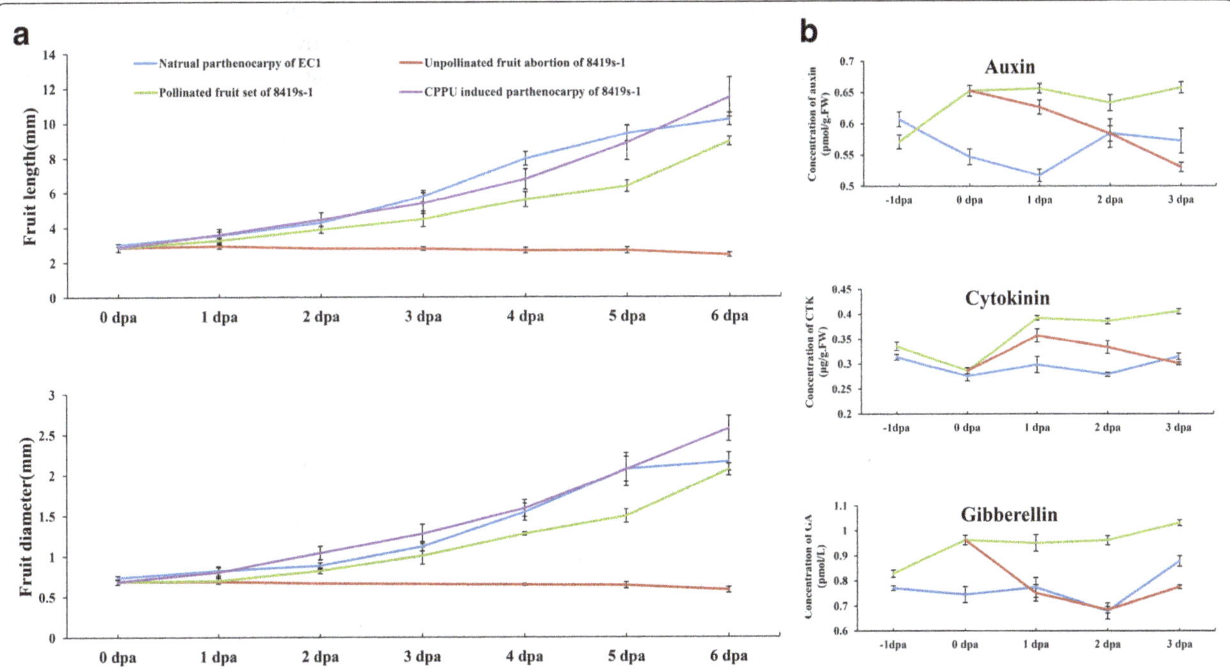

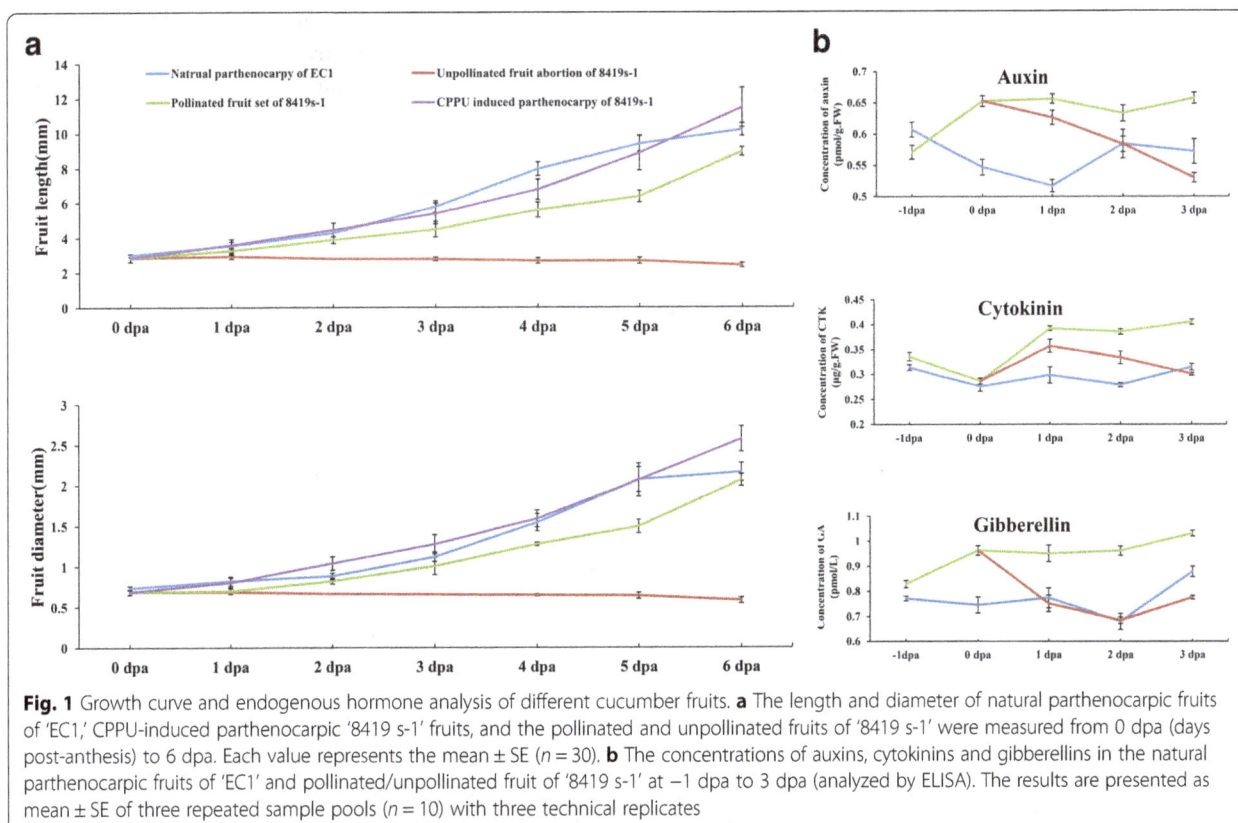

Fig. 1 Growth curve and endogenous hormone analysis of different cucumber fruits. **a** The length and diameter of natural parthenocarpic fruits of 'EC1,' CPPU-induced parthenocarpic '8419 s-1' fruits, and the pollinated and unpollinated fruits of '8419 s-1' were measured from 0 dpa (days post-anthesis) to 6 dpa. Each value represents the mean ± SE ($n = 30$). **b** The concentrations of auxins, cytokinins and gibberellins in the natural parthenocarpic fruits of 'EC1' and pollinated/unpollinated fruit of '8419 s-1' at −1 dpa to 3 dpa (analyzed by ELISA). The results are presented as mean ± SE of three repeated sample pools ($n = 10$) with three technical replicates

concentrations also decreased unexpectedly during the natural parthenocarpic fruit set of 'EC1' (Fig. 1b).

For a further comparison of the fruit developmental differences between 'EC1' and '8419 s-1,' we conducted ovary treatment experiments. The ovaries of 'EC1' and '8419 s-1' were treated with hormones, hormone inhibitors and pollen separately at anthesis. The weight, length and diameter measurement of the treated ovaries was conducted at 4dpa to reveal different phytohormone responses between 'EC1' and '8419 s-1'. In cucumber, etiolation of ovary tips is the principal identifying symbol to identify whether the fruits are set or aborted since 2 dpa. It was showed that the ovaries with etiolated tips did not grow or even wilt by comparing with the 0 dpa ovaries (Table 1). The non-etiolated tip phenotypes and growth of ovaries suggested that parthenocarpy of '8419 s-1' could be induced by all of the exogenous hormones, including NAA, CPPU, GA_3 and EBR; however, NAA, GA3 and EBR exerted weak effects on fruit growth, causing the treated ovaries to grow slightly in length and remain in a dormant-like state (Table 1, Additional file 1: Figure S1). Marcelis et al. [59] suggested that cell division in fruit of cucumber occurs about a week after anthesis whereas cell size increases markedly only after cell division begins to decline. Fruit size increases because of increase in both the number and size of cells. Therefore we speculated that NAA, GA3 and EBR might be primarily involved in cell division, rather than cell expansion during fruit development of cucumber, thus ovaries treated by these hormones are smaller than pollinated and CCPU-treated fruits. On the other hand, the weak effect of NAA, GA3 and EBR might be due to the low hormone concentration used in this study. Fu et al. [51] showed that 0.2 μM (20 mg/L) of EBR led to high efficiency of fruit growth in cucumber. But in this study we found 10 mg/L of EBR showed weak effect on fruit expansion. However, half concentration of CPPU (50 mg/L) referred to Fu's method still lead to strong effect on fruit set and growth.

The strong effect of CPPU on fruit development of cucumber was observed and documented in many other reports [34, 51, 52]. Similar observation was also found in many other species such as watermelon, apple, kiwifruit and blueberry [60–63]. Early fruit development generally consists of three stages: fruit formation, cell division and cell expansion [13]. The growth of cucumber fruit size is often mirrored by the increase in cell number and size [64]. Previous cytological observation showed that the CPPU-treated cucumber fruits was initiated with an increase of cell numbers in the pericarp and placenta tissues, and the size of pericarp cells were bigger than natural parthenocarpic fruits, although the number of cell layers was similar [53]. The findings suggested that CPPU might be involved in both processes of cell division and cell expansion. In addition, fruit growth is tightly related to the availability of carbohydrate, because fruit is a very strong metabolic sink. Many studies confirmed that CKs was demonstrated to regulate carbohydrate allocation in fruit [65, 66]. We thought that maybe another reason why CPPU induced parthenocarpic fruit was consistently bigger than the fruit induced by other PRGs.

Martínez et al. [21] have demonstrated that the inhibition of ethylene response (STS treatment) is sufficient to induce the set and early development of the fruit in absence of pollination in both the parthenocarpic and the non-parthenocarpic cultivar of zucchini squash. Coincidentally, it was showed that STS has stimulated parthenocarpy in the three non-parthenocarpic cucumber cultivars; however it had no effect on fruit development of parthenocarpic cultivar 'EC1' (Additional file 2: Table S6). Besides, diameter, length and weight of ethephon treated fruit of 'EC1' showed no significant differences to the natural parthenocarpic fruits (Additional file 2: Table S6). It indicated that neither ethephon nor STS (ethylene response inhibitor) could affect parthenocarpy of EC. Interestingly, the ethephon treated ovaries of non-parthenocarpic cucumbers displayed more severe atrophy of ovaries by comparing with their unpollinated ovaries (Additional file 2: Table S6), suggested that the fruit abortion of the non-parthenocarpic cultivars maybe accelerated by ethephon. Irradiated pollen treatment could promote stenospermocarpy in 8419 s-1, but the stenospermocarpic ovaries were much smaller than the active pollen-treated ovaries, implying that seed set may be essential for the fruit growth of non-parthenocarpic varieties. Interestingly, the seedless fruit of 'EC1' formed by parthenocarpy was much larger than its pollinated fruit. The pollination fruits of '8419 s-1' were blocked by either a hormone inhibitor mixture or individual hormone inhibitors (Table 1). Etiolation was observed in the pollen and hormone inhibitor co-treated ovaries. In contrast, the natural parthenocarpic fruit set of 'EC1' could not be blocked by hormone inhibitors, but the growth of the ovaries was suppressed (Table 1).

iTRAQ-based proteomic study of differently developed cucumber fruits

Early fruit development generally consists of three stages: fruit formation, cell division and cell expansion [13]. The earliest stage, in which the ovary is aborted or allowed to proceed with fruit development, is referred to as fruit set. In this study, we focused on the fruit set stage in cucumber, which is 0 to 2 days post-anthesis [53, 59, 64, 67, 68]. The natural parthenocarpic fruits of 'EC1,' cytokinin-induced parthenocarpic fruits of '8419 s-1,' the pollinated and unpollinated fruits of '8419 s-1' were investigated (described in the M&M section). The proteomes of these

Table 1 The measurement of ovary weight, length and diameter after treated by pollen, exogenous hormones and hormone inhibitors

		0 dpa ovary	Unpollination[b]	Pollination I[c]	Pollination II	NAA	CPPU	GA_3	EBR	Hormone Mix[d]
Ovaries of 8419 s-1	Weight (g)	0.75 ± 0.14 D[a]	0.66 ± 0.15 D	8.39 ± 1.4 B	5.05 ± 0.94 C	1.01 ± 0.12 D	12.81 ± 1.25 A	1.01 ± 0.17 D	0.86 ± 0.21 D	13.04 ± 3.14 A
	Length (mm)	28.4 ± 2.4 E	27.60 ± 3.3 E	69.33 ± 4.25 B	58.91 ± 3.96 C	34.06 ± 1.91 D	77.76 ± 5.67 A	33.31 ± 2.59 D	31.31 ± 3.1 DE	81.43 ± 6.77 A
	Diameter (mm)	6.86 ± 0.33 D	6.30 ± 0.48 D	13.46 ± 1.38 B	11.07 ± 2.8 C	6.96 ± 0.7 D	15.87 ± 1.9 A	6.98 ± 0.58 D	6.73 ± 0.33 D	14.99 ± 2.3 A
	P/A/T[e]		0/30/30	–	–	31/0/31	30/0/30	36/0/36	34/0/34	35/0/35
		TIBA	Lovastatin	Uniconazole	Brz	Inhibitor Mix[g]				
Pollinated ovaries of 8419 s-1[f]	Weight (g)	0.67 ± 0.06 D	0.67 ± 0.05 D	0.69 ± 0.04 D	0.68 ± 0.03 D	0.62 ± 0.08 D				
	Length (mm)	25.7 ± 0.55 F	26.8 ± 1.11 E	25.9 ± 1.59 E	26.9 ± 1.27 E	26.5 ± 1.40 E				
	Diameter (mm)	5.96 ± 0.24 F	5.84 ± 0.05 F	5.70 ± 0.35 F	5.69 ± 0.31 F	6.07 ± 0.19 F				
	P/A/T	0/35/35	0/29/29	0/30/30	0/32/32	0/30/30				
		0 dpa ovary	Unpollination	Pollination I		TIBA	Lovastatin	Uniconazole	Brz	Inhibitor Mix
Ovaries of EC1	Weight (g)	0.83 ± 0.07 E	8.09 ± 1.51 A	2.64 ± 0.49 B		1.56 ± 0.23 CD	1.22 ± 0.37 CDE	0.93 ± 0.13 E	1.76 ± 0.38 C	1.11 ± 0.2 DE
	Length (mm)	30.6 ± 3.1 F	58.70 ± 2.74 A	45.27 ± 1.51 B		34.82 ± 1.74 D	34.52 ± 0.58 D	33.89 ± 0.55 DE	38.06 ± 1.33 C	32.3 ± 2.53 EF
	Diameter (mm)	6.87 ± 0.27 E	10.58 ± 1.11 A	8.39 ± 1.47 BC		7.20 ± 0.38 DE	7.17 ± 0.78 DE	8.81 ± 1.33 B	9.16 ± 1.35 B	7.84 ± 0.87D
	P/A/T		30/0/30	–		30/0/30	30/0/30	30/0/30	30/0/30	37/0/37

The treatments were conducted at the anthesis day (0 dpa), and the measurements were conducted at 4 dpa. Means (±SE) of three independent experiments were calculated

[a]Letters indicate differences between the treated ovaries with statistical significance at $P \leq 0.05$ (t-test). The same letter means not significantly different; different letters means significantly different

[b]Underline words means the treatment can induce etiolation of ovary tips at 4dpa and lead to fruit abortion

[c]'Pollination I' means hand pollination with active pollens; 'Pollination II' means hand pollination with irradiated pollens (γ-ray irradiation at a dose of 200 Gy)

[d]Mixed solution of NAA (50 mg/L), CPPU (50 mg/L), GA3 (50 mg/L) and EBR (10 mg/L)

[e]P/A/T: number of parthenocarpic ovaries/number of abortive ovaries/total treated ovaries; etiolation phenotype of ovary tips is the principal identifying symbol to identify whether the fruits are set or aborted since 2dpa

[f]Ovaries of 8419 s-1 were pollinated at 0 dpa, the hormone inhibitor treatments were conducted 3 h after pollination

[g]Mixed solution of TIBA (50 mg/L), Lovastatin (50 mg/L), uniconazole (50 mg/L) and Brz (10 mg/L)

differently developed fruits were analyzed by iTRAQ with two technical replicates per sample. The strategy for analysis is shown in Additional file 3: Figure S2.

Protein homolog identification was conducted by BLASTP against the cucumber Refseq database and the *Arabidopsis thaliana* Refseq database (*E*-value <1E-10). After redundancies were removed, 683 unique proteins were identified, including 359 fruit set-related proteins (natural/cytokinin-induced parthenocarpic fruit and pollinated fruit) and 377 fruit abortion-related proteins. Gene Ontology (GO) analysis showed that the proteins detected in both set and aborted fruits were involved in similar biological processes (Fig. 2a). Proteins related to cell growth (GO:0016049), cell cycle (GO:0007049), cell death (GO:0008219) and cell communication (GO:0007154) were detected that actively expressed during fruit development, that was consistent with our previous transcriptomic study [53], suggesting the cell growth, cell cycle, cell death and cell communication related genes were actively involved in fruit development at both mRNA and protein levels.

Identification and comparison of differentially expressed proteins involved in different fruit developmental processes

Large numbers of polymorphic SNPs (Single Nucleotide Polymorphisms), InDels (Insertion-Deletion Polymorphisms) and SVs (Structural Variations) were detected between 'EC1' and '8419 s-1' by genome resequencing study [25]. Besides, 84 differentially expressed proteins were also identified between 0dpa fruits of 'EC1' and '8419 s-1' (Additional file 4: Figure S3). These findings demonstrated that the genetic background of the two varieties was significantly different with each other. In order to identifiy fruit developmental DEPs (Differentially Expressed Proteins), a proteome comparison strategy was employed (Additional file 3: Figure S2). Proteomic comparisons were not conducted between cultivars, therefore the differentially expression of identified DEPs were not

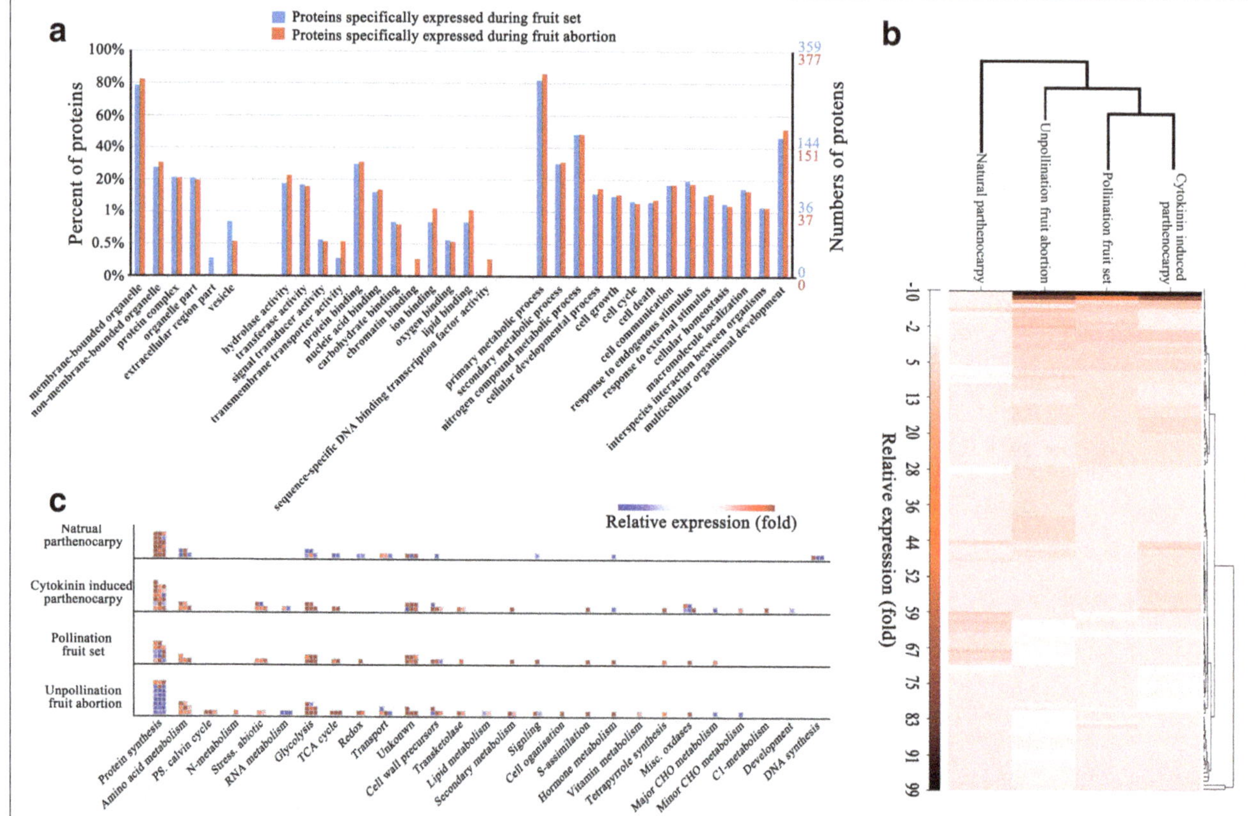

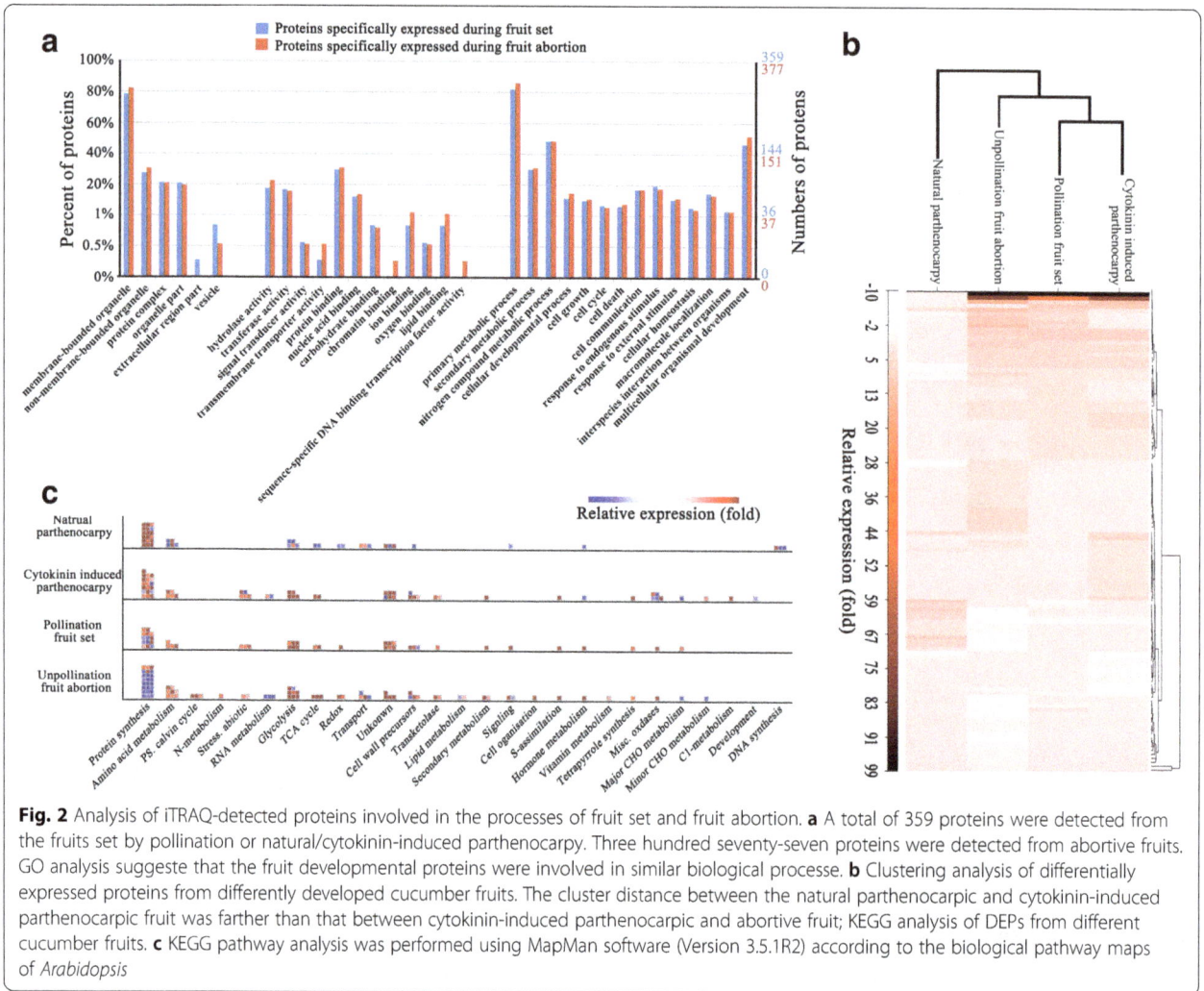

Fig. 2 Analysis of iTRAQ-detected proteins involved in the processes of fruit set and fruit abortion. **a** A total of 359 proteins were detected from the fruits set by pollination or natural/cytokinin-induced parthenocarpy. Three hundred seventy-seven proteins were detected from abortive fruits. GO analysis suggeste that the fruit developmental proteins were involved in similar biological processe. **b** Clustering analysis of differentially expressed proteins from differently developed cucumber fruits. The cluster distance between the natural parthenocarpic and cytokinin-induced parthenocarpic fruit was farther than that between cytokinin-induced parthenocarpic and abortive fruit; KEGG analysis of DEPs from different cucumber fruits. **c** KEGG pathway analysis was performed using MapMan software (Version 3.5.1R2) according to the biological pathway maps of *Arabidopsis*

caused by genetic variations but only correlated with development of fruit. The false discovery rate (FDR) method (FDR ≤ 0.05, |fold ≥ 1.5|, *P*-value < 0.05) was used to determine the significance of the differential protein expression. Consequently 138 DEPs were identified and four groups of DEPs were screened (Additional file 4: Figure S3). In pollination and CP fruits, most of the DEPs were up-regulated; in contrast, more down-regulated proteins were detected in the abortive fruit of '8419 s-1' and the natural parthenocarpic (NP) fruit. Clustering analysis indicated that the DEPs in CP and pollinated fruit showed similar protein expression patterns, but the protein expression profiles of the CP and NP fruits were clustered into two groups, and the cluster distance between CP and NP fruits was greater than that between CP and the abortive fruit (Fig. 2b).

Kyoto Encyclopedia of Genes and Genomes (KEGG) pathway analysis of the DEPs was conducted using MapMan software [69], according to the biological pathway maps of *Arabidopsis* (http://mapman.gabipd.org/web/guest/home). Large numbers of DEPs were shown to be involved in the biological process of protein synthesis and were mainly up-regulated during fruit set and down-regulated during fruit abortion (Fig. 2c). Transcriptome analysis has shown that amino acid metabolism, glycolysis and TCA cycle-related genes were actively expressed during fruit set [53], corroborating the present proteomic analysis, which also showed that the proteins related to these biological processes were actively expressed (Fig. 2c).

The interactions between the DEPs were analyzed based on the reference proteome-wide binary protein-protein interaction (PPI) map of *Arabidopsis* [70]. A core fruit developmental PPI network was revealed, which consisted of 30 DEPs and 19 'bridging' interaction proteins (Fig. 3, Additional file 2: Table S1). The proteins within the PPI network were mainly involved in the processes of protein metabolism (GO:0019538), transport (GO:0006810) and signal transduction (GO:0007165). Interactions within the PPI network occurred more

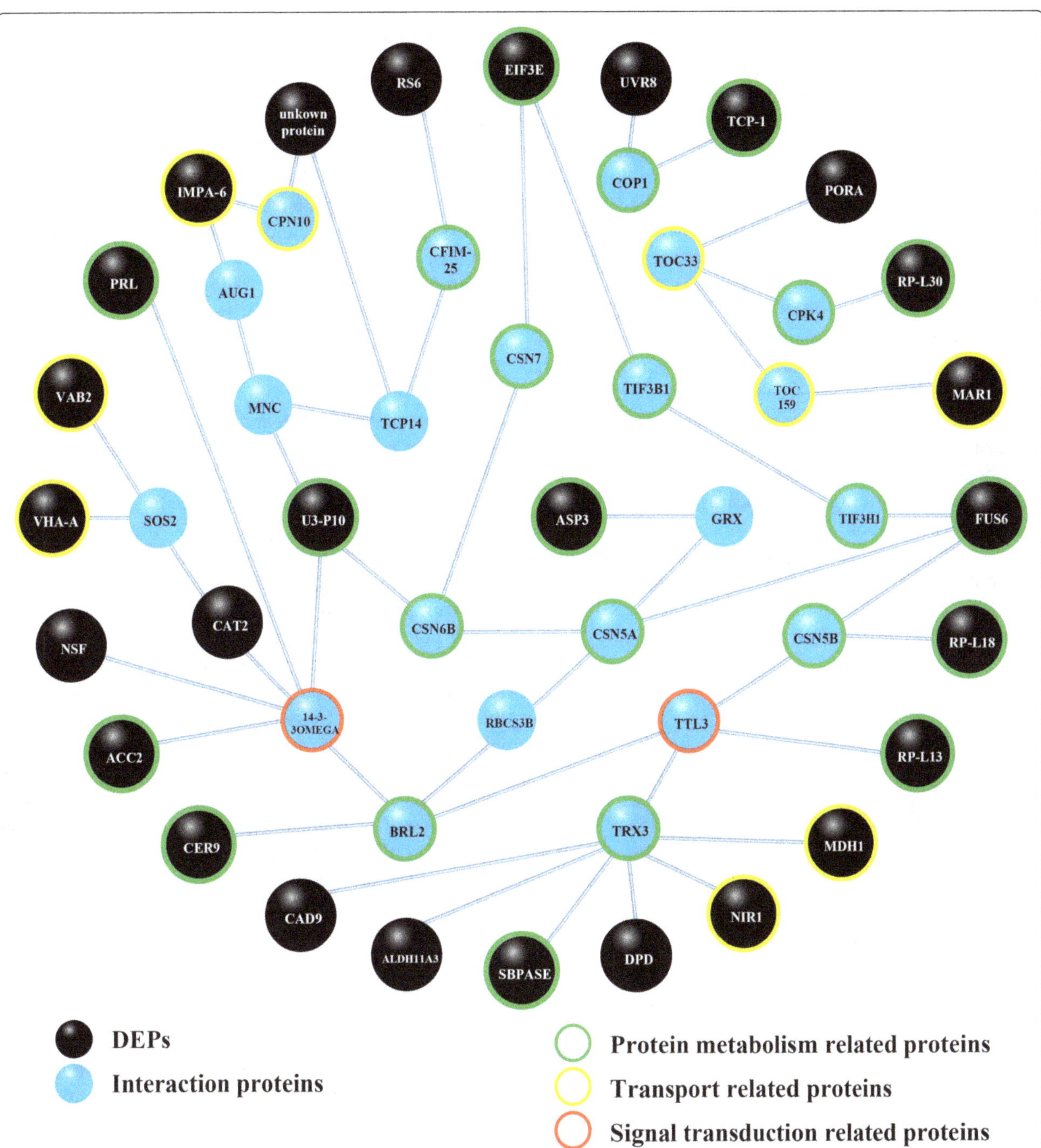

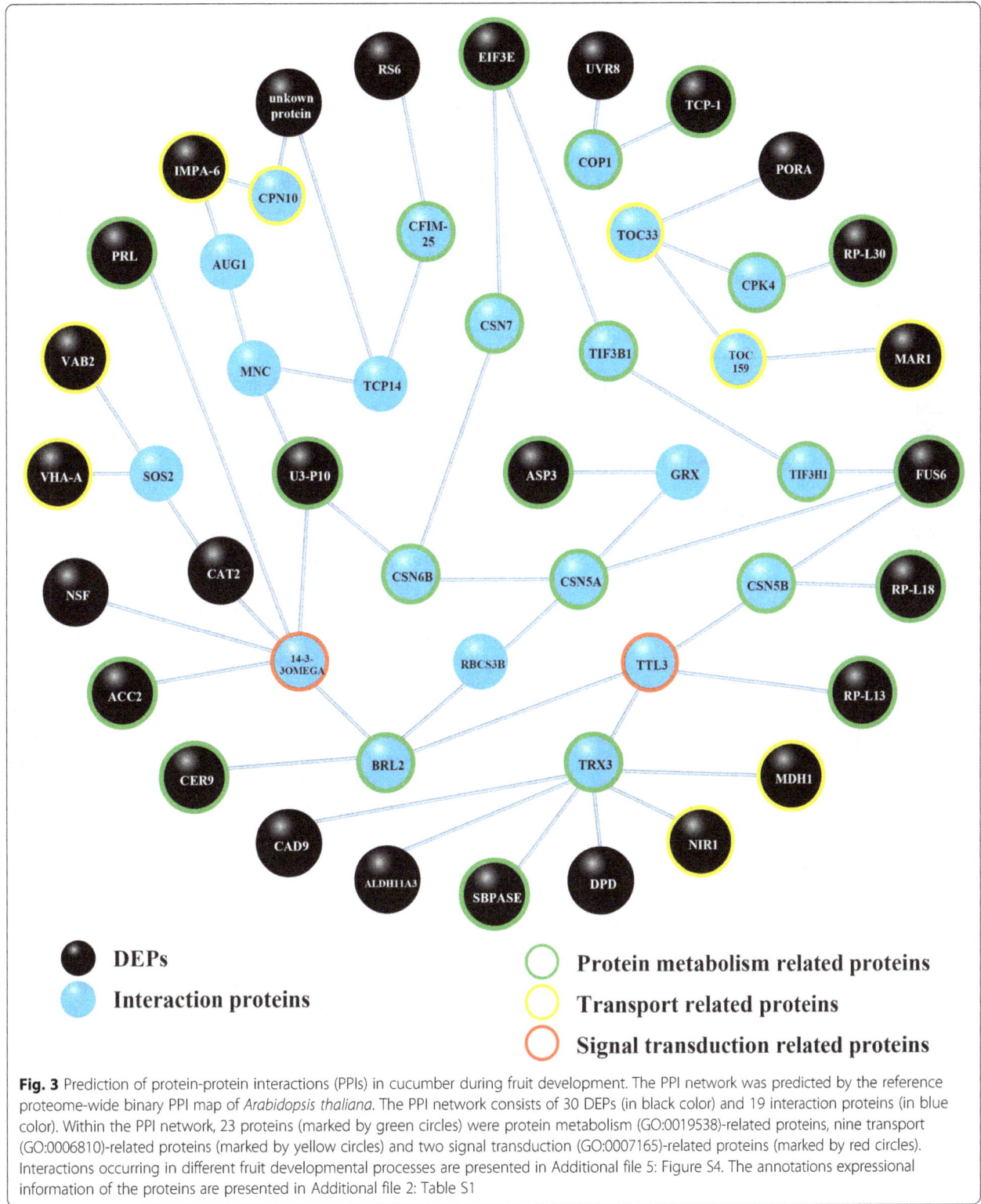

Fig. 3 Prediction of protein-protein interactions (PPIs) in cucumber during fruit development. The PPI network was predicted by the reference proteome-wide binary PPI map of *Arabidopsis thaliana*. The PPI network consists of 30 DEPs (in black color) and 19 interaction proteins (in blue color). Within the PPI network, 23 proteins (marked by green circles) were protein metabolism (GO:0019538)-related proteins, nine transport (GO:0006810)-related proteins (marked by yellow circles) and two signal transduction (GO:0007165)-related proteins (marked by red circles). Interactions occurring in different fruit developmental processes are presented in Additional file 5: Figure S4. The annotations expressional information of the proteins are presented in Additional file 2: Table S1

frequently in fruit abortion, while the smallest scale of interactions was involved in natural parthenocarpy (Additional file 5: Figure S4). Two interaction proteins exhibited specialized expression in natural parthenocarpy: CER9 (ECERIFERUM 9, Csa7M073540.1), which is involved in cuticle metabolism and the maintenance of plant water status [71], and PRL (PROLIFERA, Csa7M407650.1), which is specifically expressed in populations of dividing cells in the sporophytic tissues of the plant body [72]. TOC159 (Csa1M229500.1, a chloroplast biogenesis-related protein), IMPA-6 (Csa1M597740.1, a nuclear import protein) and RS6 (Csa1M229500.1, a putative ovule development regulator) showed specialized up-regulation in CPPU-induced parthenocarpic fruit set (Additional file 2: Table S1).

The common and specific DEPs in the differently developed cucumber fruits are shown in the Venn diagram in Fig. 4. No DEPs were commonly expressed in NP, CP and pollination fruit. Twelve common proteins were identified in the CP and pollinated fruits and showed similar expression patterns (Additional file 2: Table S2). Most of these CP and pollination-specialized DEPs are closely related to pollen and seed development. Eleven DEPs were commonly expressed in NP and abortive fruit, most of which were protein metabolism-related proteins and mainly involved in the biological processes of pollen germination, gametophyte and endosperm development as well as root morphogenesis; however, these proteins showed opposite expression trends during parthenocarpy and fruit abortion (Additional file 2: Table S2).

Forty-one parthenocarpy-specialized DEPs were identified, but only 3 DEPs are commonly found in the NP and CP fruits (Additional file 2: Table S3). The remaining 39 parthenocarpy-specialized DEPs were

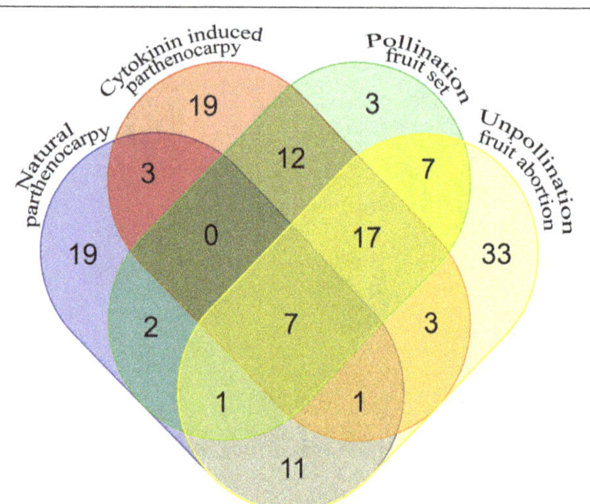

Fig. 4 Venn diagram and relative expression of DEPs in natural and cytokinin-induced parthenocarpic, pollinated and abortive cucumber fruits. The differently expressed proteins (DEPs) from the natural parthenocarpic fruit of 'EC1', cytokinin-induced parthenocarpic fruit of '8419 s-1', pollinated and unpollinated fruits of '8419 s-1' were compared. Twelve DEPs were commonly expressed in cytokinin-induced parthenocarpic and pollinated fruits (the common DEPs are annotated in Additional file 2: Table S2). Eleven DEPs were commonly expressed in natural parthenocarpic and abortive fruits and showed opposite expression trends (Additional file 2: Table S2). Three DEPs were parthenocarpy-specialized proteins that are commonly expressed in both natural and cytokinin-induced parthenocarpic fruits (Additional file 2: Table S3). The natural and cytokinin-induced parthenocarpy-specialized proteins are individually annotated in Table 2

divided into two groups, of which 19 were uniquely expressed in NP fruits, while the other 19 existed only in CP fruits (Table 2). These DEPs were mainly involved in the biological processes of pollen germination, seed and seedling development, cell proliferation and programmed cell death (Table 2). The NP-specialized DEPs Csa7M073540.1 and Csa7M450640.1, which were related to cell cycle and proliferation, and Csa1M025890.1 and Csa4M036590.1, the amino acid biosynthesis-related proteins, showed dramatically up-regulated expression during fruit set (expression fold >5) (Table 2). Moreover, Csa2M139820.1, which has the putative function of translational elongation, was the only DEP that was dramatically increased in the CP fruit (Table 2).

Expression analysis of NP- and CP-specialized proteins in response to phytohormones

The transcription profiles of the parthenocarpy-specialized DEPs in the hormone- and hormone inhibitor-treated fruits were compared (the treated fruits are described in the 1st part of the results section). Consistent with the result of iTRAQ, the specialized DEPs from both parthenocarpy groups showed inactive transcription during pollinated fruit set (Fig. 5, middle panel). The CP-specialized DEPs were actively expressed in the cytokinin-induced parthenocarpic fruits of '8419 s-1.' However, in the parthenocarpic fruits of 'EC1,' including the hormone inhibitor-treated but not blocked parthenocarpic fruits, most proteins showed decreased expression (Fig. 5, left column of the panel). In contrast, the NP-specialized DEPs showed up-regulated expression in parthenocarpic fruits of 'EC1,' whereas many of these DEPs were silenced in the cytokinin-induced parthenocarpic fruits (Fig. 5, right column of the panel). Although a few of the NP-specialized DEPs were differentially transcribed in cytokinin-induced parthenocarpic fruits, the expression levels of the genes in these fruits were the same as those in the parthenocarpic fruits of 'EC1.' Moreover, similar transcription patterns of NP-specialized DEPs between the natural parthenocarpic and unblocked natural parthenocarpic fruits were found, indicating that the transcription of these proteins could not be affected by hormone inhibitors (Fig. 5, top right of the panel). These findings, to some extent, indicated that the transcription of NP-specialized DEPs was not sensitive to hormones or hormone inhibitors.

iTRAQ showed that four NP-specialized proteins—Csa7M073540.1, Csa7M450640.1, Csa1M025890.1 and Csa4M036590.1—showed a high abundance in expression during parthenocarpy (expression >5-fold; Table 2, marked by stars). To further investigate the expression characteristics of the active parthenocarpy-specialized proteins, we conducted a western blot analysis. Consistent with the results of iTRAQ, the NP-specialized proteins

Table 2 DEPs specifically expressed in natural or Cytokinin induced parthenocarpic fruit

Protein ID	Top Hit[a]	Description[b]	Relative Biological Processes[c]	Expression fold of DEPs in parthenocarpic fruit[d]
Natural parthenocarpy specialized DEPs				
Csa1M024830.1	AT1G48630.1	RACK1B, RECEPTOR FOR ACTIVATED C KINASE 1B	Seed Germination and Early Seedling Development; shoot development (GO:0048367)	−1.97
Csa1M025890.1★	AT5G46180.1	DELTA-OAT, ORNITHINE-DELTA-AMINOTRANSFERASE	Pollen germination and tube growth [86]; cellular amino acid biosynthetic process (GO:0008652)	85.50
Csa1M025980.1	AT3G10920.1	MSD1, MANGANESE SUPEROXIDE DISMUTASE 1	Seed Germination [87]; female gametophyte development; programmed cell death	−1.94
Csa2M223140.1	AT3G04840.1	40S ribosomal protein S3a-like protein	Pollen germination and tube growth [86]	2.58
Csa2M338890.1	AT1G26880.1	60S ribosomal protein L34	Pollen germination and tube growth [86]	2.99
Csa3M002370.1	AT4G35630.1	PHOSPHOSERINE AMINOTRANSFERASE 1, PSAT1	Serine biosynthesis; responses to cytokinin	−3.16
Csa3M827370.1	AT3G01280.1	VOLTAGE DEPENDENT ANION CHANNEL 1, VDAC1	Female gametogenesis; pollen germination and tube growth [86]	−2.53
Csa4M001980.1	AT1G02780.1	EMB2386, EMBRYO DEFECTIVE 2386	Pollen germination and tube growth [86]; embryonic development (GO:0009790)	3.02
Csa4M012460.1	AT1G60710.1	Aldo/keto reductase family protein	Seed Germination and Floral Development	−2.65
Csa4M036590.1★	AT4G39660.1	AGT2, ALANINE:GLYOXYLATE AMINOTRANSFERASE 2	Responses to brassinosteroids; cellular amino acid biosynthetic process (GO:0008652)	8.87
Csa4M179090.1	AT4G33680.1	AGD2, ABERRANT GROWTH AND DEATH 2	Responses to cytokinin; pollen germination and tube growth (Wang et al. [86]); cell growth	−2.44
Csa4M290220.1	AT1G09200.1	Histone H3	Cell cycle [88]; cell expansion and proliferation; male gametogenesis	−3.25
Csa4M664520.1	AT1G76550.1	Fructose-6-phosphate 1-phosphotransferase	seed development [89]; glycolysis (GO:0006096)	2.29
Csa6M193590.1	AT1G07660.1	Histone H4	Chromatin organization (GO:0006325)	−3.16
Csa6M450410.1	AT2G36460.1	FBA6, FRUCTOSE-BISPHOSPHATE ALDOLASE 6	Seed Germination [87]; responses to cytokinin	−2.03
Csa6M451470.1	AT3G52880.1	MDAR1, MONODEHYDROASCORBATE REDUCTASE 1	Pollen germination and tube growth [86]; Cell Wall Regeneration	−1.95
Csa7M073540.1★	AT4G34100.1	CER9, ECERIFERUM 9	Seed Germination; pollen germination and tube growth [86]; Cell cycle [88]	7.44
Csa7M407650.1	AT4G02060.1	PRL, PROLIFERA	Cell cycle and division [72]; GO:0007049)	3.66
Csa7M450640.1★	AT1G67120.1	MDN1, MIDASIN 1	Seed germination and seedling development; female gametophyte development; cell proliferation	87.90
Cytokinin induced parthenocarpy specialized DEPs				
Csa1M003540.1	AT4G20360.1	RABE1B, RAB GTPASE HOMOLOG E1B	Seed Germination and development [87, 89]	2.55
Csa1M031900.1	AT1G48410.1	AGO1, ARGONAUTE 1	Fruit development; cell division	−1.67
Csa1M042700.1	AT3G18080.1	BGLU44, B-S GLUCOSIDASE 44	Female gametophyte development; Cell wall proteins	−1.80
Csa1M229500.1	AT5G20250.1	DIN10, DARK INDUCIBLE 10	Seed germination and seedling development; pollen germination and tube growth [86]	1.78
Csa1M573730.1	AT5G56680.1	EMB2755, EMBRYO DEFECTIVE 2755	Gametogenesis and embryo development (GO:0009793)	1.95
Csa1M597740.1	AT5G20720.1	CPN21, CHAPERONIN 20	Seed development [89]; pollen germination and tube growth	1.55
Csa1M604600.1	AT1G50480.1	THFS,10-FORMYLTETRAHYDROFOLATE SYNTHETASE	Seed development [89]; responses to cytokinin	2.88
Csa2M139820.1★	AT1G07920.1	EF1α, Elongation factor 1-alpha	Pollen development; translational elongation (GO:0006414)	6.08

Table 2 DEPs specifically expressed in natural or Cytokinin induced parthenocarpic fruit *(Continued)*

Protein ID	Top Hit[a]	Description[b]	Relative Biological Processes[c]	Expression fold of DEPs in parthenocarpic fruit[d]
Csa2M264020.1	AT4G34880.1	GAtA, Glutamyl-tRNA (Gln) amido transferase subunit A	Translation (GO:0006412)	2.70
Csa2M350200.1	AT1G24510.1	TCP-1/cpn60 chaperonin family protein	Plant cell death [90]	1.95
Csa4M094000.1	AT3G29360.1	UGD2, UDP-GLUCOSE DEHYDROGENASE 2	Pollen germination and tube growth [86]; cell wall organization (GO:0007047)	3.98
Csa4M496230.1	AT5G63860.1	UVR8, UVB-RESISTANCE 8	Cell cycle (GO:0007049)	−3.02
Csa5M623870.1	AT5G07030.1	Aspartic proteinase nepenthesin-1	Re-arrangements of cell wall [91]	−2.44
Csa5M644550.1	AT3G02530.1	TCP-1/cpn60 chaperonin family protein	Plant cell death [90]	1.85
Csa6M439410.1	AT5G19440.1	Cinnamoyl CoA reductase-like protein	Seed development [89]; lignin biosynthetic pathway	−4.09
Csa7M048110.1	AT3G14940.1	PPC3, PHOSPHOENOLPYRUVATE CARBOXYLASE 3,	Development of male gametophyte; Cell cycle [88]	2.75
Csa7M075590.2	AT5G42650.1	AOS, ALLENE OXIDE SYNTHASE	Floral organ development; defense response (GO:0006952)	−2.65
Csa7M390010.1	AT3G02080.1	40S ribosomal protein S19	Translation (GO:0006412)	2.55
Csa7M405310.1	AT4G02290.1	GH9B13, GLYCOSYL HYDROLASE 9B13	Root development; cell wall organization (GO:0007047)	−2.42

[a]Homologous search was conducted by BLASTP against the *Arabidopsis* Refseq database (http://www.arabidopsis.org/index.jsp). *E*-value was set to <1E-10. The *Arabidopsis* gene ID with highest score is picked for further analysis

[b]The proteins were annotated based on the public databases: *Arabidopsis* database (http://www.arabidopsis.org/index.jsp) and cucumber Refseq database (http://cucumber.genomics.org.cn/page/cucumber/index.jsp)

[c]The proposed biological processes were refer from the GO terms and related research reports of the top hit *Arabidopsis* genes

[d]The expression fold was calculated as the ratio of the protein expression in 2 dpa fruits vs. protein expression in 0 dpa fruit, *P*-value <0.05; The dramatically increased (fold >5) DEPs were marked with stars

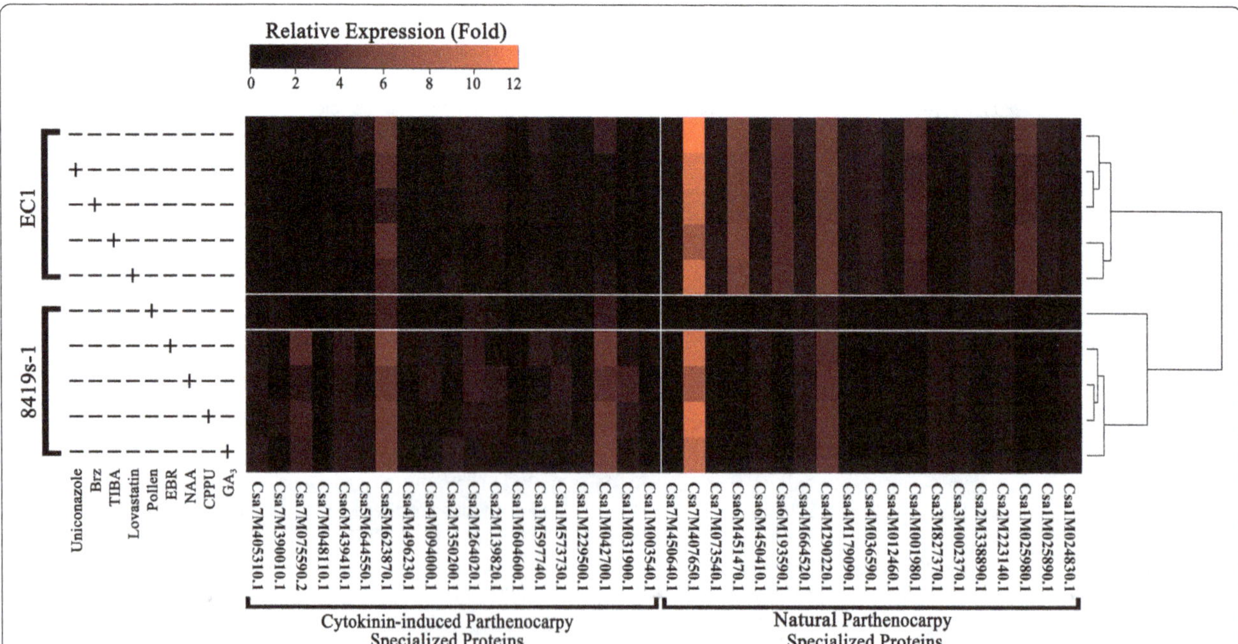

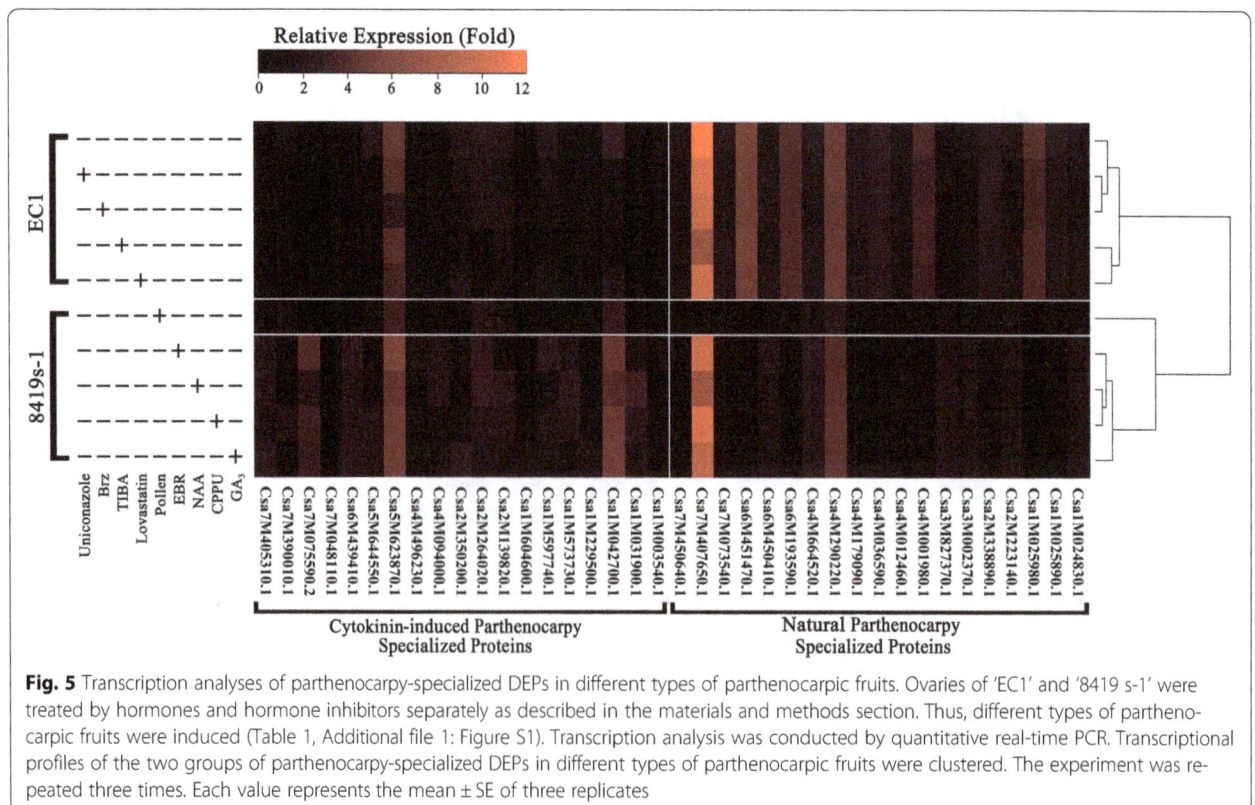

Fig. 5 Transcription analyses of parthenocarpy-specialized DEPs in different types of parthenocarpic fruits. Ovaries of 'EC1' and '8419 s-1' were treated by hormones and hormone inhibitors separately as described in the materials and methods section. Thus, different types of parthenocarpic fruits were induced (Table 1, Additional file 1: Figure S1). Transcription analysis was conducted by quantitative real-time PCR. Transcriptional profiles of the two groups of parthenocarpy-specialized DEPs in different types of parthenocarpic fruits were clustered. The experiment was repeated three times. Each value represents the mean ± SE of three replicates

were up-regulated during natural parthenocarpic fruit set but inactively expressed in the cytokinin-induced fruit, while the NP-specialized protein was not detected in the NP fruit but was actively expressed in the CP fruit (Fig. 6a). Auxins, cytokinins and gibberellins are essential fruit developmental phytohormones, which can induce parthenocarpy in cucumber. The expression patterns of the parthenocarpy-specialized proteins responding to these hormones were also investigated (Fig. 6b). Our results showed that the treatment with endogenous hormones NAA, CPPU and GA3 increased the expression of the CP-specialized protein Csa2M139820.1 but decreased the expression of Csa4M036590.1 and Csa7M450640.1 (NP-specialized proteins). Although the NP-specialized proteins Csa1M025890.1 and Csa7M073540.1 were up-regulated by GA3, they were not significantly changed by NAA and CPPU (Fig. 6b, Additional file 6: Figure S5A).

The protein expression pattern of Csa2M059750.1 was also analyzed, which was up-regulated during natural parthenocarpy and down-regulated in abortive fruit (Additional file 2: Table S3). Csa2M059750.1 was the unique DEP located in the chromosome region of the major parthenocarpic QTL *Parth2.1* of 'EC1' [25]. Western blot analysis showed that Csa2M059750.1 was degraded during fruit abortion but actively expressed during fruit set. However, in contrast to the natural parthenocarpic fruit set, the increasing expression of Csa2M059750.1 was delayed in the pollinated and CP fruits until 3 dpa (Fig. 7a). Western blot analysis also showed that the expression of Csa2M059750.1 was not significantly affected by hormones, indicating that Csa2M059750.1 may be a hormone-insensitive protein (Fig. 7b, Additional file 6: Figure S5B).

Discussion

Although studies on parthenocarpy have been conducted for over 100 years, current understanding of parthenocarpy remains at a nascent stage. Cucumber is emerging as a model for the *Cucurbitaceae* family. The rich parthenocarpic germplasm of cucumber offers an opportunity to investigate the coordination and communication of environmental factors, hormone signals and genes during parthenocarpy. In this study, we investigated the proteomes of cucumber fruits to help improve the global understanding of parthenocarpy.

Post-translational regulation of fruit development in cucumber

Proteins are executors with a vast array of functions within organisms. The translational and post-translational regulations of proteins such as protein synthesis, proteolysis, glycosylation, phosphorylation and folding are essential for plant development. Studies on hormone-dependent biological processes have revealed that post-translational regulations frequently occur during fruit development. For instance, in the absence of auxins, the function of ARFs (auxin response factors) was inhibited via heterologous dimerization with Aux/IAA. However, in the presence of auxin, the ARFs dissociated from Aux/IAA proteins, whose targeting and degradation were mediated by the E3 ubiquitin ligase $SCF^{TIR1/AFB}$, and consequently stimulated fruit set [5, 73–77]. The fruit developmental responses to ethylene were mediated by the $SCF^{EBF1/EBF2}$-dependent proteolysis of EIN3 (Ethylene insensitive 3) [78]. Furthermore, the function of EIN3 was regulated by the MAPK-dependent phosphorylation within the EPR1 domain of the transcriptional factor [79].

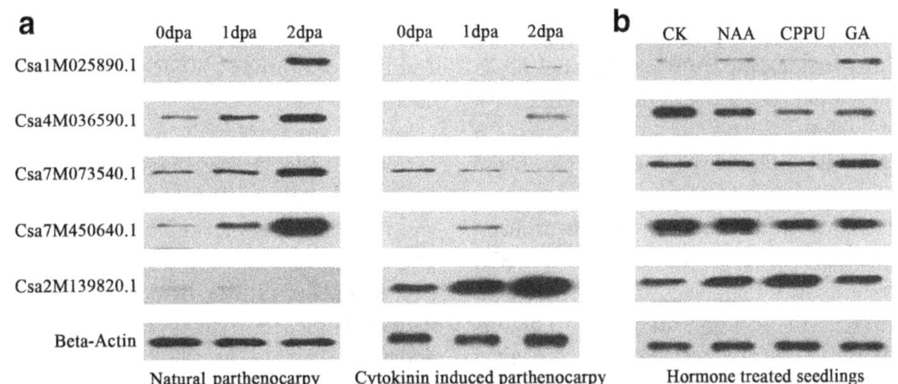

Fig. 6 Western blot analysis of the parthenocarpy-specialized proteins that were actively expressed during NP and CP fruit set. iTRAQ result showed that Csa1M025890.1, Csa4M036590.1, Csa7M073540.1, and Csa7M450640.1 were dramatically increased in natural parthenocarpic fruits, while Csa2M139820.1 was highly increased in cytokinin-induced parthenocarpic fruits (Table 2; marked by solid stars, >5-fold). The expression patterns of these proteins were further analyzed by western blotting. **a** Expression analysis of the parthenocarpy-specialized proteins during NP and CP fruit set individually. **b** Expression analysis of the parthenocarpy-specialized proteins in response to hormone treatments in seedlings. The cucumber beta-actin (Csa5M182010.1) was used as reference protein for Western blotting. The experiment was repeated three times. The band intensity analysis of western blots was conducted using ImageJ (Version 1.4), and the data are presented in Additional file 6: Figure S5A. **CK**: Seedlings without phytohormone treatment; **NAA**: treated with 50 μM NAA; **CPPU**: treated with 10 μM CPPU; **GA**: treated with 10 μM GA_3

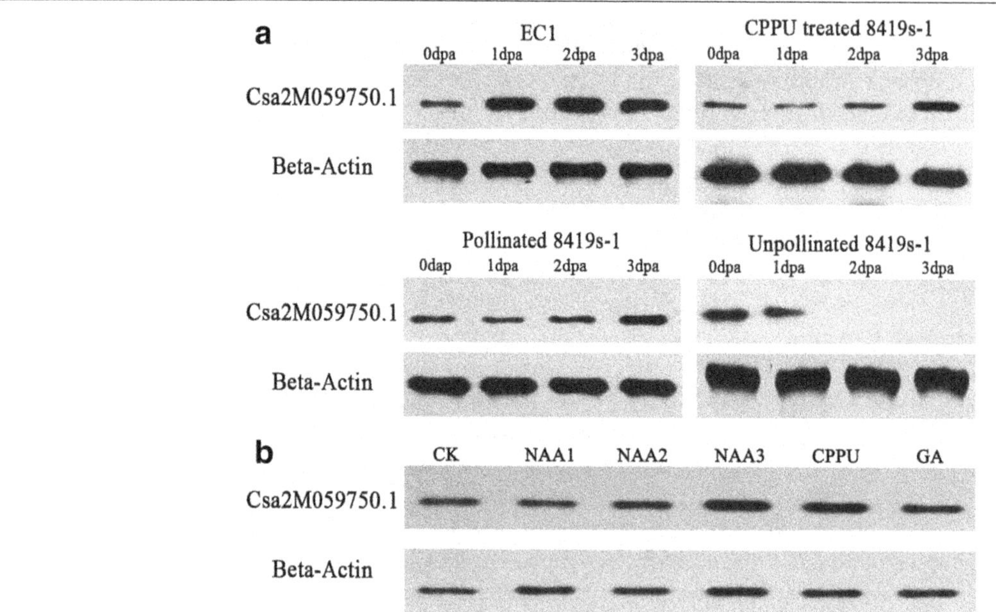

Fig. 7 Expression analysis of Csa2M059750.1 during different fruit developmental processes and the response to phytohormone treatments. Csa2M059750.1 was considered a candidate parthenocarpy regulatory protein by combined analysis of iTRAQ and genetic mapping results (Wu et al. [25]). The protein expression of Csa2M059750.1 was analyzed by western blotting. **a** The expression of Csa2M059750.1 during fruit development; **b** The expression of Csa2M059750.1 after phytohormone treatment in cucumber seedlings. The cucumber beta-actin (Csa5M182010.1) was used as a reference protein for western blotting. The experiment was repeated three times. The band intensity analysis of western blots was recorded using ImageJ (Version 1.4), for which the data are presented in Additional file 6: Figure S5B. **CK**: Seedlings without phytohormone treatment; **NAA1**: treated with 5 μM NAA; **NAA2**: treated with 10 μM NAA; **NAA3**: treated with 50 μM NAA; **CPPU**: treated with 10 μM CPPU; **GA**: treated with 10 μM GA_3

In addition, our previous transcriptome study confirmed that glycosylation reactions were dramatically active throughout fruit development in cucumber [53]. In the present proteomic study, the protein folding-related proteins were up-regulated in both pollinated and parthenocarpic fruits, including Csa1M255160.1, which was annotated as a TCP-1 (T-COMPLEX PROTEIN 1 ALPHA SUBUNIT), and was actively expressed in natural and cytokinin-induced parthenocarpic fruits (Additional file 2: Table S3). Moreover, Csa2M099450.1, also defined as a TCP-1-like protein, showed specialized expression in pollinated and cytokinin-induced parthenocarpic fruits (Additional file 2: Table S1).

Many lines of evidence indicate that protein synthesis/degradation may function during cell growth [80, 81]. The present proteomic study showed that over 30% of the differentially expressed proteins during the cucumber fruit development were related to the biological process of protein metabolism (Fig. 2c). Within the predicted IPP network, nearly half of the interaction proteins were protein metabolism-related proteins, such as the three TIF (TRANSLATION INITIATION FACTOR) proteins involved in the initiation phase of eukaryotic translation, four CSN (COP9 signalosome) multi-proteins that functioned in the ubiquitin–proteasome pathway, and three ribosomal proteins (Additional file 2: Table S1). Protein metabolism related proteins were commonly expressed in unpollinated and natural parthenocarpic fruit (Additional file 2: Table S2). The opposite expression patterns of the common proteins indicated the fate (set or abortion) of the mature ovary in cucumber, which was determined by protein metabolism pathways.

The cues of hormone-independent parthenocarpy in cucumber

Growth measurement showed that natural and cytokinin-induced parthenocarpic fruits presented similar growth curves (Fig. 1a). However, clustering analysis showed that the protein expression profiles of the CP and NP fruits were quite different from each other. Moreover, the cluster distance between the CP and NP fruits was greater than that between the CP fruit and the abortive fruit (Fig. 2b). The specialized proteins expressed in the parthenocarpic fruits were divided into two individual groups, which were separately involved in the NP and CP fruit set, as shown in the Venn diagram (Fig. 4; Table 2). These findings suggested that there may be individual parthenocarpic pathways in cucumber.

Gustafson [54, 55] proposed that plants produce parthenocarpic fruits because the ovary contains enough auxins to promote fruit initiation. Since then, many studies have confirmed that parthenocarpy is a phytohormone-dependent biological process. In cucumber, polar auxin transport-blocking experiments have shown that parthenocarpy could be triggered by the sufficient accumulation of auxin in the ovary [44]. Moreover, the application of exogenous hormones such as auxins, cytokinins, gibberellins and brassinosteroids could induce parthenocarpy [51]. In this study, hormone measurement showed that the endogenous hormone levels increased during fruit set but decreased during fruit abortion in '8419 s-1' (Fig. 1b). However, the endogenous hormone levels were relatively low and remained stable during natural parthenocarpic fruit set in 'EC1' compared with '8419 s-1.' Moreover, the NP fruits showed a broad resistance to hormone inhibitors (Fig. 1b; Table 1; Additional file 1: Figure S1), indicating the existence of a hormone-independent parthenocarpic mechanism in 'EC1.' This speculation was supported by expression analysis in the parthenocarpy-specialized proteins, whereby the NP-specialized proteins performed hormone-insensitive transcriptional and translational functions (Figs. 5, 6 and 7; Additional file 6: Figure S5).

Inhibiting the regulation of fruit abortion in cucumber

Dormant fruits, as a result of first-fruit inhibition or nutritional stress, can always be observed in the field [25, 26]. However, the dormant state of these fruits usually leads to fruit abortion in a short time (2 days at most). Although natural parthenocarpic fruits of 'EC1' could not be blocked by hormone inhibitors, these treated fruits stayed in a dormant state for a long time (more than 4 days) (Additional file 1: Figure S1). We speculated that inhibitory regulations of fruit abortion might exist in 'EC1,' causing the fruits to maintain a dormant state. Coincidentally, the common proteins that detected in the NP fruits of 'EC1' and the abortive fruit of '8419 s-1,' showed opposite expression trends. Most of these proteins were down-regulated during fruit abortion but up-regulated during NP fruit set (Additional file 2: Table S2). Besides, ethephon treating experiments showed that although fruit abortion of the non-parthenocarpic cultivars was accelerated by ethephon which had no effect on fruit development of 'EC1,' further suggesting inhibitory regulations of fruit abortion might exist in 'EC1'. Conversely, hormone inhibitor-induced dormant state in 'EC1' indicated that hormone stimuli might be required for fruit expansion in either parthenocarpic cultivars or non-parthenocarpic cultivars.

Conclusions

Based on the evidence provided in this study, a working hypothesis for the cucumber parthenocarpic fruit set was proposed (Fig. 8), whereby parthenocarpy in cucumber may be promoted by a 'parallel switch,' namely,

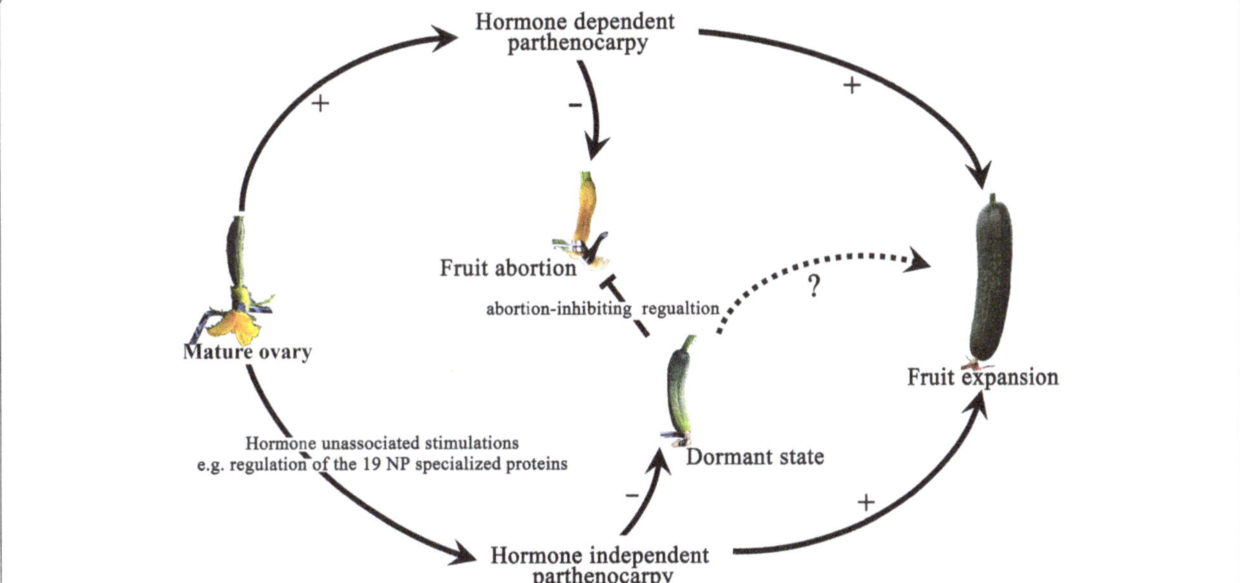

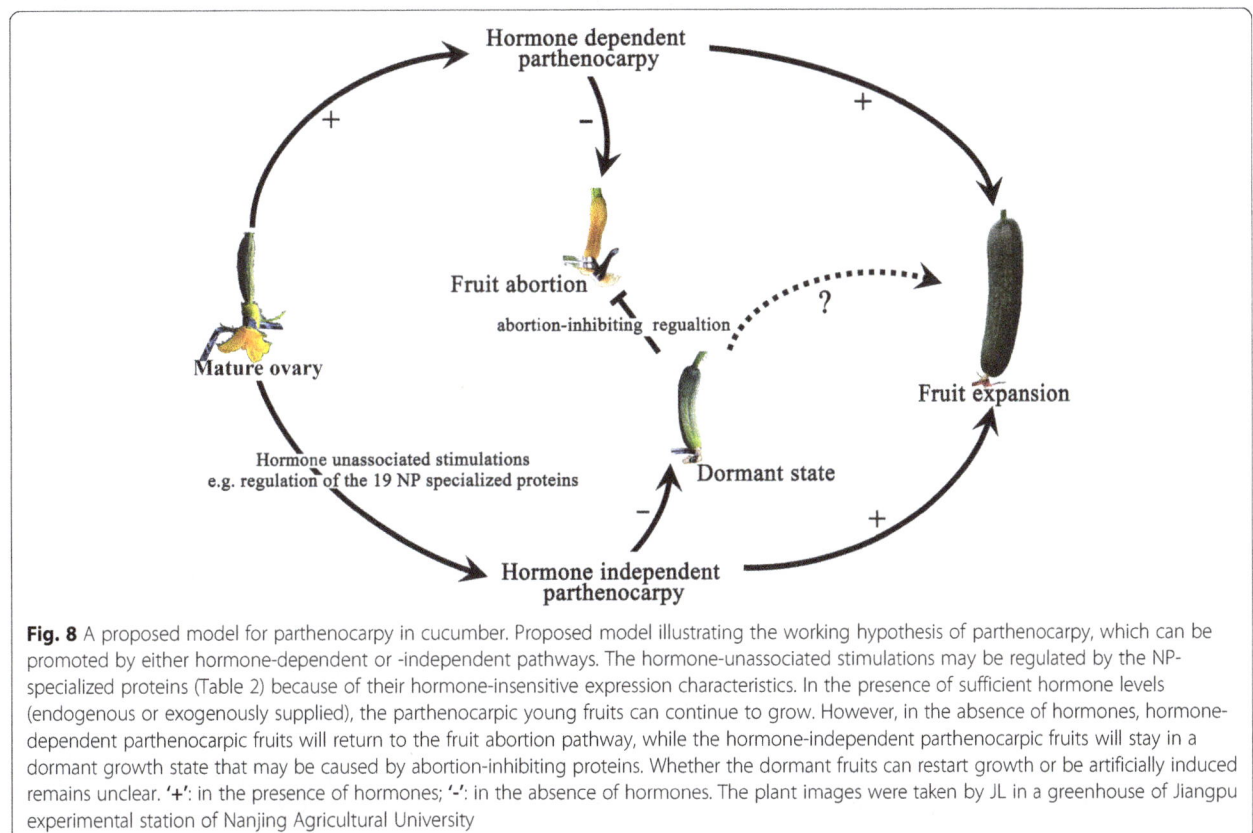

Fig. 8 A proposed model for parthenocarpy in cucumber. Proposed model illustrating the working hypothesis of parthenocarpy, which can be promoted by either hormone-dependent or -independent pathways. The hormone-unassociated stimulations may be regulated by the NP-specialized proteins (Table 2) because of their hormone-insensitive expression characteristics. In the presence of sufficient hormone levels (endogenous or exogenously supplied), the parthenocarpic young fruits can continue to grow. However, in the absence of hormones, hormone-dependent parthenocarpic fruits will return to the fruit abortion pathway, while the hormone-independent parthenocarpic fruits will stay in a dormant growth state that may be caused by abortion-inhibiting proteins. Whether the dormant fruits can restart growth or be artificially induced remains unclear. '+': in the presence of hormones; '-': in the absence of hormones. The plant images were taken by JL in a greenhouse of Jiangpu experimental station of Nanjing Agricultural University

hormone-dependent and hormone-independent pathways. During hormone-independent parthenocarpy, fruit set was promoted by hormone-insensitive regulatory proteins, such as the NP-specialized proteins in 'EC1.' In the presence of sufficient hormones, young fruits formed through both hormone-dependent and -independent pathways could continuously grow to maturity. In the absence of hormones, the development of hormone-sensitive fruits proceeds to fruit abortion, whereas the hormone-insensitive fruits remain in a dormant state because of the increasing expression of abortion-inhibiting proteins. However, the expansion of dormant fruits and their further promotion are unknown. Although the accurate regulation of parthenocarpy in cucumber remains unclear, our studies provide a theoretical framework for understanding the mechanism of parthenocarpy for its application in agricultural production.

Methods

Plant material and growth conditions

In this study, the cucumber cultivar 'EC1' was used as a parthenocarpic sample (Gynoecious inbred line, European Glasshouse type, parthenocarpic rate ≥ 95%) and '8419 s-1' as a non-parthenocarpic sample (Monoecious inbred line, European Glasshouse type, the rare occurrence of parthenocarpy is occasionally observed in the senescence phase of the cultivar). Plants were grown in a greenhouse at Nanjing Agricultural University with a 14 h photoperiod, a mean daily air temperature of 28/20 °C (day/night).

Phytohormone measurement

Phytohormones were separately analyzed through ELISA using IAA, ZR and GA3 ELISA Kits (Sangon Biotech Company) based on Weiler's method [82]. The results are presented as the mean ± SE ($n = 10$) with three technical replicates.

Ovary treatments

The female flowers (at the 12-15th node of the main stem) of the above cucumber cultivars were previously trapped with bags in order to prevent pollen contamination on the day before anthesis. When anthesis, the trapped ovaries were treated separately: keeping trapping (unpollination), hand pollination [34] and CPPU treatment. CPPU (N-(2-chloro-4-pyridyl)-N′-phenyl urea) is a kind of synthetic cytokinin which could induce parthenocarpy in cucumber. For CPPU treatment, 20 μL CPPU solution (100 mg/L) was sprayed on the surface of the ovaries. All the treated ovaries were harvested at 0, 1, 2 and 3 dpa (days-post-anthesis). Thirty ovaries of each treatment were ground into powder with liquid nitrogen and mix as a sample pool for iTRAQ and western blot analysis.

The trapped ovaries of 'EC1' and '8419 s-1' were also treated with phytohormones (NAA, 1-naphthaleneacetic acid, 50 mg/L; CPPU, 100 mg/L; GA3, Gibberellin A3, 50 mg/L; EBR, Epi-Brassinosteroids, 10 mg/L; Ethephon, 100 mg/L) and hormone inhibitors (TIBA, 2,3,5-triodobenzoic acid, inhibitor of auxin, 50 mg/L; Lovastatin, inhibitor of cytokinin, 50 mg/L; uniconazole, inhibitor of gibberellin 50 mg/L; Brz, Brassinazole, inhibitor of Brassinosteroids, 10 mg/L; STS, silver thiosulphate, inhibitor of ethylene, 0.25 mM). The exogenous phytohormones and hormone inhibitors were separately sprayed on the surface of the ovaries at 0dpa. Active pollens and irradiated pollens (γ-ray irradiation at a dose of 200Gy) also used to treat the 0dpa ovaries by hand-pollination. After spaying and hand-pollination, the ovaries were trapped again. The weight, length and diameter of the ovaries were measured at 4dpa. The experiments were repeated three times ($n = 30$). The treated ovaries were also harvested at 2dpa, of which the RNA was isolated for qRT-PCR analysis.

For western blotting, the seedlings of '8419 s-1' (at the three true leaf stage) were also treated with exogenous phytohormones (50 μM, 10 μM, or 5 μM NAA; 10 μM CPPU and 10 μM GA3) by spraying the solutions on the surface of the true leaves. After growing in the growth chamber with a 14 h photoperiod and 25 °C for 24 h, in total 30 true leaves from five individual plants by same treatment were collected and mixed by grinding in liquid nitrogen, then stored at −80 °C before protein extraction.

Protein extraction and quantization

Approximately 1 g of powdered sample was mix with 3 mL extraction buffer [500 mM Tris-HCl (pH 7.5), 150 mM NaCl, 50 mM ethylene diaminetetraacetic acid (EDTA), 1% Triton-X-100, 2 mM dithiothreitol (DTT), 2 mM phenylmethanesulfonyl fluoride (PMSF)]. Protein extraction was performed using the methods described by Omar et al. [83]. The protein precipitation was collected and washed with cold methanol containing 10 mM DTT three times, cold acetone containing 10 mM DTT twice and then dried by vacuum freeze. The extracted proteins were quantified by using the Bradford method [84].

Protein digestion and iTRAQ labeling

One hundred micrograms Proteins from each samples were precipitated with five volume of cold acetone at −20 °C for 1, centrifuged by 12,000 rpm for 15 min at 4 °C, and dried by vacuum freeze dryer (Thermo savant, USA). Pellets were dissolved in the dissolution buffer with reducing reagent described in iTRAQ Reagent 8-Plex kit (Applied Biosystems, USA), and alkylated by cysteine-blocking reagent according to the manufacturer's instructions [85]. After digestion with 50 μl of

50 ng/μl sequence grade modified trypsin (Promega, USA) solution overnight at 37 °C, the peptide samples were labeled. The samples were labeled with the iTRAQ tags as described in Additional file 1: Figure S1.

SCX chromatography and LC–MS/MS analysis

The vacuum dried iTRAQ labeled samples were re-suspended with 100 μl SCX (Strong cation exchange) buffer A (10 mM ammonium formate, 20% ACN (acetonitrile), pH 2.8) and fractionated using a Poly-SEA HPLC (High Performance Liquid Chromatography) column (2.0 × 150 mm, 5 μm particle size, 300   pore size) using at a flow rate of 0.3 ml/min on the Agilent 1200 HPLC System (Agilent, USA). The 50 min HPLC gradient consisted of 100% buffer A (10 mM ammonium formate, 20%ACN, pH 2.8) for 5 min, 0–50% buffer B (500 mM ammonium formate, 20%ACN, pH 2.8) for 25 min, then 50–80% buffer B for 15 min, followed by 80–100% buffer B for 10 min, and lastly 100% buffer B for 15 min. Chromatograms were recorded at 215 and 280 nm. All the collected fractions were vacuum dried, and re-suspended with Nano- RPLC (Reversed Phase Liquid Chromatography) buffer A (0.1% FA, folic acid; 2%ACN). Samples were desalted with C18 nanoLC trap column (100 μm ID × 3 cm, 3 μm particle size, 150   pore size) and Nano-RPLC buffer A (0.1%FA, 2%ACN) at 2 μl/min for 10 min for LC–MS/MS analysis.

The mass spectroscopy analysis was performed using a Triple TOF 5600 System (AB SCIEX, USA), coupled with the Eksigent nanoLC-Ultra™ 2D System (AB SCIEX, USA).The iTRAQ labeled peptides were separated using an analytical ChromXP C18 column (75 μm ID × 15 cm, 3 μm particle size, 120 Å pore size) (New Objectives, USA) with a nanospray emitter (2500 V, 30 PSI (pounds per square inch) curtain gas, 5 PSI nebulizer gas, 150 °C interface heater temperature) (New Objectives, USA), and analyzed by LC-MS/MS. A rolling collision energy setting was applied to all precursor ions for collision-induced dissociation (CID). For information dependent acquisition (IDA), survey scans were acquired in 250 ms and as many as 35 product ion scans were collected if they exceeded a threshold of 150 counts per second (counts/s) with a 2+ to 5+ charge-state. The total cycle time was fixed to 2.5 s. Dynamic exclusion was set for one-half of peak width (18 s), and then the precursor was refreshed off the exclusion list. The peak areas of the iTRAQ reporter ions reflect the abundance of the proteins in the samples.

Protein identification

Mass spectrometric data was processed with Protein Pilot Software v. 4.0 (AB SCIEX, USA) against *Cucumber* database using the Paragon algorithm, and further processed by a Pro Group algorithm where isoform-specific quantification was adopted to trace the differences between expressions of various isoforms which was applied to the peptide identification. Protein identification was performed with emphasis on biological modifications option. Database search parameters were the followings: instrument was TripleTOF 5600, iTRAQ 8-plex quantification, cysteine modified with iodoacetamide, biological modifications were selected as the ID focus, trypsin digestion. An automatic decoy database search strategy was employed to estimate the false discovery rate (FDR) using the Proteomics System Performance Evaluation Pipeline Software (PSPEP) t was integrated in the Protein Pilot Software. In this study, only protein quantification data with the value of global FDR ≤0.05 were chosen for further analysis, and proteins with a |fold change ≥1.5| were considered to be significantly differentially expressed.

Quantitative real time PCR

Proteins based on their differential expression patterns revealed by iTRAQ were selected for verification by Quantitative real-time PCR (qRT-PCR) with primers designed using Primer 5.0 software (Additional file 2: Table S4). Total RNA of the samples described above was extracted by Trizol (Invitrogen, USA). After extraction, total RNA was treated with DNase I (Fermentas, UK) according to the manufacturer's protocol. First-strand cDNA synthesis was carried-out using the PrimeScript™ RT-PCR Kit (TaKaRa, Japan). The real-time qRT-PCR was accomplished in a thermal cycler and analyzed by an IQ5 multicolor Real-time PCR detection system (Bio-Rad, USA). To determine relative fold differences for each sample in each experiment, the CT values were normalized using *Cs-actin* as an internal control and calculated relative to a calibrator using the formula $2^{-\Delta\Delta Ct}$. The experiment was repeated three times.

Western blot analysis

Proteins extracted from ovaries and leaves as previously described were mixed with protein lysis buffer at 4 °C. Total protein lysis was boiled at 98 °C for 10 min and separated with 10% SDS-PAGE (Sodium dodecyl sulfate-polyacrylamide gel electrophoresis), then transferred to nitrocellulose (NC) membranes (GE Hybond, USA) with semi-dry approach. After 2 h of blocking with 5% milk in TBST (Tris Buffered Saline with Tween), membranes were incubated with polyclonal antibodies against the proteins of cucumber that were raised in rabbit by synthetic peptides (Lufei, P.R.China; Additional file 2: Table S5). Polyclonal antibody against cucumber beta-actin was used as internal control. The antibodies were used at 1:300 dilutions. Membranes were incubated with goat anti-rabbit IgG (Proteintech, USA) at 1:2000 dilutions in

TBST for 2 h, after that membranes were washed with TBST for five times. Signals were detected by using enhanced electro-chemiluminescence (Beyotime, P.R. China).

Additional files

Additional file 1: Figure S1. The typical phenotypes of the treated ovaries of 'EC1' and '8419 s-1' at 4 dpa. (DOCX 196 kb)

Additional file 2: Table S1. The annotation and expressional information of the proteins within the protein-protein interactions (PPIs) which presented in Fig. 5. **Table S2.** DEPs commonly expressed in pollinated and Cytokinin induced parthenocarpic fruit (I), and DEPs commonly existed in unpollinated and natural parthenocarpic fruit (II). **Table S3.** DEPs commonly expressed in natural and cytokinin induced parthenocarpic fruits. **Table S4.** The primers of natural and cytokinin induced parthenocarpy specialized protein encoding genes for qRT-PCR. **Table S5.** The antigenic peptides of parthenocarpy specialized proteins for producing rabbit polyclonal antibodies. **Table S6.** The measurement of ovary weight, length and diameter after treated by ethephon and ethylene inhibitors.The treatments were conducted at the anthesis day (0 dpa), and the measurements were conducted at 4 dpa. Means (±SE) of three independent experiments were calculated.Letters indicate differences between the treated ovaries with statistical significance at $P \leq 0.05$ (t-test).Detail information of EC1 and 8419 s-1 was described in M&M section. Both cucumber cultivars CC3 and CCMC were non-parthenocarpic, monoecious inbred lines, Asia ecotype. (XLSX 35 kb)

Additional file 3: Figure S2. The analytical strategy of the iTRAQ based proteome analysis of cucumber fruits. Total proteins of each sample were extracted and labeled separately with tags (113 to 118). Proteomic analysis was conducted by iTRAQ. The differentially expressed proteins (DEPs) were identified by comparing proteomes of 0 and 2 dpa ovaries of each treatment. **CK1 and CK2:** Control sample 1 and Control sample 2; **NP:** natural parthenocarpic fruits of EC1; **Unp:** Unpollination fruits of 8419 s-1 (fruit abortion); **P:** pollination fruits of 8419 s-1; **CP:** Cytokinin induced parthenocarpic fruits of 8419 s-1. Bar = 20 mm. (DOCX 357 kb)

Additional file 4: Figure S3. Statistics of differentially expressed proteins during different fruit developmental processes of cucumber. **NP:** natural parthenocarpic fruits of EC1; **Unp:** Unpollination fruits of 8419 s-1 (fruit abortion); **P:** pollination fruits of 8419 s-1; **CP:** Cytokinin induced parthenocarpic fruits of 8419 s-1. (DOCX 247 kb)

Additional file 5: Figure S4. Protein interactions occurred in different fruit developmental processes based on the predicted PPIs network. The interactions marked in red color means the interactions involved in Cytokinin induced parthenocarpy (A), natural parthenocarpy (B), pollination fruit set (C) and unpollination fruit abortion (D). (DOCX 349 kb)

Additional file 6: Figure S5. Band intensity analysis of western blotting by using the software ImageJ. The relative expression fold of each parthenocarpy specialized protein was calculated by the formula: (band intensity after hormone treatments/band intensity without hormone treatment)/(band intensity of beta-actin in hormone treated sample/band intensity of beta-actin in untreated sample). Each value represents the mean ± SE of three Western blotting replicates. (DOCX 238 kb)

Abbreviations

Brz: Brassinazole, a kind of inhibitor of Brassinosteroids; CP: Cytokinin induced parthenocarpy; CPPU: N-(2-chloro-4-pyridyl)-N0-phenylurea, a diphenylurea-derived cytokinin; DEP: Differentially expressed protein; dpa: days post anthesis; EBR: Epi-Brassinosteroids; FDR: False discovery rate; iTRAQ: isobaric tags for relative and absolute quantitation; NP: NATURAL Parthenocarpy; QTL: Quantitative trait locus; STS: Silver thiosulphate; TIBA: 2,3,5-triodobenzoic acid, a kind of inhibitor of auxin

Acknowledgements

We would like to thank Professor Robert N. Trigiano (University of Tennessee, USA) for critical reading of the manuscript.

Funding

This work was supported by the National Natural Science Foundation of China [31430075, 31672168]; Special Fund for Agro-Scientific Research in the Public Interest [201403032]; National Key Research and Development Program of China [2016YFD0101705-5]; Agricultural science and Technology Innovation Fund of Jiangsu Province [CX(15)1019]. 31430075, 31672168 were involved in iTRAQ analysis. 201403032, 2016YFD0101705-5 and CX(15)1019 were involved in sample collection. The funding bodies were not involved in the design of the study and collection, analysis, and interpretation of data and in writing the manuscript.

Authors' contributions

JFC and JL conceived the study. JL and JX performed most of the experiments. QWG helped to prepare samples for iTRAQ. ZW, TZ, KJZ and CYC carried out the iTRAQ data analysis. PYZ together with JX performed the phytohormone measurement. QFL gave useful advices and contributed to experiments. JL interpreted the results and wrote the paper. All authors read and approved the final version of the manuscript.

Competing interests

The authors declare that they have no competing interests.

References

1. Pandolfini T. Seedless fruit production by hormonal regulation of fruit set. Nutrients. 2009;1(2):168–77.
2. Tiedjens VA. The relation of environment to shape of fruit in Cucumis Sativus L. and its bearing on the genetic potentialities of the plants. J Agr Res. 1928;36:794–809.
3. Denna DW. Effects of genetic parthenocarpy and gynecious flowering habit on fruit production and growth of cucumber *Cucumis Sativus* L. J Am Soc Hortic Sci. 1973;98(6):602–4.
4. Falavigna A, Soressi GP. Influence of the pat-sha gene on plant and fruit traits in tomato (*L. esculentum* mill.) In: Modern trends in tomato genetics and breeding; 1987. p. 128.
5. Wang H, Jones B, Li Z, et al. The tomato aux/IAA transcription factor IAA9 is involved in fruit development and leaf morphogenesis. Plant Cell. 2005; 17(10):2676–92.
6. Gorguet B, Eggink PM, Ocana J, et al. Mapping and characterization of novel parthenocarpy QTLs in tomato. Theor Appl Genet. 2008;116(6):755–67.
7. Goetz M, Vivian-Smith A, Johnson SD, et al. AUXIN RESPONSE FACTOR8 is a negative regulator of fruit initiation in Arabidopsis. Plant Cell. 2006;18(8): 1873–86.
8. Wittwer SH, Bukovac MJ, Sell HM, et al. Some effects of gibberellin on flowering and fruit setting. Plant Physiol. 1957;32(1):39.
9. Martí C, Orzáez D, Ellul P, et al. Silencing of DELLA induces facultative parthenocarpy in tomato fruits. Plant J. 2007;52(5):865–76.
10. Carrera E, Ruiz-Rivero O, Peres LEP, et al. Characterization of the procera tomato mutant shows novel functions of the SIDELLA protein in the control of flower morphology, cell division and expansion, and the auxin-signaling pathway during fruit-set and development. Plant Physiol. 2012;160(3):1581–96.
11. Mapelli S. Changes in cytokinin in the fruits of parthenocarpic and normal tomatoes. Plant Sci Lett. 1981;22(3):227–33.
12. Bohner J, Bangerth F. Effects of fruit set sequence and defoliation on cell number, cell size and hormone levels of tomato fruits (*Lycopersicon esculentum* mill.) within a truss. Plant Growth Regul. 1988;7(3):141–55.
13. Gillaspy G, Ben-David H, Gruissem W. Fruits: a developmental perspective. Plant Cell. 1993;5(10):1439.
14. Srivastava A, Handa AK. Hormonal regulation of tomato fruit development: a molecular perspective. J Plant Growth Regul. 2005;24(2):67–82.
15. Trueman SJ. Endogenous cytokinin levels during early fruit development of macadamia. Afr J Agric Res. 2010;5(24):3402–7.

16. Matsuo S, Kikuchi K, Fukuda M, et al. Roles and regulation of cytokinins in tomato fruit development. J Exp Bot. 2012;63(15):5569–79.
17. Ding J, Chen B, Xia X, et al. Cytokinin-induced parthenocarpic fruit development in tomato is partly dependent on enhanced gibberellin and auxin biosynthesis. PLoS One. 2013;8(7):e70080.
18. Li Y, Yu JQ. Photosynthesis and 14C-assimilate distribution as influenced by CPPU treatment on ovary. Acta Agric Nucleatae Sin. 2001;15:355–9. (in Chinese with English abstract)
19. Vriezen WH, Feron R, Maretto F, et al. Changes in tomato ovary transcriptome demonstrate complex hormonal regulation of fruit set. New Phytol. 2008;177(1):60–76.
20. Pascual L, Blanca JM, Cañizares J, et al. Transcriptomic analysis of tomato carpel development reveals alterations in ethylene and gibberellin synthesis during pat3/pat4 parthenocarpic fruit set. BMC Plant Biol. 2009;9(1):67.
21. Martínez C, Manzano S, Megías Z, et al. Involvement of ethylene biosynthesis and signalling in fruit set and early fruit development in zucchini squash (*Cucurbita pepo* L.). BMC Plant Biol. 2013;13(1):139.
22. Fos M, Nuez F, García-martínez JL. The gene pat-2, which induces natural parthenocarpy, alters the gibberellin content in unpollinated tomato ovaries. Plant Physiol. 2000;122(2):471–80.
23. Beraldi D, Picarella ME, Soressi GP, et al. Fine mapping of the parthenocarpic fruit (*pat*) mutation in tomato. Theor Appl Genet. 2004;108(2):209–16.
24. Miyatake K, Saito T, Negoro S, et al. Development of selective markers linked to a major QTL for parthenocarpy in eggplant (*Solanum melongena* L.). Theor Appl Genet. 2012;124(8):1403–13.
25. Wu Z, Zhang T, Li L, et al. Identification of a stable major-effect QTL (Parth 2.1) controlling parthenocarpy in cucumber and associated candidate gene analysis via whole genome re-sequencing. BMC Plant Biol. 2016;16(1):182.
26. Lietzow CD, Zhu H, Pandey S, et al. QTL mapping of parthenocarpic fruit set in north American processing cucumber. Theor Appl Genet. 2016;129(12): 2387–401.
27. Smith O, Cochran HL. Effect of temperature on pollen germination and tube growth in the tomato. New York: The University; 1935.
28. Rylski I. Effects of season on parthenocarpic and fertilized summer squash (*Cucumis pepo* L.). Exp Agric. 1974;10(01):39–44.
29. Sun C, Li Y, Zhao W, et al. Integration of hormonal and nutritional cues orchestrates progressive corolla opening. Plant Physiol. 2016;171(2):1209–29.
30. Cholodny N. Beiträge zur hormonalen Theorie von Tropismen. Planta. 1928; 6(1):118–34.
31. Went FW, Thimann KV. Phytohormones. NY: Macmillan; 1937.
32. Rudich J, Baker LR, Sell HM. Parthenocarpy in *Cucumis sativus* L. as affected by genetic parthenocarpy, thermo-photoperiod, and femaleness. J Am Soc Hort Sci 1977;102(2):225-8.
33. Matlob AN, Kelly WC. Growth regulator activity and parthenocarpic fruit production in snake melon and cucumber grown at high temperature. J Am Soc Hortic Sci. 1975;100:406–9.
34. Kim IS, Okubo H, Fujieda K. Endogenous levels of IAA in relation to parthenocarpy in cucumber (*Cucumis sativus* L.). Sci Hortic. 1992;52(1–2):1–8.
35. Huang S, Li R, Zhang Z, et al. The genome of the cucumber, *Cucumis sativus* L. Nat Genet. 2009;41(12):1275–81.
36. Hawthorn LR, Wellington R. Geneva, a greenhouse cucumber that develops fruit without pollination. N Y State Agric Exp Station. 1930;2:3–11.
37. Pike LM, Peterson CE. Inheritance of parthenocarpy in the cucumber (*Cucumis sativus* L.). Euphytica. 1969;18:101–5.
38. Kvasnikov BV, Rogova NT, Tarakanova SI, Ignatov SI. Methods of breeding vegetable crops under the covered ground. Trudy Prikl Bot Genet Selek. 1970;42:45–57.
39. Juldasheva L. Inheritance of the tendency towards parthenocarpy in cucumbers. Byull Vsesoyuznogo Ordena Lenina Inst Rastenievodstva Imeni NI Vavilova. 1973;32:58–9.
40. Meshcherov E, Juldasheva L. Parthenocarpy in cucumber. Trudy Prikl Bot Genet Selek. 1974;51:204–13.
41. Shawaf EI, Baker L. Inheritance of parthenocarpic yield in gynoecious pickling cucumber for once-over mechanical harvest by diallel analysis of six gynoecious lines. J Am Soc Hortic Sci. 1981;106:359–64.
42. Sun ZY, Lower RL, Staub JE. Analysis of generation means and components of variance for parthenocarpy in cucumber (*Cucumis sativus* L.). Plant Breed. 2006;125:277–80.
43. Yan LY, Lou LN, Li XL, et al. Evaluation of parthenocarpy in cucumber germplasm. Acta Hortic Sin. 2009;36:975–82. (in Chinese with English abstract)
44. Robinson RW, Cantliffe DJ, Shannon S. Morphactin-induced parthenocarpy in the cucumber. Science. 1971;171(3977):1251–2.
45. Cantliffe D. Parthenocarpy in the cucumber induced by some plant growth-regulating chemicals. Can J Plant Sci. 1972;52(5):781–5.
46. Quebedeaux B, Beyer E. Chemically-induced parthenocarpy cucumber by a new inhibitor of auxin transport. Hortscience. 1972;7:474–6.
47. Elassar GJ, Patevitch D, Kedar N. Induction of Parthenocarpic fruit development cucumber by growth regulators. Hortscience. 1974;9(3):238–9.
48. Takeno K, Ise H, Minowa H, et al. Fruit growth induced by benzyladenine in Cucumis Sativus L.: influence of benzyladenine on cell division, cell enlargement and indole-3-acetic acid content. J Jpn Soc Hortic Sci. 1992;60(4):915–20.
49. Yin Z, Malinowski R, Ziolkowska A, et al. The DefH9-iaaM-containing construct efficiently induces parthenocarpy in cucumber. Cell Mol Biol Lett. 2006;11(2):279–90.
50. Ogawa Y, Inoue N, Aoki S. Promotive effects of exogenous and endogenous gibberellins on the fruit development in *Cucumis sativus* L. J Jpn Soc Hortic Sci. 1989;58(2):327–31.
51. Fu FQ, Mao WH, Shi K, et al. A role of brassinosteroids in early fruit development in cucumber. J Exp Bot. 2008;59(9):2299–308.
52. Hikosaka S, Sugiyama N. Effects of exogenous plant growth regulators on yield, fruit growth, and concentration of endogenous hormones in Gynoecious Parthenocarpic cucumber (*Cucumis sativus* L.). Hortic J. 2015;84(4):342–9.
53. Li J, Wu Z, Cui L, et al. Transcriptome comparison of global distinctive features between pollination and parthenocarpic fruit set reveals transcriptional phytohormone cross-talk in cucumber (*Cucumis sativus* L.). Plant Cell Physiol. 2014;55(7):1325–42.
54. Gustafson FG. Further studies on artificial parthenocarpy. Am J Bot. 1938a; 25:237–44.
55. Gustafson FG. Induced parthenocarpy. Bot Gaz. 1938b;99(4):840–4.
56. Schwabe WW. Hormones and parthenocarpic fruit set, a literature survey. Hortic Abstr. 1981;51:661–98.
57. Vivian-Smith A, Koltunow AM. Genetic analysis of growth-regulator-induced parthenocarpy in Arabidopsis. Plant Physiol. 1999;121(2):437–52.
58. Spena A, Rotino GL, Bhajwani SS, Soh WY. Parthenocarpy. State of the art, Current trends in the embryology of Angiosperms. Kluwer Academic Publishers; 2001. p. 435-450.
59. Marcelis LFM, Baan Hofman-Eijer LR. Cell division and expansion in the cucumber fruit. J Hortic Sci. 1993;68(5):665–71.
60. Huitrón MV, Diaz M, Diánez F, et al. Effect of 2, 4-D and CPPU on triploid watermelon production and quality. Hortscience. 2007;42(3):559–64.
61. Bangerth F, Schröder M. Strong synergistic effects of gibberellins with the synthetic cytokinin N-(2-chloro-4-pyridyl)-N-phenylurea on parthenocarpic fruit set and some other fruit characteristics of apple. Plant Growth Regul. 1994;15(3):293–302.
62. Lewis DH, Burge GK, Hopping ME, et al. Cytokinins and fruit development in the kiwifruit (*Actinidia deliciosa*). II. Effects of reduced pollination and CPPU application. Physiol Plant. 1996;98(1):187–95.
63. NeSmith DS. Response of rabbiteye blueberry (*Vaccinium ashei* Reade) to the growth regulators CPPU and gibberellic acid. Hortscience. 2002;37(4):666–8.
64. Boonkorkaew P, Hikosaka S, Sugiyama N. Effect of pollination on cell division, cell enlargement, and endogenous hormones in fruit development in a gynoecious cucumber. Sci Hortic. 2008;116(1):1–7.
65. Fang JB, Tian LL, Li SH, et al. Influence of CPPU on the sink and source of kiwifruit. Acta Horticulturae Sin. 2000;27(6):444–6. (in Chinese with English abstract)
66. Zeng H, Yang W, Lu C, et al. Effect of CPPU on carbohydrate and endogenous hormone levels in young macadamia fruit. PLoS One. 2016; 11(7):e0158705.
67. Ando K, Grumet R. Transcriptional profiling of rapidly growing cucumber fruit by 454-pyrosequencing analysis. J Am Soc Hortic Sci. 2010;135(4):291–302.
68. Ando K, Carr KM, Grumet R. Transcriptome analyses of early cucumber fruit growth identifies distinct gene modules associated with phases of development. BMC Genomics. 2012;13(1):518.
69. Thimm O, Bläsing O, Gibon Y, et al. MAPMAN: a user-driven tool to display genomics data sets onto diagrams of metabolic pathways and other biological processes. Plant J. 2004;37(6):914–39.

70. Dreze M, Carvunis A-R, Charloteaux B, et al. Evidence for network evolution in an Arabidopsis interactome map. Science. 2011;333:601–7.
71. Lü S, Zhao H, Des Marais DL, et al. Arabidopsis ECERIFERUM9 involvement in cuticle formation and maintenance of plant water status. Plant Physiol. 2012;159(3):930–44.
72. Springer PS, Holding DR, Groover A, et al. The essential Mcm7 protein PROLIFERA is localized to the nucleus of dividing cells during the G (1) phase and is required maternally for early Arabidopsis development. Development. 2000;127(9):1815–22.
73. De Jong M, Wolters-Arts M, Feron R, et al. The Solanum Lycopersicum auxin response factor 7 (SlARF7) regulates auxin signaling during tomato fruit set and development. Plant J. 2009;57(1):160–70.
74. Goetz M, Hooper LC, Johnson SD, et al. Expression of aberrant forms of AUXIN RESPONSE FACTOR8 stimulates parthenocarpy in Arabidopsis and tomato. Plant Physiol. 2007;145(2):351–66.
75. Ren Z, Li Z, Miao Q, et al. The auxin receptor homologue in Solanum Lycopersicum stimulates tomato fruit set and leaf morphogenesis. J Exp Bot. 2011;62(8):2815–26.
76. Dharmasiri N, Dharmasiri S, Weijers D, et al. Plant development is regulated by a family of auxin receptor F box proteins. Dev Cell. 2005;9(1):109–19.
77. Kepinski S, Leyser O. The Arabidopsis F-box protein TIR1 is an auxin receptor. Nature. 2005;435(7041):446–51.
78. Guo H, Ecker JR. Plant responses to ethylene gas are mediated by SCF EBF1/EBF2-dependent proteolysis of EIN3 transcription factor. Cell. 2003; 115(6):667–77.
79. Li J, Li Z, Tang L, et al. A conserved phosphorylation site regulates the transcriptional function of ETHYLENE-INSENSITIVE3-like1 in tomato. J Exp Bot. 2012;63(1):427–39.
80. Key JL. Ribonucleic acid and protein synthesis as essential processes for cell elongation. Plant Physiol. 1964;39(3):365.
81. Faurobert M, Mihr C, Bertin N, et al. Major proteome variations associated with cherry tomato pericarp development and ripening. Plant Physiol. 2007; 143(3):1327–46.
82. Weiler EW, Jourdan PS, Conrad W. Levels of indole-3-acetic acid in intact and decapitated coleoptiles as determined by a specific and highly sensitive solid-phase enzyme immunoassay. Planta. 1981;153(6):561–71.
83. Omar AA, Song WY, Grosser JW. Introduction of Xa21, a Xanthomonas-resistance gene from rice, into 'Hamlin' sweet orange [*Citrus sinensis* (L.) Osbeck] using protoplast-GFP co-transformation or single plasmid transformation. J Hortic Sci Biotechnol. 2007;82(6):914–23.
84. Bradford MM. A rapid and sensitive method for the quantitation of microgram quantities of protein utilizing the principle of protein-dye binding. Anal Biochem. 1976;72(1–2):248–54.
85. Wisniewski JR, Zougman A, Nagaraj N, et al. Universal sample preparation method for proteome analysis. Nat Methods. 2009;6(5):359.
86. Wang Y, Zhang WZ, Song LF, et al. Transcriptome analyses show changes in gene expression to accompany pollen germination and tube growth in Arabidopsis. Plant Physiol. 2008;148(3):1201–11.
87. Rajjou L, Belghazi M, Huguet R, et al. Proteomic investigation of the effect of salicylic acid on Arabidopsis seed germination and establishment of early defense mechanisms. Plant Physiol. 2006;141(3):910–23.
88. Menges M, Hennig L, Gruissem W, et al. Cell cycle-regulated gene expression in Arabidopsis. J Biol Chem. 2002;277(44):41987–2002.
89. Hajduch M, Hearne LB, Miernyk JA, et al. Systems analysis of seed filling in Arabidopsis: using general linear modeling to assess concordance of transcript and protein expression. Plant Physiol. 2010;152(4):2078–87.
90. Hwang IS, Kim NH, Choi DS, et al. Overexpression of Xanthomonas campestris pv. Vesicatoria effector AvrBsT in Arabidopsis triggers plant cell death, disease and defense responses. Planta. 2012;236(4):1191–204.
91. Irshad M, Canut H, Borderies G, et al. A new picture of cell wall protein dynamics in elongating cells of Arabidopsis Thaliana: confirmed actors and newcomers. BMC Plant Biol. 2008;8(1):94.

7

Genetic diversity and structure of Iberian Peninsula cowpeas compared to world-wide cowpea accessions using high density SNP markers

Márcia Carvalho[1], María Muñoz-Amatriaín[2], Isaura Castro[1,3*], Teresa Lino-Neto[4], Manuela Matos[3,5], Marcos Egea-Cortines[6], Eduardo Rosa[1], Timothy Close[2] and Valdemar Carnide[1,3]

Abstract

Background: Cowpea (*Vigna unguiculata* L. Walp) is an important legume crop due to its high protein content, adaptation to heat and drought and capacity to fix nitrogen. Europe has a deficit of cowpea production. Knowledge of genetic diversity among cowpea landraces is important for the preservation of local varieties and is the basis to obtain improved varieties. The aims of this study were to explore diversity and the genetic structure of a set of Iberian Peninsula cowpea accessions in comparison to a worldwide collection and to infer possible dispersion routes of cultivated cowpea.

Results: The Illumina Cowpea iSelect Consortium Array containing 51,128 SNPs was used to genotype 96 cowpea accessions including 43 landraces and cultivars from the Iberian Peninsula, and 53 landraces collected worldwide. Four subpopulations were identified. Most Iberian Peninsula accessions clustered together with those from other southern European and northern African countries. Only one accession belonged to another subpopulation, while two accessions were 'admixed'. A lower genetic diversity level was found in the Iberian Peninsula accessions compared to worldwide cowpeas.

Conclusions: The genetic analyses performed in this study brought some insights into worldwide genetic diversity and structure and possible dispersion routes of cultivated cowpea. Also, it provided an in-depth analysis of genetic diversity in Iberian Peninsula cowpeas that will help guide crossing strategies in breeding programs.

Keywords: *Vigna unguiculata*, Single nucleotide polymorphism, Genetic diversity and variation, Population structure

Background

Cowpea (*Vigna unguiculata* L. Walp., 2n = 2× = 22) is a member of the Fabaceae family and one of the most important grain legumes growing in tropical and subtropical regions [1]. Grain-type cowpea, also known as common cowpea or African cowpea belongs to subspecies *unguiculata* while vegetable cowpea, commonly known as asparagus bean or 'yardlong' bean, belongs to subspecies *sesquipedalis* [2]. These two subspecies are differentiated mainly by their plant architecture, pod size and thickness, and end use [3, 4], but they both possess a high protein content [3, 5]. Other important characteristics of cowpea are the capacity to fix atmospheric nitrogen through symbiosis with root nodule bacteria [6], the ability to grow in low fertility soils [7], and the high tolerance to high temperatures and drought [8]. These attributes make cowpea a key crop in the context of global climate change and food security. In Southern Europe, namely the Iberian Peninsula, rainfall is projected to decrease while temperature is projected to increase [9].

* Correspondence: icastro@utad.pt
[1]Centre for Research and Technology of Agro-Environmental and Biological Sciences (CITAB), University of Trás-os-Montes and Alto Douro (UTAD), 5000-801 Vila Real, Portugal
[3]Department of Genetics and Biotechnology, University of Trás-os-Montes and Alto Douro (UTAD), 5000-801 Vila Real, Portugal
Full list of author information is available at the end of the article

Cowpea is native to Africa [10, 11] although the center of domestication is still uncertain. In the Neolithic period, cowpea was first introduced into India, which is now considered a secondary center of genetic diversity [12]. Some reports suggest that cowpea has been cultivated in Europe at least since the eighteenth century BC and possibly since prehistoric times [13, 14], while others suggest that it was introduced in Europe around 300 BC, where it still remains as a minor crop in the southern part of the continent. These two scenarios are not mutually exclusive. From Europe, more specifically from Spain, it has been speculated that cowpea was exported in the seventeenth century to the New World [15–17].

Assessment of the genetic diversity within a crop's germplasm is fundamental for crop improvement and selection [1]. Moreover, the utilization of landraces is valuable as they can contain favorable alleles for many agronomic traits [18]. Until now, Iberian Peninsula cowpeas, including landraces, have not been genetically characterized, which is a prerequisite for their full exploitation in breeding. Recently, an iSelect BeadArray which assays 51,128 SNPs has been developed for cowpea and used to generate a consensus genetic map containing 37,372 SNPs and to assess genetic diversity within West African breeding materials [19], and to better understand the genetic basis underlying pod length variation [2].

Europe has a deficit of grain legumes, including cowpea. Imports into Europe were about 1.7 million tonnes worth 1.3 billion € in 2015 [20]. The recently developed Cowpea iSelect Consortium Array [19] provides an opportunity to use this tool to understand diversity in Iberian Peninsula cowpea germplasm and to apply this knowledge to breeding varieties producing higher and stable yields in the hotter, drier summers of Southern Europe. The main objectives of this study were to: (1) understand genetic diversity and structure in a set of Iberian Peninsula cultivated cowpea accessions in comparison to a worldwide collection of cowpea accessions; and (2) infer possible dispersion routes of cultivated cowpea, focusing on the contribution of the Iberian Peninsula cowpea germplasm.

Methods

Plant material

A total of 96 cowpea accessions from twenty-four countries were used in this study. They included 33 accessions from Portugal, 10 accessions from Spain (for a total of 43 accessions representing the diversity of Iberian Peninsula germplasm), and 53 accessions from genebanks at the National Institute for Agrarian and Veterinarian Research (INIAV, Portugal), the National Plant Genetic Resources Centre-National Institute for Agricultural and Food Technology Research (CRF-INIA, Spain), the Leibniz Institute of Plant Genetics and Crop Plant Research (IPK, Gatersleben, Germany), the Botanic Garden Meise (Belgium), the University of Perugia (Italy), and the Brazilian Agricultural Research Corporation (EMBRAPA, Brazil). These 53 accessions were chosen to represent worldwide cowpea diversity (Additional file 1). From these 96 accession, 86 belonged to ssp. *unguiculata*, while 10 were part of the ssp. *sesquipedalis*.

Leaves from three individual plants of each accession were collected. Total genomic DNA from each plant was extracted from 50 mg of well-developed trifoliate leaves (two-weeks-old) with the NucleoSpin® Plant II kit (Macherey-Nagel, Düren, Germany) using the Lysis Buffer 1 (based on the CTAB method) and the standard protocol according to the manufacturer's instructions. DNA concentrations were measured using a NanoDrop 1000 (Invitrogen, California, USA). In order to verify DNA integrity, 2 μL of DNA were subjected to gel electrophoresis on 1.0% (*w*/*v*) agarose gel, stained with ethidium bromide. Equal amounts of the three DNA samples of each accession were bulked for genotyping to get a better estimation of diversity within each accession/bulk.

SNP genotyping and data curation

The 96 accessions were genotyped with the Illumina Cowpea iSelect Consortium Array containing 51,128 SNPs [19] at the University of Southern California Molecular Genomics Core Facility (Los Angeles, CA, USA). SNPs included in this iSelect array were discovered in a panel of 37 phenotypically and genetically diverse accessions of cultivated cowpea from 12 countries in Africa, China and the USA, and included four accessions of ssp. *sesquipedalis* (Muñoz-Amatriaín et al. [19]). SNP calling was performed in GenomeStudio v.2011.1 software (Illumina Inc., San Diego, CA, USA) using the same cluster file as in Muñoz-Amatriaín et al. [19]. Quality control filters were applied to both SNPs and samples: first, SNPs with missing data and/or heterozygous calls in >20% accessions were eliminated; second, accessions with >20% missing SNP calls (which may be indicative of poor DNA quality) and/or >20% heterozygous calls were removed from further analysis. The 20% heterozygosity threshold was chosen based on outcrossing rates from 1 to 15% reported for cultivated cowpea [3, 21, 22]. In addition, SNPs were used to identify potentially identical individuals in the collection by performing pair-wise comparisons.

Population structure and genetic diversity analyses

Population structure was estimated using the Bayesian model-based approach implemented in the software STRUCTURE v2.3.4 [23] and by Principal Component Analysis (PCA) in TASSEL v.5 [24] using SNPs with a minor allele frequency (MAF) >0.05. To identify the most likely number of subpopulations, STRUCTURE was run for each hypothetical number of subpopulations (K) between 1 and 8 using a burn-in period of 5000 iterations and a run length of 5000 Monte Carlo Markov Chain (MCMC) iterations. LnP(D) and ΔK values [25] were plotted with Structure Harvester [26]. After estimating the best K, a new run using a burn-in period of 100,000 and 100,000 MCMC was performed to assign accessions to subpopulations. Those accessions with a membership probability lower than 0.70 of belonging to one subpopulation were assigned to an 'admixed' group.

Principal Component Analysis (PCA) was conducted in TASSEL v.5 [24] on the same dataset and plotted using TIBCO Spotfire® 6.5.0.

A neighbor-joining (NJ) tree was generated based on Manhattan distances using the R package "Phyclust" [27].

Expected heterozygosity (*He*) and polymorphism information content (PIC) [28] were calculated for all *V. unguiculata* ssp. *unguiculat*a accessions and then separately for Iberian Peninsula accessions and for the world-wide set of accessions as in Muñoz-Amatriaín et al. [19].

SNP data were used to generate a similarity matrix between *V. unguiculata* ssp. *unguiculat*a accessions from Iberian Peninsula based on simple matching coefficient (number of common SNP alleles divided by the total number of SNPs).

Results

SNP genotyping and data curation

A high-density genotyping array containing 51,128 SNPs [19] was used to genetically characterize 43 landraces and cultivars from the Iberian Peninsula and 53 landraces collected worldwide for a total of 96 cowpea accessions. After SNP calling using GenomeStudio software (Illumina Inc., San Diego, CA, USA), quality control (QC) filtering was applied to both SNPs and accessions with the goal of removing SNPs with low performance accuracy, and accessions that failed in the SNP assay and/or were highly heterozygous (see Methods). Five accessions were eliminated, one of them (Ac61) because of its high percentage of missing calls (40%) indicating poor DNA quality, and the remaining four (Ac45, Ac46, Ac65 and Ac79) because they had high levels of "heterozygosity" (because DNAs were mixed from three plants, the apparent heterozygosity may have an alternative explanation of high heterogeneity between individuals), ranging from 22% to 33% heterozygous calls. These percentages exceeded the expected genetic variability within a cowpea landrace, where outcrossing rates from <1% to a maximum of 15% have been reported [3, 21, 22]. The remaining 91 accessions had percentages of heterozygosity from 0 to 16%, with an average of 2.7% heterozygosity.

A total of 44,056 good-quality polymorphic SNPs and 91 samples were used for further analysis. Pairwise SNP comparisons among accessions showed that Ac39 and Ac43 were potentially duplicates (100% similar SNP calls). These two accessions are members of ssp. *sesquipedalis* that were obtained from the National Plant Genetic Resources Centre-National Institute for Agricultural and Food Technology Research (CRF-INIA, Spain) genebank. This identity was also apparent at the phenotypic level (e.g. samples had the same growth habit, leaf type, flower color, seed color and shape, and hilum color).

Genetic diversity and structure in the whole population

Genetic structure in the entire population of 91 accessions was evaluated using STRUCTURE v.2.3.5 [25], principal component analysis (PCA) in TASSEL V.5.0 [24] and a Neighbor-Joining (NJ) tree generated with "Phyclust" [27].

Using STRUCTURE, the estimated log probability of the data for each given population (K), from 1 to 8, reached a maximum at $K = 4$ (Additional files 2 and 3). In addition, Evanno's ΔK also showed the highest value at $K = 4$ (Additional files 2 and 3). These results indicated that the most likely number of subpopulations in this dataset is four. A new run was performed at $K = 4$ to assign accessions to subpopulations. Accessions with membership probability lower than 0.70 of belonging to one subpopulation were assigned to an 'admixed' group (Additional file 4). Subpopulation 1 included nine accessions, all of them members of ssp. *sesquipedalis*. All other subpopulations (2, 3, and 4) consisted of ssp. *unguiculata* accessions (Fig. 1; Additional file 4). Subpopulation 2 (41 accessions) included accessions from southern Europe, North Africa and Cuba; subpopulation 3 (13 accessions) included accessions from countries in South and Southeast Africa, South America and Asia; and subpopulation 4 (4 accessions) was composed of only West African accessions (Fig. 1; Additional file 4). The remaining 24 accessions were 'admixed'.

This four major subpopulations were also distinguished by PCA (Fig. 2, upper plots): PC1 clearly separated subpopulations 2 and 3, while PC2 separated ssp. *sesquipedalis* accessions belonging to subpopulation 1 from the ssp. *unguiculata* ones. Subpopulation 4 was separated from the rest in PC3 (Fig. 2, upper plots). The NJ tree showed accessions clustered by subpopulation membership, supporting results from both STRUCTURE and PCA (Fig. 3).

PIC and *He* were calculated for the entire population and separately for each subpopulation (Table 1). Considering the whole dataset, the average PIC and *He* were 0.22 and 0.26, respectively. Average PIC values ranged

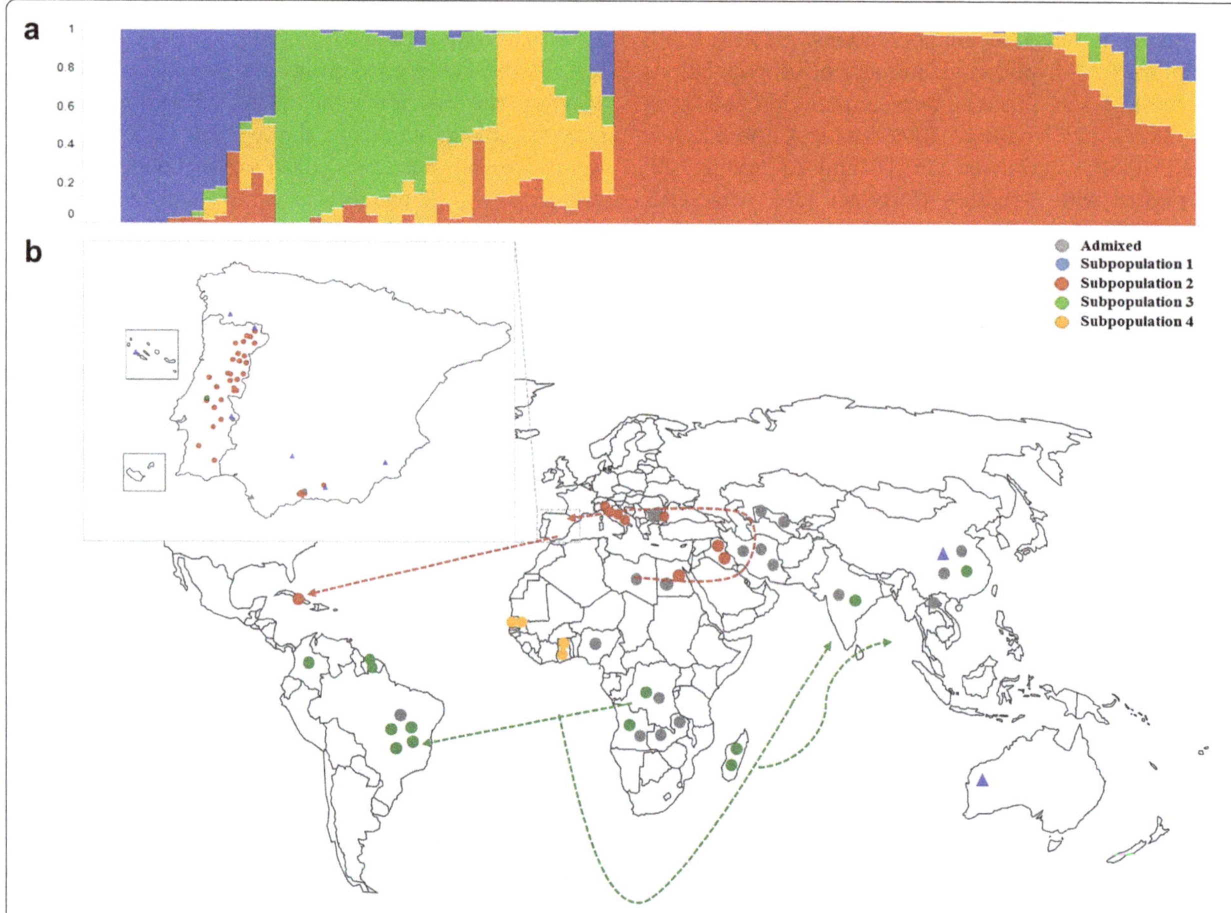

Fig. 1. Population structure for 91 cowpea accessions. **a** Plot of ancestry estimates for K = 4; **b** geographical distribution and population structure of accessions used in this study, and inferred cowpea dispersion routes. Exact locations are provided for Iberian Peninsula accessions. For genebank accessions, coordinates were slightly adjusted in cases where latitude and longitude were identical to allow a visualization of all samples in the study. Each color represents a subpopulation as inferred by STRUCTURE (blue = subpopulation 1; red = subpopulation 2; green = subpopulation 3; orange = subpopulation 4), with 'grey' being used for the 'admixed' group (membership coefficient < 0.7). Shapes are used to distinguish the two subspecies of *Vigna unguiculata* used in this study, with circles representing ssp. *unguiculata* accessions and triangles indicating ssp. *sesquipedalis* accessions

from 0.07 in subpopulation 2 to 0.18 in subpopulation 3, while average *He* ranged from 0.09 to 0.23 in subpopulations 2 and 3, respectively (Table 1). This indicates that subpopulation 3 is the most diverse genetically, while subpopulation 2 appeared the least diverse, even though it contained the highest number of accessions (Table 1).

The geographical distribution of accessions together with their subpopulation membership allowed inference of possible dispersion routes (Fig. 1). The similarity between European and northern African accessions seems to indicate that cowpeas were brought by Arabs to Europe. The accession from Cuba may have been brought by Spanish navigators because Cuba was a Spanish colony and consequently commercial exchanges were frequent. The accessions from South America and Asia belonged to the same subpopulation as those from South/East Africa (Fig. 1). It is possible that these were brought from that region in Africa to Asia and South America during the discovery period, when Portuguese had an important role in commercial routes in the southern hemisphere. If so, Iberian Peninsula people may have had an important role in the distribution of cowpea from Africa and Europe to other parts of the world.

Genetic structure and diversity of Iberian peninsula accessions from subspecies *unguiculata*

Genetic structure and diversity were explored for 35 Iberian Peninsula accessions belonging to ssp. *unguiculata* compared to 46 world-wide ssp. *unguiculata* accessions. Due to the low number of ssp. *sesquipedalis* accessions in the dataset (10 in total) and the fact that grain-type cowpea (ssp. *unguiculata*) is the most cultivated and consumed in Europe, ssp. *sesquipedalis* accessions were not included in these analyses. Most of the 35 *V. unguiculata* ssp. *unguiculata* accessions from the

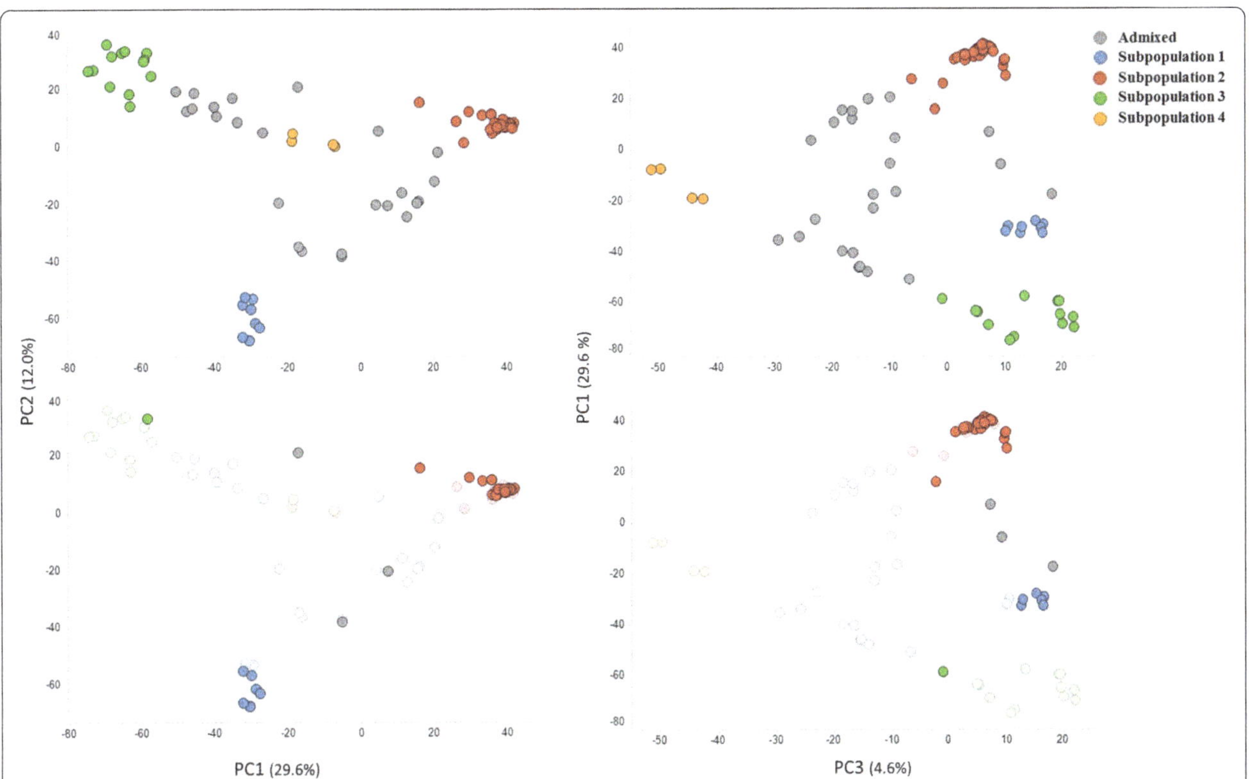

Fig. 2 Principal component analysis of cowpea accessions used in this study. The accessions are colored by subpopulation membership ($K = 4$). Upper plots display all accessions, while the lower plots highlight only cowpea accessions from Iberia Peninsula

Iberian Peninsula belonged to subpopulation 2, together with other Genebank accessions from Europe (Fig. 2, lower plots; Additional file 4). Only two accessions from Portugal (Ac5 and Ac13) and one accession from Spain (Ac38) did not belong to this subpopulation: Ac13 belonged to subpopulation 3, while accessions Ac5 and Ac38 were considered admixed (estimated proportion of subpopulation 2 = 0.43 and 0.61, respectively). These three accessions would then likely contain unique alleles not present in any other Iberian Peninsula accession studied. An examination of the SNP data from all 35 Iberian Peninsula accessions showed that, of all polymorphic SNPs (29,550) in the Iberian Peninsula dataset, 4777 were contributed only by Ac13 (16.2%). These unique alleles from Ac13 were distributed all over the linkage groups (LGs; Additional file 5). As expected, Ac5 and Ac38 contained a lower number of unique alleles, 1849 (6.3%) for Ac5 and 534 (1.8%) for Ac38. Unique alleles from Ac5 were found in all cowpea chromosomes, while those from Ac38 were mainly present on the pericentromeric region of LG3 and LG11, and towards the distal end of LG8 (Additional file 5).

PIC and *He* were calculated for the entire set of 81 *V. unguiculata* ssp. *unguiculata* accessions, and then separately for Iberian Peninsula accessions and for those from other countries (Table 2). Considering the ssp. *unguiculata* whole dataset, average PIC and *He* were 0.21 and 0.25, respectively. PIC and *He* values were quite different between accessions from the Iberian Peninsula (0.09 and 0.10, respectively) and those from the world-wide collection (0.25 and 0.31, respectively). This indicates that genetic diversity in Iberian Peninsula ssp. *unguiculata* accessions is low compared to the diversity available in the world-wide sample of cultivated cowpeas. To better understand and compare accessions from the Iberian Peninsula at the genetic level, similarity matrix was generated based on comparisons between all 35 accessions (Additional file 6). From this it was apparent that Ac13, Ac5 and Ac38 had the lowest similarity indexes with the rest of the Iberian Peninsula accessions. This was expected since they had the lowest genomic ancestry proportions of subpopulation 2, to which all other Iberian Peninsula accessions belong (Additional file 4). The other 32 accessions were very similar to each other, with percentages of similarity ranging from 77.0% to 99.9%.

Discussion

Genetic characterization of germplasm resources is essential for conservation and the sustainable use of their diversity [29]. In recent years, several studies have characterized cowpea germplasm mainly from Africa and Asia [13, 30–33]. However, there have been no studies

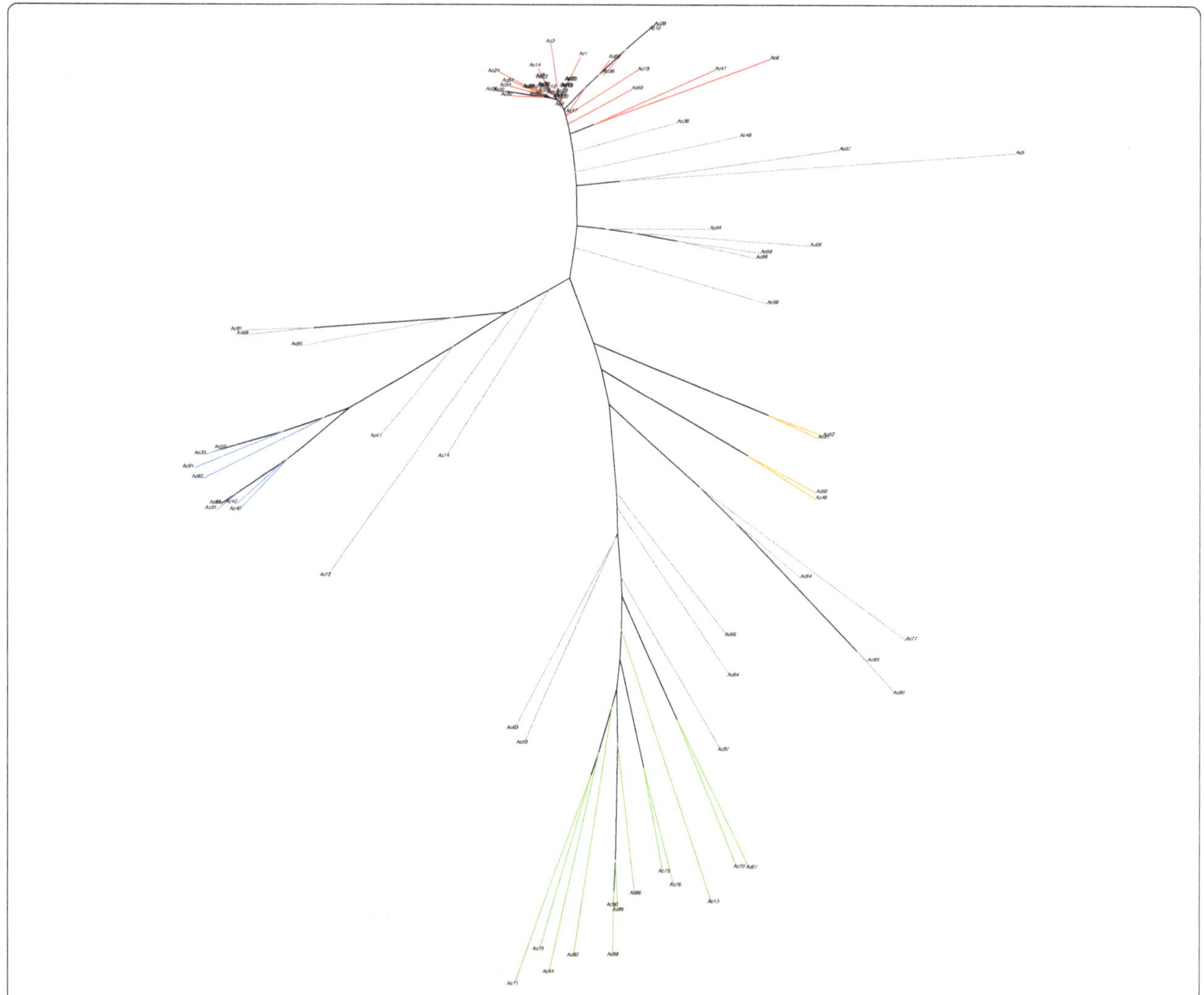

Fig. 3 Neighbor-joining tree of 91 cowpea accessions with colors representing subpopulation membership (blue = subpopulation 1; red = subpopulation 2; green = subpopulation 3; orange = subpopulation 4; and grey = admixed)

exploring in depth the genetic diversity of southern European cowpeas.

In this study, high-density SNP genotyping using the Cowpea iSelect Consortium Array [19] has provided a means to study population structure and genetic diversity in a set of 91 world-wide cowpea accessions, with a special focus on 43 accessions from the Iberian Peninsula. A high proportion of the SNPs assayed by the array were polymorphic in the dataset (44,056 of 51,128; 86%). Also PIC and *He* values obtained from the entire population are similar to those reported by Huynh et al. [34] and Muñoz-Amatriaín et al. [19] using a larger dataset, indicating that the selection of worldwide accessions in the present work provides a good representation of the diversity in cultivated cowpea.

Table 1 Polymorphism information content (PIC) and expected heterozygosity (*He*) calculated for the entire population and for each subpopulation

Data set	N° accessions	N° countries	PIC	*He*
All accessions	91	24	0.22	0.26
Subpopulation 1	9	4	0.12	0.14
Subpopulation 2	41	7	0.07	0.09
Subpopulation 3	12	8	0.18	0.23
Subpopulation 4	4	2	0.12	0.15

Table 2 Polymorphism information content (PIC) and expected heterozygosity (*He*) calculated for *V. unguiculata* ssp. *unguiculata* accessions

Data set	N° accessions	N° countries	PIC	*He*
All *V. unguiculata* ssp. *unguiculata* accessions	81	23	0.21	0.25
Iberian Peninsula accessions	35	2	0.09	0.10
Accessions from other countries	46	21	0.25	0.31

The SNP genotyping of these accessions enabled identification of one apparent duplication: Ac39 and Ac43, which are members of the subspecies *sesquipedalis.* These were provided by the National Plant Genetic Resources Centre-National Institute for Agricultural and Food Technology Research (CRF-INIA, Spain) genebank, and their passport information is limited. Ac39 and Ac43 are both from Spain, but from two different regions: Ac1 is from Cordoba (Andalucia region, south of Spain) and Ac43 from Ourense (Galicia region, north of Spain). A common cause of redundant accessions is the unwitting submission of the same accession to the genebank, then generating more than one name or designator. Identifying these redundant accessions is not possible using phenotype data alone [35]. Duplicated accessions do not contribute to genetic diversity of collections while generating unnecessary and additional costs to genebank [36].

The population structure analysis assigned the 91 accession to four subpopulations. In agreement with the results of Huynh et al. [34] and Xiong et al. [37], two of the subpopulations identified (subpopulation 3 and subpopulation 4) corresponded to the East/South Africa and the West Africa gene pools, respectively. In addition to those two genetic clusters, our study identified two more subpopulations composed of North Africa and South Europe accessions (subpopulation 2) and *V. unguiculata* ssp. *sesquipedalis* accessions (subpopulation 1). The aforementioned studies may not have identified those two populations because of a lack of accessions from these regions.

The geographic distribution of the accessions from the three ssp. *unguiculata* subpopulations enabled inference of possible dispersion routes of domesticated cowpea (Fig. 1). It has been reported that some Iberian Peninsula crops were introduced in Europe through the "Arab corridor" [38]. Our study is consistent with the idea that cowpea was one of the crops brought by Arabs from North Africa to Europe in ancient times. From the end of the fifteenth century until the middle of the seventeenth century, Portugal and Spain, which form the Iberian Peninsula, had an important role in the great discovery period. Saúco and Cubero [38] described how powers from the Iberian Peninsula had an important contribution to the exchange and acclimatization of new and old world crops, including cowpea, due to exploration voyages and commercial routes established by them. This information together with the genetic data from this study seems to indicate that the accession from Cuba (Ac62) belonging to subpopulation 2 may have been brought by the Spaniards. This island was discovered in 1492 by Christopher Columbus and belonged to Spain until 1898, so it seems plausible that the Spaniards introduced this crop to Cuba. On the other hand, Portuguese sailors explored and dominated the Southern hemisphere including South America (more specifically Brazil), Southern Africa (Angola, Guinea Bissau, Mozambique) and India. They established direct contact between Europe, South America and India, and later with Southeast Asia and China [38]. Since subpopulation 3 includes accessions from all these regions, it is possible that slaves being transported in Portuguese ships crossing the Atlantic Ocean were the ones who introduced cowpea cultivation into Brazil. Additional cowpea introduction into India and later China may also have occurred through the Portuguese sea routes as well.

Cowpea genetic diversity among countries and regions can be affected by environmental factors and customs of cowpea consumption [37]. In the Iberian Peninsula, cowpea is a minor crop, mostly based on cultivation of landraces. These landraces reflect the cultural identity of local people and are reservoirs of diversity for breeding improvement. Given the narrow genetic base found in this study for most of the Iberian Peninsula cowpea, introduction of additional diversity into the Iberian Peninsula genepool seems sensible to keep increasing yields under changing climatic conditions [29]. Three of the accessions belonging to the Iberian Peninsula were more diverse than the rest: Ac13 was the most different from the others and had mostly subpopulation 3 ancestry, while Ac5 and Ac38 had admixed ancestry from subpopulations 2 and 3, and subpopulations 1 and 2, respectively (Additional file 4). Ac5 is a variety developed by breeders at INIAV-Elvas (Portugal) and Ac38 is a landrace from Spain. Given its proportion of ancestry from subpopulation 3 (0.50), Ac5 may have resulted from crosses between accessions from the Iberian Peninsula and South/East African materials. Although Ac38 is morphologically similar to other ssp. *unguiculata* accessions, its genome has an estimated proportion of subpopulation 1 ancestry of 0.39 (Additional file 4). This accession could be the result of intentional crosses between the two cultivar-groups. The introduction of Ac13, a member of subpopulation 3, into Portugal could have occurred in the 70's. During that time, Portuguese living in Angola, Guinea and Mozambique returned to Portugal and could have brought that cowpea landrace with them. It is also possible that during the great discover period navigators brought that accession from Africa, Asia or South America (Brazil). The aforementioned accessions Ac5, Ac13 and Ac38 can be very useful for breeding programs as they can bring additional genetic diversity without compromising adaptation to the environment.

Conclusions

Higher cowpea production is needed in Europe to meet demand, and only Southern European countries possess climatic conditions that are favorable for growing this legume crop. Here we have genetically characterized a geographically diverse set of cowpeas that are cultivated in the Iberian Peninsula using a high-density genotyping array, and we have compared them to cowpea accessions collected world-wide. Our study identified four subpopulations in the whole dataset, with most Iberian Peninsula accessions of ssp. *unguiculata* belonging to the same subpopulation and having lower levels of genetic diversity than world-wide cowpea accessions. However, we identified one Iberian Peninsula landrace with ancestry from another subpopulation and two accessions having admixture of different subpopulations. These three accessions may be used to incorporate new genetic diversity into breeding programs without compromising adaptation. Possible dispersion routes of cultivated cowpea have been also inferred using the SNP data combined with passport information. In the future, favorable alleles for simple and complex traits could be mined from these accessions via genome-wide association studies.

Additional files

Additional file 1: Information on cowpea accessions used in this study. (XLSX 12 kb)

Additional file 2: Raw STRUCTURE output for all runs (left) and Δ*K* calculations for each number of ***K*** (right). (XLSX 12 kb)

Additional file 3: Exploration of the optimal number of subpopulations (*K*) in the entire dataset. Plots were generated with Structure Harvester [26]. (A) Estimated log probability of the data for each *K* between 1 and 8. (B) Δ*K* values as a function of *K*. (TIFF 75 kb)

Additional file 4: Genetic structure information on the 91 accessions. The estimated membership of each accession in the four subpopulations is shown, as well as the PCA coordinates. (XLSX 17 kb)

Additional file 5: Genomic location of unique alleles in Ac13, Ac5 and Ac38 on cowpea linkage groups (LGs). Genomic regions colored in red contain unique alleles in the corresponding accession, while regions containing non-unique alleles are represented in blue. For the figure, one marker per locus was kept, giving priority to unique alleles over non-unique ones. In white are represented regions lacking mapped SNPs. LG number and cM positions are based on the cowpea consensus genetic map available from Muñoz-Amatriaín et al. [19]. (TIFF 2302 kb)

Additional file 6: Matrix showing genetic pair-wise similarity values for Iberian Peninsula accessions. (XLSX 16 kb)

Abbreviations

BC: Before Christ; CRF-INIA: National Plant Genetic Resources Centre-National Institute for Agricultural and Food Technology Research; CTAB: cetyl trimethyl ammonium bromide; DNA: Deoxyribonucleic acid; EMBRAPA: Brazilian Agricultural Research Corporation; GWAS: Genome-wide association studies; *He*: Expected heterozygosity; INIAV: National Institute for Agrarian and Veterinarian Research; IPK: Leibniz Institute of Plant Genetics and Crop Plant Research; LG: Linkage group; MAF: Minor allele frequency; MCMC: Monte Carlo Markov Chain; NJ: Neighbor-joining; PCA: Principal component analysis; PIC: Polymorphism Information Content; QC: Quality control; QTL: Quantitative Trait Locus; SNP: Single Nucleotide Polymorphism

Acknowledgments

Authors would like to thank the seed providing namely, the National Institute for Agrarian and Veterinarian Research (INIAV, Portugal), National Plant Genetic Resources Centre-National Institute for Agricultural and Food Technology Research (CRF-INIA, Spain), Leibniz Institute of Plant Genetics and Crop Plant Research (IPK, Gatersleben, Germany), the Botanic Garden Meise (Belgium), the University of Perugia (Italy), and the Brazilian Agricultural Research Corporation (EMBRAPA, Brazil).

Funding

This study was supported by EUROLEGUME project. This project has received funding from the European Union's Seventh Framework Programme for research, technological development and demonstration under grant agreement no 613781. European Investment Funds by FEDER/COMPETE/POCI – Operational Competitiveness and Internationalization Programme, under Project POCI-01-0145-FEDER-006958 and National Funds by FCT – Portuguese Foundation for Science and Technology, under the project UID/AGR/04033/2013. MMA was partially supported by the Feed the Future Innovation Lab for Climate Resilient Cowpea (USAID Cooperative Agreement AID-OAA-A-13-00070), which is directed by TJC. The funding entities had no role in the design of the study, collection, analysis and interpretation of data, or in writing the manuscript.

Authors' contributions

IC, MEC and VC provided material. MC, IC and VC conducted the experiment. MC, MMA and TC analyzed the data. MC drafted the manuscript. MMA, IC, MEC and VC conceived and designed the whole experiment. MMA, IC, MEC, TLN, MM, ER, TC and VC revised the manuscript. All authors read and approved the final version of the manuscript.

Competing interests

The authors declare that they have no competing interests.

Author details

[1]Centre for Research and Technology of Agro-Environmental and Biological Sciences (CITAB), University of Trás-os-Montes and Alto Douro (UTAD), 5000-801 Vila Real, Portugal. [2]Department of Botany and Plant Sciences, University of California Riverside, Riverside, CA 92521-0124, USA. [3]Department of Genetics and Biotechnology, University of Trás-os-Montes and Alto Douro (UTAD), 5000-801 Vila Real, Portugal. [4]Biosystems & Integrative Sciences Institute (BioISI), Plant Functional Biology Center (CBFP), University of Minho, Campus de Gualtar, 4710-057 Braga, Portugal. [5]Biosystems & Integrative Sciences Institute (BioISI), Sciences Faculty, University of Lisbon, Campo Grande, 1749-016 Lisbon, Portugal. [6]Instituto de Biotecnología Vegetal, Universidad Politécnica de Cartagena, 30202 Cartagena, Spain.

References

1. Tan H, Tie M, Luo Q, Zhu Y, Lai J, Li H. A review of molecular makers applied in cowpea (*Vigna unguiculata* L. Walp.) breeding. J Life Sci. 2012;6:1190–9.
2. Xu P, Wu X, Muñoz-Amatriaín M, Wang B, Wu X, Hu Y, et al. Genomic regions, cellular components and gene regulatory basis underlying pod length variations in cowpea (*V. unguiculata* L. Walp). Plant Biotechnol J. 2016;15:1–11.
3. Timko MP, Ehlers JD, Roberts PA. Cowpea. In: Kole CM, editor. Genome mapping and molecular breeding in plants: pulses, sugar and tuber crops. New York: Springer-Verlag; 2007. p. 49–67.
4. Xu P, Wu X, Wang B, Liu Y, Qin D, Ehlers JD, et al. Development and polymorphism of *Vigna unguiculata* ssp. *unguiculata* microsatellite markers used for phylogenetic analysis in asparagus bean (*Vigna unguiculata* ssp. *sesquipedialis* (L.) Verdc.). Mol Breed. 2010;25(4):675–84.
5. Singh BB, Fatokun CA, Tarawali SA, Kormawa PM, Tamò M. Recent genetic studies in cowpea. In: Cowpea genetic and breeding; 2002. p. 3–13.
6. Ehlers JD, Hall AE. Genotypic classification of cowpea based on responses to heat and photoperiod. Crop Sci. 1996;36(3):673–9.
7. Eloward HO, Hall AE. Influence of early and late nitrogen fertilization on yield and nitrogen fixation of cowpea under well-watered and dry field conditions. F. Crop Res. 1987;15:229–44.

8. Hall AE. Breeding for adaptation to drought and heat in cowpea. Eur J Agron. 2004;21:447–54.
9. Kröner N, Kotlarski S, Fischer E, Lüthi D, Zubler E, Schär C. Separating climate change signals into thermodynamic, lapse-rate and circulation effects: theory and application to the European summer climate. Clim Dyn. 2017; 48(9–10):3425–40.
10. Richard A. Tentamen florae abyssinicae. Paris: Arthus Bertrand; 1847.
11. Steele WM. Cowpeas, *Vigna unguiculata* (Leguminosae Papillionatae). In: Simmonds NW, editor. Evolution of crop plants. London: Longman; 1976. p. 183–5.
12. Pant KK, Chandel K, Joshi B. Analysis of diversity in Indian cowpea genetic resources. SABRAO J. 1982;14:103–11.
13. Coulibaly S, Pasquet RS, Papa R, Gepts PAFLP. Analysis of the phenetic organization and genetic diversity of *Vigna unguiculata* L. Walp. Reveals extensive gene flow between wild and domesticated types. Theor Appl Genet. 2002;104(2–3):358–66.
14. Tosti N, Negri V. Efficiency of three PCR-based markers in assessing genetic variation among cowpea (*Vigna unguiculata* subsp. *unguiculata*) landraces. Genome. 2002;45:268–75.
15. Purseglove JW. Tropical crops - Dicotyledons. London: Longman; 1968.
16. Fang J, Chao CCT, Roberts PA, Ehlers JD. Genetic diversity of cowpea [*Vigna unguiculata* (L.) Walp.] in four West African and USA breeding programs as determined by AFLP analysis. Genet Resour Crop Evol. 2007;54:1197–209.
17. Badiane FA, Diouf M, Diouf D. Cowpea. In: Singh M, Bisht IS, Dutta M, editors. Broadening the Genetic Base of grain legumes. India: Springer; 2014. p. 95–114.
18. Sinha AK, Mishra PK. Agro-morphological characterization and morphology based genetic diversity analysis of landraces of rice variety (*Oryza sativa* L.) of Bankura district of West Bengal. Int J Curr Res. 2013;5(10):2764–9.
19. Muñoz-Amatriaín M, Mirebrahim H, Xu P, Wanamaker SI, Luo M, Alhakami H, et al. Genome resources for climate-resilient cowpea, an essential crop for food security. Plant J. 2017;89(5):1042–54.
20. CBI, Ministry of Foreign Affairs [Internet] [cited 2017 Feb 10]. Available from: https://www.cbi.eu/market-information/grains-pulses/trends/
21. Duke KA. *Vigna unguiculata* (L.) Walp. ssp *unguiculata*. In: Handbook of legumes of world economic importance Plenum Press. New York: Plenum Press; 1981.
22. Pasquet RS. Morphological study of cultivated cowpea *Vigna unguiculata* (L.) Walp. Importance of ovule number and definition of cv gr Melanophthalmus. Agronomie. 1998;18(1):61–70.
23. Pritchard JK, Stephens M, Donnelly P. Inference of population structure using multilocus genotype data. Genetics. 2000;155(2):945–59.
24. Bradbury PJ, Zhang Z, Kroon DE, Casstevens TM, Ramdoss Y, Buckler ES. TASSEL: software for association mapping of complex traits in diverse samples. Bioinformatics. 2007;23(19):2633–5.
25. Evanno G, Regnaut S, Goudet J. Detecting the number of clusters of individuals using the software STRUCTURE: a simulation study. Mol Ecol. 2005;14(8):2611–20.
26. Earl DA, von Holdt BM. STRUCTURE HARVESTER: A website and program for visualizing STRUCTURE output and implementing the Evanno method. Conserv Genet Resour. 2012;4(2):359–61.
27. Chen WC. Overlapping codon model, phylogenetic clustering, and alternative partial expectation conditional maximization algorithm. Iowa: Iowa Stat University; 2011.
28. Botstein D, White RL, Skolnick M, Davis RW. Construction of a genetic linkage map in man using restriction fragment length polymorphisms. Am J Hum Genet. 1980;32(3):314–31.
29. Govindaraj M, Vetriventhan M, Srinivasan M. Importance of genetic diversity assessment in crop plants and its recent advances: an overview of its analytical perspectives. Genet Res Int. 2015;2015:1–14.
30. Ba FS, Pasquet RS, Gepts P. Genetic diversity in cowpea [*Vigna unguiculata* (L.) Walp.] as revealed by RAPD markers. Genet Resour Crop Evol. 2004;51:539–50.
31. Lee JR, Back HJ, Yoon MS, Park SK, Cho HY, Kim CY. Analysis of genetic diversity of cowpea landraces from Korea determined by simple sequence repeats and establishment of a core collection. Korean J Breed Sci. 2009; 41(4):369–76.
32. Asare AT, Gowda BS, Galyuon IKA, Aboagye LL, Takrama JF, Timko MP. Assessment of the genetic diversity in cowpea (*Vigna unguiculata* L. Walp.) germplasm from Ghana using simple sequence repeat markers. Plant Genet Resour. 2010;8(2):142–50.
33. Badiane FA, Gowda BS, Cissé N, Diouf D, Sadio O, Timko MP. Genetic relationship of cowpea (*Vigna unguiculata*) varieties from Senegal based on SSR markers. Genet Mol Res. 2012;11(1):292–304.
34. Huynh B, Close TJ, Roberts PA, Hu Z, Wanamaker S, Lucas MR, et al. Gene pools and the genetic architecture of domesticated cowpea. Plant Genome. 2013;6(2):1–8.
35. Muñoz-Amatriaín M, Cuesta-Marcos A, Endelman JB, Comadran J, Bonman JM, Bockelman HE, Chao S, Russell J, Waugh R, Hayes PM, Muehlbauer GJ. The USDA barley core collection: genetic diversity, population structure, and potential for genome-wide association studies. PLoS One. 2014;9(4):1–13.
36. Spooner D, Van Treuren R, De Vicente MC. Molecular markers for Genebank management. In: IPGRI technical bulletin N 10. Rome, Italy: International Plant Genetic Resources Institute; 2005. p. 24–90.
37. Xiong H, Shi A, Mou B, Qin J, Motes D, Lu W, et al. Genetic diversity and population structure of cowpea (*Vigna unguiculata* L. Walp). PLoS One. 2016;11(8):1–15.
38. Saúco VG, Cubero JI. Contribution of Spain and Portugal to the exchange and acclimatization of new and old world crops. Acta Hortic. 2011;916:71–82.

8

Genome analysis of the foxtail millet pathogen *Sclerospora graminicola* reveals the complex effector repertoire of graminicolous downy mildews

Michie Kobayashi[1*], Yukie Hiraka[1], Akira Abe[1], Hiroki Yaegashi[1], Satoshi Natsume[1], Hideko Kikuchi[1], Hiroki Takagi[1], Hiromasa Saitoh[1,3], Joe Win[2], Sophien Kamoun[2] and Ryohei Terauchi[1,4*]

Abstract

Background: Downy mildew, caused by the oomycete pathogen *Sclerospora graminicola*, is an economically important disease of Gramineae crops including foxtail millet (*Setaria italica*). Plants infected with *S. graminicola* are generally stunted and often undergo a transformation of flower organs into leaves (phyllody or witches' broom), resulting in serious yield loss. To establish the molecular basis of downy mildew disease in foxtail millet, we carried out whole-genome sequencing and an RNA-seq analysis of *S. graminicola*.

Results: Sequence reads were generated from *S. graminicola* using an Illumina sequencing platform and assembled de novo into a draft genome sequence comprising approximately 360 Mbp. Of this sequence, 73% comprised repetitive elements, and a total of 16,736 genes were predicted from the RNA-seq data. The predicted genes included those encoding effector-like proteins with high sequence similarity to those previously identified in other oomycete pathogens. Genes encoding jacalin-like lectin-domain-containing secreted proteins were enriched in *S. graminicola* compared to other oomycetes. Of a total of 1220 genes encoding putative secreted proteins, 91 significantly changed their expression levels during the infection of plant tissues compared to the sporangia and zoospore stages of the *S. graminicola* lifecycle.

Conclusions: We established the draft genome sequence of a downy mildew pathogen that infects Gramineae plants. Based on this sequence and our transcriptome analysis, we generated a catalog of *in planta*-induced candidate effector genes, providing a solid foundation from which to identify the effectors causing phyllody.

Keywords: *Sclerospora graminicola*, Graminicolous downy mildew, Oomycetes, Whole genome sequence, Effector, Jacalin-like lectin, *Setalia italica*, Phyllody

Background

The oomycetes form a diverse group of filamentous eukaryotic microorganisms, also known as water molds, which include saprophytes as well as pathogens of plants, insects, crustaceans, fish, vertebrate animals, and various microorganisms [1, 2]. In plants, pathogenic oomycetes cause devastating diseases in a wide range of species including agricultural crops. Foxtail millet (*Setalia italica* (L.) Beauv.), the second most important millet in terms of global yield [3], suffers from downy mildew disease caused by *Sclerospora graminicola* (Sacc.) Schroet. in regions including India, China, Japan, and Russia.

Twenty genera of downy mildews are known, of which eight are graminicolous downy mildews [4]. Among these, *S. graminicola* (Sacc.) Schroet. is an obligate biotrophic oomycete. The likely source of the *S. graminicola* primary inoculum is oospores remaining in the soil or diseased plant residues. Fourteen graminaceous species are established hosts of *S. graminicola*, with strict host specificity observed among the various isolates of the pathogen [5]. After pathogen invasion,

* Correspondence: m-kobayashi@ibrc.or.jp; terauchi@ibrc.or.jp
[1]Iwate Biotechnology Research Center, Iwate, Japan
Full list of author information is available at the end of the article

systemically infected leaves generally show chlorosis along the veins. When the pathogen colonizes the branched inflorescences, known as panicles, the floral organs are often transformed into leafy structures, in a process termed phyllody [6]. Phyllody leads to the disease referred to as "witches' broom", "green ear disease", or "crazy top", and is caused in foxtail millet, pearl millet, maize, and finger millet by pathogens belonging to the three genera, *Peronosclerospora*, *Sclerophthora*, and *Sclerospora* [6, 7]. No induction of phyllody in dicots by downy mildews has been reported.

Whole-genome sequencing and transcriptome analyses have profoundly changed research into plant-microbe interactions in recent years [8], and draft genome sequences of oomycetes have been published for five downy mildew pathogens [9–13]. Whole-genome sequencing has revealed that obligate pathogens including the downy mildews often lose some metabolic pathways, such as for nitrate and sulfate metabolism [9, 10, 13]. In addition, sequence analyses point to conservation of a subset of the effectors that oomycetes secrete to manipulate plant physiology or suppress plant immunity [14]. Such effectors are classified as apoplastic or cytoplasmic based on their localization in the host plants. Apoplastic effectors include (1) secreted hydrolytic enzymes such as proteases, lipases, and glycosylases that can degrade plant tissue, (2) protease inhibitors that protect the oomycetes from host defense enzymes, (3) necrosis and ethylene-inducing peptide 1 (Nep1)-like proteins (NLPs), and (4) PcF-like small cysteine-rich proteins (SCRs) [14]. By contrast, RXLR domain-containing proteins and crinklers (CRNs) are characteristic cytoplasmic effectors in plant pathogenic oomycetes [15, 16]. Several genomic sequences for oomycetes and dicot downy mildews have been released; however, with the exception of the recently published transcriptome analysis of pearl millet infected with *S. graminicola* [17], there have been no genomic analyses of the graminicolous downy mildew pathogens.

Here, we perform whole-genome sequencing on the *S. graminicola* strain that infects foxtail millet. We further report RNA-seq–based gene prediction and annotation of the *S. graminicola* genome, and expression profiling of the putative secreted protein genes flagged as effector candidate genes.

Results

De novo assembly of the *S. graminicola* (*Sg*) genome

We prepared genomic DNA from a mixture of sporangia and zoospores colonized on the leaves of foxtail millet. The genome was sequenced using an Illumina platform and a paired-end library with a mean insert size of 370 bp, as well as mate-pair libraries with insert sizes of 2, 4, and 6 kbp. To check for contamination with bacterial and host plant DNA, some of the short reads were assembled using Platanus v.1.2.1 [18], and the generated contigs were used for a BLASTn search against the NCBI nt database. Of the 97 scaffolds over 200 bp in length, 11 scaffolds showed a high similarity to other oomycete or fungal sequences (Additional file 1: Table S1). The others did not show any significant similarity to the sequences in the database. From this result, we judged that the level of contamination from bacterial and host plant DNA was negligible, and proceeded to de novo assemble all the sequencing reads that had sufficient Phred quality scores.

The filtered Illumina sequencing reads were used for the de novo assembly in Platanus v.1.2.1 (Table 1). The total size of the assembled contigs was 254 Mbp, with an

Table 1 Genome statistics of *Sclerospora graminicola* (*Sg*) and other previously sequenced oomycetes[a]

Characteristic	*Sg*	*Plh*	*Hpa*	*Phi*	*Phs*
Estimated genome size	360 Mbp	100 Mbp	100 Mbp	240 Mbp	95 Mbp
Number of scaffolds	64,505	3162	3408	4921	1810
N50 scaffold length	24.3 kbp	1540 kbp	332 kbp	1570 kbp	463 kbp
Total scaffold length	254 Mbp	75 Mbp	78.4 Mbp	228.5 Mbp	86.0 Mbp
GC content	46.4%	45.3%	47%	51.0%	54.4%
Repeat (%)	73%	40%	42%	74%	39%
Number of genes	16,736	15,469	14,321	17,787	18,969
Secreted protein genes	1220	631	762	1568	1701
CEGMA					
Group 1[b]	92.42%	93.94%	89.5%	96.97%	96.97%
Group 2 [b]	94.64%	96.43%	96.5%	96.43%	98.21%
Group 3 [b]	98.36%	98.36%	98.5%	96.72%	100.00%
Group 4 [b]	96.92%	100.00%	97.0%	96.92%	98.46%

[a]*Plh Plasmopara halstedii, Hpa Hyaloperonospora arabidopsidis, Phi Phytophthora infestans, Phs Phytophthora sojae*

[b]The CEGs are split into 4 groups with Group 1 being the least conserved between organisms, and Group 4 being the most conserved between organisms

N50 scaffold length of 24.3 kbp. The longest contig was 279 kbp. The completeness of the assembled genome was analyzed using the CEGMA pipeline [19]. Complete and partial mapping identified 95.56% and 98.39% of the 248 core eukaryotic genes (CEGs) in the *Sg* sequence, respectively, suggesting that our *Sg* draft genome sequence was of sufficient quality for further analysis and gene prediction. Phylogenetic analysis using the CEGs of available oomycete genomes revealed that *Sg* is closely related to *Plasmopara halstedii* (*Plh*), which infects sunflower (Fig. 1).

Sg has a large and heterozygous genome

Analysis of the k-mer frequency using paired-end reads showed two peaks, possibly derived from heterozygous and homozygous DNA sequences (Fig. 2a). To estimate the ploidy level of the *Sg* genome, we analyzed the distribution of the biallelic SNP call rate (Fig. 2b). The SNP counts had a single mode around 0.5, suggesting that the genome was diploid. The number of heterozygous SNPs, with a call rate of between 0.4 to 0.6, was 226,400 (Fig. 2c). The total genome size, estimated from the k-mer frequency at the peak corresponding to the putative homozygous DNA, was approximately 360 Mbp.

Sg has a highly repetitive genome

Gene prediction was carried out using Trinity/PASA, Tophat2/Cufflinks/PASA, MAKER2, AAT, based on the RNA-seq data and the in silico method [20–25]. By combining multiple types of evidence using EvidenceModeler [22], we identified a total of 16,736 genes supported by RNA-seq data. Analysis of repeated sequences using RepeatModeler [26] and RepeatMasker [27] revealed that approximately 73% of the assembled genome was repetitive, with more than half composed of long terminal repeat (LTR)-elements (Additional file 2: Table S2).

The *Sg* genome encodes proteins comparable to those of other downy mildews

To compare the *Sg* genome with those of other oomycetes, we performed clustering analyses of orthologs and paralogs from three downy mildew pathogens (DMs) (*Sg*, *Plh*, and *Hyaloperonospora arabidopsidis*; *Hpa*) and two *Phytophthora* species (*Ph. infestans*; *Phi* and *Ph. sojae*; *Phs*) based on the OMA orthology database [28]. There were 3548 and 2725 common orthologous groups in the DMs and in the five genomes (three DMs plus the two *Phytophthora* species), respectively (Additional file 3: Table S3). A total of 2055 groups were conserved in the *Phytophthora* species but not in the DMs, while only 128 groups were conserved among the DMs but not in *Phytophthora*. Some obligate biotrophs have lost the nitrogen and sulfate metabolic pathways [9, 10, 13]; an ortholog search revealed that *Sg* similarly lacked nitrate reductase, nitrite reductase, nitrate transporter, glutamine synthetase, and cysteine synthetase (Additional file 4: Table S4).

To gain insights into the unique features of the *Sg* genome, we compared the frequency of the protein domains encoded in the five oomycete genomes. In *Sg*, 11 domains were overrepresented (Fisher's exact test, $p < 0.05$), compared with two in the DMs and/or *Phytophthora* species (Additional file 5). In particular, the Jacalin-like lectin domain was overrepresented among the putative secreted proteins. Although no domains were underrepresented in *Sg* alone, 85 domains were underrepresented in the three DMs in comparison with the *Phytophthora* species. Of these 85 domains, 20 were associated with cellular transporters and 11 were linked to plant cell wall degradation. Several protein families related to plant defense, such as elicitin and cellulose-binding elicitor lectin, were also less common in the DM genomes than in *Phytophthora* (Additional file 5).

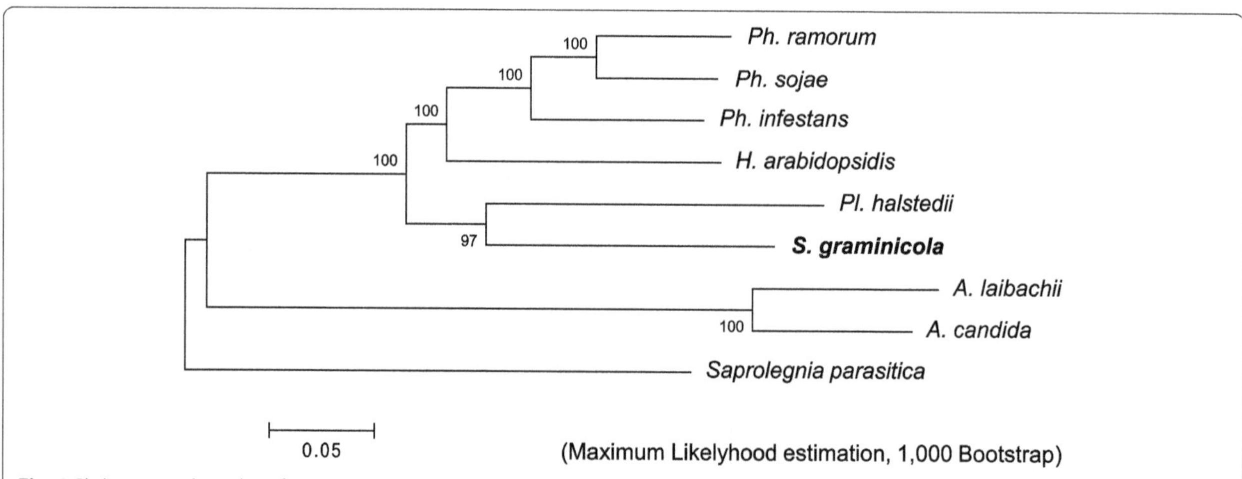

Fig. 1 Phylogenetic relationship of oomycete genomes. The tree was generated based on the nucleotide sequences of orthologous genes predicted by CEGMA pipeline using the Maximum Likelihood method implemented in MEGA6.06-mac. Bootstrap values from 1000 replicates are indicated on the branches

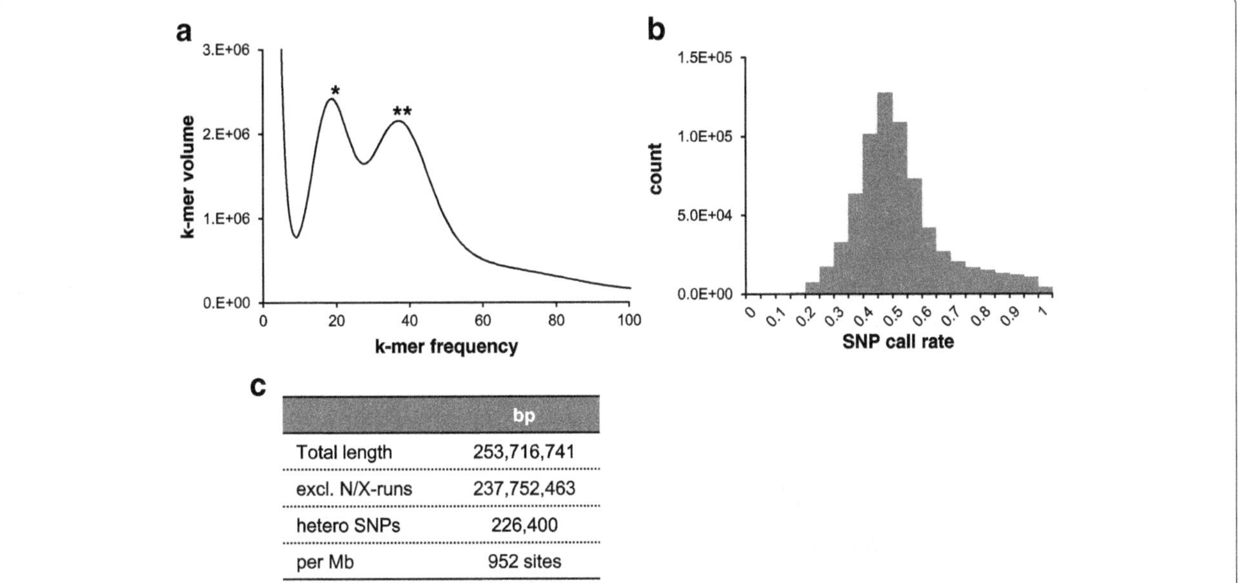

	bp
Total length	253,716,741
excl. N/X-runs	237,752,463
hetero SNPs	226,400
per Mb	952 sites

Fig. 2 *Sg* has a large and diploid genome with high heterozygosity. **a** K-mer distribution and coverage of sequencing reads at K = 15. Peaks with single and double asterisks were estimated as k-mer species derived from heterozygous (k-mer frequency = 19) and homozygous (k-mer frequency = 37) sequences, respectively. **b** Ploidy analysis displaying the distribution of the SNP call rate. **c** Heterozygosity was evaluated by counting the SNPs based on the alignment of genome sequence reads

Sg expresses conserved effector-like protein genes during infection

A total of 1220 *Sg* proteins were classified as putative secreted proteins based on the presence of signal peptides, predicted by SignalP4.1 [29], and the absence of transmembrane domains. This total was greater than those of *Plh* and *Hpa*, but fewer than that of *Phi* (Table 1). The number of proteins related to pathogenicity in *Sg* was comparable to that in other DMs, except for the RXLR-like proteins, of which *Sg* had more than *Plh* but fewer than those in the *Phytophthora* species (Table 2).

To search for effector candidates involved in *Sg* infection, an RNA-seq analysis was performed using total RNA extracted from sporangia/zoospores (inocula) and infected leaves. The foxtail millet leaves were inoculated with a spray containing a mixture of sporangia and zoospores. Primary penetration hyphae appeared 16–18 h after inoculation, and haustoria were formed one

Table 2 Summary of putative pathogenicity genes in *Sclerospora graminicola* and related oomycetes

Genes encoding	*Sg*	*Plh*	*Hpa*	*Phi*	*Phs*
Serine protease [a]	32 (5)	30	28	34	31
Aspartic protease [a]	6 (1)	5	5	6	6
Cysteine protease	14 (4)	15	16	18	17
Metalloprotease [a]	26 (3)	30	30	32	29
Kazal-like serine protease inhibitor [b]	10 (9)	16	5	34	23
Cystatin-like cysteine protease inhibitor [b]	1 (1)	2	0	3	2
Cutinase [b]	2 (2)	2	2	4	16
Pectate lyase [b]	8 (5)	3	12	46	46
Pectin lyase [b]	11 (7)	5	4	11	19
CAP domain [b, e]	20 (12)	22	15	30	40
NPP1-like [b]	24 (17)	19	21	27	74
Elicitin-like [b]	17 (10)	16	16	44	56
RXLR-like [c]	355 (355)	274 [d]	134 [d]	563 [d]	396 [d]
CRN-like	45 (4)	77 [d]	20 [d]	196 [d]	100 [d]

[a]PANTHER11.0 classification database, [b]: Interproscan, [c]: presence of N-terminal putative secretion signal and RXLR motif. [d]: reported in previous papers. [e]: CAP domain indicates Cysteine-rich secretory protein. Numbers in parentheses indicate numbers of putative secreted protein genes

day after inoculation. We analyzed the gene expression profiles at five time points (stage 1: SPO (sporangia and zoospores), stages 2, 3, 4, and 5: 16 hpi (hpi; hours post infection), and 1, 2, and 3 dpi (dpi; days post infection), respectively). Distribution of maximum transcripts per million (TPM) value of all genes in five data points indicated that 54% of the genes were lower than 20, 31% were from 20 to 100, 14% were from 100 to 1000, and 1.6% were higher than 1000. From differentially expression gene (DEG) analysis using edgeR [30], expression of 91 putative secreted protein genes significantly changed during infection. The maximum value of TPM of all DEGs was more than 20.

Ninety one DEGs were classified into four clusters based on their expression patterns using ward's method (Fig. 3, Additional file 6: Table S5). Representative genes of each cluster were validated by quantitative reverse transcription PCR (qRT-PCR) (Additional file 7: Fig. S1). Cluster I included genes expressed in sporangia or zoospores, but not during infection. The expression of genes belonging to cluster II increased in late stage of infection, suggesting that they include components contributing to pathogen expansion into leaves and the absorption of nutrition from host cells. Genes belonging to clusters III and IV were induced during stage 2 when the primary penetration hyphae developed, after which the expression of genes in clusters III gradually returned to basal levels. To determine the gene families overrepresented in each cluster, an enrichment analysis of protein domains predicted by InterProScan was performed (Additional file 8). CAP domain (CAP: the cysteine-rich secretory proteins, antigen 5, and pathogenesis-related 1 proteins superfamily proteins) and CUB domain which is related to Trypsin-like peptidase were enriched in cluster I. Jacalin-like lectin domain and Necrosis inducing protein domain were significantly enriched in cluster III, indicating that these domain could function in the early stages of *Sg* infection.

Different clustering methods could provide different results. We additionally performed clustering analyses using two methods, logFC-Cosine method using the cosine similarity of the vectors of their log-fold-change (logFC) values (Additional file 9: Figure S2) and model-based clustering method [31] (MBCluster; Additional file 10: Figure S3). Cluster I was separated into two clusters and some genes of cluster III and IV were classified into the same cluster by logFC-Cosine and MBCluster, however, most of genes showed similar clustering patterns by multiple clustering methods (Additional file 6: Table S5). Interproscan domain enrichment analysis indicated that Jacalin-like lectin domain and Necrosis inducing protein domain were also enriched in cluster 4 of logFC-Cosine method and cluster 2 of MBCluster that contain genes induced in early infection phase (Additional file 8).

To reveal features of *Sg* secretome, putative secreted proteins of *Sg* and 11 oomycetes (*Plh, Hpa, Phi, Phs, Ph. ramorum, Ph. capsici, Ph. parasitica, Albugo candida, A. laibachii, Pythium ultimum, Saprolegnia parasitica*) were clustered using TribeMCL protein family clustering algorithm [32]. 13,328 proteins were clustered into 1252 families (each family contains at least two sequences) and 1862 singletons. Of the 1252 familes, 230 contained *Sg* and other oomycete proteins and 78 were *Sg* specific families. *Sg*-specific families consisted of 39 RXLR-like families, 4 Jacalin-like domain-containing protein families, one leucine-rich repeat domain-containing family, one Mitochondrial carrier domain-containing family, and 33 unknown protein families (Additional file 11). Of these *Sg*-specific Tribes, Jacalin-like domain-containing families included genes those have high TPM levels, especially in stage 2 and 3 (Additional file 12: Fig. S4).

Jacalin-like lectin domain proteins

Jacalin-like lectin domain-containing proteins belong to a subgroup of lectins with binding specificity to mannose or galactose, and are involved in multiple biological processes. Jacalin-like proteins were overrepresented in the *Sg* genome (Additional file 5), and a phylogenetic analysis indicated many were specific to *Sg* (Fig. 4a). Among the jacalin-like protein genes of *Plh, Hpa*, and *Phi*, the closest to the *Sg*-specific clade was PITG_22899. Intriguingly, most of the *Sg*-jacalin-like proteins, including proteins with putative secreted signals and significant expression levels, belonged to the *Sg*-specific clade (Fig. 4a, red filled circles, Additional file 13). Effector genes are distributed in gene-sparse regions of the *Phi* genome [33, 34]. From the analysis of intergenic distance, jacalin-like protein genes appeared to distribute in gene-sparse regions (Fig. 4c, Wilcoxon rank sum test, 5′-intergenic length; p-value = 0.03721, 3′-intergenic length; p-value = 0.01161), however, most of jacalin-like protein genes were located near the scaffold border and were not possible to determine intergenic distance.

Nep1-like proteins (NLPs)

NLPs are a widespread effector family among filamentous and bacterial pathogens that show very different lifestyles [35]. Oomycetes have two types of NLPs: type 1 NLPs with a cation-binding pocket required for cytotoxicity, and type 1a NLPs with amino acid substitutions in their cation-binding pocket [35]. The *Sg* genome contained 24 NLP-encoding genes, 17 of which had an N-terminal secretion signal peptide (Additional file 14). One NLP, SG00816, was classified as a type 1 NLP with a TRAP repeat and the other 23 were type 1a NLPs.

Six of the 24 *SgNLP*s were DEGs (Additional file 14). The type 1 NLP, SG00816, was not significantly

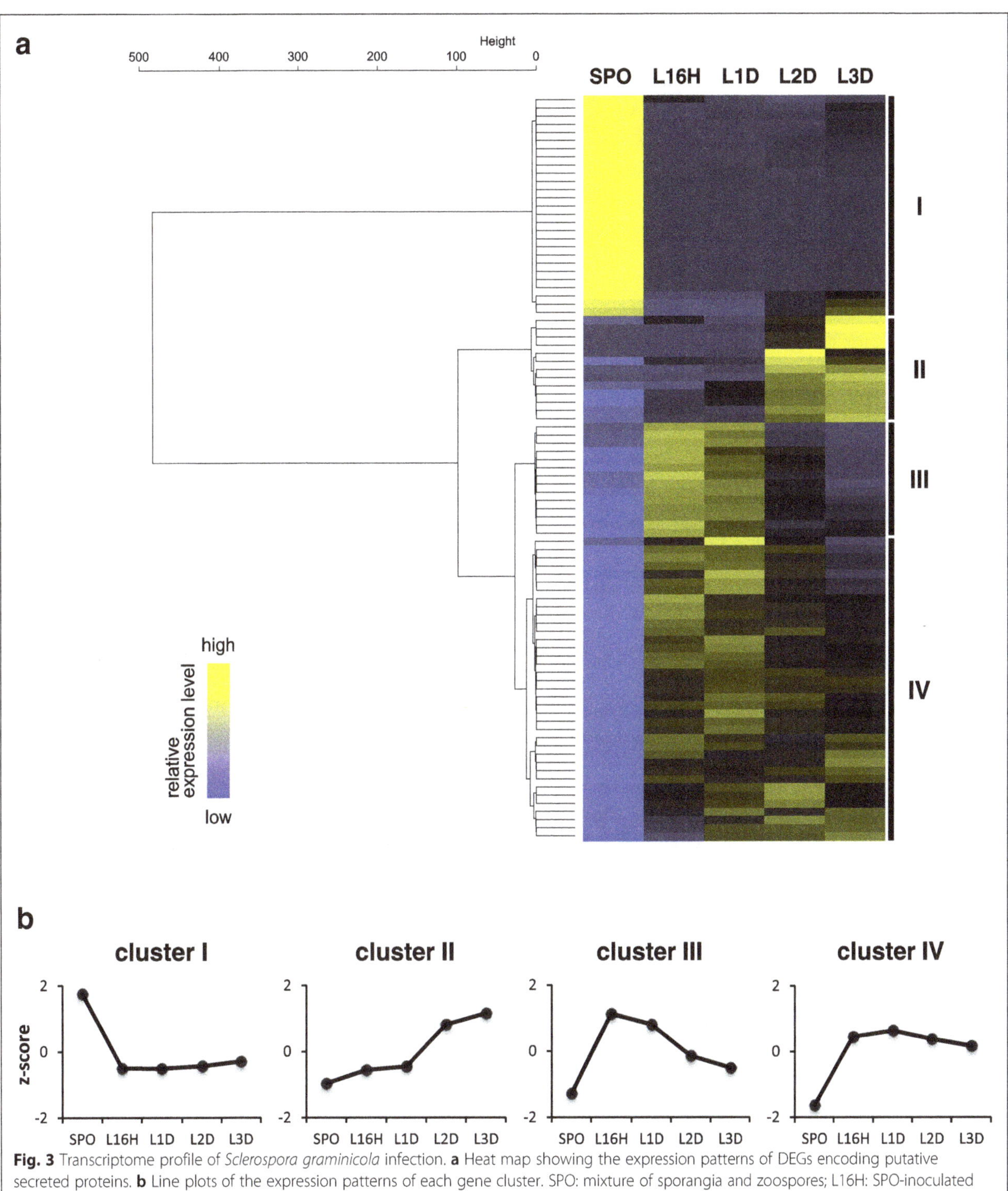

Fig. 3 Transcriptome profile of *Sclerospora graminicola* infection. **a** Heat map showing the expression patterns of DEGs encoding putative secreted proteins. **b** Line plots of the expression patterns of each gene cluster. SPO: mixture of sporangia and zoospores; L16H: SPO-inoculated leaves 16 h after inoculation; L1D, L2D, and L3D: SPO-inoculated leaves at one, two, and three days after inoculation, respectively

expressed at any stage (Additional file 14). Intriguingly, these DEGs of NLPs were in one clade of the *Sg*-specific expansion groups (see asterisk in Fig. 5). All of six differentially expressed NLPs were classified into cluster III and IV (Additional file 14).

Crinklers (CRNs)

CRNs are cytoplasmic effectors originally identified in *Phi* as secreted proteins that have a conserved LFLAK motif in the 50 amino acid residues of the N-terminal [36]. We identified 45 CRN-like genes in *Sg* (Table 2).

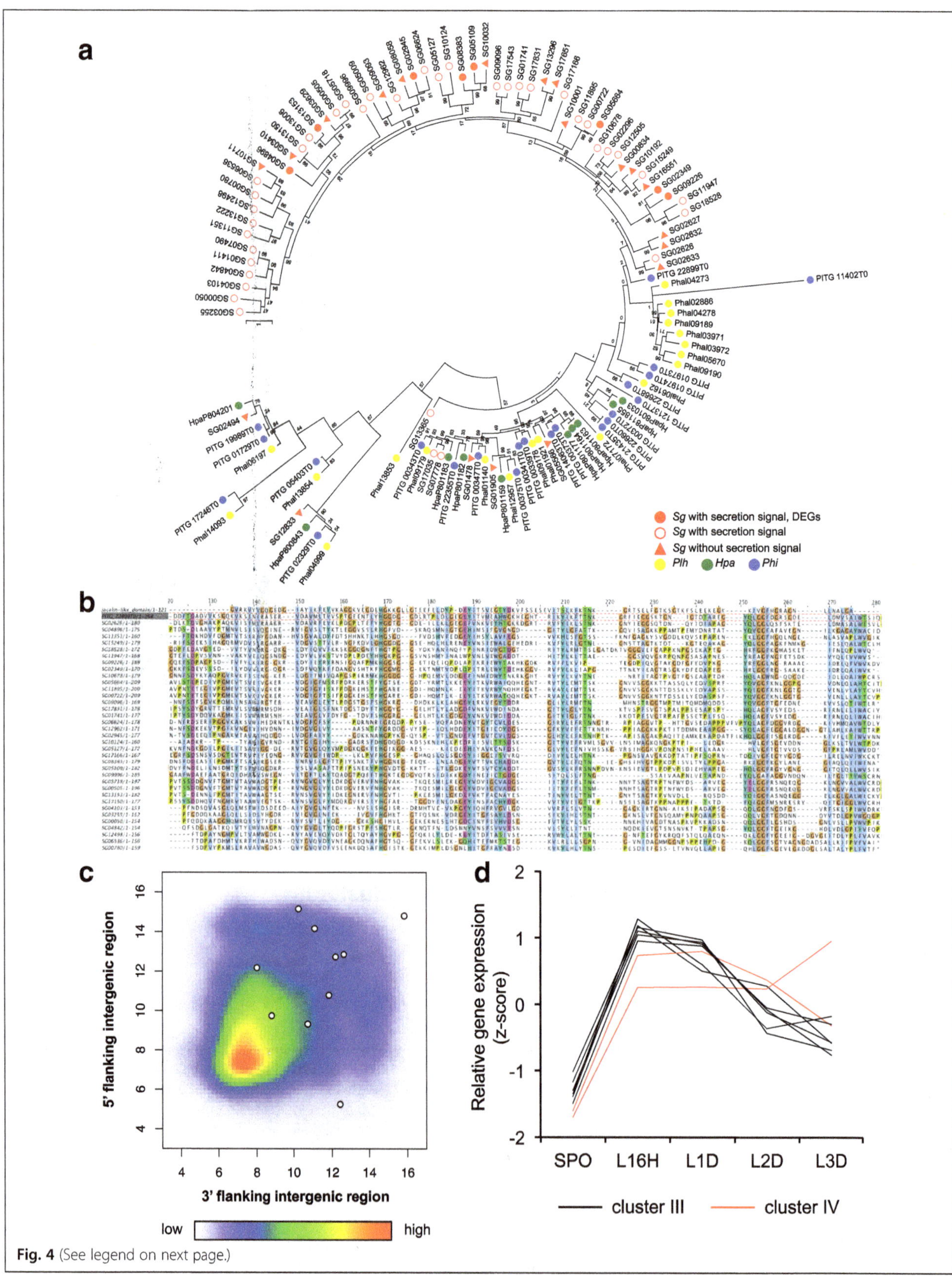

Fig. 4 (See legend on next page.)

(See figure on previous page.)
Fig. 4 Features of jacalin-like lectin domain-containing protein genes. **a** Phylogeny of the jacalin-like lectin domain-containing proteins of *Sg*, *Plh*, *Hpa*, and *Phi*. The tree was conducted using the Maximum Likelihood method implemented in MEGA6.06-mac, with 1000 bootstrap replicates. **b** Multiple sequence alignment showing the sequence similarity between PITG22899T0 and the jacalin-like lectin domains of the Sg proteins. **c** Distribution of intergenic region length of Sg genes. All predicted genes are represented by a heatmap and the jacalin-like protein genes are represented by white circles. **d** Relative expression of DEGs of jacalin-like protein during infection. Clusters III and IV are defined in Fig. 3

Only four of these had a signal peptide at the N-terminus. *SgCRNs*, including four putative secreted *CRN* genes, were not significantly expressed during infection (Additional file 15).

RXLR-like proteins

The RXLR domain is a putative host-targeting motif [37] and is highly conserved among plant-pathogenic oomycetes. We predicted RXLR-like protein genes by searching for a RXLR(−EER) sequence following the N-terminal putative signal peptide. Proteins showing high similarity to known RXLR-like proteins were also included as RXLR-like protein candidates. A total of 355 RXLR-like proteins were found, among which 165 had the exact RXLR-EER motif and 60 had the RXLR motif, while 130 were predicted to be RXLR(−EER) variants (Fig. 6a). Some RXLR effectors contain a core α-helical fold known as the WY-fold [38]. We explored whether our identified RXLR-like proteins had the WY-fold using HMMER, and found a total of 38 proteins with at least one WY-fold (Additional file 16). In the gene expression profile and expression pattern

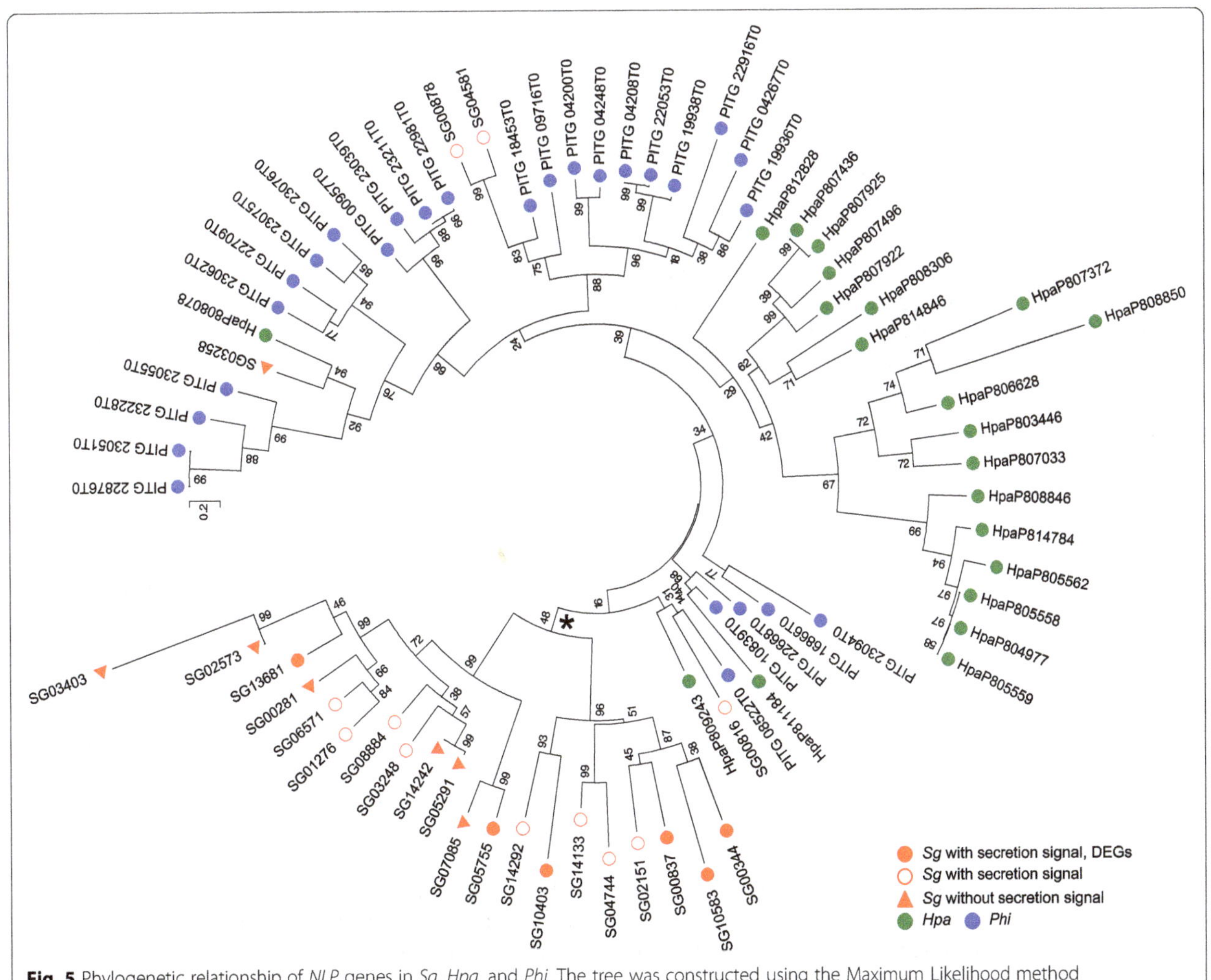

Fig. 5 Phylogenetic relationship of *NLP* genes in *Sg*, *Hpa*, and *Phi*. The tree was constructed using the Maximum Likelihood method implemented in MEGA6.06-mac, with 1000 bootstrap replicates. The asterisk indicates the *Sg*-specific expansion group

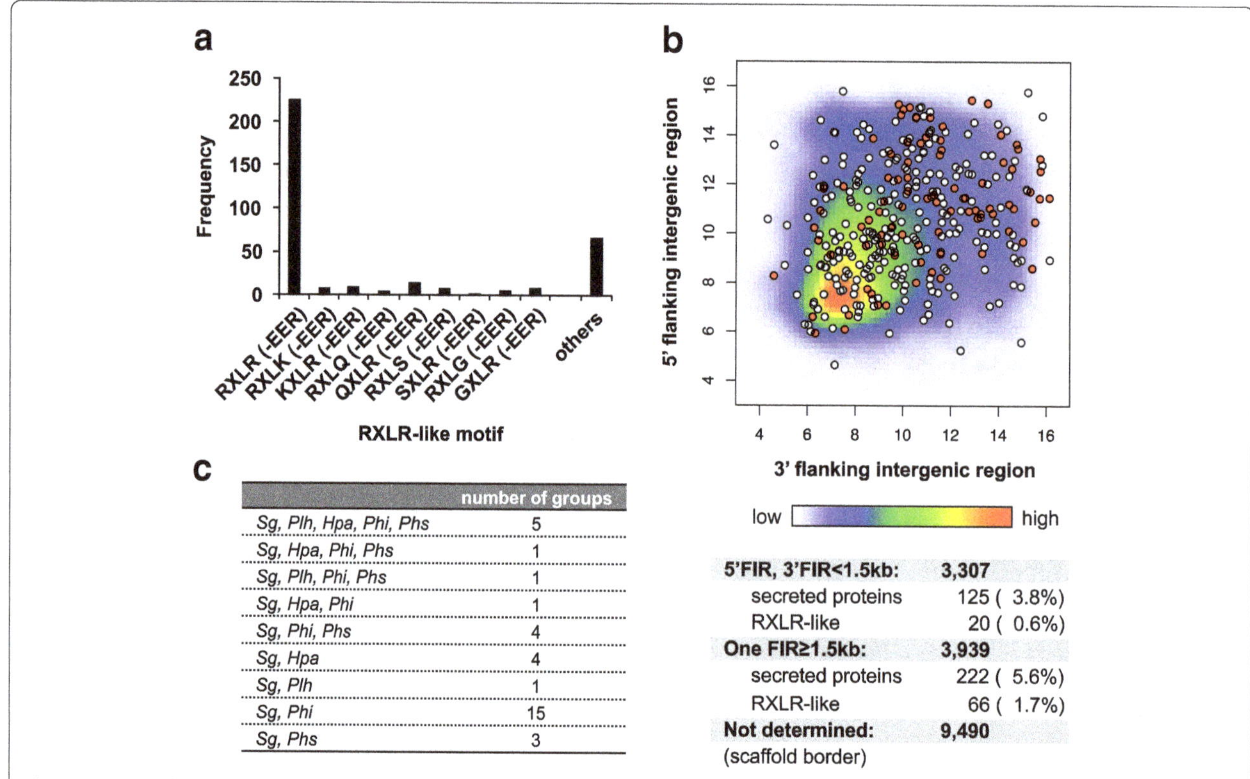

Fig. 6 Features of RXLR-like protein genes. **a** Distribution of the conserved sequence patterns of putative RXLR-like proteins. **b** Distribution of *Sg* genes according to the length of their 5′ and 3′ flanking intergenic regions. The density of genes in each positional bin is indicated by a heatmap. Putative secreted proteins (white) and RXLR-like proteins (red) genes are represented by circles. (C) Orthologous groups of SgRXLR-like proteins within the putative secreted proteins of four oomycetes

clustering, RXLR-like protein genes were not enriched in any clusters; however, 22 of these genes were induced during infection (Additional files 8 and 15).

Effector genes are distributed in gene-sparse regions of the *Phi* genome [33, 34]. In the *Sg* genome, secreted protein genes, in particular RXLR-like protein genes, were distributed in relatively gene-sparse regions compared with all of the predicted genes (Fig. 6b). Wilcoxon rank sum test indicated that distribution of intergenic length of RXLR-like genes was significantly different from that of all predicted genes (5′-intergenic length; p-value = 9.244e-05, 3′-intergenic length; p-value = 1.225e-08). We searched for orthologs of SgRXLR-like proteins among the putative secreted proteins of five oomycetes (*Sg, Plh, Hpa, Phi,* and *Phs*) and compared them using the OMA orthology database [28]. There were 35 ortholog groups that contained SgRXLR-like proteins (Fig. 6c), with most *Sg* orthologs found in the *Phi* genome.

Discussion

S. graminicola (*Sg*) has a large and highly heterozygous genome

Our analysis of Illumina sequencing paired-end reads suggested that the genome size of *Sg* is approximately 360 Mbp. This is 1.3 times larger than the genome of *Phytophthora mirabilis*, the largest among the previously sequenced oomycete plant pathogen genomes [39]. Phylogenetic analyses indicated that *Sg* is closely related to *Plh*, which has a 100-Mbp genome (Table 1), suggesting that expansion of the *Sg* genome probably occurred after its divergence from *Plh*. A broad range of genome sizes among closely related oomycetes is also found in *Phytophthora*; the smallest genome among the deeply sequenced *Phytophthora* species is 65 Mbp (in *Ph. ramorum*), while the largest genome is 240 Mbp (in *Ph. infestans*) [33]. Genome expansion occurred in *Ph. infestans* with an increase in repetitive regions such as the Gypsy elements. We found that at least 73% of *Sg* and 40% of *Plh* genomes comprised repeat regions, respectively. The number of protein-coding genes in the *Sg* genome was comparable to that in *Plh,* indicating that the larger genome size in *Sg* is not caused by an increased number of genes but by the expansion of the repetitive elements.

Proteins encoded by the *Sg* genome are mostly comparable to those of dicot downy mildews

A total of 2055 orthologous gene groups were conserved in the *Phytophthora* species but not the DMs. By contrast, the number of groups conserved among the DMs but not in *Phytophthora* was only 128. This suggests two possibilities: either the *Phytophthora* species are more phylogenetically closely related while the DMs are more diversified, or the obligate biotrophs have lost substantial numbers of genes in comparison with non-obligate microbes. Indeed, the DMs, including *Sg*, lack part of the nitrogen and sulfate metabolic pathways. When we compared the protein-coding domain frequency between the DM and *Phytophthora* genomes, we found fewer genes encoding transporters, cell wall degrading enzymes, and elicitin in the DMs than in *Phytophthora*. These results suggest that DMs have adapted to their hosts and developed their obligate biotroph lifestyles by losing components that might induce the host defense response.

Expression patterns of putative secreted protein genes

We performed expression profiling of putative secreted protein genes during infection and classified them into five clusters. Cluster I included genes expressed only in sporangia and zoospores, which likely having no direct influence on *Sg* infection of foxtail millet leaves. By contrast, the expression of genes belonging to clusters II, III, and IV increased during *Sg* infection in foxtail millet leaves. Genes of cluster II gradually increased with development of internal hyphae, suggesting that these genes contribute to the haustorial development of *Sg* and might be involved in the induction of phyllody in the *Sg*-infected foxtail millet plants. The expression of genes belonging to clusters III and IV were induced in stage 2 of infection, during the development of the primary penetration hyphae, then subsequently returned gradually to their basal expression levels. We hypothesize that *Sg* genes belonging to these clusters have roles in overcoming the host defense responses in foxtail millet, and that the effector candidate genes determining host specificity are included in clusters II, III, and IV.

Jacalin-like lectin domain proteins

We found that jacalin-like lectin domain-containing protein genes were specifically overrepresented in the *Sg* genome in comparison with *Plh*, *Hpa*, *Phi*, and *Phs* (Additional file 5). Additionally, clustering of *Sg* and 11 oomycetes secretomes using TribeMCL showed that *Sg* has four *Sg*-specific families which include 36 genes of jacalin-like domain proteins. PITG_22899, the closest gene to the *Sg*-specific clade, is induced in *Phi* during plant infection stages, and has been reported as an effector candidate by an in silico analysis (Fig. 4a) [34]. These findings imply that jacalin-like genes play a role in infection and have specifically diversified in the *Sg* genome. Our clustering analysis of the *Sg* gene expression patterns indicated that eight jacalin-like protein genes were found as DEGs (Additional file 13, in cluster III and IV). Many of jacalin-like genes other than DEGs also indicated high level of TPM (Additional file 13) implying that jacalin-like genes play roles during early infection.

If jacalin-like proteins are a novel class of effectors in *Sg*, it would be reasonable to expect the jacalin-like genes to be distributed in gene-sparse regions. While this did appear to be the case (Fig. 4c) [33, 34], the assembled scaffolds in this study were too short to determine genetic distances for a large number of genes. The use of long sequencing reads to improve the assembly will be required to determine the genetic distances of all genes, in particular the effector candidates located in gene-sparse regions.

Previous reports suggested that plant jacalin-like proteins play a role in the defense response; for example, a jacalin-related lectin-like gene in wheat positively regulates resistance to fungal pathogens [40]. The authors reported that Ta-JA1 and OsJAC1 function in bacterial and fungal resistance in wheat and rice plants, respectively. Both proteins belong to a Poaceae-specific protein family, the members of which contain jacalin-related lectin and dirigent domains [41, 42]. Analysis of the separated domains of OsJAC1 indicated that the jacalin-related lectin domain is important for its targeting to the site of pathogen attack [42]. Another study revealed that six of eight grass species have nucleotide-binding leucine-rich repeat (NLR) protein genes including jacalin domain-encoding sequences [43]. These reports imply that jacalin-like lectin domains play a role in defense responses in the Gramineae plants. The foxtail millet genome contained one NLR-jacalin fusion protein gene and four jacalin-like protein genes with a dirigent domain (Additional file 17). Taken together, the previous reports and the results of the present study suggest the possibility that *Sg* secretes jacalin-like proteins to disturb host immune signaling, enabling it to successfully establish an infection. Future studies should determine the function of *Sg* effectors containing the jacalin-like lectin domain.

Nep1-like proteins (NLPs)

Oomycetes have cytotoxic-type (type 1) and non-cytotoxic-type (type 1a) NLPs. [35]. In *Hpa* and *Plh*, most NLPs were classified as type 1a [9, 13]. Although the *Sg* genome contained 24 NLP-encoding genes, only one, SG00816, was a type 1 NLP. The expression of SG00816 was very low, suggesting that this protein plays only a minor role in infection (Additional file 14). In hemibiotrophs such as *Phytophthora* and *Colletotrichum higginsianum*, cytotoxic NLPs are believed to control the transition from the biotrophic phase to the necrotrophic phase by inducing cell death in the host plants [44, 45].

DMs are biotrophic pathogens, and cytotoxic NLPs are presumably not required for their lifecycles.

Non-cytotoxic NLPs are expressed in the biotrophic phase of hemibiotrophic pathogens [45–48]; therefore, they are believed to play a role in host penetration or the establishment of infection [35]. Intriguingly, expression of 12 *SgNLPs* including six DEGs peeked at 16 hpi (Additional file 14). These results suggest that the SgNLPs also play a role in the establishment of *Sg* infection in foxtail millet.

CRNs and RXLR proteins

Oomycetes have cytoplasmic effectors belonging to the RXLR and CRN protein families, which comprise many members [9–13]. A total of four CRNs and 355 RXLR protein genes with putative secreted signals were predicted in the *Sg* genome. In addition to four CRNs in putative secrete proteins, there are 41 CRN-like proteins without N-terminal secretion signals (Additional file 15), in agreement with a previous report that a large number of non-secreted CRNs are present in the *Plh* genome [13]. Our RNA-seq analysis revealed that 21 RXLR protein genes were found as DEGs during infection, and could have roles as effectors in *Sg*. Clustering analysis based on protein sequence using TribeMCL indicated that there are 39 RXLR families belonging to the *Sg*-specific tribes. However, expression patterns were not similar among RXLR-like genes and RXLR-like genes were not enriched in any clusters like as jacalin-like genes (Additional file 8). These results suggest that roles of RXLRs are not correlated with sequence similarities. By contrast, the expression levels of the four CRN genes were very low, suggesting that they may only have minor roles in *Sg*-foxtail millet interactions. Of the 355 RXLRs, 165 had the exact RXLR-EER motif. This contrasted with the situation in *Plh*, the most closely phylogenetically related oomycete to *Sg*, in which only 34 of 274 RXLRs had a typical RXLR-EER motif [13]. An ortholog search indicated low numbers of orthologs among the related oomycetes (Fig. 6c). Considering the above findings, the RXLRs may have evolved separately in each species, depending on the process of interaction with their host plants.

Conclusions

In this study, we report the first genome sequence of a graminicolous downy mildew pathogen, *S. graminicola*. Although the relatively large *Sg* genome showed high heterozygosity and was repetitive, it encoded a similar number of genes to other oomycete genomes. A phylogenetic analysis indicated that *Sg* was most closely related to *Plh* among the oomycetes for which genome sequences are available; however, the significantly smaller genome of *Plh* suggested that the genome expansion of *Sg* occurred after its divergence from *Plh*. Gene prediction and transcriptome analysis revealed that the *Sg* genome had several of the common effectors conserved throughout the oomycetes. In addition, *Sg* had a species-specific clade of jacalin-like lectin protein genes that were distributed in gene-sparse regions of the genome. Further analyses are needed to address the function of these jacalin-like genes and to determine whether other graminicolous downy mildews have homologous jacalins. The resources provided in this study will be invaluable for future advances in understanding the pathology of *S. graminicola*, and to determine how this pathogen perturbs host development.

Methods

Plant and oomycete materials

The foxtail millet (*Setaria italica* (L.) P. Beauv, cultivar 'Ootsuchi-10'), obtained from the experimental field of Iwate Agricultural Research Center (IARC), Karumai, Iwate, Japan with a permission, was used in this study. The single zoospore isolated strain of *Sclerospora graminicola* (Sacc.) Schroet. was derived from the isolate collected in the IARC field with a permission in 2013. Plants were grown in an artificial climate chamber at 20–25 °C with 15 h light. Four-week-old plants were infected with *S. graminicola* by spraying them with a mixture of sporangia and zoospores ($1–5 \times 10^5$ per mL). Seven days after inoculation, the leaves were harvested, incubated in 70% ethanol for 30 s, rinsed with distilled water, and used for inoculum preparation. Sporulation was induced by incubating the infected leaves at 100% humidity at 20 °C for 5–6 h. Mixtures of sporangia and zoospores were collected by rinsing the sporulated leaves with chilled sterile water.

DNA extraction

Genomic DNA was prepared from spores. The spores were ground in liquid nitrogen, to which CTAB buffer (140 mM sorbitol, Tris-HCl (pH 8.0), 22 mM Na-EDTA, 800 mM NaCl, 1% sarkosyl, and 0.8% CTAB (hexadecyltrimethylammonium bromide)) was added, before being mixed and incubated at 65 °C for 10 min. The lysate was then mixed with chloroform and centrifuged at 20,000×*g* for 5 min, after which the upper phase was transferred and precipitated using isopropanol. The DNA pellet was washed with 70% ethanol then dried and dissolved in RNase solution (0.5 x TE, 20 μg/mL RNaseA) and incubated at 37 °C for 30 min. Genomic DNA was purified using Genomic-tip (Qiagen, Germany) according to the manufacturer's protocol.

RNA extraction

Four-week-old leaves were sprayed with spores (10^6 per mL) and incubated at 22 °C in 100% humidity in darkness.

Leaves were harvested at time points of 16 h, and 1, 2, and 3 d after inoculation. Spores were sampled as a time point 0. Total RNA was prepared using PureLink Plant RNA Reagent (Thermo Fisher Scientific, USA), according to the manufacturer's protocol. The RNA samples were treated with TURBO DNase (Thermo Fisher Scientific) to remove contamination from genomic DNA.

Library preparation and sequencing

Libraries for paired-end reads and mate-pair reads of various insert sizes, including 2, 4, and 6 kbp, were constructed using the TruSeq DNA LT Sample Prep Kit and the Nextera Mate Pair Sample Prep Kit (both Illumina, USA), respectively. The paired-end library was sequenced on the Illumina MiSeq platform, while the mate-pair libraries were sequenced on the HiSeq 2500 platform (Illumina). For RNA-seq analysis, 4 μg total RNA was used to construct cDNA libraries using the TruSeq Stranded Total RNA Library Prep Kit (Illumina), according to the manufacturer's instructions. The libraries were used for paired-end sequencing in 2 × 75 cycles on the NextSeq 500 platform (Illumina) in the high output mode. The sequencing reads were filtered for their Phred quality score, and reads with a quality score of ≥30, comprising ≥90% of the reads, were retained.

Genome size estimation by k-mer distribution

Genome size was estimated by analyzing the k-mer frequency using the paired-end short reads. The peak of the k-mer frequency (M) of the reads is correlated with the real sequencing depth (N), read length (L), and k-mer length (K), and their relationships can be expressed by the following formula: $M = N \times (L - K + 1) / L$ [49]. The peak of the 15-mer frequency from the paired-end reads of *S. graminicola* was 37 (Fig. 2a). We divided the total sequence length (14,257,601,560 bp) by the real sequencing depth (39.398) to obtained an estimated the genome size of 361,885,068 bp (approx. 360 Mbp).

Genome assemblies

All sequence reads in the FASTQ format were filtered for quality using the FASTX-Toolkit version 0.0.13 [50]. The paired-end reads from Miseq were processed by removing 10 bp of the 3′-end of the second reads, and then the first reads and the trimmed second reads with a Phred quality score of ≥20, comprising ≥80% of the reads, were retained. For mate-pair reads, only those sequence reads with a Phred quality score of ≥30, comprising ≥90% of the reads, were retained. Adaptor trimming and the removal of mate-pair reads with the wrong insert sizes were performed using an in-house pipeline of scripts written in Perl and C++. Finally, the paired-end and mate-pair reads were assembled using Platanus v.1.2.1 [18].

Repeat element masking

Repeat elements were masked using RepeatModeler v1.0.8 [26]. RECON v1.08 [51] and RepeatScout v1.0.5 [52] were used to perform de novo repeat element prediction. Repbase library version 20,140,131 [53] was imported to RepeatModeler for reference-based repeat element searches. The final set of predicted repeat elements were then masked in the genome assembly using RepeatMasker v4.0.5 [27].

Gene predictions

Genes were predicted based on ab initio and RNA-seq data. RNA-seq reads were assembled and mapped to the assembled genome using the Trinity/PASA pipeline. Redundant cDNA and protein sequences were merged using cd-hit and cd-hit-est., respectively, with a 90% sequence identity level. RNA-seq reads were also mapped to the assembled genome using the TopHat2/Cufflinks/PASA pipeline, and redundant cDNA and protein sequences were merged using cd-hit and cd-hit-est., respectively, with a 90% identity level. Predicted genes from Trinity/PASA and TopHat2/Cufflinks/PASA were merged, and redundant genes were merged with a 100% sequence identity level. The results were used as evidence for an expressed gene. A SNAP HMM, trained using the CEGMA output, and GeneMark-ES were used to generate sets of gene models. We ran MAKER2 [23] (first round) using these expressed genes, and the outputs from SNAP HMM, GeneMark-ES, and RepeatMasker. A SNAP HMM was then trained using the MAKER2 first-round output, and was used to re-run MAKER2. The intron-exon boundaries were predicted by AAT [20] using RepeatMasker output and the list of putative expressed genes. Finally, the results of the MAKER2 second round were merged with the evidence of gene expression and the AAT output using EvidenceModeler. Genes encoding complete protein sequences, whose expression was determined in the RNA-seq analysis, were defined as predicted genes.

Phylogenetic analyses

Phylogenetic analyses were conducted using the orthologous genes predicted by CEGMA pipeline or annotated proteins using the Maximum Likelihood method implemented in MEGA6.06-mac [54], with 1000 bootstrap replicates.

Orthology analyses

Orthology analyses were performed with the OMA [28] software, using a minimum score cut-off of 180 to define orthologous proteins among the five oomycete genomes. Genomic and protein sequences of *Plasmopara halstedii* [13] were obtained from their local server [http://dataportal-senckenberg.de/database/metacat/rsharma.26.4/bikf], and other oomycete species were obtained from Ensembl database [http://www.ensembl.org/index.html].

TribeMCL analysis

Protein sequences of the putative secreted proteins from Sg and 11 oomycetes were clustered into families by TribeMCL algorithm [32] using BLASTp with an e-value cut-off of 1.0e-10. Protein sequences of 11 oomycetes were obtained from local server [https://www.dropbox.com/s/q37suzp15jkzshk/oomycetes_11species_secretomes.faa.zip?dl=0].

Secreted protein predictions

Signal peptides were predicted using SignalP4.1 [29]. Mature proteins lacking signaling peptides were checked for transmembrane domains using TMHMM [55].

Functional annotations

Functional annotations of predicted genes were added using InterProScan 5.15–54.0 [56] and the PANTHER classification system [57]. Protein family mapping was performed using pantherScore v.1.03, with the PANTHER database v11.

Crinkler (CRN) protein predictions

First, CRN pre-candidates were identified by their sequence similarity to known CRN proteins using BLASTp. The resulting 12 proteins with a LF/YLAK motif in their N-terminal 120 amino acids (aa) were used in a manual HMM search. The HMM was trained from the N-terminal 120 aa of these genes, and the pre-candidates were searched using HMMER v3.1 [58] with an e-value cut-off of 1e-3. The resultant proteins were identified as CRN-like proteins.

RXLR protein predictions

Candidate RXLR-like proteins were extracted from predicted secreted proteins using Perl regular expressions, HMM, and a BLASTp search. An initial set of proteins were searched using Perl regular expressions as described previously [33] and in HMM using the hmm profile [59]. The following approaches and criteria were used to extract exact RXLR proteins: (1) signal peptides within residues 1–30 followed by an RXLR motif [33, 59]; (2) Regex: allowing for a signal peptide between residues 10–40, followed by the RXLR motif within the next 100 residues, followed by the EER motif, allowing D and K [33]; (3) HMM search using Win's hmm profile.

To complement the above approach, the predicted secreted proteins were scanned using HMM and a BLASTp search to extract RXLR-like proteins: (4) an HMM was trained on 40 aa sequences including the RXLR-EER motif from the exact RXLR proteins, and putative secreted proteins were searched for using HMMER v3.1 [58] with an e-value cut-off of 1e-3. (5) Putative secreted proteins with sequence similarity to known RXLR proteins were searched using BLASTp with an e-value cut-off of 1e-10.

The results for approaches 1–5 above were merged and the non-overlapping set of proteins were defined as RXLR-like protein genes (Additional file 18).

WY-domain predictions

The WY-domains of predicted RXLR-like proteins were extracted using a pfam search, MEME [60], PSIPRED [61], and HMM, as described previously [38]. First, conserved motifs annotated as RXLR by the pfam search (Additional file 19) were searched using MEME with following parameters: -protein -oc. -nostatus -time 18,000 -maxsize 60,000 -mod zoops -nmotifs 5 -minw 6 -maxw 50. The protein secondary structure was predicted using PSIPRED (http://bioinf.cs.ucl.ac.uk/psipred/). From the MEME results, motif 1 included repeating WLY sequences and spanned an a-helical fold (Additional file 20: Fig. S5). We used sequences including motif 1 for the manual HMM search as a WY-domain. After training the HMM, the RXLR-like proteins were searched using HMMER v3.1 [58] with an e-value cut-off of 0.05 (Additional file 18).

Expression profiling

Expression levels of predicted genes were determined using the TopHat2/Cufflinks pipeline [24, 25]. Differential expression was evaluated by the Fisher's exact test using the edgeR package (version 3.18.1) [30]. TPM was calculated by the following formula: TPM = (FPKM / (sum of FPKM over all transcripts)) * 10^6. Clustering by the ward's method was performed using R Commander [62]. Clustering by logFC-Cosine method was performed using the cosine similarity of the vectors of their logFC values calculated by edgeR. Clustering by model-based clustering method was performed using MBCluster.Seq package (version 1.0) [31]. Expression levels of putative pathogenicity genes were indicated in Additional files 14, 15, 16, and 21.

qRT-PCR analysis

cDNA was synthesized using ReverTra Ace® (Toyobo, Osaka, Japan). The qRT-PCR was performed using StepOne ™ real-time PCR instrument (Applied Biosystems, Foster city, CA, USA) with 10 μL reaction mixtures containing 0.5 μL cDNA, 5 μL the KAPA SYBR FAST Universal 2X qPCR Master Mix (Kapa Biosystems, Wilmington, MA, USA), 0.3 μL of each gene-specific primer (0.1 mM), and 1.9 μL ddH_2O under the following reaction conditions: 95 °C for 20 s, followed by cycling for 40 cycles of denaturation at 95 °C for 3 s, and annealing and extension at 60 °C for 30 s. Finally, melt curve analyses (from 60 to 95 °C) were included at the end to ensure the consistency of the amplified products. A comparative CT (ΔΔCT) experiment used an endogenous control to determine the quantity of target in a sample relative to the quantity of target in a reference sample. Histone H2A gene (SG05345) was used as internal control. The primer sequences are provided in Additional file 22.

Ploidy analysis

The ploidy level was estimated as described previously [63]. Paired-end reads were mapped to the assembled genome using BWA. SNPs with at least 10 × coverage were counted using samtools v0.1.18.

Heterozygosity

To calculate heterozygosity, paired-end reads were mapped to the assembled genome using BWA. The SNPs were counted using samtools v0.1.18. SNPs with an allele frequency of between 0.4 and 0.6 were counted as heterozygous.

Domain search for *S. italica* jacalin-like proteins

S. italica proteins were downloaded from the foxtail millet database of the Beijing Genome Initiative [64]. Jacalin-like domain-containing proteins were identified using InterProScan 5.15–54.0 [56] and the *S. italica* jacalin-like proteins were annotated using the HMMER web server [65].

Additional files

Additional file 1: Table S1. BLASTn results of the assembled scaffolds against the nt NCBI database. (XLSX 54 kb)

Additional file 2: Table S2. Putative transposable elements in the *Sclerospora graminicola* genome sequence. (XLSX 51 kb)

Additional file 3: Table S3. Number of ortholog groups within the oomycete genomes. (XLSX 46 kb)

Additional file 4: Table S4. Gene IDs for nitrogen and sulfur assimilation enzymes in *Sclerospora graminicola* and related oomycetes. (XLSX 52 kb)

Additional file 5: Enrichment analysis of InterProScan domains between *Sclerospora graminicola* and related oomycetes. (XLSX 41 kb)

Additional file 6: Table S5. TPM values of DEGs encoding putative secreted proteins and cluster numbers from clustering analyses. (XLSX 68 kb)

Additional file 7: Figure S1. qRT-PCR analyses of differentially expression genes. (PDF 92 kb)

Additional file 8: Table S6.1 Summary of interproscan domain enrichment of DEGs encoding putative secreted proteins. Clusters I to IV correspond to the expression profiles given in Figure 3. **Table S6.2** Summary of interproscan domain enrichment of putative secreted proteins clustered using logFC-Cosine method. **Table S6.3** Summary of interproscan domain enrichment of putative secreted proteins clustered using MBCluster method. (XLSX 13 kb)

Additional file 9: Figure S2. Heat map showing the expression patterns of DEGs encoding putative secreted proteins. Genes were clustered by logFC-Cosine method. Line plots of the expression patterns of each gene cluster. L16H: *Sg*-inoculated leaves 16 h after inoculation; L1D, L2D, and L3D: *Sg*-inoculated leaves at one, two, and three days after inoculation, respectively. (PDF 204 kb)

Additional file 10: Figure S3. Heat map showing the expression patterns of all genes. Genes were clustered by model-based clustering method. Line plots of the expression patterns of each gene cluster. SPO: mixture of sporangia and zoospores; L16H: SPO-inoculated leaves 16 h after inoculation; L1D, L2D, and L3D: SPO-inoculated leaves at one, two, and three days after inoculation, respectively. (PDF 124 kb)

Additional file 11: Summary of clustering results of TribeMCL. (XLSX 2667 kb)

Additional file 12: Figure S4. Distribution of gene expression values. Box plot of TPM of putative secreted protein genes (A) and genes clustered in *Sg*-specific Tribe of jacalin-like domain-containing proteins by TribeMCL. (B) (PDF 444 kb)

Additional file 13: Jacalin-like domain protein genes predicted in genome of *Sclerospora graminicola* and their expression levels during infection. (XLSX 49 kb)

Additional file 14: NLP genes predicted in genome of *Sclerospora graminicola* and their expression levels during infection. (XLSX 44 kb)

Additional file 15: CRN genes predicted in genome of *Sclerospora graminicola* and their expression levels during infection. (XLSX 43 kb)

Additional file 16: RXLR-like genes predicted in genome of *Sclerospora graminicola* and their expression levels during infection. (XLSX 82 kb)

Additional file 17: Jacalin-like domain containing protein genes of *Setaria italica.* (XLSX 58 kb)

Additional file 18: Candidate RXLR-like effectors of *Sclerospora graminicola*, predicted based on gene models. (XLSX 68 kb)

Additional file 19: Conserved motifs of *Sclerospora graminicola* protein sequences annotated as RXLR by the pfam search. (TXT 5 kb)

Additional file 20: Figure S5. Prediction of WY-motifs in SgRXLR-like proteins. (PDF 162 kb)

Additional file 21: Putative effector-like protein genes in *Sclerospora graminicola* and their expression during infection. (XLSX 46 kb)

Additional file 22: Primers sequences for qRT-PCR. (XLSX 56 kb)

Abbreviations

CEGMA: Core eukaryotic gene mapping approach; CEGs: Core eukaryotic genes; CRN: Crinkler; CTAB: Hexadecyltrimethylammonium bromide; DM: Downy mildew pathogens; FPKM: Fragments per kilobase of transcript per million mapped reads; HMM: Hidden markov model; Kbp: Kilobase pair; LINE: Long interspersed elements; LTR: Long terminal repeat; Mbp: Megabase pair; NCBI: National center for biotechnology information; NLPs: Necrosis and ethylene-inducing peptide 1 (Nep1)-like proteins; NLR: Nucleotide-binding leucine-rich repeat; OMA: Orthologous MAtrix; SCRs: PcF-like small cysteine-rich proteins; SINE: Short interspersed nuclear element;; SNP: Single Nucleotide Polymorphism

Acknowledgements

We thank N. Urasaki of Okinawa Agricultural Research Center and H. Matsumura of Shinshu University for genome and RNA sequence. We also thank I. Chuma and K. Yoshida of Kobe University, H. Sakai and K. Naito of the National Institute of Agrobiological Sciences, and M. Sato of Hokkaido University for technical advice. Computations were mostly performed on the NIG supercomputer at ROIS National Institute of Genetics.

Funding

This work was supported by JSPS KAKENHI Grant Number 15 K18650 and the Ministry of Education, Culture, Sports, Science and Technology of Japan (Grant-in-Aid for Scientific Research on Innovative Areas 23,113,009).

Authors' contributions

MK conceived and performed experiments, computer analyses, interpreted the data, and developed the draft of the manuscript; YH performed experiments and single zoospore isolation of *S. graminicola*; AA procured seeds of *S. italica* and diseased *S. italica* samples infected with *S. graminicola*;

HS interpreted the data and contributed significantly in drafting the manuscript; HY, SN, and HK performed computer analyses; HT assisted in genome and RNA sequence; SK analyzed, interpreted the results and wrote the paper. JW analyzed and interpreted the results; RT supervised the entire project. All authors read and approved the final manuscript.

Competing interests

The authors declare that they have no competing interests.

Author details

[1]Iwate Biotechnology Research Center, Iwate, Japan. [2]The Sainsbury Laboratory, Norwich, UK. [3]Department of Molecular Microbiology, Tokyo University of Agriculture, Tokyo, Japan. [4]Kyoto University, Kyoto, Japan.

References

1. Lamour K, Kamoun S. Oomycete genetics and genomics: diversity, interactions, and research tools. New Jersey: John Wiley & Sons; 2009.
2. Thines M, Kamoun S. Oomycete-plant coevolution: recent advances and future prospects. Curr Opin Plant Biol. 2010;13:427–33.
3. http://exploreit.icrisat.org/profile/Small%20millets/187. ICRISAT, Retrieved 20 May 2017.
4. Thines M, Telle S, Choi YJ, Tan YP, Shivas RG. *Baobabopsis*, a new genus of graminicolous downy mildews from tropical Australia, with an updated key to the genera of downy mildews. IMA FUNGUS. 2015;6(2):483–91.
5. Singh SD, King SB, Werder J. Downy mildew disease of pearl millet. Information Bulletin No. 37. Patancheru, AP 502 324, India: International Crops Research Institute for the Semi Arid Tropics. 1993;36pp.
6. Das IK, Nagaraja A, Tonapi VA. Diseases of millets- a ready reckoner. Indian Institute of Millets Research, Rajendranagar, Hyderabad 500030, Telangana. 2016;67pp. ISBN: 81-89-335-59-6.
7. Jegera MJ, Gilijamsea E, Bockb CH, Frinking HD. The epidemiology, variability and control of the downy mildews of pearl millet and sorghum, with particular reference to Africa. Plant Pathol. 1998;47:544–69.
8. Dong S, Raffaele S, Kamoun S. The two-speed genomes of filamentous pathogens: waltz with plants. Curr Opin Genet Dev. 2015;35:57–65.
9. Baxter L, Tripathy S, Ishaque N, Boot N, Cabral A, Kemen E, Thines M, Ah-Fong A, Anderson R, Badejoko W, Bittner-Eddy P, Boore JL, Chibucos MC, Coates M, Dehal P, Delehaunty K, Dong S, Downton P, Dumas B, Fabro G, Fronick C, Fuerstenberg SI, Fulton L, Gaulin E, Govers F, Hughes L, Humphray S, Jiang RHY, Judelson H, Kamoun S, Kyung K, Meijer H, Minx P, Morris P, Nelson J, Phuntumart V, Qutob D, Rehmany A, Rougon-Cardoso A, Ryden P, Torto-Alalibo T, Studholme D, Wang Y, Win J, Wood J, Clifton SW, Rogers J, Van den Ackerveken G, Jones JDG, McDowell JM, Beynon J, Tyler BM. Signatures of adaptation to obligate biotrophy in the *Hyaloperonospora arabidopsidis* genome. Science. 2010;330(6010):1549–51.
10. Links MG, Holub E, Jiang RH, Sharpe AG, Hegedus D, Beynon E, et al. *De novo* sequence assembly of *Albugo candida* reveals a small genome relative to other biotrophic oomycetes. BMC Genomics. 2011;12:503.
11. Kemen E, Gardiner A, Schultz-Larsen T, Kemen AC, Balmuth AL, Robert-Seilaniantz A, et al. Gene gain and loss during evolution of obligate parasitism in the white rust pathogen of *Arabidopsis thaliana*. PLoS Biol. 2011;9(7):e1001094.
12. Derevnina L. Chin-Wo-Reyes S, Martin F, wood K, Froenicke L, spring O, Michelmore R. Genome sequence and architecture of the tobacco downy mildew pathogen *Peronospora tabacina*. Mol Plant-Microbe Interact. 2015;28:1198–215.
13. Sharma R, Xia X, Cano LM, Evangelisti E, Kemen E, Judelson H, Oome S, Sambles C, van den Hoogen DJ, Kitner M, et al. Genome analyses of the sunflower pathogen *Plasmopara halstedii* provide insights into effector evolution in downy mildews and *Phytophthora*. BMC Genomics. 2015;16:741.
14. Kamoun SA. Catalogue of the effector secretome of plant pathogenic oomycetes. Annu Rev Phytopathol. 2006;44:41–60.
15. Torto TA, Li S, Styer A, Huitema E, Testa A, Gow NAR, van West P, Kamoun SEST. Mining and functional expression assays identify extracellular effector proteins from the plant pathogen *Phytophthora*. Genome Res. 2003;13:1675–85.
16. Morgan W, Kamoun SRXLR. Effectors of plant pathogenic oomycetes. Curr Opin Microbiol. 2007;10:332–8.
17. Kulkarni KS, Zala HN, Bosamia TC, Shukla YM, Kumar S, Fougat RS, Patel MS, Narayanan S, Joshi CG. *De novo* transcriptome sequencing to dissect candidate genes associated with pearl millet-downy mildew (*Sclerospora graminicola* Sacc.) interaction. Front Plant Sci. 2016;22(7):847.
18. Kajitani R, Toshimoto K, Noguchi H, Toyoda A, Ogura Y, Okuno M, Yabana M, Harada M, Nagayasu E, Maruyama H, Kohara Y, Fujiyama A, Hayashi T, Itoh T. Efficient de novo assembly of highly heterozygous genomes from whole-genome shotgun short reads. Genome Res. 2015;24:1384–95.
19. Parra G, Bradnam K, Korf ICEGMA. A pipeline to accurately annotate core genes in eukaryotic genomes. Bioinformatics. 2007;23(9):1061–7.
20. Huang X, Adams MD, Zhou H, Kerlavage AA. Tool for analyzing and annotating genomic sequences. Genomics. 1997;46:37–45.
21. Haas BJ, Delcher AL, Mount SM, Wortman JR, Smith RK Jr, Hannick LI, Maiti R, Ronning CM, Rusch DB, Town CD, Salzberg SL, White O. Improving the *Arabidopsis* genome annotation using maximal transcript alignment assemblies. Nucleic Acids Res. 2003;31:5654–66.
22. Haas BJ, Salzberg SL, Zhu W, Pertea M, Allen JE, Orvis J, White O, Buell CR, Wortman JR. Automated eukaryotic gene structure annotation using EVidenceModeler and the program to assemble spliced alignments. Genome Biol. 2008;9:R7.
23. Holt C, Yandell M. MAKER2: an annotation pipeline and genome-database management tool for second-generation genome projects. BMC Bioinformatics. 2011;12:491.
24. Kim D, Pertea G, Trapnell C, Pimentel H, Kelley R, Salzberg SL. TopHat2: accurate alignment of transcriptomes in the presence of insertions, deletions and gene fusions. Genome Biol. 2013;14:R36.
25. Trapnell C, Hendrickson DG, Sauvageau M, Goff L, Rinn JL, Pachter L. Differential analysis of gene regulation at transcript resolution with RNA-seq. Nat Biotechnol. 2013;31:46–53.
26. http://www.repeatmasker.org/RepeatModeler/, v1.0.8. Accessed 4 June 2015.
27. http://www.repeatmasker.org/, v4.0.5. Accessed 4 June 2015.
28. Altenhoff AM, Škunca N, Glover N, Train CM, Sueki A, Piližota I, Gori K, Tomiczek B, Müller S, Redestig H, Gonnet GH, Dessimoz C. The OMA orthology database in 2015: function predictions, better plant support, synteny view and other improvements. Nucleic Acids Res. 2015;43(Database issue):D240–9.
29. Nielsen H, Engelbrecht J, Brunak S, von Heijne G. Identification of prokaryotic and eukaryotic signal peptides and prediction of their cleavage sites. Protein Eng. 1997;10(1):1–6.
30. Robinson MD, McCarthy DJ, Smyth GK. edgeR: a bioconductor package for differential expression analysis of digital gene expression data. Bioinformatics. 2010;26:139–40.
31. Si Y, Liu P, Li P, Brutnell TP. Model-based clustering for RNA-seq data. Bioinformatics. 2014;30:197–205.
32. Enright AJ, Van Dongen S, Ouzounis CA. An efficient algorithm for largescale detection of protein families. Nucleic Acids Res. 2002;30: 1575–84.
33. Haas BJ, Kamoun S, Zody MC, Jiang RH, Handsaker RE, Cano LM, et al. Genome sequence and analysis of the Irish potato famine pathogen *Phytophthora infestans*. Nature. 2009;461(7262):393–8.
34. Raffaele S, Win J, Cano LM, Kamoun S. Analyses of genome architecture and gene expression reveal novel candidate virulence factors in the secretome of *Phytophthora infestans*. BMC Genomics. 2010;11:637.
35. Oome S, Van den Ackerveken G. Comparative and functional analysis of the widely occurring family of Nep1-like proteins. Mol Plant-Microbe Interact. 2014;27(10):1081–94.
36. Schornack S, van Damme M, Bozkurt TO, Cano LM, Smoker M, Thines M, Gaulin E, Kamoun S, Huitema E. Ancient class of translocated oomycete effectors targets the host nucleus. Proc Nat Acad Sci USA. 2010;107:17421–6.
37. Whisson SC, Boevink PC, Moleleki L, Avrova AO, Morales JG, Gilroy EM, Armstrong MR, Grouffaud S, van West P, Chapman S, Hein I, Toth IK, Pritchard L, Birch PRJ. A translocation signal for delivery of oomycete effector proteins into host plant cells. Nature. 2007;450:115–8.
38. Boutemy LS, King SR, Win J, Hughes RK, Clarke TA, Blumenschein TM, et al. Structures of Phytophthora RXLR effector proteins: a conserved but adaptable fold underpins functional diversity. J Biol Chem. 2011;286(41):35834–42.
39. Raffaele S, Kamoun S. Genome evolution in filamentous plant pathogens: why bigger can be better. Nat Rev Microbiol. 2012:1–14.
40. Xiang Y, Song M, Wei Z, Tong J, Zhang L, Xiao L, Ma Z, Wang YA. Jacalin-related lectin-like gene in wheat is a component of the plant defence system. J Exp Bot. 2011;62(15):5471–83.
41. Ma QH, Zhen WB, Liu YC. Jacalin domain in wheat jasmonate-regulated protein ta-JA1 confers agglutinating activity and pathogen resistance. Biochimie. 2013;95:359e365.

42. Weidenbach D, Esch L, Möller C, Hensel G, Kumlehn J, Höfle C, Hückelhoven R, Schaffrath U. Polarized defense against fungal pathogens is mediated by the Jacalin-related lectin domain of modular Poaceae-specific proteins. Mol Plant. 2016;9:514–27.
43. Sarris PF, Cevik V, Dagdas G, Jones JDG, Krasileva KV. Comparative analysis of plant immune receptor architectures uncovers host proteins likely targeted by pathogens. BMC Biol. 2016;14:8.
44. Qutob D, Kamoun S, Gijzen M. Expression of a *Phytophthora sojae* necrosis-inducing protein occurs during transition from biotrophy to necrotrophy. Plant J. 2002;32:361–73.
45. Kleemann J, Rincon-Rivera LJ, Takahara H, Neumann U, van Themaat EV, Van der does HC, Hacquard S, Stüber K, Will I, Schmalenbach W, Schmelzer E, O'Connell RJ. Sequential delivery of host-induced virulence effectors by appressoria and intracellular hyphae of the phytopathogen *Colletotrichum higginsianum*. PLoS Pathog. 2012;8:e1002643.
46. Kanneganti TD, Huitema E, Cakir C, Kamoun S. Synergistic interactions of the plant cell death pathways induced by *Phytophthora infestans* Nep1-like protein PiNPP1.1 and INF1 elicitin. Mol Plant-Microbe Interact. 2006;19:854–63.
47. Cabral A, Oome S, Sander N, Küfner I, Nürnberger T, Van den Ackerveken G. Nontoxic Nep1-like proteins of the downy mildew pathogen *Hyaloperonospora arabidopsidis*: repression of necrosis-inducing activity by a surface-exposed region. Mol Plant-Microbe Interact. 2012;25:697–708.
48. Dong S, Kong G, Qutob D, Yu X, Tang J, Kang J, Dai T, Wang H, Gijzen M, Wang Y. The NLP toxin family in *Phytophthora sojae* includes rapidly evolving groups that lack necrosis-inducing activity. Mol Plant-Microbe Interact. 2012;25:896–909.
49. Li R, Fan W, Tian G, Zhu H, He L, Cai J, et al. The sequence and *de novo* assembly of the giant panda genome. Nature. 2010;463(21):311–7.
50. http://hannonlab.cshl.edu/fastx_toolkit/, version 0.0.13, release date: 2 Feb 2010.
51. Bao Z, Eddy SR. Automated de novo identification of repeat sequence families in sequenced genomes. Genome Res. 2002;12(8):1269–76.
52. Price AL, Jones NC, Pevzner PA. *De novo* identification of repeat families in large genomes. Bioinformatics. 2005;21(Suppl 1):i351–8.
53. Jurka J, Kapitonov VV, Pavlicek A, Klonowski P, Kohany O, Walichiewicz J. Repbase update, a database of eukaryotic repetitive elements. Cytogenet Genome Res. 2005;110(1-4):462–7.
54. Tamura K, Stecher G, Peterson D, Filipski A, Kumar S. MEGA6: Molecular Evolutionary Genetics Analysis Version 6.0. Mol Biol Evol. 2013;30(12):2725–9.
55. Krogh A, Larsson B, von Heijne G, Sonnhammer EL. Predicting transmembrane protein topology with a hidden Markov model: application to complete genomes. J Mol Biol. 2001;305(3):567–80.
56. Hunter S, Apweiler R, Attwood TK, Bairoch A, Bateman A, Binns D, Bork P, Das U, Daugherty L, Duquenne L et al. InterPro: the integrative protein signature database. Nucleic Acids Res 2009;37(Database issue):D211–D215.
57. Mi H, Muruganujan A, Casagrande JT, Thomas PD. Large-scale gene function analysis with the PANTHER classification system. Nat Protoc. 2013;8:1551–66.
58. Eddy SRA. New generation of homology search tools based on probabilistic inference. Genome informatics international conference on. Genome Informatics. 2009;23(1):205–11.
59. Win J, Morgan W, Bos J, Krasileva KV, Cano LM, Chaparro-Garcia A, Ammar R, Staskawicz BJ, Kamoun S. Adaptive evolution has targeted the C-terminal domain of the RXLR effectors of plant pathogenic oomycetes. Plant Cell. 2007;19:2349–69.
60. Bailey TL, Elkan C. Fitting a mixture model by expectation maximization to discover motifs in biopolymers. Proc Int Conf Intell Syst Mol Biol. 1994;2:28–36.
61. Buchan DWA, Minneci F, Nugent TCO, Bryson K, Jones DT. Scalable web services for the PSIPRED Protein Analysis Workbench. Nucleic Acids Res. 2013;41(W1):W340–48.
62. Fox J. The R commander: a basic statistics graphical user interface to R. J Stat Softw. 2005;14(9):1–42.
63. Yoshida K, Schuenemann V, Cano C, Pais P, Mishra B, Sharma R, Lanz C, Martin F, Kamoun S, Krause J, Thines M, Weigel D, Burbano H. The rise and fall of the *Phytophthora infestans* lineage that triggered the Irish potato famine. elife. 2013;2:e00731.
64. http://foxtailmillet.genomics.org.cn/page/species/download.jsp, Foxtail millet database, the Beijing Genome Initiative. Accessed 7 Oct 2016.
65. Finn RD, Clements J, Eddy SR. HMMER web server: interactive sequence similarity searching. Nucleic Acids Res. 2011;39(Web Server issue):W29–W37.

9

Comparative transcriptome analysis of a lowly virulent strain of *Erwinia amylovora* in shoots of two apple cultivars – susceptible and resistant to fire blight

Joanna Puławska*, Monika Kałużna, Wojciech Warabieda and Artur Mikiciński

Abstract

Background: *Erwinia amylovora* is generally considered to be a homogeneous species in terms of phenotypic and genetic features. However, strains show variation in their virulence, particularly on hosts with different susceptibility to fire blight. We applied the RNA-seq technique to elucidate transcriptome-level changes of the lowly virulent *E. amylovora* 650 strain during infection of shoots of susceptible (Idared) and resistant (Free Redstar) apple cultivars.

Results: The highest number of differentially expressed *E. amylovora* genes between the two apple genotypes was observed at 24 h after inoculation. Six days after inoculation, only a few bacterial genes were differentially expressed in the susceptible and resistant apple cultivars. The analysis of differentially expressed gene functions showed that generally, higher expression of genes related to stress response and defence against toxic compounds was observed in Free Redstar. Also in this cultivar, higher expression of flagellar genes (FlaI), which are recognized as PAMP (pathogen-associated molecular pattern) by the innate immune systems of plants, was noted. Additionally, several genes that have not yet been proven to play a role in the pathogenic abilities of *E. amylovora* were found to be differentially expressed in the two apple cultivars.

Conclusions: This RNA-seq analysis generated a novel dataset describing the transcriptional response of the lowly virulent strain of *E. amylovora* in susceptible and resistant apple cultivar. Most genes were regulated in the same way in both apple cultivars, but there were also some cultivar-specific responses suggesting that the environment in Free Redstar is more stressful for bacteria what can be the reason of their inability to infect of this cultivar. Among genes with the highest fold change in expression between experimental combinations or with the highest transcript abundance, there are many genes without ascribed functions, which have never been tested for their role in pathogenicity. Overall, this study provides the first transcriptional profile by RNA-seq of *E. amylovora* during infection of a host plant and insights into the transcriptional response of this pathogen in the environments of susceptible and resistant apple plants.

Keywords: Fire blight, RNA-seq, Virulence

Background

Erwinia amylovora is the causal agent of fire blight, occurring on over 130 plant species belonging to 40 genera, mainly from the family *Rosaceae* [1]. It is a serious bacterial pathogen, causing severe loses in production of apples and pears worldwide. The symptoms of fire blight can be observed on all above-ground parts of the plant. The most common are wilt and death of flowers; dieback of shoots, twigs, leaves, and fruits; and cankers of branches and the trunk, which can cause the dieback of the whole plant.

The pathogenic abilities of *E. amylovora* are determined by several factors. Based on present knowledge, the most important are the type III secretion system (T3SS) and biosynthesis of exopolysaccharides (EPS) amylovoran and levan. *E. amylovora,* as for many other pathogenic bacteria, uses T3SS to deliver effector proteins (T3E) into the cytosol of host plants. In the host cell, T3Es exert a

* Correspondence: joanna.pulawska@inhort.pl
Research Institute of Horticulture, ul. Konstytucji 3 Maja 1/3, 96-100 Skierniewice, Poland

number of effects that help the pathogen to survive and to escape immune response [2]. Exopolysaccharides play a role in bypassing the plant defence system, in blocking the vascular system of the plant and in protecting the bacteria against water and nutrient loss during dry conditions and the toxic effect of reactive oxygen species (ROS) [3, 4]. Additionally, they are crucial in the formation of biofilm, which is essential for attachment to several surfaces and for pathogenicity of bacteria [5].

To establish a pathogenic relationship with a host plant, *E. amylovora* uses complex regulatory systems that sense environmental signals and induce virulence genes. These systems include two component signal transduction systems (TCSTs), regulating amylovoran biosynthesis and swarming motility [6, 7], c-di-GMP, which positively regulates the secretion of the main exopolysaccharide in *E. amylovora*, amylovoran, leading to increased biofilm formation and quorum sensing [8, 9], the bacterial alarmone ppGpp, crucial for T3SS regulation [10], and small RNAs (sRNAs) [11]. For successful infection other important factors are: i/ motility [12], ii/ biofilm formation [5], iii/ adhesion [13], iv/ stress responses including efficient expulsion of a wide range of compounds toxic to bacteria [14, 15], v/ resistance towards host plant toxins, such as phytoalexins [16], vi/ adaptation to the environmental niche via catabolism of available carbohydrates, such as sucrose [17] and sorbitol [18], vii/ production of siderophores for iron acquisition [19], viii/ the production of metalloproteases, which are important for tissue colonization [20].

The *E. amylovora* strains collected worldwide have been found to be very similar in terms of phenotypic and genetic features, as reviewed by Puławska and Sobiczewski [21]; however, they are quite different in their levels of virulence [22]. The difference in virulence of particular strains was observed mostly on hosts with different susceptibility to fire blight, e.g., different apple cultivars. Some strains are able to infect only susceptible cultivars, while others can also infect cultivars that are found to be resistant to fire blight [23, 24]. One qualitative difference between strains responsible for overcoming resistance to fire blight of *Malus* × *robusta* 5 is that the single nucleotide polymorphism (SNP) resulting in an exchange of cysteine to serine was detected in type 3 effector (T3E) $avrRpt2_{EA}$ [25]. The difference in virulence between *E. amylovora* strains can have also quantitative background e.g., the amount of amylovoran produced and the expression of genes crucial for pathogenicity [23, 26]. However, no complex studies revealing the differences at the transcriptome level have been performed to date.

The infection of apple plants by *E. amylovora* elicits several mechanisms related to plant defence. These plant defence responses include various molecular, physiological and cellular processes, activation of expression of multiple genes, and accumulation of secondary metabolites. These processes involve a hypersensitivity response, which leads to building systemic acquired resistance, an oxidative burst, cutin formation and callose deposition [27, 28].

The available data show that resistance to fire blight in apples is based on several mechanisms involving various pathways. Several QTLs (Quantitative Trait Loci) related to resistance to fire blight in different apple genetic backgrounds and in response to different *E. amylovora* strains have been found [reviewed in [29]. Comparative studies of the reaction of sensitive and resistant apple cultivars to *E. amylovora* infection have revealed higher expression of a gene encoding vacuolar processing enzyme (VPE) - a caspase-like protease active during programmed cell death [30], BAX inhibitor and HIR proteins involved in hypersensitivity reactions and controlled cell death, and proteins involved in signal transduction, especially serine/threonine kinase and β-1,3-glucanase (PR-2 protein) [31]. Milcevičová et al. [32] indicated that the resistant plants might represent a less favourable environment for bacterial growth and have higher levels of some defence-related compounds, such as salicylic acid, or increased activities of these compounds, such as the PAL enzyme. Additionally, the levels of phenolic compounds, which are potential inhibitors of *E. amylovora*, are higher in resistant plants [33].

The aim of our study was to decipher differences in the response of a lowly virulent *E. amylovora* strain to infection of susceptible and resistant apple cultivars at the transcriptome level. For this purpose, we applied an RNA-seq technique to see the global changes in gene expression of *E. amylovora* while interacting with two apple cultivars at two time points after inoculation of shoots. We believed to find differences resulting in the inability to infect the resistant apple cultivar. Additionally, we compared transcriptomes of *E. amylovora* growing on a microbiological medium and *in planta* to elucidate transcriptional changes in bacterial cells induced by the host plant environment. Until now, no detailed studies on the mechanism of action of *E. amylovora* on hosts with different susceptibility levels have been carried out. This is also the first study to apply an RNA-seq technique for analysis of *E. amylovora* gene expression.

Results

E. amylovora virulence test

The analysis of virulence of *E. amylovora* strain 650 revealed differences in its ability to infect the different apple genotypes. The most intensive disease symptoms – 94.1% – were observed on the cv. Idared, known to be susceptible to fire blight. On the middle susceptible cv. Elstar the virulence was 57.2%, while on the cv. Free Redstar, known to be resistant to fire blight, the virulence of strain 650 was estimated to be 2.6% (Fig. 1). In the view of this results and earlier studies [34], strain 650 could be classified as lowly virulent strain.

Overview of RNA-seq results

For each biological replicate, the library was constructed and sequenced on the MiSeq (Illumina). For each sample, from 4,628,510 to 17,099,456 reads were obtained, and 1,320,224 to 10,215,328 reads were mapped to genes of *E. amylovora* CFBP 1430 genome (Additional file 1: Table S1). Mapping to rRNA operons showed that one replicate, FR-650-6d-3, possessed over 72.95% of reads complementary to rRNA (Additional file 1: Table S1), although analysis of this sample on a 2100 Bioanalyzer (Agilent) before sequencing did not show any traces of rRNA. This sample was eliminated from further analysis. The rest of the biological replicates showed a very high level of correlation ($r \geq 0.99$). The Principal Component Analysis (PCA) of the log2-transformed normalized expression values highlighted the variability among the samples and revealed the influence of different environmental conditions on bacterial gene expression (Fig. 2).

The accuracy of the RNA-seq data was verified using reverse transcription quantitative real-time polymerase chain reaction, RT-qPCR. Fold changes in expression values under different experimental conditions obtained with these two techniques were plotted on a scatter graph, with fold change values obtained from RT-qPCR on the X-axis and those obtained from RNA-seq on the Y-axis (Fig. 3). A high value for the Pearson correlation coefficient ($r = 0.954$; $p < 0.001$; $R^2 = 0.909$) indicated a positive correlation between the two variables.

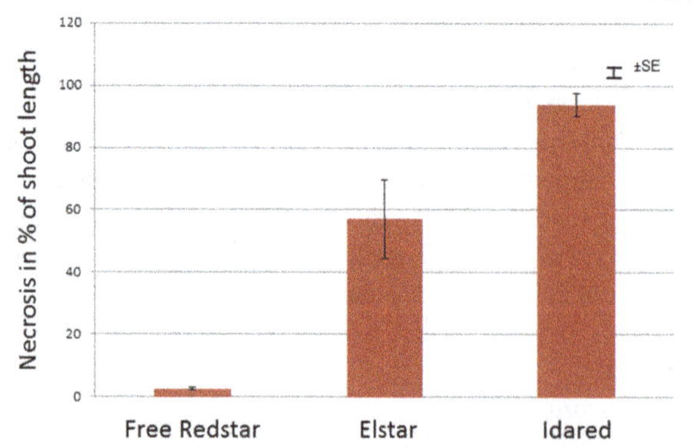

Fig. 1 Mean virulence rating of *E. amylovora* 650 strain used for inoculation of actively growing shoots of three apple cultivars of different susceptibility. Virulence was measured 6 weeks post-inoculation and it is expressed as a percent of the length of a shoot exhibiting necrosis divided by the entire length of shoot. Mean virulence ratings were separated with Tukey's test at a significance level of $P = 0.05$. The vertical bars represent standard error

Expression of *E. amylovora* 650 genes *in planta*

Over 50% of the *E. amylovora* 650 genes were differentially expressed *in planta* both at 24 h and 6 days after inoculation of two apple cultivars compared to the transcriptome of bacteria in pure culture in liquid TY medium. A total of 640 down-regulated genes and a set of another 698 up-regulated genes were found for both apple cultivars at the two time points after inoculation (Table 1, Fig. 4, Fig. 5).

The 640 down- and 698 up-regulated genes were classified into the same 19 eggNOG/COG categories. In the case of down-regulated genes, six eggNOG/COG categories were over-represented. The over-represented categories corresponding to the lowest *p*-values in increasing order included the following: translation (J), energy production and conversion (C), cell wall/membrane/envelope biogenesis (M), intracellular trafficking and secretion (U), cell motility (N), and lipid transport and metabolism (I) (Fig. 5a). Among the 698 up-regulated genes, eight eggNOG/COG categories were over-represented. The over-represented categories with the lowest p-values in increasing order included the following: amino acid transport and metabolism (E), carbohydrate transport and metabolism (G), inorganic ion transport and metabolism (P), transcription (K), signal transduction mechanisms (T), energy production and conversion (C), coenzyme transport and metabolism (H), and secondary metabolite biosynthesis (Q) (Fig. 5b, Additional file 2: Table S2).

Introduction of *E. amylovora* cells to apple tree tissue also influenced the metabolic pathways of the bacteria. The genes of metabolic pathways in the general categories of metabolism, genetic information processing and environmental information processing were found among the differentially expressed genes. Among the up-regulated genes, over-representation of genes playing roles in the pathways of xenobiotic biodegradation and metabolism, signal transduction, energy and amino acid metabolism, and membrane transport was observed. On the other hand, among the down-regulated pathways, translation and transcription were mostly over-represented (Additional file 3: Table S3).

Out of all *E. amylovora* genes located on the chromosome and plasmid, in all experimental combinations, the highest abundance of transcripts was observed for the gene EAMY_3112, annotated to code a hypothetical protein, followed by the genes *ompA*, coding for outer membrane protein A precursor, and *lpp*, coding for a major outer membrane lipoprotein precursor. Out of 1633 genes that were differentially expressed (DEGs) in Idared 24 h after inoculation, expression of 288 genes (17.63%) was not found to be changed in Free Redstar. However, out of 1910 DEGs found in Free Redstar, 622 (32.56%) had unchanged expression in Idared. In the case of both apple cultivars, most down-regulated genes belonged to the group of genes responsible for siderophore

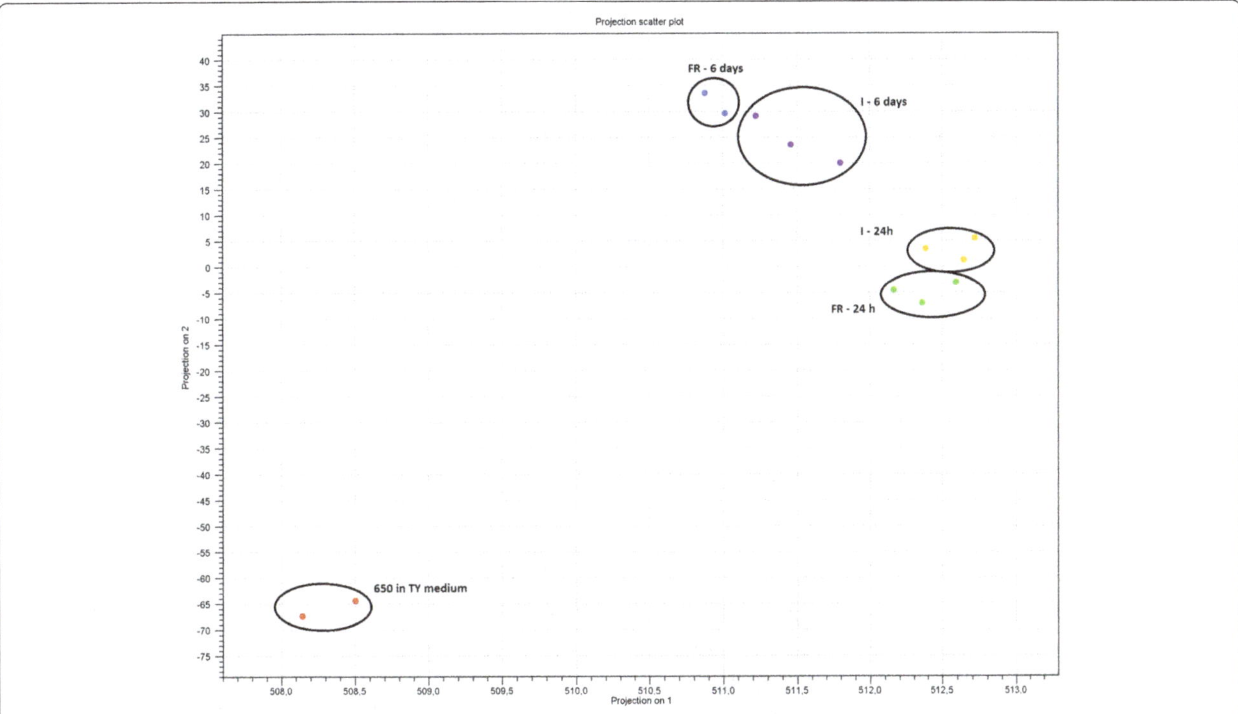

Fig. 2 The Principal Component Analysis (PCA) of the $\log_2$ - transformed normalized expression values highlighted the variability between the samples. FR – Free Redstar, I – Idared, 24 h – 24 h after inoculation, 6 days – 6 days after inoculation

biosynthesis, *dfoJAC*, and a few genes of glp regulon involved in glycerol catabolism. The most up-regulated gene in the two apple cultivars was *fldX;* in the case of Free Redstar, two other flavodoxin (electron-transfer protein) genes, *fldA3* and *fldZ*, were the most up-regulated. Among other highly up-regulated genes, some genes of the *ssuEADCB* cluster and gene *cbl* were found. The *ssu* genes are required for the utilization of sulfur from aliphatic sulfonates in *E. coli* and are regulated by the transcriptional regulator *cbl* [35]. Other genes that were significantly up-regulated included *cysGDN*, which also plays a role in sulfur metabolism. Among the most up- and down-regulated genes, several genes coding for hypothetical proteins were found.

The differences in expression of the most important genes involved in pathogenicity of *E. amylovora* were very

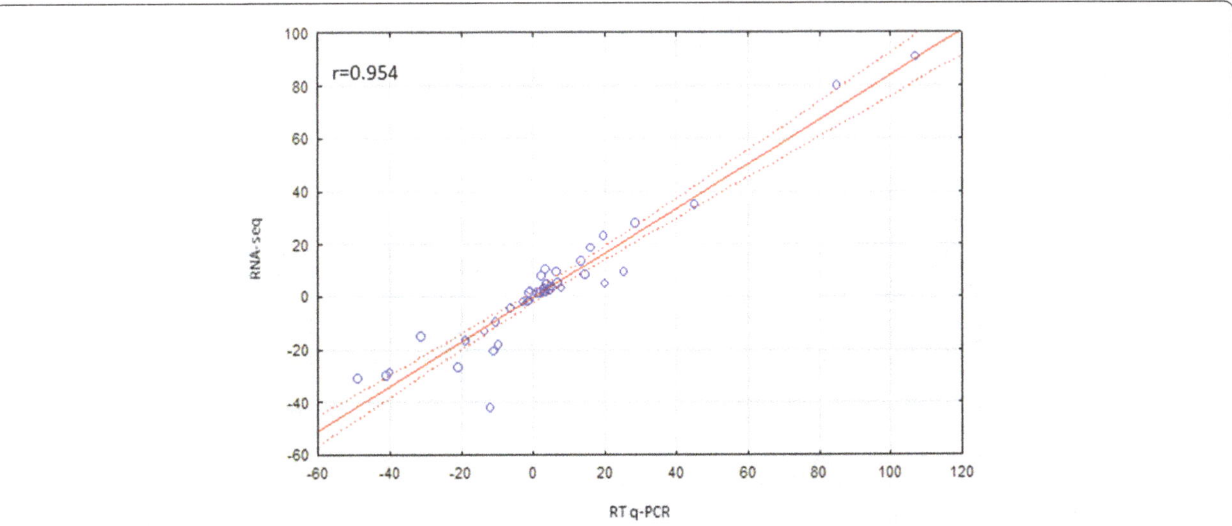

Fig. 3 Validation of RNA-seq data using RT-qPCR. Fold changes of gene expression detected by RNA-seq were plotted against the data of qPCR. The reference line indicates the linear relationship between the results of RNA-seq and qPCR

Table 1 Number of differentially expressed genes between experimental combinations

	650-bact	I-24 h	I-6d	FR-24 h	FR-6d
650-bact	–	1057 ↑ 922 ↓	1063 ↑ 1014 ↓	1020 ↑ 817↓	1053 ↑ 1086 ↓
I-24 h	922 ↑ 1057 ↓	–	519 ↑ 494 ↓	142 ↑ 150 ↓	na
I-6d	1014 ↑ 1063 ↓	494 ↑ 519 ↓	–	na	6 ↑ 4 ↓
FR-24 h	817 ↑ 1020 ↓	150 ↑ 142 ↓	na	–	572 ↑ 510 ↓
FR-6d	1086 ↑ 1053 ↓	na	4 ↑ 6 ↓	510 ↑ 572 ↓	–

650 – *E. amylovora* strain used in the study; bact – RNA isolated from pure bacterial culture; I – Idared; FR – Free Redstar; 24 h – sample collected 24 h after inoculation; 6d - sample collected 6 days after inoculation; na – not analysed
↑ - up-regulated and ↓- down-regulated genes of the samples listed in the first row in relation to the samples listed in the first column

similar between all combinations of pure bacterial culture vs. *in planta* transcriptomes. All or almost all genes involved in amylovoran biosynthesis, T3SS (hrp – PAI-1), sucrose and sorbitol metabolism, and biosynthesis of 6-thioguanine and c-di-GMP were up-regulated. Additionally, up-regulation *in planta* of a gene described on the genome of *E. amylovora* strain ATCC49946 as EAM_2938 and localized in the position 569,413 ... 569,255 of the strain CFBP 1430 genome was observed. This gene, putatively coding for a membrane protein, was found by [36] to be up-regulated by *hrpL* – the alternative sigma factor that positively regulates transcription of T3SS components. Two sets of flagellar genes localized in different regions of genome (FlaI and FlaII), T3SS genes (PAI-2 and PAI-3), iron uptake genes (*foxR, dfoJAC*), and genes involved in T1SS - metalloprotease synthesis and secretion (*prtADEF*) were mostly down-regulated, or their expression was not differential (Additional file 4: Fig. S1). The main differences in expression change between pure bacterial culture vs. *in planta* were observed in FlaI genes. Almost all FlaI genes were down-regulated in Idared; half of these genes showed no change in the expression level in Free Redstar, and *fliOPQR* genes coding for inner membrane proteins involved in flagellar biosynthesis pathways were up-regulated, or their expression was not changed. Thiamin biosynthesis genes (*thiFGSO)* located on pEA29 plasmid were also up-regulated *in planta*.

In Free Redstar, up-regulation of the multidrug efflux pump genes *acrAB*, genes coding for components of other multidrug efflux pumps, such as *aaeA, mdtB, norM, emrAB3*, genes coding for some permeases involved in the transport of metabolites or resistance to toxic substances, such as *ydfJ, ydhC, ccmB, eamA, rhaT,* and *yrbE*, and several other membrane proteins was observed. Expression of these genes was not changed in Idared. Genes *srfABC*, annotated to be putative virulence factors and known to be responsible for the biosynthesis of surfactin, a surface cyclic lipopeptide in *Bacillus subtilis* [37], were down-regulated in Idared, but their expression was unchanged in Free Redstar.

We also tested the difference in expression of 40 sRNAs identified by Zeng et al. [11]. Differences in expression between bacterial culture and *in planta* were observed only for sRNA *gcvB* and hrs17. These two sRNAs were down-regulated in apple shoots (Additional file 5: Table S4).

The majority of the genes of the type VI secretion system T6SS – cluster 1 and cluster 3 [38] were down-regulated, or no expression changes were observed *in planta* compared to in pure bacterial culture, with the exception of the EAMY_3224 gene, coding for a membrane protein, whose expression was 2.66 and 3.99 times higher in Idared and Free Redstar, respectively. Genes of the T6SS – cluster 2 were up-regulated, or their expression was not changed.

Difference in *E. amylovora* transcriptome response in apple trees of different susceptibility to fire blight 24 h after inoculation

At 24 h post inoculation, compared to the transcriptome of bacteria grown in TY medium, in Idared and Free Redstar, 1057 and 1020 genes were up-regulated, and

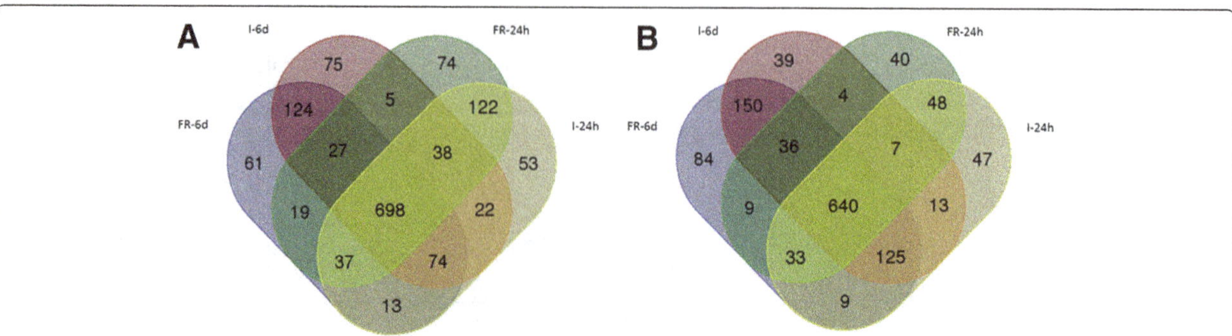

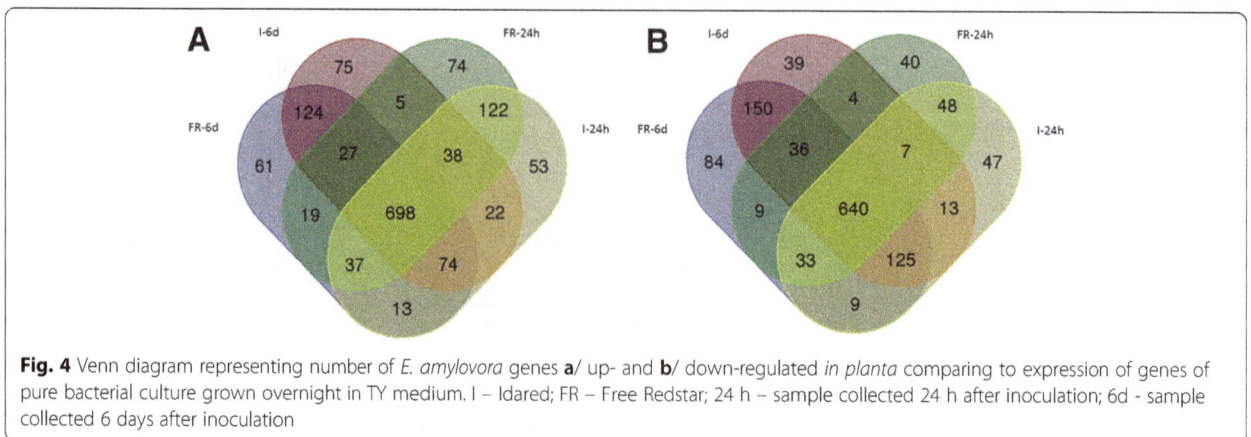

Fig. 4 Venn diagram representing number of *E. amylovora* genes **a**/ up- and **b**/ down-regulated *in planta* comparing to expression of genes of pure bacterial culture grown overnight in TY medium. I – Idared; FR – Free Redstar; 24 h – sample collected 24 h after inoculation; 6d - sample collected 6 days after inoculation

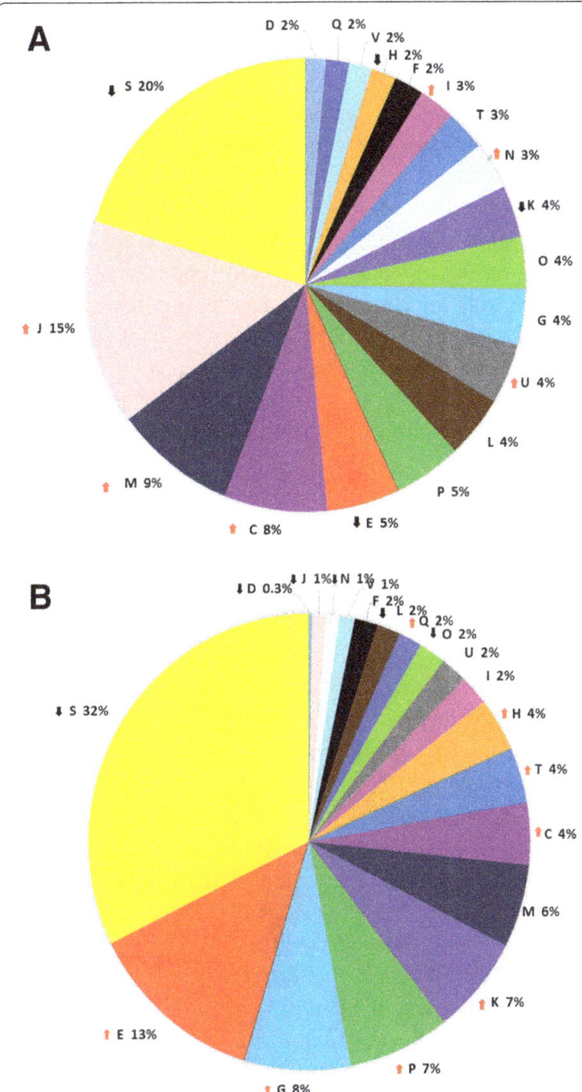

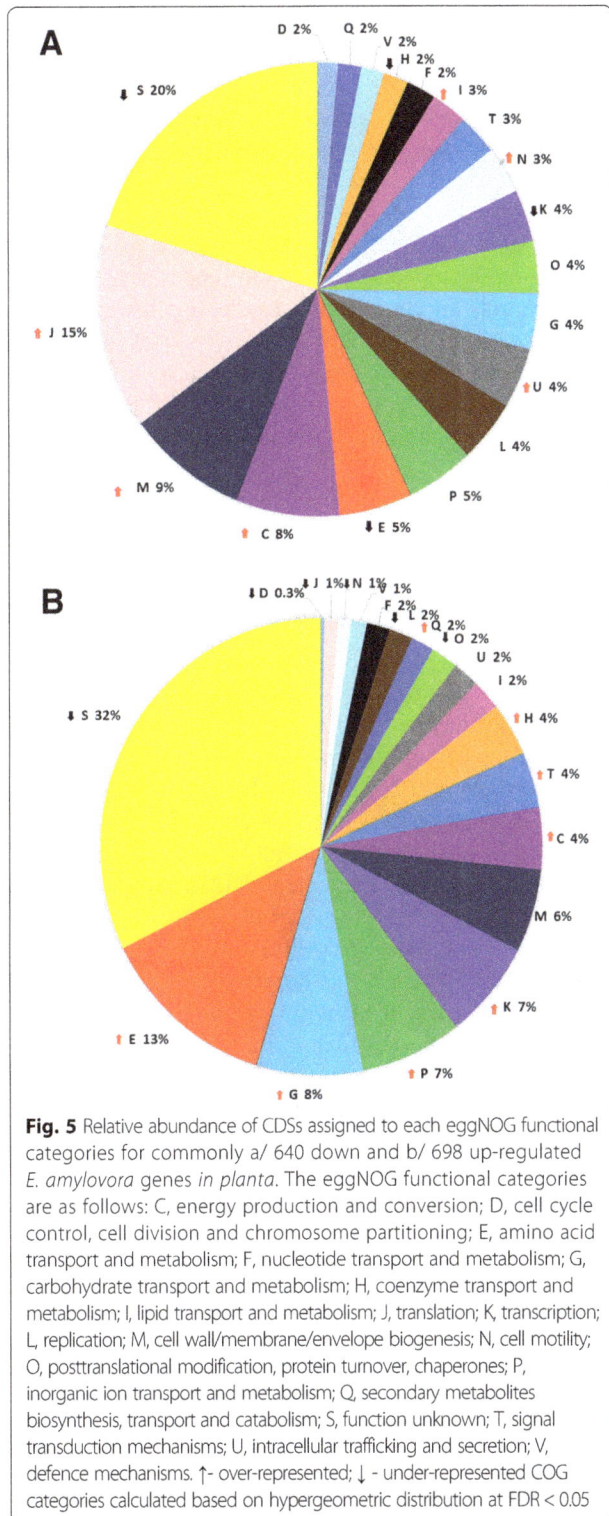

Fig. 5 Relative abundance of CDSs assigned to each eggNOG functional categories for commonly a/ 640 down and b/ 698 up-regulated *E. amylovora* genes *in planta*. The eggNOG functional categories are as follows: C, energy production and conversion; D, cell cycle control, cell division and chromosome partitioning; E, amino acid transport and metabolism; F, nucleotide transport and metabolism; G, carbohydrate transport and metabolism; H, coenzyme transport and metabolism; I, lipid transport and metabolism; J, translation; K, transcription; L, replication; M, cell wall/membrane/envelope biogenesis; N, cell motility; O, posttranslational modification, protein turnover, chaperones; P, inorganic ion transport and metabolism; Q, secondary metabolites biosynthesis, transport and catabolism; S, function unknown; T, signal transduction mechanisms; U, intracellular trafficking and secretion; V, defence mechanisms. ↑- over-represented; ↓ - under-represented COG categories calculated based on hypergeometric distribution at FDR < 0.05

922 and 817 genes were down-regulated, respectively, (Table 1, Additional file 6: Table S5abcd). Among these genes, in Idared and Free Redstar, 162 and 125 were uniquely up-regulated, while 194 and 84 were uniquely down-regulated, respectively.

At 24 h after inoculation, compared to in Free Redstar, in Idared, 150 genes of *E. amylovora* 650 had significantly higher expression, and 142 genes had significantly lower expression (Additional file 7; Table S6, Additional file 8: Table S7). The most differentially expressed genes were those coding for hypothetical proteins, based on the fact that they are not classified to any COG/eggNOG category. Generally, genes coding for hypothetical, putative or uncharacterized proteins constituted half of all calculated DEGs.

For the COG/eggNOG categories that were differentially expressed in the two apple cultivars, the highest differences were observed for categories C (energy production and conversion), E (amino acid transport and metabolism - including genes involved in methionine biosynthetic process (*metAEFR*) and glycine cleavage genes *gcvTHP*) and G (carbohydrate transport and metabolism including glycogen *glgBXCAP* operon), which were uniquely or more often represented among genes of higher expression in Idared than in Free Redstar. Among categories of genes of higher expression in Free Redstar than in Idared, M (cell wall/membrane/envelope biogenesis), N (cell motility), O (posttranslational modification, protein turnover, chaperones) and V (defence mechanisms) were all prevalent.

The most differentially expressed *E. amylovora* genes between the two apple cultivars are annotated to code for hypothetical proteins. Gene EAMY_3203, which was found to be 80.64 times up-regulated in Idared compared to in Free Redstar, is placed on the *E. amylovora* CFBP 1430 genome in the group of genes annotated to code for hypothetical proteins and located between T6SS genes. The three most up-regulated genes in Free Redstar compared to in Idared were EAMY_0674, EAMY_0946, EAMY_2509, also coding for hypothetical proteins. Gene EAMY_0674 is located on the *E. amylovora* chromosome in the group of genes related to heme utilization or adhesion, while the functions of the rest of genes in this group are unknown. The other two genes are also located in groups of genes with unidentified functions.

Among *E. amylovora* genes with higher expression in Idared than in Free Redstar, genes involved in biotin synthesis, *bioABFCD* and *ynfK*, were found, as well as two genes possibly involved in adhesion, *csuB* and *csuE*, and one biofilm regulator *yceP*. Genes *yhcN1* and *yhcN3*, coding for members of the family of proteins functioning as acid stress adaptation factors in *Yersinia pestis* [39] or described as a biofilm modulation genes in *Escherichia coli* [40] were also more highly expressed in Idared than in Free Redstar, as well as the *ygiW* gene coding for a protein that is important in stress responses, including resistance to H_2O_2, cadmium and acid, and that is important in biofilm formation in *E. coli* [41], the *alsSD* gene, coding for a-acetolactate synthase/decarboxylase, and *yidQ*, which is described as a hyperadherence gene helping to colonize host tissue in the genome of *Pantoea ananatis* [42]. In Idared, there were also up-regulated genes known to function in cellular metabolism under anaerobic conditions, namely, *adhE1, sfsA, frdC,* and *yceJ* [43], as well as genes that code alcohol dehydrogenases, *yghA, adhB, adhP,* and *adhE1*, and *ftnA*, which codes ferritin, a universal protein that stores iron and releases it in a controlled fashion.

In Free Redstar, up-regulated genes included the *suhB* gene, which has not been studied in *E. amylovora* but has been found to be an essential gene for T3SS gene expression in *Pseudomonas aeruginosa* [44], *ampC*, which is one of the beta-lactamase coding genes localized in the *E. amylovora* genome, *inlA*, coding for a precursor of intenalin-A, a surface protein known to be used by *Listeria monocytogenes* to invade mammalian cells [45], and *srfABC*, surfactin biosynthesis genes [37].

In the genome of CFBP 1430, four *pqq* operon genes coding for pyrroloquinoline quinone (PQQ) were annotated (*pqqBCDE*). In this study, we found a fifth one, *pqqA*, located between positions 3,173,704 and 3,173,784 of the CFBP 1430 genome. Three of these *pqqECB* genes were found to be up-regulated in Free Redstar. PQQ is known to be a small, redox-active molecule that serves as a cofactor for several bacterial dehydrogenases, introducing pathways for carbon utilization that confer a growth advantage, but it was also shown to be essential for antibacterial activity of the biocontrol agent *Rahnella aquatilis* [46].

Among genes known to be important for the pathogenicity of *E. amylovora,* one of two sets of flagellar genes, FlaI, was found to be expressed more in Free Redstar 24 h after infection. In Free Redstar, 25 FlaI genes were up-regulated 24 h after inoculation compared to in Idared. A similar observation was made for ten chemotaxis, motility and biofilm formation genes located within FlaI (*cheZ1, cheY1, cheB1, cheR1, tap3, tsr3, cheW1, cheA1, notB1,* and *motA1*) and the gene *aer* located in another region of the genome.

We found other genes known to play or possibly play a role in pathogenicity that were more highly expressed in Idared, including *hrpA1*, belonging to T3SS, *galF,* a precursor of amylovoran formation, *srlA*, sorbitol permease, three (EAMY_1021, EAMY_1023, EAMY_1024) out of five genes involved in biosynthesis of 6-thioguanine and *argD,* a gene coding for the N-acetylornithine aminotransferase enzyme, which is involved in the production of the amino acid arginine, and mutation in this gene causes arginine auxotrophy, nonpathogenicity in apples, and reduced virulence in pears [47].

In Free Redstar, up-regulation of the multidrug efflux pump gene *acrA*, which is a part of the multidrug efflux pump AcrAB required for virulence of *E. amylovora,* resistance towards apple phytoalexins and successful colonization of the host plant [14, 16], was observed. Up-regulation was also noted for another set of multidrug efflux pumps, *emrA* and *emrB3,* protecting the cell from several chemically unrelated antimicrobial agents [48], and for genes *aaeA* and *aaeB,* which are subunits of the p-hydroxybenzoic acid efflux pump.

Among the two-component signal transduction system (TCST) genes present in the *E. amylovora* genome, differences in expression between two apple cultivars were observed only for genes related to the motility of bacterial cells: *cheA1, cheB1, cheY1,* which were up-regulated in Free Redstar, and *baeR,* which was down-regulated in this cultivar. *baeR* may play a role in the virulence of *E. amylovora* because its overexpression significantly increased amylovoran biosynthesis [49].

In Idared, we also observed higher expression of the *rmsA* (*csrA)* gene, which has been the subject of contradictory reports. Ancona et al. [50] used knock-out mutants to find that *rsmA* (*csrA*) positively regulates virulence factors, such as motility, amylovoran production, and T3SS, while Ma et al. [51] showed that the presence of many copies of the *rsmA* gene in an *E. amylovora* cell supresses motility and EPS production. The protein RsmA creates a regulatory system with a nontranslatable RNA regulator *rmsB (csrB).* Ancona et al. [50] and Ma et al. [51] showed opposite roles of *rmsB*. In studies by Ancona et al. [50], ΔcsrB mutants were hypermotile, overproduced EPS, and showed increased expression of T3SS, while according to Ma et al. [51], multiple copies of *rsmB* in *E. amylovora* cell induced the same effect. In our studies, no difference in expression of *rmsB* was observed between cultivars, but a generally smaller amount of *rmsB* was detected in *E. amylovora in planta* than in pure bacterial culture.

All stress-related genes in the CFBP 1430 genome had different expression *in planta* than in pure bacterial culture. Of these, 17 genes, whose products participate in stress responses, were identified to be differentially expressed in Idared and Free Redstar 24 h after infection. The increased expression in Free Redstar was observed for 11 genes: genes related to heat shock (*clpB3, dnaJ, dnaK, grpE, htrB, htpG, hslU, ibpA*, EAMY_0674) and cold shock (*deaD, cspA*), while in Idared increased expression was observed for only six genes: *cspD,* related to cold shock, *yedU,* related to heat shock, and genes playing roles in the general stress response: *dps, yfiA, A* and *uspB.* This result indicates that the environment in Free Redstar is more stressful for bacterial cells than that in Idared because a higher number of stress-related genes were more intensively expressed; primarily heat shock-related genes are involved in defence against stress.

Differences in expression of a few genes coding for outer membrane proteins were detected between Idared and Free Redstar. Outer membrane proteins create a selective barrier and protect the bacteria from the environment by preventing the entry of many toxic molecules into the cell; additionally, they are members of transport systems. Out of 307 genes in the *E. amylovora* CFBP 1430 genome annotated to code for membrane proteins, 24 were differentially expressed in the two apple cultivars, and 11 and 13 were up-regulated in Idared and in Free Redstar, respectively. Among membrane protein genes with higher expression in Idared, genes responsible for the transport of amino acids, inorganic ions and coenzymes and for envelope biogenesis were observed. The up-regulated membrane protein genes in Free Redstar included genes playing a role in intracellular trafficking and secretion, defence mechanisms, lipid and inorganic ion transport and metabolism, envelope biogenesis, and cell cycle control (Additional file 9: Table S8).

Differences in expression of *E. amylovora* genes 24 h and 6 days after inoculation of Idared and free Redstar shoots

A similar number of *E. amylovora* genes were differentially expressed 24 h and 6 days after inoculation of Idared and Free Redstar shoots: 1013 and 1082, respectively. In both cases, about half of the DEGs were up-regulated and down-regulated (Table 1, Additional file 10: Table S9, Additional file 11: Table S10). Taking into consideration the COG/eggNOG categories, a different number of genes belonging to some of the categories were up-regulated and down-regulated in Idared compared to in Free Redstar (Additional file 12: Fig. S2). Among the up-regulated categories, carbohydrate transport and metabolism genes (G) and transcription genes (K) were over-represented in Free Redstar, while intracellular trafficking and secretion genes (U) were under-represented in Idared. Among the down-regulated categories, genes belonging to the cell motility category (N) were over-represented in Free Redstar and under-represented in Idared. Additionally, over-representation of the categories of energy production and conversion (C), amino acid, nucleotide, carbohydrate and lipid transport and metabolism (E, F, G, I) were observed in Idared, and cell wall/membrane/envelope biogenesis (M) in Free Redstar.

Three genes were up-regulated in Free Redstar but down-regulated in Idared at both time points: EAMY_0930, coding for ornithine utilization regulator, EAMY_1210, a putative ABC transport system, and EAMY_1750, coding for a putative flavoprotein monooxygenase. Differences were observed in the most down- and up-regulated genes in two apple cultivars. In Free Redstar, the gene *pagC* was the most down-regulated gene (166.36 times down-regulated), while on Idared, it was down-regulated only by 2.36 times. PagC protein is a well described *Enterobacteriaceae* virulence membrane protein belonging to the family Ail/OmpX/PagC/Lom. The members of this family are responsible for conferring resistance to complement-mediated killing, survival in macrophages, and adhesion and invasion of host cells in *Yersinia* and *Salmonella* strains [52]. These proteins could be important for virulence by neutralizing host defence mechanisms. Comparative genomic analysis revealed that the *pagC* gene is present in pathogenic *E. amylovora* and *E. pyrifoliae* but not in the non-pathogenic *E. tasmaniensis* [53]. Among the down-regulated genes in Free Redstar, several genes coding for hypothetical proteins were found, but no change in their expression was observed in Idared and vice versa. A similar observation was made for the groups of the most up-regulated genes in both cultivars – they consist mostly of genes coding for hypothetical proteins (Additional file 10: Table S9, Additional file 11: Table S10).

Differences in expression of virulence-related genes were observed between Idared and Free Redstar while comparing samples 24 h and 6 days after inoculation. Comparing the expression of genes at 6 days to that at 24 h after inoculation, a higher number of genes involved in amylovoran and metalloprotease biosynthesis was down-regulated in Idared than in Free Redstar. In Idared, the expression of the majority of FlaI genes and motility genes was similar at the two time points, while in Free Redstar, more genes were down-regulated 6 days after infection compared to 24 h; no expression differences between apple cultivars were observed 6 days after inoculation. Two genes (*edcB* and *edcE*) involved in c-di-GMP biosynthesis were up-regulated in Idared, while in Free Redstar, their expression was not different between the two time points after inoculation. Almost all *hrp* T3SS genes were down-regulated at 6 days compared to at 24 h after inoculation, except the T3E avirulence gene *avrRpt2*, whose expression increased over time in both apple cultivars (Additional file 10: Table S9, Additional file 11: Table S10, Additional file 13: Fig. S3).

No change in expression was observed for genes of inv./spa-type T3SSs PAI-2, with the exception of two genes, *spaK* and *spaM1*, which were significantly down-regulated in Free Redstar 6 days after inoculation compared to at 24 h (11.33 and 48.67 times, respectively). Similarly, for T3SS PAI-3, *prgK3* and *spaN3* genes were down-regulated in Free Redstar (fold changes of 3.09 and 73.55, respectively), while in Idared, over time, the *invB* gene was up-regulated, and the *sipC3* gene was down-regulated. The *srfABC* gene involved in surfactin biosynthesis, the multidrug efflux pump *acrAB* and the majority of chemotaxis and motility genes were down-regulated in Free Redstar while no expression changes were observed in Idared between the two time-points of the experiment.

Differences in expression of *E. amylovora* genes in Idared and free Redstar 6 days after inoculation

Only 11 genes were found to be differentially expressed in the two apple cultivars 6 days after inoculation. In Idared, five genes were up-regulated compared to in Free Redstar: siderophore biosynthetic genes *dfoJAC,* one gene coding for the hypothetical protein EAMY_1906, the gene EAM_2938, which is newly annotated in the CFBP 1430 genome and known to contribute to the virulence of *E. amylovora* [36] (Additional file 5: Table S4, Additional file 14: Table S11).

In Free Redstar, six genes were up-regulated: *vanA*, a putative vanillate O-demethylase oxygenase subunit, EAMY_1750, a putative flavoprotein monooxygenase, *pucI*, a putative NCS1-family allantoin permease, and three hypothetical proteins, EAMY_1683, EAMY_1948 and EAMY_3440 (Additional file 14: Table S11). The VanA protein sequence of *E. amylovora* showed over 80% similarity to proteins in human and animal pathogens of the *Enterobacteriaceae* family. In these pathogens, VanA and VanB are responsible for resistance to glycopeptides – a group of antimicrobial compounds [54] while *pucI* is a gene coding for an allantoin transport protein. Allantoin is a naturally occurring compound and a major metabolic intermediate in most living organisms, including bacteria; it often accumulates in stressed plants and may also activate stress responses [55].

Discussion

We used RNA-seq technology to analyse differences in the transcriptome of the lowly virulent *E. amylovora* strain 650 in apple shoots of two apple cultivars differing in their susceptibility to fire blight. The susceptible cultivar, Idared, could be easily infected by this strain, while the resistant one, Free Redstar, exhibited almost no disease symptoms after inoculation. The results of transcriptome analysis show clear differences between *E. amylovora* gene expression in the two apple cultivars. However, the only significant differences in expression of previously recognized genes crucial for pathogenesis were observed for flagellar genes (FlaI), which had higher expression in Free Redstar, and *hrpA,* three out of five genes involved in the biosynthesis of 6-thioguanine, which were more intensively expressed in Idared 24 h after inoculation. Six days after inoculation, siderophore biosynthetic genes *dfoJAC* were up-regulated in Idared.

The transcriptome analysis showed that expression of two sets of flagellar genes located in the *E. amylovora* genome, FlaI and FlaII, was differentially regulated. Compared to in the bacterial culture, the majority of FlaI genes were down-regulated *in planta*; no change in expression was observed for the majority of FlaII genes. From the studies of Zhao et al. [56], it is known that operon deletion of FlaII does not influence the motility of the tested strain. Moreover, a phylogenetic analysis based on concatenation of 14 conserved flagellar protein sequences revealed that both FlaI and FlaII are clustered with enterobacteria, but the phylogenetic position of the FlaI system is much closer to the phylogeny of *E. amylovora* species than that of FlaII, which is more closely related to those of *Sodalis glossinidius* – an insect endosymbiont [57]. The same phylogenetic origin was found for two non-flagellar T3SS pathogenicity islands, PAI-2 and PAI3, which were mostly down-regulated *in planta* in our studies, in contrast to hrp T3SS. However, they were previously reported to be uninvolved in *E. amylovora* virulence in plants [58] but involved in insect cell invasion by *S. glossinidius* [57]. These results indicate that PAI2, PAI3 and FlaII may be acquired from the same source by horizontal gene transfer [58].

Flagellum-based motility is important for the virulence of bacterial pathogens. In our experiment, we observed general down-regulation of FlaI genes *in planta* in Idared 24 h after inoculation and in both cultivars 6 days after inoculation compared to in bacterial culture. This is in agreement with the observations of Raymundo and Ries [59], who found that *E. amylovora* cells isolated directly from apple shoots are not motile. However, almost all FlaI genes were up-regulated 24 h after inoculation in Free Redstar compared to in Idared. The higher expression of *E. amylovora* flagellar genes in Free Redstar can explain why strain 650 cannot effectively attack Free Redstar trees. The conserved part of the flagellin polypeptide, the flg22-domain, which faces the inside of the flagellar tube, is recognized as PAMP (pathogen-associated molecular pattern) by the innate immune systems of plants [60], and as was found in the proteomic studies performed by Holtappels et al. [61], lower virulent strains have more flagellin- and motility-associated proteins. However, the question is why is the expression of *E. amylovora* flagellar genes higher in a resistant apple cultivar than in a susceptible one? Flagellum synthesis undergoes transcriptional and posttranscriptional regulation. At the transcriptional level, genes involved in flagellum synthesis are expressed in a hierarchical fashion. At the top of this hierarchy is the master regulator *flhDC,* as reviewed by Chilcott et al. [62]. At 24 h after inoculation, we observed that expression of *flhC1* was higher in Free Redstar, while the expression of *flhD1* was unchanged. The operon *flhDC* is sensitive to environmental and cell state sensors and is controlled by numerous regulators, including cAMP-CRP, H-NS, *EnvZ/OmpR, barA/uvrY* (*gacA/gacS*), *lrhA* and the phosphorelay system RcsCDB [6, 63], but these genes were not differentially expressed between the two apple genotypes. However, their levels of expression are unlikely to reflect the type of environmental signals they sense, which may be different between two apple cultivars. The second level of flagellar gene expression regulation includes the positive regulator σ^{28} factor encoded by the *fliA1* gene and a negative one, anti-σ^{28} protein, coded by *flgM*. We observed an up-regulation of *fliA1* in Free Redstar and no difference in expression of *flgM* between the two apple cultivars 24 h after inoculation, which can explain the higher expression of flagella synthesis genes in Free Redstar. Comparing expression of regulation genes 24 h and 6 days after inoculation, *fliA1* was down-regulated in both cultivars, and *flgM* was up-regulated in both cultivars at the later time point, while the *flhC1* gene was down-regulated in Free Redstar 6 days after inoculation, resulting in down-regulation of the majority of FlaI genes.

One of the clear differences in transcription of *E. amylovora* genes between the two apple genotypes was the transcription of stress-related genes. They were generally more highly expressed in the Free Redstar cultivar; most products were classified as heat shock proteins, which are a group of proteins that repress the denaturation of molecules by various stressful circumstances, such as heat, cold, UV light, oxygen, and Ca2+. A difference in abundance of these proteins was also observed among lowly and highly virulent *E. amylovora* strains based on proteomic studies performed on the leaves of a susceptible apple clone. More heat shock proteins were produced by a more virulent strain in a susceptible apple cultivar [64], while in the case of our study, the same was observed for a lowly virulent strain in a resistant cultivar. This type of protein was also induced during *E. amylovora* infection of immature pears [65].

Another group of genes that were more highly expressed in Free Redstar are genes of multidrug efflux pumps and permeases involved in the transport of metabolites or resistance to toxic substances. During infection of the plant, bacteria are exposed to a variety of antimicrobial compounds produced by the host; these protein structures are able to recognize and efficiently expel a wide range of structurally diverse compounds from the bacterial cell and play a very important role in the success of the pathogen [66]. This observation could suggest that Free Redstar produces more antimicrobial compounds, and therefore, expression of genes coding for proteins involved in detoxification of bacterial cells is higher in the more resistant cultivar.

Several genes, such as surfactin biosynthesis genes *srfABC*, *csuBE*, involved in adhesion, the biofilm regulator *yceP*, *suhB*, found to be important for T3SS in *P. aeruginosa*, and stress response genes that have not yet been shown to play a role in the pathogenic abilities of *E. amylovora* were found to be differentially expressed in the two apple cultivars. Based on this fact and on their function, structure or reports of their roles in other bacterial species, detailed studies are required to elucidate their role in the pathogenicity of *E. amylovora*. One of the most significant observations during this study is the fact that among genes with the highest fold change in expression between experimental combinations or the highest transcript abundance, there are several genes without ascribed functions. This fact suggests that although their role is unknown, their function could be important during interactions with a host plant. The importance of genes coding for hypothetical proteins was observed even during a study with a minimal cell concept, where an experimental design of a minimal synthetic genome revealed a surprising number of genes of unknown function – ca. 30% of the genome essential for bacterial life [67]. This is the general problem in genomics. At present, numerous genome projects are adding thousands of nucleotide sequences to public databases each day. The challenge is in translating sequence into function. The most common approach is to search databases for well-characterized proteins that have similar amino acid sequences to the protein encoded by a new gene and employ a method to explore the gene's function from there. Using this approach, only a fraction of predicted genes will have annotated products and functions. In the genome of *E. amylovora* CFBP1430, over 850 predicted genes are annotated to be putative proteins, proteins of unknown function or hypothetical proteins. Approximately 40% of genes cannot be classified to any COG category or are classified to category S: function unknown. This is related to the fact that although studies on genes and their function have been conducted for many years by many teams, even using new challenging techniques, there is still much work ahead for scientists to fully understand all the processes in bacterial and eukaryotic cells. Additionally, some mistakes can be generated in RNA-seq data analyses because of weak points of the algorithms applied for data normalization and gene expression fold change calculation [68]. Particularly for the extremal values of fold change, differences can be observed depending on the algorithm applied; genes of particular interest should be additionally analysed with other techniques, e.g., real-time PCR.

Only a few genes were differentially expressed 6 days after inoculation in two apple cultivars, although clear differences in disease symptoms were observed. On Idared, four genes of known relation to pathogenicity were up-regulated – genes coding for siderophore desferrioxamine biosynthesis, *dfoJAC* and EAM_2938. Desferrioxamine plays a dual role in iron acquisition and protecting bacterial cells against lethal doses of hydrogen peroxide [19]. In Idared, extracellular development of the pathogen and a rapid host cell death likely lead to iron deficiency or higher antimicrobial activity of reactive oxygen species (ROS) because ROS are closely associated with lesion development after inoculation of apple leaves with *E. amylovora* [30]. In Free Redstar, genes known to play a role in stress responses and bacterial cell defence in other bacterial species were more intensively expressed, as well as a group of genes coding for proteins of unknown function.

Zhao et al. [65] identified *E. amylovora* genes induced during infection of immature pear tissue. We found that only approximately 30% of up-regulated genes listed by Zhao et al. [65] were also up-regulated in our tests on both apple cultivars. The different plant tissue, different experimental conditions, e.g., the microbiological medium used to grow bacteria prior to the inoculation, or different types of techniques used for gene expression analysis may be the reason for these discrepancies.

To elucidate the background of the differences in virulence of *E. amylovora* strains, Holtappels et al. [64] applied a proteomics approach. After separate inoculations of apple leaves with highly and lowly virulent strains, they identified a group of 154 proteins that were differentially expressed in these two tested strains. Only a few genes coding for these proteins were found to be differentially expressed in Idared and Free Redstar; proteins identified by Holtappels et al. [64] to be more abundant in lowly virulent strain were not found to be down-regulated in the more sensitive apple cultivar or vice versa. Out of the proteins that were more abundant in the lower virulence strain, four (*mdh, yedU, dps, yfiA*) were up-regulated, and three were down-regulated (*htpG, grpE, cspA*), while among the proteins that were more highly expressed in the more virulent strain, four (*argD, mdh, galF,* EAMY_3259) were up-regulated, and three were down-regulated (*dnaK, clpB3, rho*) in Idared compared to in Free Redstar 24 h after inoculation. However, as shown by Hack [69], the proteome and transcriptome data are quite often contradictory. The poor correlation between of mRNA and protein amounts is considered as a result of different factors. One of them is weak complementarity between Shine Dalgarno sequence on the transcript and rRNA what results in lower translation level. Additionally there is important role of a secondary structure of RNA, which can be changed in certain conditions and also influences the efficiency of translation [70]. Regulatory proteins [71] and sRNA [72] which act as translational modulators as well as other factors like half-life of protein, its location and interaction with other proteins play also a role in the efficiency of translation [73].

Conclusions

This RNA-seq analysis generated a novel dataset describing the transcriptional response of the lowly virulent strain of *E. amylovora* in susceptible and resistant apple cultivar. The genes known as important for the *E. amylovora* pathogenicity were only slightly differentially expressed between apple cultivars. However, the higher expression of *E. amylovora* flagellar genes (recognized as PAMP) in Free Redstar can explain why strain 650 cannot effectively attack Free Redstar trees. Also higher expression of stress related genes and genes of multidrug efflux pumps and permeases can suggest that the environment in Free Redstar is more stressful for bacteria what can be the barrier for the efficient infection of this cultivar. Among genes with the highest fold change in expression between experimental combinations or with the highest transcript abundance, there are many genes without ascribed functions, which have never been tested for their role in pathogenicity. This fact suggests that although their role is unknown, their function could be important during interactions with a host plant.

Methods

E. amylovora virulence test

Strain 650 was isolated from a hawthorn with fire blight symptoms in central Poland and kept in the collection of the Laboratory of Bacteriology at the Research Institute of Horticulture, Skierniewice, Poland. To check its virulence, the test on apple cultivars of different susceptibility was performed. The shoots of three apple genotypes: Idared (susceptible), Elstar (middle susceptible) and Free Redstar (resistant), were used for inoculation by shoot tip cutting with scissors immersed in bacterial solution of strain 650. Fifteen trees were tested for each genotype. The virulence of *E. amylovora* strains was expressed as a percentage of shoot necrosis in relation to the entire length of the shoot measured 6 weeks after inoculation. The results were analysed with ANOVA, and means were separated with Tukey's test at $P = 0.05$.

Sample collection and RNA isolation

One-year-old, potted apple trees cultivars Idared and Free Redstar grafted on M.26 were inoculated with *E. amylovora* strain 650 in greenhouse conditions. Inoculation was made on actively growing shoots, punctured with a sterile needle on approximately 7 cm of their length and covered by droplets of bacterial suspension grown overnight in TY (Bacto Tryptone 0.5%, Yeast Extract 0.3%, $CaCl_2$ 0.065%) medium. After 24 h and 6 days from the inoculation time (Fig. 6), samples were processed, and total RNA was isolated with a Total RNA Purification Kit (Norgen Biotek), as described by Kałużna et al. [74]. At each time point, RNA was isolated separately from at least six shoots of each apple cultivar. Additionally, RNA was isolated from the pure culture of *E. amylovora* 650 grown overnight in TY medium – the same used for bacterial growth for inoculation purposes. DNA was removed from samples by DNAse treatment (Deoxyribonuclease I, ThermoScientific, Lithuania). The efficiency of DNA removal was tested by nested-PCR with the primers peant1/peant2 and AJ75/AJ76 [75, 76] complementary to plasmid pEA29. Determination of the quality and concentration of obtained RNA free from DNA was tested on an Agilent 2100 Bioanalyzer using the

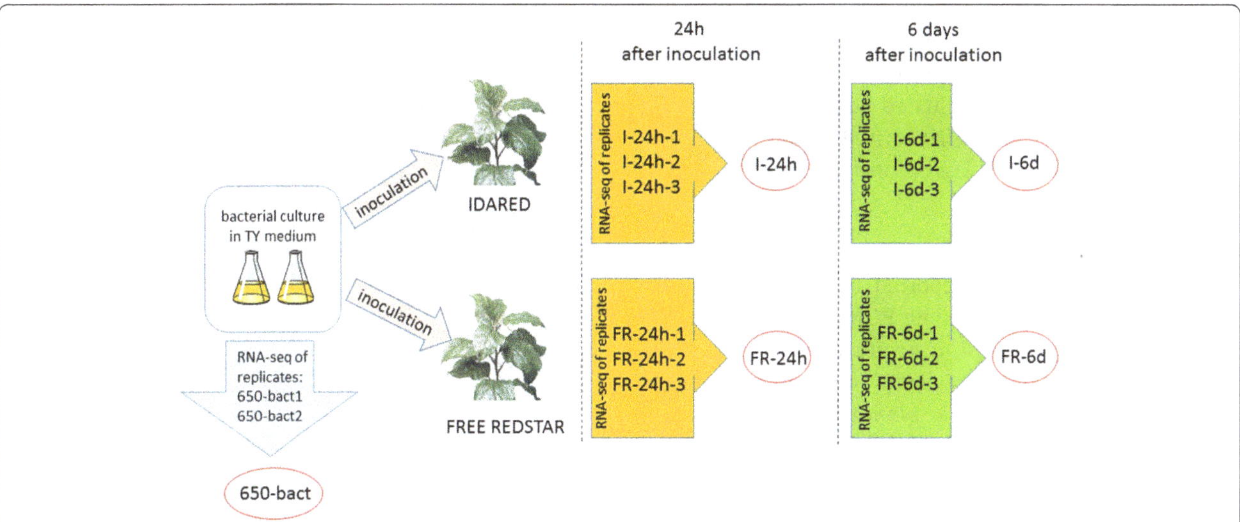

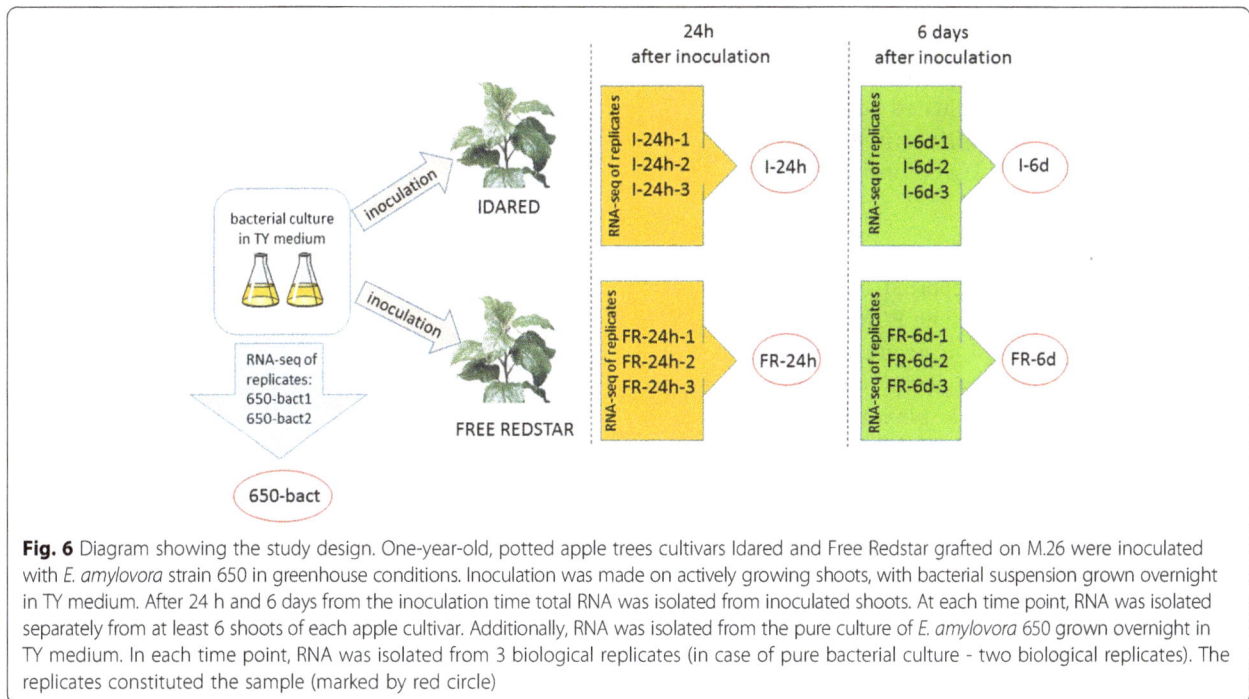

Fig. 6 Diagram showing the study design. One-year-old, potted apple trees cultivars Idared and Free Redstar grafted on M.26 were inoculated with *E. amylovora* strain 650 in greenhouse conditions. Inoculation was made on actively growing shoots, with bacterial suspension grown overnight in TY medium. After 24 h and 6 days from the inoculation time total RNA was isolated from inoculated shoots. At each time point, RNA was isolated separately from at least 6 shoots of each apple cultivar. Additionally, RNA was isolated from the pure culture of *E. amylovora* 650 grown overnight in TY medium. In each time point, RNA was isolated from 3 biological replicates (in case of pure bacterial culture - two biological replicates). The replicates constituted the sample (marked by red circle)

Agilent RNA 6000 Nano Kit according to the manufacturer's instructions. Three samples of the best quality (RIN) of each apple cultivar were subjected to rRNA depletion using a Ribo-Zero™ Magnetic Kit (Gram-Negative Bacteria); they constituted three biological replicates for each experimental combination.

Library preparation and sequencing

The rRNA depleted sample concentration was measured using the 2100 Bioanalyzer (Agilent) and an RNA 6000 Pico Kit (Agilent, 5067-1513). Since the RNA concentration was low, the maximum allowable volume (6 μl) was used for the library construction using NEBNext Ultra Directional RNA Library Preparation Kit for Illumina (New England Biolabs, E7420S). The libraries have been sequenced on the MiSeq (Illumina) using the MiSeq Reagent Kit v2 (500-cycles) (Illumina, MS-102-2003) in the PE250 read mode. The resulting reads were additionally trimmed with Cutadapt [77]. These sequence data have been submitted to the ArrayExpress (EMBL) databases under accession number E-MTAB-5630.

Bioinformatic analysis

Low quality sequence ends (ambiguous base limit: 2, quality limit: 0.05) were trimmed using CLC Genomics Workbench (v. 8.1) (Qiagen) Trim Sequences tool. For mapping, *E. amylovora* strain CFBP 1430 genome (FN434113, FN434114), which consists of chromosome (3,805,573 bp) and pEA29 plasmid (28,259 bp) and carries 3706 and 28 CDS on chromosome and plasmid, respectively [35] was used. High quality sequences were aligned to this genome using the CLC RNA-seq reference mapping algorithm with settings appropriate for Prokaryotic genomes (mapping to gene regions only). A quality control to check whether the overall variability of the samples reflected their grouping and the reproducibility between repetitions was performed with Principal Component Analysis (PCA). For the differentially expressed gene (DEG) analysis, expression values were normalized using the Trimmed Mean of M values (TMM) [78], and DEGs were analysed using the Empirical Analysis of DGE tool based on Exact Test incorporated in the EdgeR Bioconductor package and implemented in CLC Genomics Workbench. A gene was considered to be differentially regulated between two conditions when the gene showed a total read number larger than five, a > 1.5-fold absolute fold-change ratio and an FDR-adjusted p value <0.05.

For the newly annotated genes and non-coding RNAs, the expression values were normalized as a percent of total reads mapped to the *E. amylovora* CFBP 1430 genome. The results were subjected to ANOVA, and significance of differences between means were tested using the Newman-Keuls test at $p < 0.05$.

Gene annotation based on Gene Ontology (GO) terms in biological process, molecular function and cellular component categories for the *E. amylovora* CFBP 1430 coding sequences were downloaded from the UniProt database (http://www.uniprot.org). To summarize the pathway information protein sequence, fasta files were submitted to KAAS (KEGG automatic annotation server) [79], and KEGG orthology assignments were obtained (Additional file 15: Table S12). The eggNOG

4.5 database [80] was used to annotate genes with common denominators or functional categories (i.e., derived from the original COG categories). Enrichment of COG and KEGG terms was evaluated by a hypergeometric distribution at FDR < 0.05. FDR derived significance thresholds was calculated using classical one-stage method [81].

qPCR validation

Transcription expression reported in the present study was validated through real time PCR using 11 candidate genes selected out of most up-regulated, down-regulated and of similar expression genes comparing different experimental combination *in planta* to in TY medium. Real time PCR was conducted with newly designed primers (Additional file 16: Table S13). Herein, three biological replicates were used to evaluate the transcription expression of *E. amylovora* strain 650 in each apple cultivar at each time point. For gene amplification, total RNA was isolated, reverse transcribed and amplified with real-time PCR, as described by Kałużna et al. [82] The qPCR runs were performed on a Bio-Rad CFX96 thermocycler with SsoAdvanced SYBR Green Supermix (Bio-Rad, Hercules, CA) under the conditions described by Kałużna et al. [82] using the comparative 2 – ΔΔCt method.

Three genes previously reported to be the most stable in expression were used for normalization of RT-qPCR expression analysis of tested genes: *proC* (DNA-directed RNA polymerase subunit beta), *recA* (recombinase A), and *ffh* (signal recognition particle protein) [82]. For assessment of the association between RNA-seq and RT-qPCR, the Pearson's correlation method was used.

Additional files

Additional file 1: Table S1. Summary of RNA-seq data. (DOCX 15 kb)

Additional file 2: Table S2. Enriched COG/eggNOG categories among commonly up- and down-regulated genes of *E. amylovora in planta* vs. in pure bacterial culture (DOCX 14 kb)

Additional file 3: Table S3. Enriched secondary KEGG pathways among commonly up- and down-regulated genes of *E. amylovora in planta* vs. in pure bacterial culture (DOCX 14 kb)

Additional file 4: Fig. S1. Change of expression of known genes involved in pathogenicity of *Erwinia amylovora* between bacteria in TY medium (650-bact) and *in planta* (I-24 h and FR-24 h) 24 h after inoculation. (PDF 339 kb)

Additional file 5: Table S4. Expression of newly annotated and non-coding RNAs (DOCX 19 kb)

Additional file 6: Table S5. Differentially expressed genes of *E. amylovora* 650 between bacterial culture in TY medium(650-bact) and in apple shoots in two time points after inoculation (I-24 h, I-6d, FR-24 h, FR-6d). (XLSX 401 kb)

Additional file 7: Table S6. *Erwinia amylovora* 650 genes up-regulated in Free Redstar (FR-24 h) in comparison to Idared (I-24 h) 24 h after inoculation (XLSX 23 kb)

Additional file 8: Table S7. *Erwinia amylovora* 650 genes up-regulated in Idared (I-24 h) in comparison to Free Redstar (FR-24 h) 24 h after inoculation (XLSX 23 kb)

Additional file 9: Table S8. Genes of *E. amylovora* 650 coding for membrane proteins and differentially expressed in two apple cultivars – Idared and Free Redstar (I-24 h and FR-24 h) 24 h after inoculation. (DOCX 16 kb)

Additional file 10: Table S9. Differentially expressed genes of *E. amylovora* 650 genes between two time point of experiment - 24 h (FR-24 h) and 6 days (FR-6d) after inoculation of Free Redstar shoots (XLSX 121 kb)

Additional file 11: Table S10. Differentially expressed genes of *E. amylovora* 650 genes between two time point of experiment - 24 h (I-24 h) and 6 days (I-6d) after inoculation of Idared shoots (XLSX 111 kb)

Additional file 12: Fig. S2. The content of different COG/eggnog categories among genes of different expression between two time points after inoculation–24 h and 6 days. UP –up-regulated, DOWN – down-regulated, I –Idared, FR –Free Redstar, 6d –6 days. (PDF 122 kb)

Additional file 13: Fig. S3. Change of expression of known genes involved in pathogenicityof Erwinia amylovorabetween24 h (FR-24 h) and 6 days (FR-6d) after inoculation on Idared and on Free Redstar (PDF 322 kb)

Additional file 14: Table S11. *Erwinia amylovora* 650 genes differentially regulated in Free Redstar (FR-6d) and in Idared (I-6d) 6 days after inoculation (XLSX 12 kb)

Additional file 15: Table S12. KEGG orthology assignments for genes located on *E. amylovora* CFBP 1430 genome (XLSX 121 kb)

Additional file 16: Table S13. Primers used for qRT-PCR validation of RNAseq data. (DOCX 14 kb)

Abbreviations

ANOVA: Analysis of variance; COG: Clusters of Orthologous Groups; DEGs: Differentially expressed genes; EPS: Exopolysaccharides; FDR: False discovery rate; GO: Gene onthology; KAAS: KEGG automatic annotation server,; KEGG: Kyoto encyclopedia of genes and genomes; PAL: Phenylalanine ammonia lyase; PAMP: Pathogen-associated molecular pattern; PCA: Principal component analysis; PQQ: Pyrroloquinoline quinone; QTL: Quantitative trait Loci; RT-qPCR: Reverse transcribed quantitative polymerase chain reaction; SNP: Single nucleotide polymorphism; T3E: Type 3 effector; T3SS: Type 3 secretion system; T6SS: Type 6 secretion system; TCSTs: Two component signal transduction systems; TMM: Trimmed mean of m values; TY: Tryptone yeast; VPE: Vacuolar processing enzyme

Acknowledgements

The authors would like to thank Mrs. Halina Kijańska for excellent technical help.

Funding

This work was funded by the National Science Centre, Poland, Grant UMO-2012/05/B/NZ9/03455. The funding body played no role in the design of the study and collection, analysis, interpretation of the data, or in the writing of the manuscript.

Authors' contributions

JP is author of the conception of the study, its design, partial analysis of the data and their interpretation. MK designed and performed RNA isolation and RNA-seq validation. AM worked out and performed virulence tests and effective inoculation of plants for RNA isolation. WW analysed data statistically. The manuscript was written by JP and commented by all authors. All authors read and approved the manuscript.

Competing interests

The authors declare that they have no competing interests.

References

1. Van der Zwet T, Keil HL. Fire blight—a bacterial disease of rosaceous plants. Agricultural handbook 510, U.S. Washington: Department of Agriculture; 1979.
2. Oh CS, Beer SV. Molecular genetics of *Erwinia amylovora* involved in the development of fire blight. FEMS Microbiol Lett. 2005;253:185–92.
3. Ordax M, Marco-Noales E, Lopez MM, Biosca EG. Exopolysaccharides favor the survival of *Erwinia amylovora* under copper stress through different strategies. Res Microbiol. 2010;161:549–55.
4. Venisse JS, Gullner G, Brisset MN. Evidence for the involvement of an oxidative stress in the initiation of infection of pear by *Erwinia amylovora*. Plant Physiol. 2001;125:2164–72.
5. Koczan JM, McGrath MJ, Zhao Y, Sundin GW. Contribution of *Erwinia amylovora* exopolysaccharides amylovoran and levan to biofilm formation: implications in pathogenicity. Phytopathology. 2009;99:1237–44.
6. Zhao Y, Wang D, Nakka S, Sundin GW, Korban SS. Systems level analysis of two-component signal transduction systems in *Erwinia amylovora*: role in virulence, regulation of amylovoran biosynthesis and swarming motility. BMC Genomics. 2009;10:245.
7. Wang DP, Korban SS, Zhao YF. The Rcs phosphorelay system is essential for pathogenicity in *Erwinia amylovora*. Mol Plant Pathol. 2009;10:277–90.
8. Castiblanco LF, Edmunds AC, Waters CM, Sundin GW. Characterization of quorum sensing and cyclic-di-GMP signaling systems in *Erwinia amylovora*. Phytopathology. 2011;101:S2.
9. Edmunds AC, Castiblanco LF, Sundin GW, Waters CM. Cyclic Di-GMP modulates the disease progression of *Erwinia amylovora*. J Bacteriol. 2013; 195:2155–65.
10. Ancona V, Lee JH, Chatnaparat T, Oh J, Hong J-I, Zhao Y. The bacterial alarmone (p)ppGpp activates the type III secretion system in *Erwinia amylovora*. J Bacteriol. 2015;197:1433–43.
11. Zeng Q, McNally RR, Sundin GW. Global small RNA chaperone Hfq and regulatory small RNAs are important virulence regulators in *Erwinia amylovora*. J Bacteriol. 2013;195:1706–17.
12. Koczan JM, Sundin GW. Deletion of *Erwinia amylovora* flagellar motor protein genes motab alters biofilm formation and virulence in apple. Acta Hort. 2011;896:203–9.
13. Koczan JM, Lenneman BR, McGrath MJ, Sundin GW. Cell surface attachment structures contribute to biofilm formation and xylem colonization by *Erwinia amylovora*. Appl Environ Microbiol. 2011;77:7031–9.
14. Al-Karablieh N, Weingart H, Ullrich MS. The outer membrane protein TolC is required for phytoalexin resistance and virulence of the fire blight pathogen *Erwinia amylovora*. Microb Biotechnol. 2009;2:465–75.
15. Pletzer D, Schweizer G, Weingart H. AraC/XylS family stress response regulators rob, SoxS, PliA, and OpiA in the fire blight pathogen *Erwinia amylovora*. J Bacteriol. 2014;196(17):3098–110.
16. Burse A, Weingart H, Ullrich MS. The phytoalexin-inducible multidrug efflux pump AcrAB contributes to virulence in the fire blight pathogen, *Erwinia amylovora*. Mol Plant Microbe In. 2004;17:43–54.
17. Bogs J, Geider K. Molecular analysis of sucrose metabolism of *Erwinia amylovora* and influence on bacterial virulence. J Bacteriol. 2000;182:5351–8.
18. Aldridge P, Metzger M, Geider K. Genetics of sorbitol metabolism in *Erwinia amylovora* and its influence on bacterial virulence. Mol Gen Genet. 1997;256:611–9.
19. Dellagi A, Brisset MN, Paulin JP, Expert D. Dual role of desferrioxamine in *Erwinia amylovora* pathogenicity. Mol Plant Microbe In. 1998;11:734–42.
20. Zhang Y, Bak DD, Heid H, Geider K. Molecular characterization of a protease secreted by *Erwinia amylovora*. J Mol Biol. 1999;289(5):1239–51.
21. Puławska J, Sobiczewski P. Phenotypic and genetic diversity of *Erwinia amylovora*: the causal agent of fire blight. Trees-Struct Funct. 2012;26:3–12.
22. Cabrefiga J, Montesinos E. Analysis of aggressiveness of *Erwinia amylovora* using disease-dose and time relationships. Phytopathology. 2005;95:1430–7.
23. Lee SA, Ngugi HK, Halbrendt NO, O'Keefe G, Lehman B, Travis JW, Sinn JP, McNellis TW. Virulence characteristics accounting for fire blight disease severity in apple trees and seedlings. Phytopathology. 2010;100:539–50.
24. Norelli JL, Aldwinckle HS. Differential susceptibility of *Malus* spp cultivars Robusta 5, novole and Ottawa 523 to *Erwinia amylovora*. Plant Dis. 1986;70:1017–9.
25. Vogt I, Wöhner T, Richter K, Flachowsky H, Sundin GW, Wensing A, Savory EA, Geider K, Day B, Hanke MV, Peil A. Gene-for-gene relationship in the host–pathogen system *Malus*× *robusta* 5–*Erwinia amylovora*. New Phytol. 2013;197(4):1262–75.
26. Wang D, Korban SS, Zhao Y. Molecular signature of differential virulence in natural isolates of *Erwinia amylovora*. Phytopathology. 2010;100:192–8.
27. Dangl JL, Jones JDG. Plant pathogens and integrated defence responses to infection. Nature. 2001;411:826–33.
28. Malnoy M, Martens S, Norelli JL, Barny MA, Sundin GW, Smits THM, Duffy B. Fire blight: applied genomic insights of the pathogen and host. Annu Rev Phytopathol. 2012;50:475–94.
29. Khan MA, Zhao Y, Korban SS. Molecular mechanisms of pathogenesis and resistance to the bacterial pathogen *Erwinia amylovora*, causal agent of fire blight disease in *Rosaceae*. Plant Mol Biol Rep. 2012;30:247–60.
30. Iakimova ET, Sobiczewski P, Michalczuk L, Wegrzynowicz-Lesiak E, Mikicinski A, Woltering EJ. Morphological and biochemical characterization of *Erwinia amylovora*-induced hypersensitive cell death in apple leaves. Plant Physiol and Bioch. 2013;63:292–305.
31. Markiewicz M, Michalczuk L. Molecular response of resistant and susceptible apple genotypes to *Erwinia amylovora* infection. Eur J Plant Pathol. 2015; 143:515–26.
32. Milcevičová R, Gosch C, Halbwirth H, Stich K, Hanke M-V, Peil A, Flachowsky H, Rozhon W, Jonak C, Oufir M, Hausman JF, Matusikova I, Fluch S, Wilhelm E. Erwinia amylovora-induced defense mechanisms of two apple species that differ in susceptibility to fire blight. Plant Sci. 2010;179:60–7.
33. Roemmelt S, Plagge J, Treutter D, Gutmann M, Feucht W, Zeller W. Defense reaction of apple against fire blight: histological and biochemical studies. Acta Hort. 1999;489:335–6.
34. Norelli JL, Aldwinckle HS, Beer SV. Differential host x pathogen interactions among cultivars of apple and strains of *Erwinia amylovora*. Phytopathology. 1984;74(2):136–9.
35. van der Ploeg JR, Iwanicka-Nowicka R, Bykowski T, Hryniewicz MM, Leisinger T. The *Escherichia coli ssuEADCB* gene cluster is required for the utilization of sulfur from aliphatic sulfonates and is regulated by the transcriptional activator Cbl. J Biol Chem. 1999;274:29358–65.
36. McNally RR, Toth IK, Cock PJA, Pritchard L, Hedley PE, Morris JA, Zhao Y, Sundin GW. Genetic characterization of the HrpL regulon of the fire blight pathogen *Erwinia amylovora* reveals novel virulence factors. Mol Plant Pathol. 2012;13:160–73.
37. Nakano MM, Zuber P. Cloning and characterization of *srfB*, a regulatory gene involved in surfactin production and competence in *Bacillus subtilis*. J Bacteriol. 1989;171:5347–53.
38. Smits THM, Rezzonico F, Kamber T, Blom J, Goesmann A, Frey JE, Duffy B. Complete genome sequence of the fire blight pathogen *Erwinia amylovora* CFBP 1430 and comparison to other *Erwinia* spp. Mol Plant Microbe In. 2010;23:384–93.
39. Vadyvaloo V, Viall AK, Jarrett CO, Hinz AK, Sturdevant DE, Hinnebusch BJ. Role of the PhoP-PhoQ gene regulatory system in adaptation of *Yersinia pestis* to environmental stress in the flea digestive tract. Microbiol-SGM. 2015;161:1198–210.
40. Hou Z, Fink RC, Sugawara M, Diez-Gonzalez F, Sadowsky MJ. Transcriptional and functional responses of *Escherichia coli* O157:H7 growing in the lettuce rhizoplane. Food Microbiol. 2013;35:136–42.
41. Lee J, Hiibel SR, Reardon KF, Wood TK. Identification of stress-related proteins in *Escherichia coli* using the pollutant cis-dichloroethylene. J Appl Microbiol. 2010;108:2088–102.
42. Megias E, Megias M, Ollero FJ, Hungria M. Draft genome sequence of *Pantoea ananatis* strain AMG521, a rice plant growth-promoting bacterial endophyte isolated from the guadalquivir marshes in southern Spain. Genome Announc. 2016;4:e01681–15.
43. Babujee L, Apodaca J, Balakrishnan V, Liss P, Kiley PJ, Charkowski AO, Glasner JD, Perna NT. Evolution of the metabolic and regulatory networks associated with oxygen availability in two phytopathogenic enterobacteria. BMC Genomics. 2012;13:110.
44. Li K, Xu C, Jin Y, Sun Z, Liu C, Shi J, Chen G, Chen R, Jin S, Wu W. SuhB is a regulator of multiple virulence genes and essential for pathogenesis of *Pseudomonas aeruginosa*. MBio. 2013;4(6):e00419–3.
45. Lecuit M, Ohayon H, Braun L, Mengaud J, Cossart P. Internalin of *Listeria monocytogenes* with an intact leucine-rich repeat region is sufficient to promote internalization. Infect Immun. 1997;65:5309–19.
46. Li L, Jiao Z, Hale L, Wu W, Guo Y. Disruption of gene *pqqA* or *pqqB* reduces plant growth promotion activity and biocontrol of crown gall disease by *Rahnella aquatilis* HX2. PLoS One. 2014;9(12):e115010.
47. Ramos LS, Lehman BL, Peter KA, McNellis TW. Mutation of the *Erwinia amylovora argD* gene causes arginine auxotrophy, nonpathogenicity in apples, and reduced virulence in pears. Appl Environ Microbiol. 2014;80:6739–49.

48. Lomovskaya O, Lewis K, Matin A. EmrR is a negative regulator of the *Escherichia coli* multidrug-resistance pump EmrAB. J Bacteriol. 1995;177: 2328–34.
49. Pletzer D, Stahl A, Oja AE, Weingart H. Role of the cell envelope stress regulators BaeR and CpxR in control of RND-type multidrug efflux pumps and transcriptional cross talk with exopolysaccharide synthesis in *Erwinia amylovora*. Arch Microbiol. 2015;197:761–72.
50. Ancona V, Lee JH, Zhao Y. The RNA-binding protein CsrA plays a central role in positively regulating virulence factors in *Erwinia amylovora*. Sci Rep. 2016;6:37195.
51. Ma WL, Cui Y, Liu Y, Dumenyo CK, Mukherjee A, Chatterjee AK. Molecular characterization of global regulatory RNA species that control pathogenicity factors in *Erwinia amylovora* and *Erwinia herbicola* pv. *gypsophilae*. J Bacteriol. 2001;183:1870–80.
52. Kolodziejek AM, Hovde CJ, Minnich SA. *Yersinia pestis* ail: multiple roles of a single protein. Front Cell Infect Microbiol. 2012;2:103.
53. Kube M, Migdoll AM, Gehring I, Heitmann K, Mayer Y, Kuhl H, Knaust F, Geider K, Reinhardt R. Genome comparison of the epiphytic bacteria *Erwinia billingiae* and *E. tasmaniensis* with the pear pathogen *E. pyrifoliae*. BMC genomics. 2010;11:393.
54. Beceiro A, Tomas M, Bou G. Antimicrobial resistance and virulence: a successful or deleterious association in the bacterial world? Clin Microbiol Rev. 2013;26:185–230.
55. Takagi H, Ishiga Y, Watanabe S, Konishi T, Egusa M, Akiyoshi N, Matsuura T, Mori IC, Hirayama T, Kaminaka H, Shimada H, Sakamoto A. Allantoin, a stress-related purine metabolite, can activate jasmonate signaling in a MYC2-regulated and abscisic acid-dependent manner. J Exp Bot. 2016;67: 2519–32.
56. Zhao YF, Qi MS, Wang DP. Evolution and function of flagellar and non-flagellar type III secretion systems in *Erwinia amylovora*. Acta Hort. 2011;896: 177–84.
57. Dale C, Welburn SC. The endosymbionts of tsetse flies: manipulating host-parasite interactions. Int J Parasitol. 2001;31:628–31.
58. Zhao Y, Sundin GW, Wang D. Construction and analysis of pathogenicity island deletion mutants of *Erwinia amylovora*. Can J Microbiol. 2009;55:457–64.
59. Raymundo AK, Ries SM. Motility of *Erwinia amylovora*. Phytopathology. 1980; 70.1062–5.
60. Felix G, Duran JD, Volko S, Boller T. Plants have a sensitive perception system for the most conserved domain of bacterial flagellin. Plant J. 1999; 18:265–76.
61. Holtappels M, Vrancken K, Schoofs H, Deckers T, Remans T, Noben JP, Valcke R. A comparative proteome analysis reveals flagellin, chemotaxis regulated proteins and amylovoran to be involved in virulence differences between *Erwinia amylovora* strains. J Proteome. 2015;123:54–69.
62. Chilcott GS, Hughes KT. Coupling of flagellar gene expression to flagellar assembly in salmonella enterica serovar typhimurium and *Escherichia coli*. Microbiol Mol Biol Rev. 2000;64:694–708.
63. Francez-Charlot A, Laugel B, Van Gemert A, Dubarry N, Wiorowski F, Castanie-Cornet MP, Gutierrez C, Cam K. RcsCDB his-asp phosphorelay system negatively regulates the flhDC operon in *Escherichia coli*. Mol Microbiol. 2003;49:823–32.
64. Holtappels M, Vrancken K, Noben JP, Remans T, Schoofs H, Deckers T, Valcke R. The in planta proteome of wild type strains of the fire blight pathogen, *Erwinia amylovora*. J Proteome. 2016;139:1–12.
65. Zhao YF, Blumer SE, Sundin GW. Identification of *Erwinia amylovora* genes induced during infection of immature pear tissue. J Bacteriol. 2005;187: 8088–103.
66. Martinez JL, Sanchez MB, Martinez-Solano L, Hernandez A, Garmendia L, Fajardo A, Alvarez-Ortega C. Functional role of bacterial multidrug efflux pumps in microbial natural ecosystems. FENS Microbiol Rev. 2009;33:430–49.
67. Hutchison CA 3rd, Chuang R-Y, Noskov VN, Assad-Garcia N, Deerinck TJ, Ellisman MH, Gill J, et al. Design and synthesis of a minimal bacterial genome. Science (New York, N.Y.) 2016;351:aad6253. http://science.sciencemag.org/content/351/6280/aad6253.
68. Dillies M-A, Rau A, Aubert J, Hennequet-Antier C, Jeanmougin M, Servant N, Keime C, et al. A comprehensive evaluation of normalization methods for Illumina high-throughput RNA sequencing data analysis. Brief Bioinform. 2013;14:671–83.
69. Hack CJ. Integrated transcriptome and proteome data: the challenges ahead. Brief Funct Genomics Proteomic. 2004;3:212–9.
70. Grossman AD, Zhou YN, Gross C, Heilig J, Christie GE, Calendar R. Mutations in the *rpoH* (*htpR*) gene of *Escherichia coli* K-12 phenotypically suppress a temperature-sensitive mutant defective in the sigma 70 subunit of RNA polymerase. J Bacteriol. 1985;161:939–43.
71. Jinks-Robertson S, Nomura M. Ribosomal protein S4 acts in trans as a translational repressor to regulate expression of the alpha operon in *Escherichia coli*. J Bacteriol. 1982;151(1):193–202.
72. Gottesman S. The small RNA regulators of Escherichia Coli: roles and mechanisms. Annu Rev Microbiol. 2004;58:303–28.
73. Maier T, Güell M, Serrano L. Correlation of mRNA and protein in complex biological samples. FEBS Lett. 2009;583(24):3966–73.
74. Kałużna M, Kuras A, Mikicinski A, Pulawska J. Evaluation of different RNA extraction methods for high-quality total RNA and mRNA from *Erwinia amylovora* in planta. Eur J Plant Pathol. 2016;146:893–9.
75. Llop P, Bonaterra A, Penalver J, Lopez MM. Development of a highly sensitive nested-PCR procedure using a single closed tube for detection of *Erwinia amylovora* in asymptomatic plant material. Appl Environ Microbiol. 2000;66:2071–8.
76. McManus PS, Jones AL. Detection of *Erwinia amylovora* by nested PCR and PCR-dot-blot and reverse blot hybridizations. Phytopathology. 1995;85:618–23.
77. Martin M. Cutadapt removes adapter sequences from high-throughput sequencing reads. EMBnetjournal, [S.l.], v. 17, n. 1, p. pp. 10-12, May. 2011. ISSN 2226-6089. Available at: <http://journal.embnet.org/index.php/embnetjournal/article/view/200/479>. Date accessed: 07 Mar. 2017. doi: http://dx.doi.org/10.14806/ej.17.1.200.
78. Robinson MD, Oshlack A. A scaling normalization method for differential expression analysis of RNA-seq data. Genome Biol. 2010;11:R25.
79. Moriya Y, Itoh M, Okuda S, Yoshizawa A, Kanehisa M. KAAS: an automatic genome annotation and pathway reconstruction server. Nucleic Acids Res. 2007;35:W182–5.
80. Huerta-Cepas J, Szklarczyk D, Forslund K, Cook H, Heller D, Walter MC, Rattei T, Mende DR, Sunagawa S, Kuhn M, Jensen LJ, von Mering C, Bork P. eggNOG 4.5: a hierarchical orthology framework with improved functional annotations for eukaryotic, prokaryotic and viral sequences. Nucleic Acids Res. 2016;44(D1):D286–93.
81. Benjamini Y, Hochberg Y. Controlling the false discovery rate: a practical and powerful approach to multiple testing. J R Stat Soc Ser B-Stat Methodol. 1995;57:289–300.
82. Kałużna M, Kuras A, Puławska J. Validation of reference genes for the normalization of the RT-qPCR gene expression of virulence genes of *Erwinia amylovora* in apple shoots. Sci Rep. 2017;7(1):2034.

10

Genome-wide analysis and transcriptomic profiling of the auxin biosynthesis, transport and signaling family genes in moso bamboo (*Phyllostachys heterocycla*)

Wenjia Wang[1†], Lianfeng Gu[1†], Shanwen Ye[1], Hangxiao Zhang[1], Changyang Cai[1], Mengqi Xiang[1], Yubang Gao[1], Qin Wang[1], Chentao Lin[1,2] and Qiang Zhu[1*]

Abstract

Background: Auxin is essential for plant growth and development. Although substantial progress has been made in understanding auxin pathways in model plants such as Arabidopsis and rice, little is known in moso bamboo which is famous for its fast growth resulting from the rapid cell elongation and division.

Results: Here we showed that exogenous auxin has strong effects on crown and primary roots. Genes involved in auxin action, including 13 *YUCCA (YUC)* genes involved in auxin synthesis, 14 *PIN-FORMED/PIN-like* (*PIN/PILS*) and 7 *AUXIN1/LIKE-AUX1* (*AUX1/LAX*) members involved in auxin transport, 10 auxin receptors (*AFB*) involved in auxin perception, 43 *auxin/indole-3-aceticacid* (*AUX/IAA*) genes, and 41 auxin response factors (*ARF*) involved in auxin signaling were identified through genome-wide analysis. Phylogenetic analysis of these genes from Arabidopsis, *Oryza sativa* and bamboo revealed that auxin biosynthesis, transport, and signaling pathways are conserved in these species. A comprehensive study of auxin-responsive genes using RNA sequencing technology was performed, and the results also supported that moso bamboo shared a conserved regulatory mechanism for the expression of auxin pathway genes; meanwhile it harbors its own specific properties.

Conclusions: In summary, we generated an overview of the auxin pathway in bamboo, which provides information for uncovering the precise roles of auxin pathway in this important species in the future.

Keywords: Moso bamboo, Auxin biosynthesis, Auxin transport, Auxin signaling, Gene expression

Background

Auxin acts as a central organization hub in controlling plant growth and development [1, 2]. Auxin action can be achieved through different levels, mainly auxin concentration pathways including auxin biosynthesis and directional auxin transport, and auxin signaling pathways including auxin perception and signal transduction [3–5].

Although auxin biosynthesis is not fully understood in plants, genetic and biochemical studies have demonstrated that the endogenous plant auxin indole-3-acetic acid (IAA) is mainly synthesized by a two-step reaction: Trp is first converted to indole-3-pyruvate (IPA) by TRYPTOPHAN AMINOTRANSFERASE OF ARABIDOPSIS (TAA) and then IAA is produced by YUC flavin-containing monooxygenase family proteins [6]. YUC proteins that encode flavin monooxygenases catalyze a rate-limiting step of the IPA pathway [7]. In Arabidopsis, 11 YUC family proteins act redundantly and cooperatively at various growth and developmental stages [4, 6].

IAA mainly exists in the protonated form (IAA-H) in the apoplast, whereas the deprotonated form (IAA^-) becomes dominant inside the plant cell due to the pH changes [8]. The IAA-H form of auxin freely enters the cell via diffusion, and the transport of its anionic form

* Correspondence: zhuqiang@fafu.edu.cn
†Equal contributors
[1]Basic Forestry and Proteomics Center (BFPC), Fujian Provincial Key Laboratory of Haixia Applied Plant Systems Biology, Haixia Institute of Science and Technology, Fujian Agriculture and Forestry University, Fujian 350002, China
Full list of author information is available at the end of the article

(IAA⁻) is mediated mainly by auxin influx transporters (AUX1/LAX proteins) and efflux transporters (ATP binding cassette B and PIN/PILS proteins) [5, 9]. In Arabidopsis, 4 amino acid permease-like family members (AtAUX1/LAX1–3) regulate auxin uptake from the apoplast [10], whereas 8 PIN proteins (AtPIN1-PIN8) and 7 PILS members (AtPILS1–7) are responsible for the polar pump-off of auxin and determine the direction of auxin flow through tissues [5]. The AtPIN-like family of proteins (AtPILS) was identified based on predicted topological similarities with AtPIN proteins [11]. As AtPIN proteins, AtPILS proteins contain the so-called InterPro auxin carrier domain, which is predicted to have auxin transport function in silico. In Arabidopsis, AtPILS proteins mainly localize in the endoplasmic reticulum (ER), and regulate intracellular auxin accumulation and the rate of auxin conjugation [11, 12]. The proper concentration of auxin throughout the plant body is achieved by auxin biosynthesis and polar auxin transport.

Synthesized auxin is distributed to the site of its action in a directional manner [5, 9]. The auxin regulatory module TIR1/AFB receptors-AUX/IAAs-ARFs, which is considered as the key components of the auxin signaling pathway in plant cells stimulates diverse auxin responses by coordinately controlling the expression of downstream genes [13]. In Arabidopsis, auxin binds to its receptor Transport Inhibitor Response 1 (TIR1), which belongs to a small gene family containing 5 additional members (AFB1–5). All of these 6 proteins redundantly function as nuclear auxin receptors [14–16]. TIR1/AFBs act as the specificity determiners for the SCF class of E3 ubiquitin ligases, which target substrate proteins for polyubiquitylation and subsequent degradation [17]. AUX/IAAs are negative regulators of the auxin signaling pathway [18]. Typically, AUX/IAA proteins have 4 conserved domains: domain I contains an ethylene responsive factor (ERF) associated amphiphilic repression motif and is required to recruit the transcriptional corepressor TOPLESS [19]; domain II is essential for auxin-induced AUX/IAA proteolysis [18]; and domains III and IV are involved in the homo- and hetero dimerization and interaction with the downstream auxin response factors (ARF) [20]. ARF proteins are a class of plant-specific B3-type transcription factors which mediate auxin-dependent transcriptional regulation [21]. In Arabidopsis, 23 ARF proteins act as either activators or repressors of the downstream auxin-responsive genes by binding to the auxin-responsive *cis*-element (AuxRE: ['TGTCTC']) [21]. The auxin receptor-AUX/IAA-ARF module precisely and sensitively controls the response of plant cells to auxin: without auxin, AUX/IAA proteins negatively regulate the abundance of ARFs, and subsequently the expression of auxin responsive genes; whereas upon elevated auxin level, the auxin-dependent ubiquitination and degradation of Aux/IAA proteins mediated by SCF^{TIR1} (TIR1 perceives auxin) releases ARF proteins to activate the transcription of auxin-responsive genes [13, 22].

Moso bamboo is one of the most important non-timber forest products in the world, due to its great economic, cultural, and environmental value [23]. Moso bamboo is famous for its fast-growing culms which were controlled by cell division and cell elongation [24]. Although the mechanisms that control plant cell number and size are not fully understood, phytohormones, especially auxin have important roles in the regulation and coordination of plant cell proliferation and elongation [25]. However, to our knowledge, no systematic study of the auxin pathway has been reported in moso bamboo. The genomic sequence of moso bamboo was recently released [26], providing an excellent opportunity to perform a comprehensive genome-wide analysis of the gene families related to auxin action. Here, a genome-wide search was carried out to identify auxin action-related genes in moso bamboo. A total of 13 *YUC* genes involved in auxin biosynthesis, 13 *PIN/PILS* and 7 *AUX1/LAX* family members involved in auxin transport, 10 putative auxin receptors involved in auxin perception, and 43 *AUX/IAA*s and 41 *ARF*s involved in auxin signaling were identified from the moso bamboo genome. Next we analyzed the phylogenetic relationships of auxin pathway orthologs from moso bamboo, Arabidopsis, and rice. Moreover, to generate a general overview of the auxin-response transcriptome in moso bamboo, we treated the plants with exogenous auxin and performed RNA-Sequencing (RNA-Seq) analyses. Our data provide a foundation for further investigations of the role of auxin in regulating moso bamboo development at the genetic and biotechnological levels.

Methods

Isolation of gene families related to auxin action in moso bamboo

The gene annotations and genomic sequences of moso bamboo were downloaded from the bamboo genome database (BambooGDB, http://www.bamboogdb.org/). To identify the members involved in auxin pathway in moso bamboo, we retrieved proteins of 11 AtYUCs, 8 AtPINs, 6 AtPILSs, 4 AtAUX1/LAXs, 6 AtTIR1/AFBs, 29 AtAUX/IAAs and 23 AtARFs from the Arabidopsis genome database, and used as query sequences against the moso bamboo database using basic local alignment search tool (BLAST) search. All sequences with an e-value $\leq 10^{-11}$ and a score value $\geq$100 were used as new queries for the 2nd cycle's search to avoid missing additional orthologs. In addition, the derived sequences were used for further protein domain analysis using the Pfam program (http://

pfam.xfam.org) and SMART (http://smart.embl-heidelberg.de/). To exclude the duplicated genes, all the candidates were aligned using the ClustalW program [27]. Information about the coding sequences, full-length sequences, and amino sequences were obtained from the bamboo genome database using the BLAST program.

Gene structure, conserved motif, and protein information analyses

The genomic and cDNA sequences of each predicted gene were downloaded from the moso bamboo genome database, and their intron distribution patterns and intro-exon boundaries were analyzed using JBrowse software as previously described [28]. The conserved motif was derived using the NCBI conserved domain search (https://www.ncbi.nlm.nih.gov/Structure/cdd/wrpsb.cgi) or the online tool Pfam. The protein sequence was analyzed using DNAman software.

Phylogenetic tree building and prediction of amino acid composition

Multiple sequence alignments of the auxin action-related proteins (YUC family, PIN/PILS family, AUX1/LAX family, AFB family, AUX/IAA family and ARF family) were performed using the ClustalW program with the default parameters [27]. The results were visualized using Editplus (https://www.editplus.com/). The phylogenetic tree was constructed from the protein sequences given above using MEGA6 program (http://www.megasoftware.net/mega.php) with the neighbor-joining (NJ) method, and the boot strap analysis was performed using 1000 replicates as described previously [29].

Plant materials and auxin treatment

To determine the effects of auxin on the growth of moso bamboo, we germinated and grew bamboo seeds in the soil in a greenhouse at 26 °C with the photoperiod of 16 h light/8 h dark for 1 month. Various concentrations of naphthaleneacetic acid (NAA, 100 nM, 500 nM, and 5 μM) were sprayed the above-ground parts and watered the roots of the seedlings for 2 weeks at 2-day intervals, and pictures were taken for further statistical analysis using Image J software. At least 3 independent biological repeats were performed.

For materials used in RNA-Seq analysis, seeds were sterilized with chlorine gas for 3–5 h and were then put on Murashige and skoog (MS) agar to germinate and grow vertically. 1-month-old seedlings were sprayed with 5 μM NAA, 5 μM IAA or 5 μM Indole-3-butytric acid (IBA) for 4 h at 1-h intervals, and the root parts were dissected for RNA extraction. For the time course experiments, 1-month-old seedlings were sprayed with 5 μM NAA, and the roots were harvested at different time points (0, 2, 4, 6, 8 and 12 h) for further qPCR analysis.

RNA extraction and RNA-Seq analysis

Total RNA from moso bamboo roots were extracted using the RNeasy Mini Kit (QIAGEN, China) according to the manufacturer's instructions. 10 μg total RNA were used for RNA-Seq analysis using the Illumina Hiseq2500 Sequencer platform (BerryGenomics, China) with paired-end sequencing, and 3 independent biological repeats were performed. In total, 6 strand-specific RNA-Seq libraries were sequenced in this study using the deoxyuridine triphosphate (dUTP) method [30]. The paired-end reads were aligned to the moso bamboo genome using tophat-2.0.11 with anchor length more than 8 nt for spliced alignments [31]. Only reads that could be uniquely aligned were retained for subsequent analysis. The expression levels of each gene were normalized as fragments per kilobase of transcript per million mapped reads (FPKM) [32]. The p-value and false discovery rate (FDR) were calculated using the edgeR package developed in Bioconductor [33]. A fold change in the expression >1.5 and an FDR <0.01 were considered to be the threshold for differentially expressed genes (DEGs). To verify the RNA-Seq results, qPCR was used as previously described [34]. The primers used for qPCR are listed in the Additional file 1: Table S7. We used DAVID [35] to test the statistical enrichment of the differentially expressed genes in the KEGG pathways.

Results

Effects of exogenous auxin on moso bamboo

Auxin controls nearly every aspect of plant growth and development, but until now no reports showed its role in bamboo growth. To test its effects on the growth of bamboo seedlings, we treated the 1-month-old seedlings with various concentration of synthetic auxin NAA for another 2 weeks. Our results showed that bamboo root is sensitive to the exogenous NAA treatment, and NAA affects the root architecture in a dose-dependent manner, low concentration (100 nM and 500 nM) of NAA promoted the formation and growth of crown roots, while an inhibition of primary root and lateral root growth was observed at a higher concentration of auxin (5 μM) (Fig. 1a and b). Under our experimental conditions, we did not find significant differences in the above-ground plant architecture under our experimental conditions (Fig. 1a and b).

Identification of gene families related to auxin action in moso bamboo

Extensive evidences showed that the auxin plays its function through controlling its homeostasis, transport and signaling (Additional file 2: Figure S1) [13, 36, 37]. As a first step to reveal the role of auxin in affecting bamboo root growth, we performed a genome-wide analysis to identify the auxin related genes in moso bamboo,

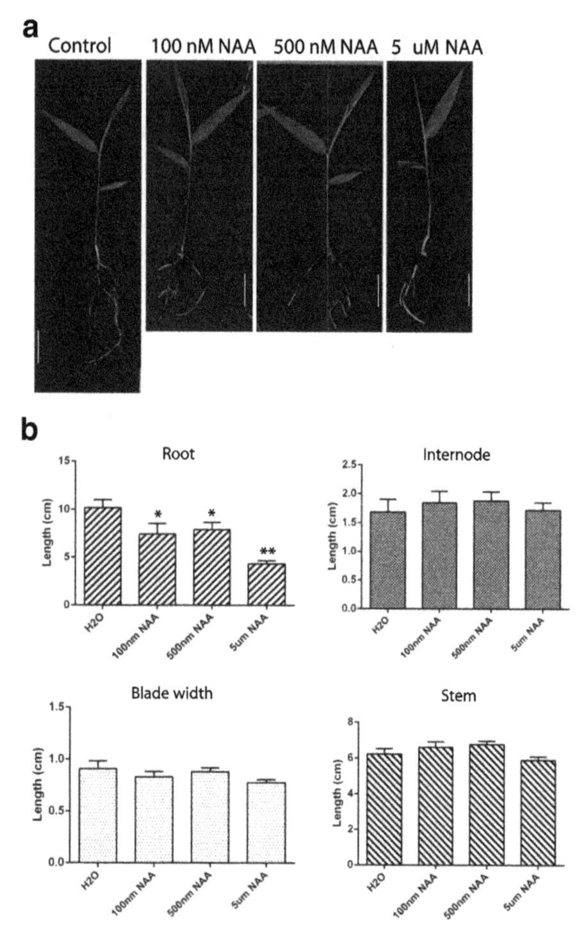

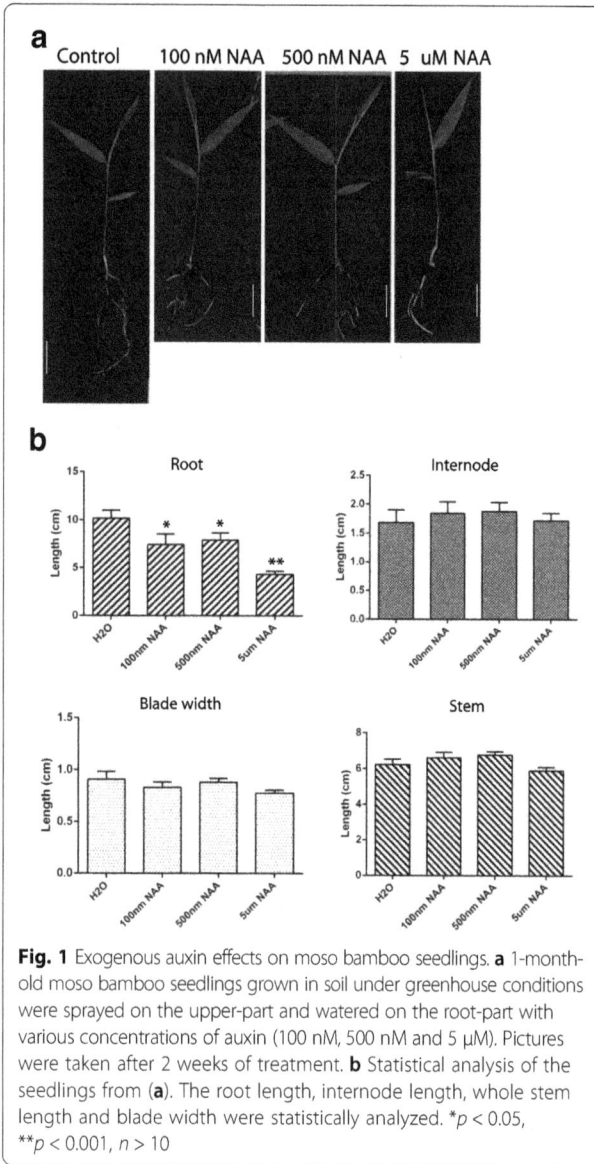

Fig. 1 Exogenous auxin effects on moso bamboo seedlings. **a** 1-month-old moso bamboo seedlings grown in soil under greenhouse conditions were sprayed on the upper-part and watered on the root-part with various concentrations of auxin (100 nM, 500 nM and 5 μM). Pictures were taken after 2 weeks of treatment. **b** Statistical analysis of the seedlings from (**a**). The root length, internode length, whole stem length and blade width were statistically analyzed. $^{*}p < 0.05$, $^{**}p < 0.001$, $n > 10$

including *YUC* family genes for auxin biosynthesis, *PIN/PILS* and *AUX1/LAX* families for polar auxin transport, *AFB* family for auxin perception, and *AUX/IAA* and *ARF* families for auxin signaling.

Auxin biosynthesis

Plant cells exhibit concentration-dependent auxin responses by tightly controlling cellular auxin levels [38]. Local auxin biosynthesis regulated by *YUC* family genes plays key roles in auxin accumulation [5, 9].

To identify *YUC* genes in moso bamboo, we searched the moso bamboo genome database using the Arabidopsis YUC proteins sequences as the queries. Combined with conserved domain analysis using Pfam or SMART, 13 *YUC* genes were identified in moso bamboo genome, and the phylogenetic analysis of predicted full-length YUC protein sequences were performed using neighbor-joining method (Fig. 2a). Our further search in the moso bambooGDB with these 13 *YUCs* did not identify additional *YUC* genes. We named these genes *PhYUC1-PhYUC13* based on the scaffold number (Fig. 2a, Additional file 1: Table S1). Analyses of the protein domains using the Pfam program showed that all of these proteins contained the conserved flavin-binding monooxygenase-like domain as reported in other species [39], whereas some PhYUC proteins have their own specific structures. For example, all PhYUC proteins except PhYUC3 contain FAD-binding motif ['GxGxxG']. PhYUC1, 4, 9 and 11 have the classical ATG-containing motif 1 ['Y(x)7ATGEN(x)5P'], while PhYUC5, 6, 7, 8, 10, 12, and 13 share a conserved motif ['D(x)4CI/NG(x)5P] in this region. The FMO identifying motif ['FxGxxxHxxxY/F'], NADPH-binding motif ['GxGxxG'], and ATG-containing motif 2 ['(F/L) ATGY'] are conserved in all PhYUC proteins except PhYUC1 and PhYUC2 (Additional file 2: Figure S2A). These structural differences suggest the possible functional diversity among the PhYUC proteins.

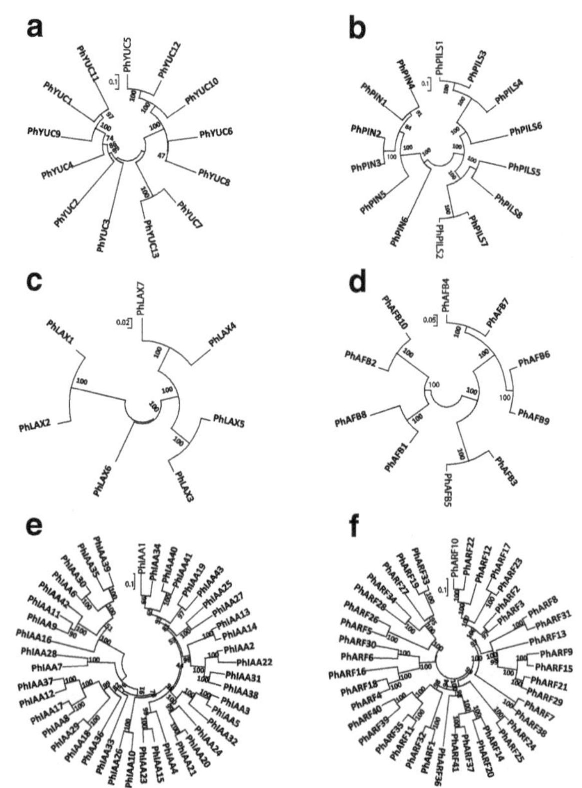

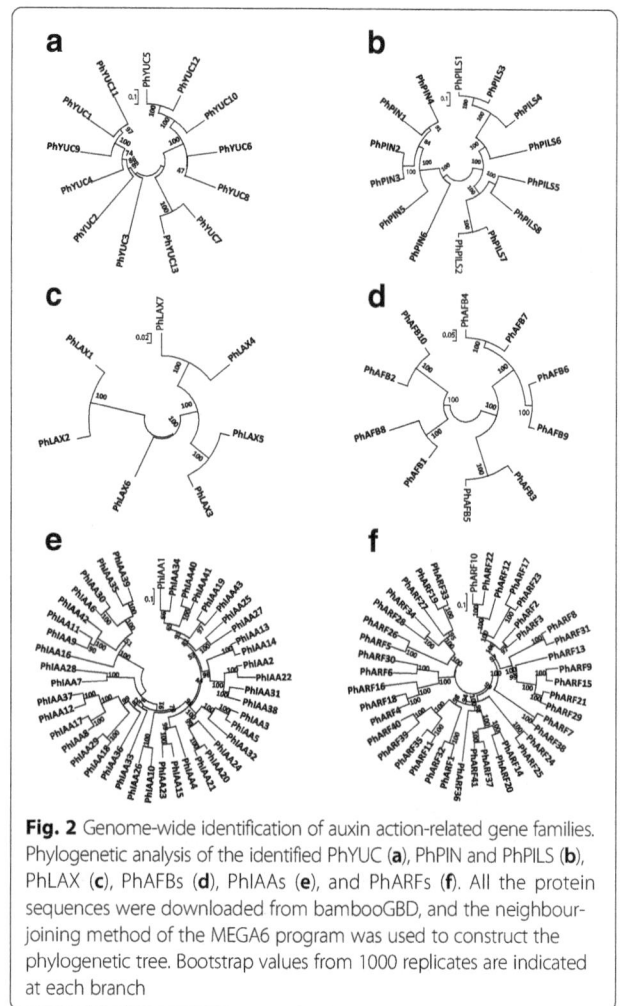

Fig. 2 Genome-wide identification of auxin action-related gene families. Phylogenetic analysis of the identified PhYUC (**a**), PhPIN and PhPILS (**b**), PhLAX (**c**), PhAFBs (**d**), PhIAAs (**e**), and PhARFs (**f**). All the protein sequences were downloaded from bambooGBD, and the neighbour-joining method of the MEGA6 program was used to construct the phylogenetic tree. Bootstrap values from 1000 replicates are indicated at each branch

The sizes of the deduced PhYUC proteins varied markedly. The open reading frame (ORF) lengths of the *PhYUC* genes range from 258 bp (*PhYUC11*) to 1656 bp (*PhYUC12*) with the predicted molecular masses varied from 9.4 to 170.6 kDa (Additional file 1: Table S1). To understand the gene structures of the *PhYUC* genes, the organization of exons and introns for each gene was obtained by comparing the cDNA sequences with the corresponding genomic sequences. The results showed that all members of the *PhYUC* family contained introns, ranging from 2 to 6 in numbers (Additional file 2: Figure S2B). The detailed information of the *PhYUCs* including the number of exons and introns, open reading frame lengths, translation lengths, molecular weights, conserved domains, and their putative subcellular localizations were listed in the Additional file 1: Table S1.

Auxin transport

After synthesis, auxin establishes the auxin gradient at the site for its function through auxin transport, a process which is controlled mainly by auxin efflux transporter PIN/PILS and influx transporter AUX1/LAX [5, 9]. To identify the *PIN* and *AUX1/LAX* members in moso bamboo, we used AtPIN/PILS and AtAUX1/LAX proteins from Arabidopsis as BLASTP queries to search for the moso bamboo orthologs in the bambooGDB. The Hidden Markov model (HMM) profiles (Pfam 01490: transmembrane amino acid transporter protein; Pfam 03547: membrane transport protein) were also employed to identify the PhLAX, PhPIN, and PhPILS protein families. A total of 6 *PhPIN* genes, 8 *PhPILS* and 7 *PhLAX* genes were identified and we further performed the phylogenetic analysis of their predicted full-length protein sequences using neighbor-joining method (Fig. 2b and c). All the *PIN* genes have similar intron/exon distribution patterns, and most of them have a large first exon that encodes the N-terminal transmembrane segments [5]. *PhPIN* genes contain 5–7 exons (Additional file 2: Figure S2C), and similar conserved exon/intron organization patterns have also been found in other plant species [40, 41]. As the case in Arabidopsis, PIN transporters in moso bamboo were classified into long PINs and short PINs based on the length of predicted proteins. The typical long PINs contained 547–611 aa (PhPIN1-PhPIN4), while the short PINs contain only 332 aa or 470 aa (PhPIN6 and PhPIN5, respectively) (Additional file 1: Table S2). Multiple sequence alignment results showed that the sequences of the N- and C-terminal transmembrane regions were highly conserved (Additional file 2: Figure S2D).

The ORFs of *PhPILS* genes range from 1023 bp (*PhPILS7*) to 1296 bp (*PhPILS1*) (Additional file 1: Table S2). The gene structures of the *PhPILS* genes showed that their exon/intron numbers were different: *PhPILS1*, *PhPILS3*, and *PhPILS4* contained 10 exons, *PhPILS2* and *PhPILS7* contained 9 exons, and *PhPILS5* and *PhPILS8* contained 8 exons (Additional file 2: Figure S2E). Interestingly, *PhPILS6* contained only 1 exon (Additional file 2: Figure S2E), similarly to the *ZmPILS5* gene in maize [40]. As PhPIN proteins, multiple sequence alignment confirmed the high similarity among the PhPILS proteins. The results also revealed that the sequences of the N- and C- terminal transmembrane regions were highly conserved while the central hydrophilic region was variable (Additional file 2: Figure S2F), indicating the functional similarity among the members of the *PhPILS* family.

AUX/LAX proteins are required to establish the auxin gradient by mediating auxin influx transport [42]. In total, 7 auxin influx transporters (*PhLAX1*-*PhLAX7*) were identified in moso bamboo genome (Fig. 2c). The exon numbers of *PhLAX* ranged from 6 (*PhLAX6*) to 8 (*PhLAX4* and *PhLAX7*) (Additional file 2: Figure S2G), and the ORF lengths varied from 1431 bp (*PhLAX6*) to 1593 bp (*PhLAX3*) (Additional file 1: Table S3). Based on the multiple sequence alignment results, all members of this family shared conserved sequences (Additional file 2: Figure S2H). Moreover, all of these proteins were predicted to localize to the cell membrane (Additional file 1: Table S3), indicating that they have similar functions. The detailed information of these auxin transport-related genes is listed in Additional file 1: Tables S2 and S3.

Auxin signaling

In plants, the auxin receptor-AUX/IAA-ARF module consists of the key components involved in auxin signaling [3]. In moso bamboo, 10 putative auxin receptors that showed highly similar protein sequences were identified, and the phylogenetic tree of their predicted full-length protein sequences was generated using neighbor-joining method (Fig. 2d). The exon number of these genes ranged from 3 to 4 (Additional file 2: Figure S3A), the ORFs varied from 1632 bp (*PhAFB5*) to 1881 bp (*PhAFB6*) (Additional file 1: Table S4).

AUX/IAA proteins are the direct targets of auxin receptors [14, 43]. 43 AUX/IAA proteins were identified in moso bamboo genome, and the corresponding phylogenetic tree was generated using neighbor-joining method (Fig. 2e). The ORF lengths ranged from 252 bp (*PhIAA19*) to 2523 bp (*PhIAA30*) (Additional file 1: Table S5). The number of exons ranged from 2 to 18 (Additional file 2: Figure S3C). Sequence alignment analysis showed that most of the PhIAA proteins had typical domains I-IV. However, the diversity in the conserved domains was also observed (Additional file 2: Figure S3D). For example, the classical domain I ['TELRLGLPG'] was found in PhIAA1, 34, 43, 40, 41, 25, and 27, while a similar domain ['LR/K/T/ELXLXXPG'] was found in PhIAA2, 4, 8, 12–15, 17–18, 20–24, 29, 31, and 36–38. Moreover, in this region, a conserved domain ['MRFK/RMR/KFRFEG'] was found in PhIAA6, 9, 11, 30, 35, 39, and 42. Whether these three domains have similar functions to that of the typical domain I needs to be further investigated (Additional file 2: Figure S3D). The other PhIAAs either lacked the domain I (PhIAA19 and PhIAA3) or had a poorly conserved domain I (PhIAA3, 5, 7, 10, 26, 28, and 32). Domain II ['VGWPP'] was found in most of the PhIAAs, however, some members did not contain domain II (PhIAA19 and PhIAA34) or had poorly conserved domain II sequences. Moreover, we found a novel conserved domain in this region ['LFGIXL'] with unknown function (Additional file 2: Figure S3D). Some members lacked either domain III (PhIAA19 and PhIAA22) or domain IV (PhIAA1, 34, and 28) (Additional file 2: Figure S3D).

Once the AUX/IAA protein was degraded, the ARF activities were released to activate downstream genes by binding to the AuxRE *cis*-elements ['TGTCTG'] in their promoters. A total of 41 *ARF* members were identified in the bamboo genome, and the phylogenetic tree was built with the sequences of their proteins using neighbor-joining method (Fig. 2f). The ORF lengths of the *PhARFs* ranged from 1284 bp (PhARF3) to 3774 bp (PhARF10) (Additional file 1: Table S6). To understand the structural components of the *PhARF* genes, the exon and intron organizations of the genes were determined by comparing the cDNA sequences with the corresponding genomic DNA sequences. The coding sequences of all family members contain introns, and the number of exons ranged from 2 (*PhARF33*) to 17 (*PhARF1*) (Additional file 2: Figure S3E). To explore the structural diversity and predict the functions of ARFs in moso bamboo, a motif analysis was performed using the NCBI conserved domain search engine (https://www.ncbi.nlm.nih.gov/Structure/cdd/wrpsb.cgi). The results showed that all 41 putative PhARFs contained the highly conserved B3 DNA binding domain (DBD) (Additional file 2: Figure S3F and G). In Arabidopsis, the AtARF3, 13 and 17 proteins do not contain the carboxy-terminal dimerization domain (CTD domain) that is involved in the protein-protein interaction by dimerizing with AUX/IAA or with ARFs [21, 22]. Similarly to those proteins from Arabidopsis, the 12 ARF proteins in moso bamboo (PhARF2, 4, 6, 16, 18, 19, 25, 26, 28, 30, 33, 34, and 36) also lacked this conserved domain (Additional file 2: Figure S3F and G). In summary, the moso bamboo ARF proteins have the typical B3 DBD domains that were required for binding to the AuxRE *cis*- elements, while their structural variations implied that the moso bamboo genome changed significantly during its evolutionary history, indicating the functional diversities of these ARF proteins.

Phylogenetic analysis of the gene families related to auxin action

To explore the possible roles of auxin action-related genes in moso bamboo and to understand their phylogenetic relationships, we constructed a phylogenetic tree using the proteins from rice, Arabidopsis, and moso bamboo. In general, the auxin action-related genes from bamboo showed high similarity to their orthologs from Arabidopsis and rice (Additional file 2: Figure S4).

The 38 members of YUC families in Arabidopsis, rice, and bamboo could be grouped into 3 classes. Of these proteins, PhYUC1, PhYUC9 and PhYUC11 belonged to class I (Additional file 2: Figure S4A) and were homologous to the genes *OsYUC5*, *OsYUC7*, and *OsYUC1* respectively. PhYUC2 and PhYUC4 belonged to class II, and the *PhYUC4* gene has two orthologs in rice *OsYUC9* and *OsYUC10*. In rice, *OsYUC9*, *OsYUC10*, and *OsYUC11* are highly expressed in the developing grains, and these genes are important for increases in the levels of IAA during grain development [44]. The highly close phylogenetic relationships of these genes indicated their potential function in bamboo grain development. Notably, the PhYUCs belonging to class III were not closely related to any rice or Arabidopsis *YUC* genes (Additional file 2: Figure S4A). This result probably reflects a diverging trend during the evolution of the YUC family members across different plant species.

Of the auxin receptor-like genes, the numbers of auxin receptor family in moso bamboo (10 members) was slightly expanded compared with that in Arabidopsis (6 members) or rice (8 members), and most of these genes are high closely related to the AFBs from these other two species (Additional file 2: Figure S4B). All the AFBs members can be grouped into 3 classes, and all AFBs from Arabidopsis and bamboo are grouped into class I and class II (Additional file 2: Figure S4B). Based on the phylogenetic tree analysis, *PhAFB10* and *PhAFB2* are closed to the Arabidopsis *TIR1*, which is well-characterized as the first-identified auxin receptor. It will be very interesting to determine the functions of these two genes in moso bamboo.

Although the PIN and PILS proteins are closely related, they belong to two separate groups in moso bamboo as in Arabidopsis and other species [11]. All *PIN* and *PILS* genes could be further divided into 3 subclasses (I-III) (Additional file 2: Figure S4C). In the PILS family, three pairs of PILS orthologs were identified between bamboo and rice: PhPILS8/OsPILS7a, PhPILS6/OsPILS2, and PhPILS4/OsPILS6

(Additional file 2: Figure S4C). In the PIN family, PhPIN1 belonged to the second group and was phylogenetically close to OsPIN10a and OsPIN10b, which are monocot-specific and have a long central hydrophilic domain. Based on the expression patterns of these genes in rice, they are involved in tillering [41, 45]. PhPIN2 and PhPIN3 are closely related to OsPIN1a, which is crucial for the negative phototropic curvature of the rice root [46]. PhPIN4 is the orthologs of the OsPIN1c and OsPIN1d from rice and *AtPIN1* from Arabidopsis, which have broad effects on plant development [45], indicating the important role of PhPIN4 in moso bamboo development. Interestingly, compared with the PIN subfamily, PhPIN6 is more closely related to the PILS subfamily (Additional file 2: Figure S4C).

The AUX1/LAX proteins could be divided into 2 major classes: PhLAX3, 4, 5, 6, and 7 belonged to class I, while PhLAX1 and 2 belonged to class II (Additional file 2: Figure S4D). *PhLAX3* and *PhLAX5* are closely related to *OsLAX2*, whereas *PhLAX4* and *PhLAX7* are closely related to *OsLAX4*; the *PhLAX1* and *PhLAX2* genes were closely related to *OsLAX1* (Additional file 2: Figure S4D). These results may reflect the occurrence of a whole-genome duplication event from moso bamboo and rice [26].

The AUX/IAA family is significantly expanded in moso bamboo (43 members) compared with those of Arabidopsis (29 members) and rice (24 members). As reported in other species [47, 48], AUX/IAA proteins in moso bamboo can be divided into 6 major classes (I-VI). Group I consisted of 12 PhIAA proteins, 6 OsIAA proteins, and 10 AtIAA proteins that form 12 sister pairs (4 PhIAA-OSIAA pairs, 4 PhIAA-PhIAA pairs, 4 AtIAA-AtIAA pairs). Group II contained 9 PhIAA proteins, 7 OsIAA proteins and 5 AtIAA proteins, which structure 7 sister pairs. Group III-VI contained 12 combined sister pairs. In total, 31 sister pairs were identified, and interestingly, no PhIAA-AtIAA pair was found (Additional file 2: Figure S4E). PhIAA15 is closely related to OsIAA1 in rice, and the overexpression of *OsIAA1* effectively inhibits root elongation and shoot growth [49]. PhIAA13 and PhIAA14 are phylogenetically close to OsIAA11, the overexpression of which leads to the loss of lateral roots in rice [50].

In regards to AUX/IAAs, the moso bamboo genome contains much more *PhARF* members (41 members) compared to the genomes of Arabidopsis (23 members) and rice (25 members) (Additional file 2: Figure S4F). The phylogenetic distribution results showed that the ARF genes from these three species could be divided into 5 major classes (I-V). Thirty-one members were clustered into class I (12 members from moso bamboo), 21 members were clustered into class II (9 members from moso bamboo), and 11 members (4 PhARF) and 15 members (7 PhARF) were clustered into classes III and IV respectively. Only two members from Arabidopsis (AtARF2) and rice (OsARF20) were classified into class V (Additional file 2: Figure S4F).

Auxin treatment induces broad changes in transcriptional activity

Under our experimental conditions, the roots of moso bamboo were sensitive to exogenous NAA treatment (Fig. 1). Our next step was to check the global gene expression changes in response to exogenous auxin by RNA-Seq. To optimize the conditions for this experiment, first we treated the bamboo seedlings with 5 μM NAA and the roots were harvested at different time points (0, 2, 4, 6, 8, and 12 h) for testing the expression patterns of selected auxin responsive genes. The genes we checked were:*Ph01003158G0110* and *Ph01000099G0730* for *Gretchen Hagen3* (*GH3* family); *Ph01000788G0760* for *the lateral organ boundaries domain* (*LBD family*); *Ph01004534G0130* for *small auxin-up RNA* (*SAUR* family); *Ph01000025G1600, Ph01000554G0550, Ph01000592G0620* and *Ph01000075G0200* for *AUX/IAA* family, whose orthologs in model plants such as Arabidopsis and rice were used as the markers for plant response to exogenous auxin treatment [51]. Our results showed that 5 μM NAA effectively changed the expression of these marker genes, and in most cases 4 h' treatment had the strongest effects (Additional file 2: Figure S5). Therefore we treated bamboo seedlings with 5 μM NAA for 4 h and harvested the roots for RNA-Seq analysis, with the DMSO treated roots as the control. In general, with a cutoff of $P < 0.05$ and a fold change >1.5, we identified 991 down-regulated genes and 1288 up-regulated genes in the root 4 h after treatment with 5 μM NAA (Fig. 3, Additional file 3: Table S8). Our results from the Pearson correlation analysis showed that the correlations among all samples were quite high (>0.95), and the biological replicates could cluster well together (Additional file 2: Figure S6). To determine the reliability of the RNA-Seq results, we selected 28 differentially expressed genes related to auxin action and performed quantitative real time PCR (qRT-PCR). The results indicated a close correlation between RNA-Seq and qRT-PCR data (Additional file 2: Figure S7).

The upregulated- and downregulated-genes were classified into many functional categories, including stress response, developmental processes, cell organization and biogenesis, signal transduction, and other processes. Moreover, a significant number of the upregulated and downregulated genes were distributed across many cellular components, such as cell wall, membrane and chloroplast, which have a broad molecular function (Fig. 3). These results suggested that auxin stimulates broad transcriptional changes in the root of moso bamboo as in other species.

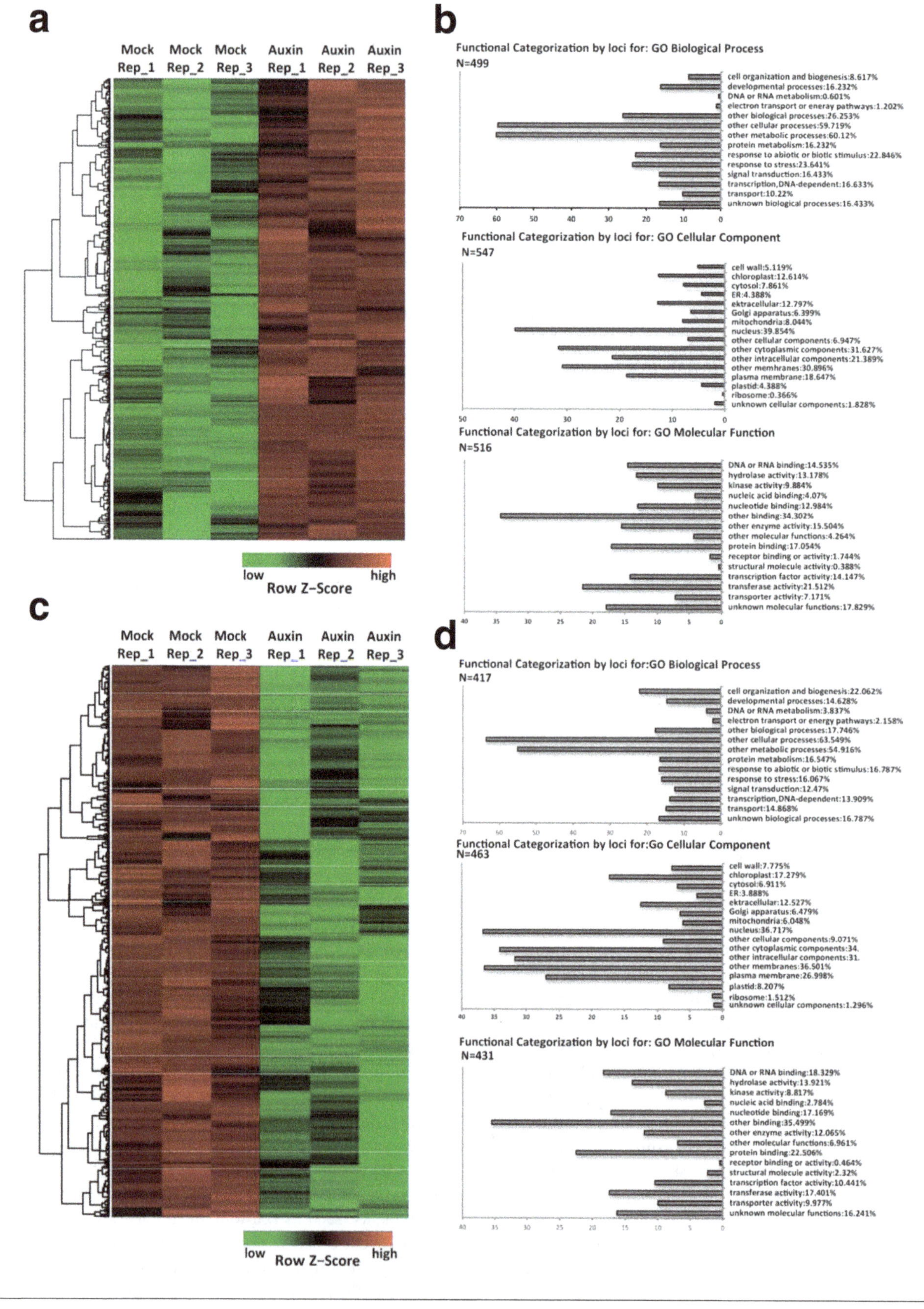

a
Mock Rep_1
Mock Rep_2
Mock Rep_3
Auxin Rep_1
Auxin Rep_2
Auxin Rep_3
low
high
Row Z-Score
b
Functional Categorization by loci for: GO Biological Process
N=499
cell organization and biogenesis:8.617%
developmental processes:16.232%
DNA or RNA metabolism:0.601%
electron transport or energy pathways:1.202%
other biological processes:26.253%
other cellular processes:59.719%
other metabolic processes:60.12%
protein metabolism:16.232%
response to abiotic or biotic stimulus:22.846%
response to stress:23.641%
signal transduction:16.433%
transcription,DNA-dependent:16.633%
transport:10.22%
unknown biological processes:16.433%
70 60 50 40 30 20 10 0
Functional Categorization by loci for: GO Cellular Component
N=547
cell wall:5.119%
chloroplast:12.614%
cytosol:7.861%
ER:4.388%
extracellular:12.797%
Golgi apparatus:6.399%
mitochondria:8.044%
nucleus:39.854%
other cellular components:6.947%
other cytoplasmic components:31.627%
other intracellular components:21.389%
other membranes:30.896%
plasma membrane:18.647%
plastid:4.388%
ribosome:0.366%
unknown cellular components:1.828%
50 40 30 20 10 0
Functional Categorization by loci for: GO Molecular Function
N=516
DNA or RNA binding:14.535%
hydrolase activity:13.178%
kinase activity:9.884%
nucleic acid binding:4.07%
nucleotide binding:12.984%
other binding:34.302%
other enzyme activity:15.504%
other molecular functions:4.264%
protein binding:17.054%
receptor binding or activity:1.744%
structural molecule activity:0.388%
transcription factor activity:14.147%
transferase activity:21.512%
transporter activity:7.171%
unknown molecular functions:17.829%
40 35 30 25 20 15 10 5 0
c
Mock Rep_1
Mock Rep_2
Mock Rep_3
Auxin Rep_1
Auxin Rep_2
Auxin Rep_3
low
Row Z-Score
high
d
Functional Categorization by loci for:GO Biological Process
N=417
cell organization and biogenesis:22.062%
developmental processes:14.628%
DNA or RNA metabolism:3.837%
electron transport or energy pathways:2.158%
other biological processes:17.746%
other cellular processes:63.549%
other metabolic processes:54.916%
protein metabolism:16.547%
response to abiotic or biotic stimulus:16.787%
response to stress:16.067%
signal transduction:12.47%
transcription,DNA-dependent:13.909%
transport:14.868%
unknown biological processes:16.787%
Functional Categorization by loci for:Go Cellular Component
N=463
cell wall:7.775%
chloroplast:17.279%
cytosol:6.911%
ER:3.888%
extracellular:12.527%
Golgi apparatus:6.479%
mitochondria:6.048%
nucleus:36.717%
other cellular components:9.071%
other cytoplasmic components:34.
other intracellular components:31.
other membranes:36.501%
plasma membrane:26.998%
plastid:8.207%
ribosome:1.512%
unknown cellular components:1.296%
40 35 30 25 20 15 10 5 0
Functional Categorization by loci for: GO Molecular Function
N=431
DNA or RNA binding:18.329%
hydrolase activity:13.921%
kinase activity:8.817%
nucleic acid binding:2.784%
nucleotide binding:17.169%
other binding:35.499%
other enzyme activity:12.065%
other molecular functions:6.961%
protein binding:22.506%
receptor binding or activity:0.464%
structural molecule activity:2.32%
transcription factor activity:10.441%
transferase activity:17.401%
transporter activity:9.977%
unknown molecular functions:16.241%
40 35 30 25 20 15 10 5 0

(See figure on previous page.)**Fig. 3** Overview of the auxin response genes in moso bamboo. Six samples from auxin treated and mock treated moso bamboo roots collected and were subjected to RNA-Seq analysis. Heatmap representation and hierarchical clustering of the differentially expressed transcript clusters for the up-regulated (**a**) or down-regulated (**c**) in response to auxin treatment are shown. Numbers on the nodes display major gene groups based on expression patterns. Color scale shows log2 signal intensity values. GO enrichment analyses for up-regulated genes (**b**) and down-regulated genes (**d**) in response to auxin. Best hits were aligned to the GO database, and most consensus sequences were grouped into three major functional categories: biological process, cellular component, and molecular function. Red indicates up-regulated genes; green indicates down-regulated genes

Expression changes of genes related to auxin action in response to treatment with exogenous auxin

We further examined the auxin response of auxin biosynthesis, transport, and signaling pathway genes from RNA-Seq datasets, and the expressions of some randomly selected genes from different families were further confirmed by qRT-PCR (Additional file 2: Figure S8). Analysis of gene expression level (FPKM) showed that all these family genes have expressions in the root (Additional file 2: Figure S9A). Furthermore, results from our qRT-PCR analysis of selected DEGs from various families of auxin pathways in different tissues showed that these auxin-related genes were expressed in root, shoot and leaf, with different extent of expression in these tissues (Additional file 2: Figure S9B). In Arabidopsis, the *YUC1, 2, 4* and *6* genes were downregulated in response to the exogenous NAA treatment through the feedback pathway [52]. In moso bamboo, 5 of the 13 *PhYUC* members responded to elevated auxin levels: *PhYUC4, 5, 9* and *13* were significantly upregulated, and *PhYUC3* was downregulated (Fig. 4, Additional file 4: Table S9).

In Arabidopsis, the expression of most auxin transporter genes are positively regulated by auxin, contributing to faster auxin transport when endogenous auxin is elevated [53]. In bamboo, we found that 3 *PIN* genes (*PhPIN2, 4, 5*) and 3 *PILS* genes (*PhPILS1, 5, 8*) changed their expression in response to exogenous auxin treatment. Auxin stimulated the expression of *PhPIN2, 4, 5* and *PhPILS1, 8,* while downregulated the expression of *PhPILS5* (Fig. 4, Additional file 5: Table S10).

In Arabidopsis roots, all *LAX* genes are upregulated by auxin treatment [51]. However, we found that in moso bamboo, the expression of 2 *PhLAX* genes (*PhLAX4* and *PhLAX6*) were downregulated by exogenous auxin treatment, whereas the transcriptional level of *PhLAX3* was upregulated (Fig. 4, Additional file 6: Table S11).

For auxin signaling pathway, auxin receptor genes did not changed at transcriptional level (Additional file 7: Table S12). The most pronounced effects of the plant response to auxin are the upregulation of *AUX/IAA* genes, and auxin signaling is activated by auxin-dependent degradation of proteins from this family [18]. In total, 21 *PhIAA* transcripts were upregulated by auxin in moso bamboo. Importantly, the transcription of *PhIAA23*, an ortholog of *AtIAA19* in Arabidopsis and acts as the key regulator in Arabidopsis lateral root formation [54], was strongly induced in response to exogenous auxin. This result is quite similar to the data from Arabidopsis and other plants, indicating that auxin-activated *AUX/IAA* gene upregulation is conserved in the plant kingdom (Fig. 4; Additional file 8: Table S13).

ARF proteins bind to the promoter elements and act as activators or repressors of downstream auxin responsive genes [21]. Of the 41 members of the ARF family identified in moso bamboo, 11 *PhARFs* were upregulated and 1 *PhARF* was downregulated by auxin (Fig. 4, Additional file 9: Table S14). Among the auxin responsive genes in Arabidopsis, *AtARF19*, an activator of auxin-dependent transcription is most sensitive to auxin at the transcriptional level. *AtARF7* and *AtARF19*, which are phylogenetically close, are considered to be the only ARF factors that are necessary and sufficient for auxin signaling in 7-d-old light-grown seedlings [55]. Supporting this point, *PhARF11*, the ortholog of *AtARF19*, and *PhARF13*, the ortholog of *AtARF7*, were strongly induced by auxin (Fig. 4, Additional file 9: Table S14). *AtARF2* acts as the communication node that links the ethylene and auxin signaling pathways [21], and its ortholog *PhARF42* in bamboo was strongly induced by auxin. All the auxin pathway genes that changed expression at transcription level were listed in the Additional file 10: Table S15.

It is likely that moso bamboo possesses similar conserved signaling pathways that stimulated the downstream genes as other species, such as Arabidopsis. On the other hand, different mechanisms in controlling the auxin concentration exist between moso bamboo and Arabidopsis, for example, the expression of some auxin transporter genes in bamboo went down, whereas all their orthologs in Arabidopsis were upregulated after auxin treatment. Future analyses of the expression patterns and subcellular localizations, as well as the functional characterization of these genes will unveil the specific mechanisms of controlling the auxin-concentration related pathways in moso bamboo.

Identification of additional important auxin-responsive transcription factors

Except for the AUX/IAA family, 3 other gene families participate in the primary auxin response at the

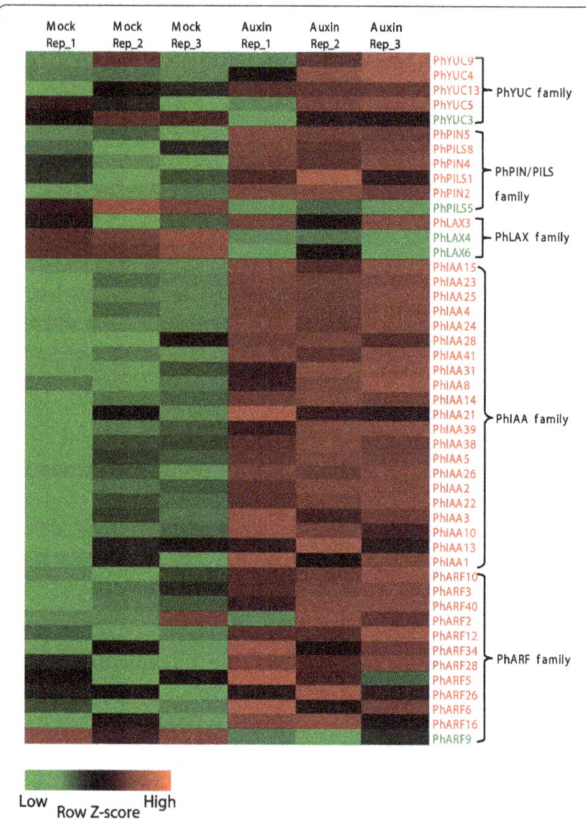

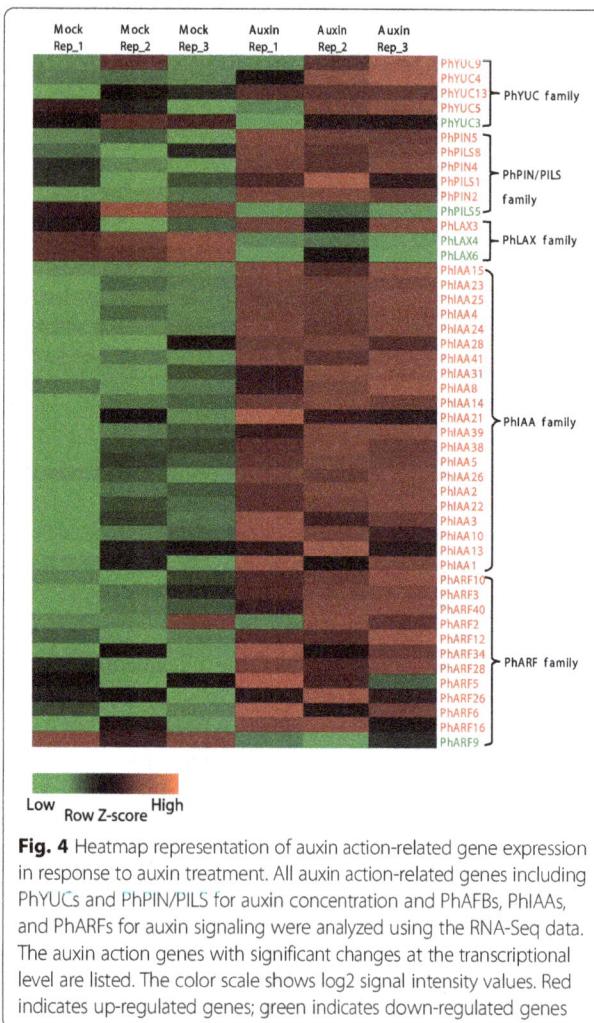

Fig. 4 Heatmap representation of auxin action-related gene expression in response to auxin treatment. All auxin action-related genes including PhYUCs and PhPIN/PILS for auxin concentration and PhAFBs, PhIAAs, and PhARFs for auxin signaling were analyzed using the RNA-Seq data. The auxin action genes with significant changes at the transcriptional level are listed. The color scale shows log2 signal intensity values. Red indicates up-regulated genes; green indicates down-regulated genes

transcriptional level in Arabidopsis: *Gretchen Hagen3* (*GH3* family), *small auxin-up RNA* (*SAUR* family) and *the lateral organ boundaries domain* (*LBD* family) [51, 56, 57]. *GH3* genes encode a class of auxin-induced conjugating enzymes and are auxin responsive genes involved in regulating auxin homeostasis and response [58]. At least 7 *GH3* genes are strongly and rapidly induced by auxin in Arabidopsis [58]. Studies from Arabidopsis and the moss *Physcomitrella patens* suggested the evolutionarily conserved role for *GH3* proteins in regulating auxin homeostasis [59]. In the bamboo genome, we found 5 putative *GH3* genes with increased transcription in response to auxin (Additional file 11: Table S16). Among the early auxin response genes, the *SAUR* gene family has the largest number of member; it is estimated that around half of the Arabidopsis *SAUR* gene transcripts are rapidly upregulated by auxin, while a small number of this family members appear to be repressed by auxin [60, 61]. Recently, a genome-wide analysis of *SAUR* genes was performed in moso bamboo [62]. Of the 38 *SAUR* genes identified, 10 *SAUR* genes changed their expressions under our auxin treatment conditions, the expression levels of 7 of those genes increased, whereas 3 genes decreased (Additional file 12: Table S17). The *LATERAL ORGAN BOUNDARIES* (LOB) proteins, which contain a conserved LOB domain (LBD), are rapidly and specifically up-regulated by auxin treatment [55]. Our RNA-Seq results also showed that in moso bamboo, at least 9 putative *LBD* genes increased their transcripts in response to exogenous auxin treatment (Additional file 13: Table S18).

Based on the acid growth hypothesis, auxin stimulates protons pumping into the cell wall matrix, thereby helping to loosen the cell wall [63]. On the other hand, auxin may modulate cell wall properties by regulating the transcription of genes related to cell wall remodeling [64]. Consistent with this hypothesis, we found that a number of cell wall-related genes encoded structural cell wall proteins or enzymes changed their expressions after auxin treatment (Additional file 3: Table S8). In total, 51 genes, including expansins, xyloglucan endotransglycosylases (XTHs), and pectinmethylesterases (PMEs), were identified. Of these genes, 24 were upregulated and 27 were downregulated (Additional file 14: Table S19). Expansins were identified as the major cell wall-loosening agents [63]. In bamboo, at least 5 expansin-like genes were upregulated at the transcriptional level (Additional file 14: Table S19). Xyloglucan endotransglycosylase /hydrolases that cut and paste xyloglucans, and endoglucanases which hydrolyze glucosidic bonds, also contribute to cell wall loosening and cell expansion by modifying cell wall properties and by integrating new wall materials, respectively [63]. Supporting to this point, 3 glucanase-related genes changed in expression after auxin treatment (Additional file 14: Table S19). Some glycanases that were transcriptionally increased by auxin catalyze the hydrolyses of cell wall polysaccharides, and are involved in the auxin-induced changes in the mechanical properties of cell walls [65]. We found that 4 genes encoding glucanases changed in expression after auxin treatment (Additional file 14: Table S19). Class III peroxidases induce cell wall loosening and growth by elongation as well as cross-linking of cell wall components. It should be noted here that the expression levels of two putative peroxidase genes were also changed (Additional file 14: Table S19). These results suggest that auxin induces a broad range of transcriptional changes in cell wall property-related genes.

Crosstalk of auxin and other phytohormones in moso bamboo

The crosstalk among hormone-regulated pathways was widely present in plant cells [37]. We further examined the interaction between auxin and other phytohormones. Cytokinin and auxin have long been recognized as crucial signaling molecules controlling plant growth and development [66]. Cytokinin is degraded by side chain cleavage through the action of the cytokinin oxidase/dehydrogenase (CKX) enzymes which are induced by auxin at the transcriptional level [67]. In moso bamboo, at least two putative cytokinin oxidases genes (*PH01000072G1090* and *PH01001279G0410*) increased their transcriptional expression in the presence of auxin (Fig. 5a, Additional file 15: Table S20). Cytokinin conjugation is an important process that maintains the subcellular levels of active cytokinin. For example, the interactions between glycosides and cytokinins which are catalyzed by cytokinin-*O*-glucosylation inactivate cytokinins. We found that the expression of *PH01021243G0010*, a putative ortholog of *cytokinin-O-glucosyltransferase 1* in maize, was upregulated. Type-A ARRs are negative regulators of cytokinin signaling in bamboo, one type-A ARR gene (*PH01000999G0470*) was upregulated by auxin (Fig. 5a, Additional file 15: Table S20). *PH01000007G1150* is an ortholog of *HAT22* which belongs to a family of cytokinin-regulated homeodomain zip (HD zip) class II transcription factors in Arabidopsis [68], and was activated by auxin at the transcriptional level in moso bamboo (Fig. 5a, Additional file 15: Table S20). Therefore auxin signaling pathways affect cytokinin synthesis, homeostasis, and other signaling pathways at the transcriptional level in moso bamboo as in other model plants like Arabidopsis and rice.

In total, 14 gibberellin acid (GA) related genes were responsive to auxin treatment in the moso bamboo root. Of these genes, 3 putative gibberellin-2-oxidases which are involved in GA degradation were downregulated. One putative Gibberellin-20-oxidase-2 which controls GA biosynthesis was upregulated (Fig. 5b, Additional file 16: Table S21). These results suggested that exogenous auxin controls GA levels through transcriptional regulation. A previous study in Arabidopsis showed that the effects of auxin on GA biosynthesis and signaling are complicated and depend on the concentration of exogenous auxin, duration of treatment or tissue type [51]. Therefore, more detailed analyses are needed to fully illustrate the crosstalk between auxin and GA.

The importance of auxin and ethylene crosstalk is well-established [69]. In total, 59 ethylene-related genes were found to be transcriptionally altered in response to auxin in the moso bamboo root (Fig. 5c, Additional file 17: Table S22). The application of auxin normally upregulates the transcription of genes encoding ethylene biosynthetic enzymes and leads to increased ethylene biosynthesis [69]. In accordance with this point, we found that 1 putative aminocyclopropane-1-carboxylic acid synthase (*ACS*) gene (*PH01000000G2120*), which catalyzes the rate-limiting step in ethylene synthesis, was upregulated under our exogenous auxin treatment condition. The second group of auxin-responsive ethylene genes is involved in a signaling pathway. Twenty-five *ERFs*, including one putative ethylene receptor gene (*PH01005673G0050*), were auxin responsive, and most of these genes were upregulated (Fig. 5c, Additional file 17: Table S22). In general, the enhanced expression of ethylene biosynthesis and signaling genes after auxin treatment showed the positive crosstalk between these two plant hormones.

The auxin signaling pathway not only controls various aspects of plant growth and development but also plays roles in plant environmental adaptation via crosstalk with some stress-related phytohormones, such as abscisic acid (ABA), jasmonate (JA) or salicylic acid (SA) [37]. Our results showed that 20 genes involved in the ABA-related pathways are auxin-responsive (Fig. 5d, Additional file 18: Table S23). Of these, *PH01000029G2070*, the putative ABA-responsive element binding protein 3 (*AREB3*) transcription factor in moso bamboo, was strongly induced by exogenous auxin treatment. We further identified 32 and 30 auxin-responsive genes that are involved in the JA and SA pathways, respectively (Fig. 5e and f, Additional file 19: Table S24 and Additional file 20: Table S25).

Discussion

Bamboo is well known for its rapid growth and high level of woodiness. At the cellular level, the fast growth of the bamboo culm is mainly due to cell elongation and division [24]. Auxin is one of the well-known phytohormones that controls cell division and expansion [13]. However, the genes involved in auxin pathway and the global auxin-response profiling in moso bamboo still need to be investigated.

Here, we showed that exogenous NAA inhibit bamboo root growth (Fig. 1). We identified the key gene families involved in auxin biosynthesis (*PhYUC* family; 13 members), auxin transport (*PhPIN*, *PhPILS* and *PhLAX* families; 6, 8, and 7 members respectively), auxin receptors (*PhAFB* family; 10 members) and auxin signaling (*PhIAA* and *PhARF* families; 43 and 41 members, respectively) in moso bamboo (Fig. 2). Our analysis indicated the importance of compartmentalized auxin homeostasis and auxin signaling throughout the plant kingdom. Compared with Arabidopsis and rice, most of the auxin action- related gene families in moso bamboo have extensively expended their numbers of members (Additional file 2: Figure S4), suggesting that the auxin action in moso bamboo is more complicated and diverse.

Until now, no auxin-related gene in moso bamboo was functionally characterized. The phylogenetic relationships of the auxin pathway genes may suggest their putative

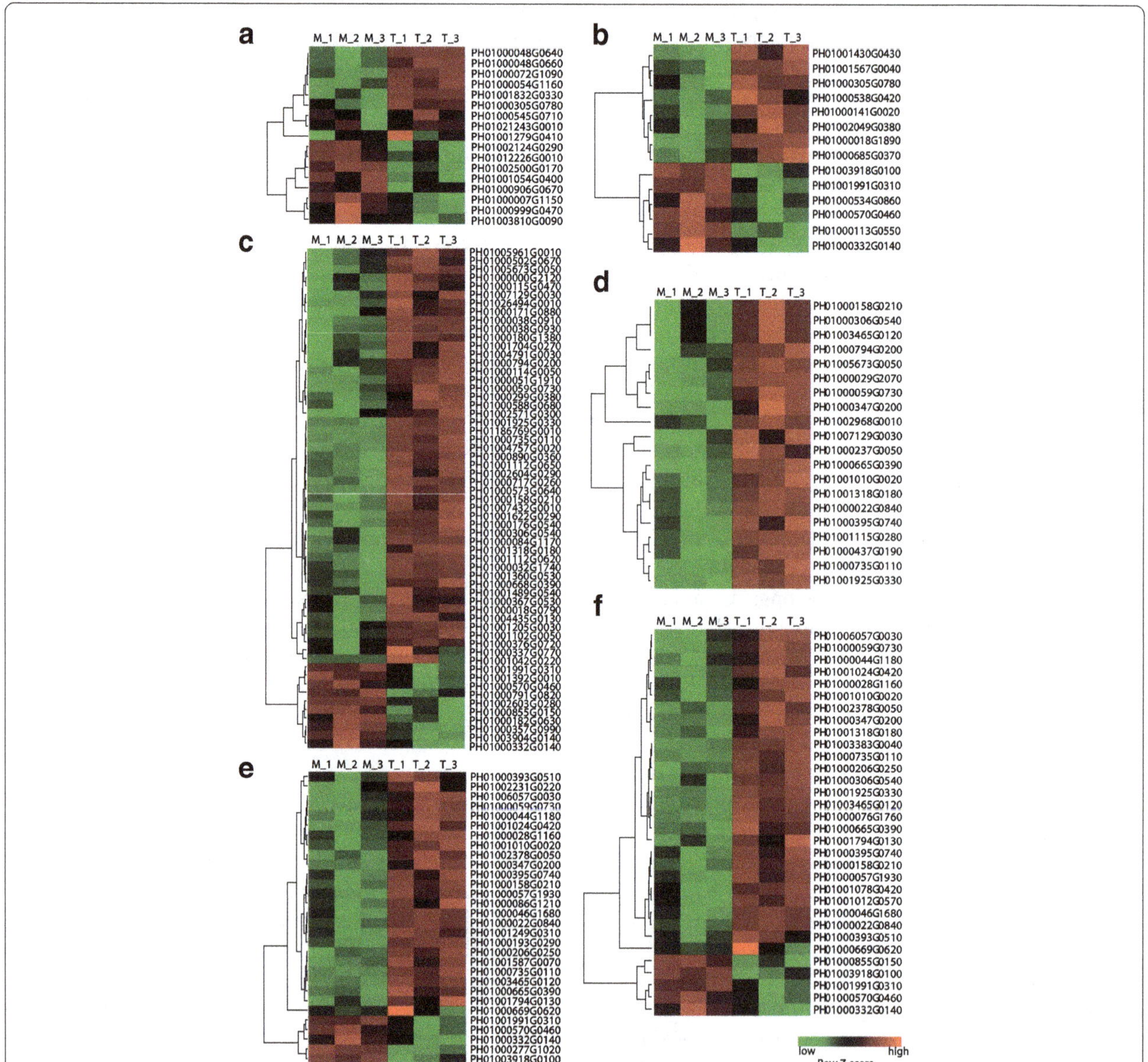

Fig. 5 Cross-talk between auxin and other phytohormones. The hierarchical clustering of auxin-responsive genes that are associated with cytokinin (**a**), ethylene (**b**), Gibberellin (**c**), ABA (**d**), SA (**e**), and JA (**f**) were shown. M_1–3, mocked sample with 3 repeats; A_1–3, auxin treated samples with 3 repeats. Color scale shows log2 signal intensity values. Red indicates up-regulated genes; green indicates down-regulated genes

roles in moso bamboo. For example, *OsYUC1* encodes the key enzyme contributing to IAA biosynthesis in rice [70], and it will be very interesting to determine the function of its ortholog in moso bamboo *PhYUC11* (Additional file 2: Figure S4). *PhPIN4* is most closely related to *OsPIN10a* and *OsPIN10b* (Additional file 2: Figure S4), which are supposed to be monocot-specific [45]. In Arabidopsis, *AtLAX3* is reported to promote the initiation of lateral root primordia by increasing the expression of a selection of cell-wall-remodeling enzymes [71]. In bamboo, we identified its closest ortholog *PhLAX6* (Additional file 2: Figure S4). Heterodimerization between Aux/IAA and ARF proteins are crucial for their unique biological functions in different tissues in Arabidopsis [22]. The studies of the protein-protein interactions and gene co-expression maps in Arabidopsis, together with our phylogenetic analysis (Additional file 2: Figure S4), will facilitate future studies of the interaction maps of PhIAAs and PhARFs. This analysis is a key step to understanding the auxin action network in moso bamboo.

Previous studies have shown that elevated cellular auxin levels stimulate the transcriptional changes of genes involved in auxin biosynthesis, conjugation, transport, and signaling [51]. We performed the KEGG analysis of the differentially expressed genes, and our results showed that those differentially expressed genes were classified into 12 pathways, and the genes involved in plant hormone signal transduction were predominantly enriched (33 DEGs) (Additional file 21: Table S26). Our results also showed that a number of auxin pathway genes changed their expression, which is consistent with a previous report (Fig. 4, Additional file 10: Table S15). While there are some exceptions, for example, all *AUX1/LAX* genes in Arabidopsis increased their transcripts after auxin treatment, but in moso bamboo, only 3 genes changed their expression levels: 2 *PhLAX* genes (*PhLAX4* and *PhLAX6*) were down-regulated, whereas the *YUC* genes involved in auxin biosynthesis (*PhYUC4, 5, 9* and *13*) were mostly upregulated (Fig. 4). The latter result is opposite to the results in Arabidopsis, in which the *YUC* genes are downregulated through the feedback pathways [72]. These results indicate that the mechanisms that control auxin concentrations in moso bamboo are more complicated, and the expression of related genes may tightly and precisely controlled by the properties of external stimuli, different tissues, or various developmental stages. Interestingly, *PhYUC3* is the only *YUC* gene that was downregulated by exogenous auxin in the moso bamboo root (Fig. 4), suggesting that PhYUC3 may encode one of the rate-limiting enzymes for auxin biosynthesis in the bamboo root. Further genetic and biochemical experiments are needed to characterize the potentially important function of these genes.

NAA, IAA and IBA are three most commonly used compounds in auxin research. Although they were transported in different ways in plant cells, they cause similar physiological response by changing overlapping downstream gene expressions [3, 73]. As a preliminary test on the effects of various auxins in moso bamboo, we treated the seedlings with 5 μM IAA or 5 μM IBA for 4 h (similar conditions as the NAA treatment we used), and randomly selected genes whose expressions were changed after NAA treatment for qRT-PCR analysis. The genes we checked include: Ph01003158G0110 and Ph01000099G0730 for GH3 family; Ph01000001G1450 and Ph01004534G0130 for SAUR family; Ph01001249G0310 for LBD family; Ph01000025G1600, Ph01000554G0550, Ph01003159G0070 and Ph01001154 G0590 for AUX/IAA family. Our results showed that these genes had similar expression patterns after IAA or IBA treatments, while the extents of the changes were different (Additional file 2: Figure S10), these results support the previous findings that different auxins have both common and specific efficacies to activate various auxin signaling pathways [74]. Moreover, previous reports also showed that different auxins cause distinct but overlapping changes in gene expression, probably due to the differences in their metabolism, transport or interaction with the signaling components [75, 76]. Thereby, it is needed to perform more detailed and systematic studies to unveil the effects of various auxins on the expression of auxin-related genes in moso bamboo in the future.

We also identified the genes that were regulated by exogenous auxin (Fig. 5). Notably, our results showed that a significant number of auxin-responsive genes are involved in the biosynthesis, metabolism or signaling pathway of other phytohormone like cytokinin, ethylene, Gibberellin, ABA, SA and JA (Fig. 5). Various plant hormones affect similar cellular processes through complicated interactions. Our results provide insights into how auxin pathways crosstalk with other plant hormone pathways to regulate moso bamboo growth and development, which should be investigated in the future.

Conclusion

In summary, we established a general overview of the main pathways involves in auxin synthesis, transport, receptor and signaling, and the global transcriptional profiling of auxin response in moso bamboo. The results from this study provided information for the elucidation of the possible functions of auxin action-related genes in bamboo. In the future, more experimental and bioinformatics work is needed to fully understand the functions of these important candidate genes and the regulatory mechanisms of some important auxin action-related proteins in this particular species.

Additional files

Additional file 1: Table S1. List of putative PhYUC family genes in maso bamboo. **Table S2.** List of putative PhPIN/PILS genes in maso bamboo. **Table S3.** List of putative PhLAX genes in maso bamboo. **Table S4.** List of putative auxin binding factors (AFB) in bamboo. **Table S5.** List of putative PhAUX/IAA genes in maso bamboo. **Table S6.** List of putative PhARF genes in maso bamboo. **Table S7.** List of primers used for qPCR analysis. (DOCX 28 kb)

Additional file 2: Figure S1. General overview of auxin biosynthesis, transport, and signaling pathway. **Figure S2.** Multiple sequence alignment, conserved motif and gene structure analysis of members involved in auxin concentration. **Figure S3.** Multiple sequence alignment, conserved motif and gene structure analysis of members involved in auxin signaling. **Figure S4.** Phylogenetic analysis of families related to auxin action in moso bamboo, rice, and Arabidopsis. **Figure S5.** qRT-PCR analysis of auxin responsive marker genes in response to exogenous auxin treatment. **Figure S6.** Heat-map of Euclidean distance among the 6 bamboo RNA-Seq libraries used in this study. **Figure S7.** Validation of RNA-Seq results by qPCR. **Figure S8.** qRTPCR confirmation of the expressions of auxin-related genes. **Figure S9.** Tissue expression patterns of the auxin related genes. **Figure S10.** qRT-PCR analysis of auxin responsive marker genes in response to exogenous IAA and IBA treatment. (PDF 3914 kb)

Additional file 3: Table S8. List of the genes in the RNA-Seq analysis. (TXT 4491 kb)

Additional file 4: Table S9. Expression of PhYUC family members in response to auxin treatment. (TXT 1 kb)

Additional file 5: Table S10. Expression of PhPIN/PhPILS family members in response to auxin treatment. (TXT 1 kb)

Additional file 6: Table S11. Expression of PhLAX family members in response to auxin treatment. (TXT 1 kb)

Additional file 7: Table S12. Expression of PhAFB family members in response to auxin treatment. (TXT 1 kb)

Additional file 8: Table S13. Expression of PhIAA family members in response to auxin treatment. (TXT 5 kb)

Additional file 9: Table S14. Expression of PhARF family members in response to auxin treatment. (TXT 5 kb)

Additional file 10: Table S15. List of the auxin action related genes with expression change in response to exogenous auxin treatment. (TXT 5 kb)

Additional file 11: Table S16. Expression of putative PhGH3 members in response to auxin treatment. (TXT 824 bytes)

Additional file 12: Table S17. Expression of PhSAUR family members in response to auxin treatment. (TXT 1 kb)

Additional file 13: Table S18. Expression of putative PhLBD members in response to auxin treatment. (TXT 1 kb)

Additional file 14: Table S19. Expression changes of cell wall related genes in response to exogenous auxin treatment. (TXT 6 kb)

Additional file 15: Table S20. Expression changes of cytokinin related genes in response to exogenous auxin treatment. (TXT 1 kb)

Additional file 16: Table S21. Expression changes of GA related genes in response to exogenous auxin treatment. (TXT 1 kb)

Additional file 17: Table S22. Expression changes of ethylene related genes in response to exogenous auxin treatment. (TXT 7 kb)

Additional file 18: Table S23. Expression changes of ABA related genes in response to exogenous auxin treatment. (TXT 2 kb)

Additional file 19: Table S24. Expression changes of JA related genes in response to exogenous auxin treatment. (TXT 3 kb)

Additional file 20: Table S25. Expression changes of SA related genes in response to exogenous auxin treatment. (TXT 3 kb)

Additional file 21: Table S26. The statistical enrichment of DEGs in the KEGG pathways. (TXT 4 kb)

Abbreviations

ABA: Abscisic acid; ACS: Aminocyclopropane-1-carboxylic acid synthase; AFB: Auxin signaling F-BOX protein; AREB: ABA-responsive element binding protein; ARFs: Auxin response factors; AUX/IAA: auxin/indole-3-aceticacid; AUX1/LAX: Auxin 1/like-AUX1; AuxRE: Auxin-responsive *cis*-element; BLAST: Basic local alignment search tool; CKX: Cytokinin oxidase/ dehydrogenase; CTD domain: Carboxy-terminal dimerization domain; DBD: DNA binding domain; DEGs: Differentially expressed genes; dUTP: Deoxyuridine triphosphate; ER: Endoplasmic reticulum; ERFs: Ethylene responsive factors; FDR: False discovery rate; FPKM: Fragment per kilobase of transcript per million mapped reads; GA: Gibberellin acid; GH3: Gretchen Hagen 3; HD zip: Homeodomain zip; HMM: Hidden Markov model; IAA: Indole-3-acetic acid; IBA: Indole-3-butytric acid; IPA: Indole-3-pyruvate; JA: Jasmonic acid; LBD: Lateral organ boundaries domain; LOB: Lateral organ boundaries; MS: Murashige and skoog; NAA: Naphthaleneacetic acid; ORF: Open reading frame; PIN/PILS: PIN-FORMED/PIN-like; PMEs: Pectinmethylesterases; qRT-PCR: Quantiative real time PCR; RNA-Seq: RNA-Sequencing; SA: Salicylic acid; SAUR: Small auxin-up RNA; SE: Standard error; TAA: Tryptophan aminotransferase; TIR1: Transport inhibitor response 1; XTHs: Xyloglucan endotransglycosylases; YUC: YUCCA

Acknowledgements

The authors thank the International Magnesium Institute for its financial support to W.J.W., and Prof. Yoshito Oka, Prof. Markus V. Kohnen, Prof. Liangsheng Zhang and Prof. Xu Chen for the critical reading of the manuscript.

Funding

This work was supported by Fujian Innovative Center for Germplasm Resources and Cultivation of Woody plants (no. 125/KLA15001E), and a grant from the Natural Science Foundation of Fujian province (no. 2016J01099) to Q.Z.. The National Natural Science Foundation of China grant (no. 31570674) to L.F.G.. The funding bodies were not involved in the design of the study or in any aspect of the data collection, analysis and interpretation of data and in paper writing.

Authors' contributions

QZ and LFG conceived this project; QZ, LFG and CTL designed experiments and interpreted the results. WJW and SWY performed the experiments, and HXZ, CYC, MQX, YBG and QW helped to collect and analyze the data. QZ wrote the manuscript. All authors read and approved the final manuscript.

Competing interests

The authors declare that they have no competing interests.

Author details

[1]Basic Forestry and Proteomics Center (BFPC), Fujian Provincial Key Laboratory of Haixia Applied Plant Systems Biology, Haixia Institute of Science and Technology, Fujian Agriculture and Forestry University, Fujian 350002, China. [2]Department of Molecular, Cell & Developmental Biology, University of California, Los Angeles, California 90095, USA.

References

1. Balzan S, Johal GS, Carraro N. The role of auxin transporters in monocots development. Front Plant Sci. 2014;5:393.
2. Gallavotti A. The role of auxin in shaping shoot architecture. J Exp Bot. 2013; 64(9):2593–608.
3. Chapman EJ, Estelle M. Mechanism of auxin-regulated gene expression in plants. Annu Rev Genet. 2009;43:265–85.
4. Kasahara H. Current aspects of auxin biosynthesis in plants. Biosci Biotechnol Biochem. 2015;80(1):34–42.
5. Adamowski M, Friml J. PIN-dependent auxin transport: action, regulation, and evolution. Plant Cell. 2015;27(1):20–32.
6. Zhao Y. Auxin biosynthesis. Arabidopsis Book. 2014;12:e0173.
7. Mashiguchi K, Tanaka K, Sakai T, Sugawara S, Kawaide H, Natsume M, Hanada A, Yaeno T, Shirasu K, Yao H, et al. The main auxin biosynthesis pathway in Arabidopsis. Proc Natl Acad Sci U S A. 2011;108(45):18512–7.
8. Vieten A, Sauer M, Brewer PB, Friml J. Molecular and cellular aspects of auxin-transport-mediated development. Trends Plant Sci. 2007;12(4):160–8.
9. Grones P, Friml J. Auxin transporters and binding proteins at a glance. J Cell Sci. 2015;128(1):1–7.
10. Kramer EM, Bennett MJ. Auxin transport: a field in flux. Trends Plant Sci. 2006;11(8):382–6.
11. Barbez E, Kubes M, Rolcik J, Beziat C, Pencik A, Wang B, Rosquete MR, Zhu J, Dobrev PI, Lee Y, et al. A novel putative auxin carrier family regulates intracellular auxin homeostasis in plants. Nature. 2012;485(7396):119–22.
12. Feraru E, Vosolsobe S, Feraru MI, Petrasek J, Kleine-Vehn J. Evolution and structural diversification of PILS putative Auxin carriers in plants. Front Plant Sci. 2012;3:227.
13. Benjamins R, Scheres B. Auxin: the looping star in plant development. Annu Rev Plant Biol. 2008;59(1):443–65.
14. Dharmasiri N, Dharmasiri S, Weijers D, Lechner E, Yamada M, Hobbie L, Ehrismann JS, Jürgens G, Estelle M. Plant development is regulated by a family of Auxin receptor F box proteins. Dev Cell. 2005;9(1):109–19.
15. Dharmasiri N, Dharmasiri S, Estelle M. The F-box protein TIR1 is an auxin receptor. Nature. 2005;435(7041):441–5.
16. Yamada M, Greenham K, Prigge MJ, Jensen PJ, Estelle M. The TRANSPORT INHIBITOR RESPONSE2 gene is required for auxin synthesis and diverse aspects of plant development. Plant Physiol. 2009;151(1):168–79.
17. Mockaitis K, Estelle M. Auxin receptors and plant development: a new signaling paradigm. Annu Rev Cell Dev Biol. 2008;24:55–80.

18. Gray WM, Kepinski S, Rouse D, Leyser O, Estelle M. Auxin regulates SCF(TIR1)-dependent degradation of AUX/IAA proteins. Nature. 2001;414(6861):271–6.
19. Szemenyei H, Hannon M, Long JA. TOPLESS mediates auxin-dependent transcriptional repression during Arabidopsis embryogenesis. Science. 2008; 319(5868):1384–6.
20. Hardtke CS, Ckurshumova W, Vidaurre DP, Singh SA, Stamatiou G, Tiwari SB, Hagen G, Guilfoyle TJ, Berleth T. Overlapping and non-redundant functions of the Arabidopsis auxin response factors MONOPTEROS and NONPHOTOTROPIC HYPOCOTYL 4. Development. 2004;131(5):1089–100.
21. Guilfoyle TJ, Hagen G. Auxin response factors. Curr Opin Plant Biol. 2007;10(5): 453–60.
22. Piya S, Shrestha SK, Binder B, Stewart CN Jr, Hewezi T. Protein-protein interaction and gene co-expression maps of ARFs and aux/IAAs in Arabidopsis. Front Plant Sci. 2014;5:744.
23. Buckingham KC, Wu L, Lou Y. Can't see the (bamboo) forest for the trees: examining bamboo's fit within international forestry institutions. Ambio. 2014;43(6):770–8.
24. Wang HY, Cui K, He CY, Zeng YF, Liao SX, Zhang JG. Endogenous hormonal equilibrium linked to bamboo culm development. Genet Mol Res. 2015; 14(3):11312–23.
25. Perrot-Rechenmann C. Cellular responses to auxin: division versus expansion. Cold Spring Harb Perspect Biol. 2010;2(5):a001446.
26. Peng Z, Lu Y, Li L, Zhao Q, Feng Q, Gao Z, Lu H, Hu T, Yao N, Liu K, et al. The draft genome of the fast-growing non-timber forest species moso bamboo (Phyllostachys Heterocycla). Nat Genet. 2013;45(4):456–61.
27. Larkin MA, Blackshields G, Brown NP, Chenna R, McGettigan PA, McWilliam H, Valentin F, Wallace IM, Wilm A, Lopez R, et al. Clustal W and Clustal X version 2.0. Bioinformatics. 2007;23(21):2947–8.
28. Skinner ME, Uzilov AV, Stein LD, Mungall CJ, Holmes IH. JBrowse: a next-generation genome browser. Genome Res. 2009;19(9):1630–8.
29. Tamura K, Stecher G, Peterson D, Filipski A, Kumar S. MEGA6: molecular evolutionary genetics analysis version 6.0. Mol Biol Evol. 2013;30(12):2725–9.
30. Parkhomchuk D, Borodina T, Amstislavskiy V, Banaru M, Hallen L, Krobitsch S, Lehrach H, Soldatov A. Transcriptome analysis by strand-specific sequencing of complementary DNA. Nucleic Acids Res. 2009; 37(18):20.
31. Kim D, Pertea G, Trapnell C, Pimentel H, Kelley R, Salzberg SL. TopHat2: accurate alignment of transcriptomes in the presence of insertions, deletions and gene fusions. Genome Biol. 2013;14(4):R36.
32. Trapnell C, Williams BA, Pertea G, Mortazavi A, Kwan G, van Baren MJ, Salzberg SL, Wold BJ, Pachter L. Transcript assembly and quantification by RNA-Seq reveals unannotated transcripts and isoform switching during cell differentiation. Nat Biotechnol. 2010; 28(5):511–5.
33. Robinson MD, McCarthy DJ, Smyth GK. edgeR: a bioconductor package for differential expression analysis of digital gene expression data. Bioinformatics. 2010;26(1):139–40.
34. Zhu Q, Dugardeyn J, Zhang C, Takenaka M, Kuhn K, Craddock C, Smalle J, Karampelias M, Denecke J, Peters J, et al. SLO2, a mitochondrial PPR protein affecting several RNA editing sites, is required for energy metabolism. Plant J. 2012;71(5):836–49.
35. Huang da W, Sherman BT, Tan Q, Kir J, Liu D, Bryant D, Guo Y, Stephens R, Baseler MW, Lane HC, et al. DAVID bioinformatics resources: expanded annotation database and novel algorithms to better extract biology from large gene lists. Nucleic Acids Res. 2007; 35(Web Server issue):W169–75.
36. Teale WD, Paponov IA, Palme K. Auxin in action: signalling, transport and the control of plant growth and development. Nat Rev Mol Cell Biol. 2006; 7(11):847–59.
37. Vanstraelen M, Benkova E. Hormonal interactions in the regulation of plant development. Annu Rev Cell Dev Biol. 2012;28:463–87.
38. De Smet I, Lau S, Voss U, Vanneste S, Benjamins R, Rademacher EH, Schlereth A, De Rybel B, Vassileva V, Grunewald W et al: Bimodular auxin response controls organogenesis in Arabidopsis. Proc Natl Acad Sci U S A 2010, 107(6):2705–2710.
39. Schlaich NL. Flavin-containing monooxygenases in plants: looking beyond detox. Trends Plant Sci. 2007;12(9):412–8.
40. Mohanta TK, Mohanta N, Bae H. Identification and expression analysis of PIN-like (PILS) gene family of Rice treated with Auxin and Cytokinin. Genes. 2015;6(3):622–40.
41. Yue R, Tie S, Sun T, Zhang L, Yang Y, Qi J, Yan S, Han X, Wang H, Shen C. Genome-wide identification and expression profiling analysis of ZmPIN, ZmPILS, ZmLAX and ZmABCB auxin transporter gene families in maize (Zea Mays L.) under various abiotic stresses. PLoS One. 2015;10(3):e0118751.
42. Swarup R, Peret B. AUX/LAX family of auxin influx carriers-an overview. Front Plant Sci. 2012;3:225.
43. Kepinski S, Leyser O. The Arabidopsis F-box protein TIR1 is an auxin receptor. Nature. 2005;435(7041):446–51.
44. Abu-Zaitoon YM, Bennett K, Normanly J, Nonhebel HM. A large increase in IAA during development of rice grains correlates with the expression of tryptophan aminotransferase OsTAR1 and a grain-specific YUCCA. Physiol Plant. 2012;146(4):487–99.
45. Wang JR, Hu H, Wang GH, Li J, Chen JY, Wu P. Expression of PIN genes in rice (Oryza Sativa L.): tissue specificity and regulation by hormones. Mol Plant. 2009;2(4):823–31.
46. Xu H-w, Y-w M, Wang W, Wang H, Wang Z. OsPIN1a gene participates in regulating negative phototropism of Rice roots. Rice Sci. 2014;21(2):83–9.
47. Paul P, Dhandapani V, Rameneni JJ, Li X, Sivanandhan G, Choi SR, Pang W, Im S, Lim YP. Genome-wide analysis and characterization of aux/IAA family genes in Brassica Rapa. PLoS One. 2016;11(4):e0151522.
48. Singh VK, Jain M. Genome-wide survey and comprehensive expression profiling of aux/IAA gene family in chickpea and soybean. Front Plant Sci. 2015;6:918.
49. Song Y, Wang L, Xiong L. Comprehensive expression profiling analysis of OsIAA gene family in developmental processes and in response to phytohormone and stress treatments. Planta. 2009;229(3):577–91.
50. Zhu ZX, Liu Y, Liu SJ, Mao CZ, Wu YR, Wu P. A gain-of-function mutation in OsIAA11 affects lateral root development in rice. Mol Plant. 2012;5(1):154–61.
51. Paponov IA, Paponov M, Teale W, Menges M, Chakrabortee S, Murray JA, Palme K. Comprehensive transcriptome analysis of auxin responses in Arabidopsis. Mol Plant. 2008;1(2):321–37.
52. Suzuki M, Yamazaki C, Mitsui M, Kakei Y, Mitani Y, Nakamura A, Ishii T, Soeno K, Shimada Y. Transcriptional feedback regulation of YUCCA genes in response to auxin levels in Arabidopsis. Plant Cell Rep. 2015; 34(8):1343–52.
53. Zažímalová E, Murphy AS, Yang H, Hoyerová K, Hošek P. Auxin transporters—why so many? Cold Spring Harb Perspect Biol. 2010;2(3): a001552.
54. Tatematsu K, Kumagai S, Muto H, Sato A, Watahiki MK, Harper RM, Liscum E, Yamamoto KT. MASSUGU2 encodes aux/IAA19, an auxin-regulated protein that functions together with the transcriptional activator NPH4/ARF7 to regulate differential growth responses of hypocotyl and formation of lateral roots in Arabidopsis Thaliana. Plant Cell. 2004;16(2):379–93.
55. Okushima Y, Overvoorde PJ, Arima K, Alonso JM, Chan A, Chang C, Ecker JR, Hughes B, Lui A, Nguyen D, et al. Functional genomic analysis of the AUXIN RESPONSE FACTOR gene family members in Arabidopsis Thaliana: unique and overlapping functions of ARF7 and ARF19. Plant Cell. 2005;17(2):444–63.
56. Okushima Y, Fukaki H, Onoda M, Theologis A, Tasaka M. ARF7 and ARF19 regulate lateral root formation via direct activation of LBD/ASL genes in Arabidopsis. Plant Cell. 2007;19(1):118–30.
57. Hagen G, Guilfoyle T. Auxin-responsive gene expression: genes, promoters and regulatory factors. Plant Mol Biol. 2002;49(3–4):373–85.
58. Staswick PE, Serban B, Rowe M, Tiryaki I, Maldonado MT, Maldonado MC, Suza W. Characterization of an Arabidopsis enzyme family that conjugates amino acids to indole-3-acetic acid. Plant Cell. 2005;17(2):616–27.
59. Ludwig-Muller J, Julke S, Bierfreund NM, Decker EL, Reski R. Moss (Physcomitrella Patens) GH3 proteins act in auxin homeostasis. New Phytol. 2009;181(2):323–38.
60. Jain M, Tyagi AK, Khurana JP. Genome-wide analysis, evolutionary expansion, and expression of early auxin-responsive SAUR gene family in rice (Oryza Sativa). Genomics. 2006;88(3):360–71.
61. Ren H, Gray WM. SAUR proteins as effectors of hormonal and environmental signals in plant growth. Mol Plant. 2015;8(8):1153–64.
62. Bai Q, Hou D, Li L, Cheng Z, Ge W, Liu J, Li X, Mu S, Gao J. Genome-wide analysis and expression characteristics of small auxin-up RNA (SAUR) genes in moso bamboo (Phyllostachys Edulis). Genome. 2016;15(10):2016–0097.
63. Cosgrove DJ. Growth of the plant cell wall. Nat Rev Mol Cell Biol. 2005;6(11):850–61.

64. Lewis DR, Olex AL, Lundy SR, Turkett WH, Fetrow JS, Muday GK. A kinetic analysis of the Auxin Transcriptome reveals Cell Wall remodeling proteins that modulate lateral root development in Arabidopsis. Plant Cell. 2013; 25(9):3329–46.
65. Kotake T, Nakagawa N, Takeda K, Sakurai N. Auxin-induced elongation growth and expressions of cell wall-bound exo- and endo-beta-glucanases in barley coleoptiles. Plant Cell Physiol. 2000;41(11):1272–8.
66. Zdarska M, Dobisova T, Gelova Z, Pernisova M, Dabravolski S, Hejatko J. Illuminating light, cytokinin, and ethylene signalling crosstalk in plant development. J Exp Bot. 2015;66(16):4913–31.
67. Schmulling T, Werner T, Riefler M, Krupkova E, Bartrina y Manns I. Structure and function of cytokinin oxidase/dehydrogenase genes of maize, rice, Arabidopsis and other species. J Plant Res. 2003;116(3):241–52.
68. Brenner WG, Romanov GA, Kollmer I, Burkle L, Schmulling T: Immediate-early and delayed cytokinin response genes of Arabidopsis thaliana identified by genome-wide expression profiling reveal novel cytokinin-sensitive processes and suggest cytokinin action through transcriptional cascades. Plant J 2005, 44(2):314–333.
69. Muday GK, Rahman A, Binder BM. Auxin and ethylene: collaborators or competitors? Trends Plant Sci. 2012;17(4):181–95.
70. Yamamoto Y, Kamiya N, Morinaka Y, Matsuoka M, Sazuka T. Auxin biosynthesis by the YUCCA genes in rice. Plant Physiol. 2007;143(3):1362–71.
71. Swarup K, Benkova E, Swarup R, Casimiro I, Peret B, Yang Y, Parry G, Nielsen E, De Smet I, Vanneste S, et al. The auxin influx carrier LAX3 promotes lateral root emergence. Nat Cell Biol. 2008;10(8):946–54.
72. Zhao Y, Dai X, Blackwell HE, Schreiber SL, Chory J. SIR1, an upstream component in auxin signaling identified by chemical genetics. Science. 2003;301(5636):1107–10.
73. Vanneste S, Friml J. Auxin: a trigger for change in plant development. Cell. 2009;136(6):1005–16.
74. Simon S, Kubes M, Baster P, Robert S, Dobrev PI, Friml J, Petrasek J, Zazimalova E. Defining the selectivity of processes along the auxin response chain: a study using auxin analogues. New Phytol. 2013;200(4):1034–48.
75. Pufky J, Qiu Y, Rao MV, Hurban P, Jones AM. The auxin-induced transcriptome for etiolated Arabidopsis seedlings using a structure/function approach. Funct Integr Genomics. 2003;3(4):135–43.
76. Woodward AW, Bartel B. Auxin: regulation, action, and interaction. Ann Bot. 2005;95(5):707–35.

11

Comprehensive genomic analysis of the *CNGC* gene family in *Brassica oleracea*: novel insights into synteny, structures, and transcript profiles

Kaleem U. Kakar[1,2†], Zarqa Nawaz[2,3†], Khadija Kakar[4], Essa Ali[1], Abdulwareth A. Almoneafy[5], Raqeeb Ullah[6], Xue-liang Ren[2,8*] and Qing-Yao Shu[1,7*]

Abstract

Background: The cyclic nucleotide-gated ion channel (CNGC) family affects the uptake of cations, growth, pathogen defence, and thermotolerance in plants. However, the systematic identification, origin and function of this gene family has not been performed in *Brassica oleracea*, an important vegetable crop and genomic model organism.

Results: In present study, we identified 26 *CNGC* genes in *B. oleracea* genome, which are non-randomly localized on eight chromosomes, and classified into four major (I-IV) and two sub-groups (i.e., IV-a and IV-b). The *BoCNGC* family is asymmetrically fractioned into the following three sub-genomes: least fractionated (14 genes), most fractionated-I (10), and most fractionated-II (2). The syntenic map of *BoCNGC* genes exhibited strong relationships with the model *Arabidopsis thaliana* and *B. rapa CNGC* genes and provided markers for defining the regions of conserved synteny among the three genomes. Both whole-genome triplication along with segmental and tandem duplications contributed to the expansion of this gene family. We predicted the characteristics of BoCNGCs regarding exon-intron organisations, motif compositions and post-translational modifications, which diversified their structures and functions. Using orthologous *Arabidopsis* CNGCs as a reference, we found that most CNGCs were associated with various protein–protein interaction networks involving CNGCs and other signalling and stress related proteins. We revealed that five microRNAs (i.e., bol-miR5021, bol-miR838d, bol-miR414b, bol-miR4234, and bol-miR_new2) have target sites in nine *BoCNGC* genes. The *BoCNGC* genes were differentially expressed in seven *B. oleracea* tissues including leaf, stem, callus, silique, bud, root and flower. The transcript abundance levels quantified by qRT-PCR assays revealed that *BoCNGC* genes from phylogenetic Groups I and IV were particularly sensitive to cold stress and infections with bacterial pathogen *Xanthomonas campestris* pv. *campestris*, suggesting their importance in abiotic and biotic stress responses.

Conclusion: Our comprehensive genome-wide analysis represents a rich data resource for studying new plant gene families. Our data may also be useful for breeding new *B. oleracea* cultivars with improved productivity, quality, and stress resistance.

Keywords: Abiotic and biotic stress, Ion channels, CNGC, Expression pattern, *Brassica oleracea*, Evolution, RNA-seq, qRT-PCR analysis

* Correspondence: renxuel@126.com; qyshu@zju.edu.cn
†Equal contributors
[2]Molecular Genetics Key Laboratory of China Tobacco, Guizhou Academy of Tobacco Science, Guiyang 550081, China
[1]State Key Laboratory of Rice Biology, Institute of Crop Science, Zhejiang University, Hangzhou 310058, China
Full list of author information is available at the end of the article

Background

Calcium is a universal secondary messenger that participates in multiple eukaryotic signalling pathways [1]. In plants, Ca^{2+} signal transduction via calcium-conducting channels is an important mechanism for transducing the signals derived from diverse environmental and developmental stimuli [2, 3]. Additionally, signal transductions contribute to growth, plant biotic interactions, and responses to hormones, light, and salt stress [4]. Cyclic nucleotide-gated ion channels (CNGCs) are components of Ca^{2+}-conducting signal transduction pathways [5]. They are Ca^{2+}-permeable cation-conducting channels that transport sodium, calcium, and potassium cations across membranes. Localized in the plasma membrane [6, 7], vacuole membrane [8], or nuclear envelope [9], CNGCs are controlled from inside the cell by secondary messengers such as Ca^{2+}/calmodulin (CaM) and cyclic nucleotide monophosphates (cNMPs; 3′,5′-cAMP and 3′,5′-cGMP) [3, 6, 10, 11]. The CNGCs are hypothesized to be involved in the uptake of both essential and toxic cations, Ca^{2+} signalling, development, pollen fertility and tip growth, gravitropism, leaf senescence, innate immunity, pathogen defence, and abiotic stress tolerance [6, 12–15].

The application of bioinformatics tools (for genes/proteins prediction and phylogenetic analysis), and experimental approaches (gene expression, mutant analysis and overexpression in yeast/Escherichia coli) have led to the identification, characterization, and functional analysis (in exceptional cases) of *CNGC* family genes in important plant species, including *Arabidopsis thaliana* [5], rice [16], tomato [17], pear [18], and *Physcomitrella patens* [19]. Researchers have only recently started to investigate the evolution, function (and underlying regulatory mechanism) of plant CNGCs, as well as their phylogenetic relationships with other channels. Briefly, plant CNGCs are characterised by conserved structural components, including a short cytosolic N-terminus, six transmembrane helices (S1–S6) with a pore-forming region between S5 and S6, and a cytosolic C-terminus containing a cNMP-binding domain (CNBD). The CNBD is the most conserved region of CNGCs carrying a plant CNGC-specific motif spanning the phosphate-binding cassette (PBC) and hinge region, which mediates channel gating by cAMP and/or cGMP [3, 20]. A latest study of the *A. thaliana CNGC12* gene suggested that plant CNGCs have multiple CaM-binding domains (CaMBDs) at cytosolic N- and C-termini [3]. Moreover, channel functionality depends on CaM binding to the conserved isoleucine–glutamine (IQ) motif in the C-terminus of the channel, indicating CaM positively and negatively regulates CNGCs [3]. Studies on individual isoforms and the *A. thaliana CNGC* family revealed that plant *CNGC* genes may be functionally distinguished in a group-dependent manner. For example, *AtCNGC19* and *AtCNGC20*, which belong to Group IV-a, are involved in salt stress responses [8]. Additionally, *AtCNGC2* and *AtCNGC4*, which are Group IV-b members, affect disease resistance against various pathogens and thermotolerance [21, 22]. Mumtaz et al. [4, 17] recently concluded that Group IV-b *SlCNGC* genes regulate different types of resistance against diverse pathogens in tomato. It is unclear whether this also applies to other plant species.

Brassica oleracea (2n = 18) is a member of the family *Brassicaceae* (approximately 338 genera and 3709 species), which consists of many important vegetable and oilseed crops, including brussels sprout, kohlrabi, and kale [23]. Among the cultivated species, *B. oleracea* exhibits the largest genetic and morphological diversity, making it highly adaptable to different environments. Sexually compatible *B. oleracea* crops, such as cabbage, cauliflower, and broccoli, are valued for their economic, nutritional, and potent anticancer properties [24]. The whole-genome sequence of this plant species was recently published [24], which enabled us to study the *B. oleracea CNGC* family. We used in silico and experimental approaches to identify, characterise, and functionally verify *CNGC* gene family members. We applied multiple tools and programs to complete in-depth analyses of each *CNGC* gene family member, including an analysis of the physiological and biochemical properties of the encoded proteins. Our objective was to elucidate the diversification, expansion, and evolution of the *CNGC* gene family. Furthermore, we investigated *CNGC* expression patterns to clarify the mechanisms underlying their responses to biotic and abiotic stresses, and to identify novel genes potentially useful for breeding.

Results

Genome-wide identification of *CNGC* genes in *Brassica oleracea*

For a complete overview of the *B. oleracea CNGC* gene family, we used the 20 *A. thaliana CNGC* genes as queries in BLAST searches of the Ensembl Plants database. Out of the 34 non-redundant putative gene sequences retrieved, eight gene accessions with truncated sequences or lacking *CNGC*-specific domains (CNBD and transmembrane) were eliminated from analyses (Additional file 1). Finally, 26 *CNGC* genes containing both essential domains (PF00520/PF07885 and PF00027) and a CNGC-specific motif were identified in the *B. oleracea* genome (i.e., *BoCNGC1–26*). Of the 26 *BoCNGC* genes identified in the latest genome assembly version in Ensembl Plants, 16 and 24 were detected in earlier versions from Bolbase (v.1.3) and GenBank (v.2.1) respectively (Table 1).

The physiological and biochemical properties of the 26 BoCNGC proteins were determined by computing different parameters, and are tabulated in Table 1. These proteins varied in length from 558 to 789 amino acids,

Table 1 Properties of 29 *BoCNGC* genes identified in *Brassica oleracea*

Name	Genes ID (Ensemble v.2.1)	brassicadb (v.1.3)	GeneBank	Transcript (bp)	Protein length (aa)	Mol.Wt. (kDa)	pI	GRAVY	Aliphatic index	II	Total atoms	Ave. residue Wt. (g/mol)	Net charge
BoCNGC1	Bo3g054400	Bol010786	LOC106335861	2115	704	81.33	9.18	−0.212	90.88	47.95	11,473	115.5	22
BoCNGC2	Bo4g009240	Bol000930	LOC106342665	2112	703	81.09	9.78	−0.139	87.92	37.21	11,462	115.3	33.5
BoCNGC3	Bo4g158880	Bol026503	LOC106336663	1950	649	75.08	9.89	−0.023	99.14	39.96	10,654	115.7	39
BoCNGC4	Bo8g100190	Bol045794	LOC106312466	2214	737	84.58	9.74	−0.177	92.51	49.74	11,989	114.8	31
BoCNGC5	Bo4g154120	Bol041958	LOC106342509	2238	745	85.27	9.85	−0.193	89.69	50.69	12,073	114.5	36.5
BoCNGC6	Bo7g114580	Bol033673	LOC106306102	2226	741	84.49	9.53	−0.169	91.61	48.38	11,959	114.0	25.5
BoCNGC7	Bo2g028960	Bol015575	LOC106326694	2166	721	82.52	9.35	−0.122	89.28	50.45	11,653	114.4	26
BoCNGC8	Bo5g021460	Bol038188	LOC106294829	2226	741	85.08	9.17	−0.266	86.22	51.02	11,997	114.8	19.5
BoCNGC9	Bo4g071760	Bol013404	LOC106337043	2055	684	79.43	10.08	−0.11	91.23	49.17	11,253	116.1	38
BoCNGC10	Bo1g013250	Bol001429	LOC106308171	2184	727	84.17	8.8	−0.187	91.21	45.73	11,876	115.8	17
BoCNGC11	Bo4g154790	Bol041907	LOC106340005	2187	728	83.84	9.12	−0.096	93.48	46.34	11,863	115.2	21.5
BoCNGC12	Bo8g099660	Bol044680	LOC106309975	2202	733	84.52	9.29	−0.13	94.69	48.3	11,982	115.3	23
BoCNGC13	Bo9g165600	Bol030411	LOC106316400	2106	701	79.63	8.23	−0.149	84.86	45.08	11,147	113.6	12
BoCNGC14	Bo8g076590	Bol008733	LOC106311631	2121	706	81.58	8.36	−0.247	86.08	52.08	11,403	115.6	15
BoCNGC15	Bo2g042020	Bol037445	LOC106327817	2088	695	80.24	8.25	−0.176	91.12	55.15	11,285	115.5	13.5
BoCNGC16	Bo9g116160	Bol038844	LOC106319027	2097	698	80.78	8.19	−0.192	90.6	55.83	11,357	115.7	12
BoCNGC17	Bo9g164130		LOC106317764	2151	716	81.93	9.81	0.018	94.61	54.04	11,569	114.4	38.5
BoCNGC18	Bo3g070140		LOC106335359	2160	674	76.69	8.38	−0.121	89.96	50.37	10,797	114.0	13.5
BoCNGC19	Bo3g070160			1677	558	63.94	7.26	−0.187	87.63	45.22	8953	114.6	5
BoCNGC20	Bo5g122720		LOC106343117	2262	753	85.98	9.8	−0.075	92.19	52.29	12,121	114.2	32
BoCNGC21	Bo1g119310		LOC106344554	2280	759	86.21	9.81	−0.067	90.59	47.96	12,147	113.6	32.5
BoCNGC22	Bo1g119340			2205	734	82.99	9.2	0.033	93.37	47.16	11,692	113.1	28
BoCNGC23	Bo1g079060		LOC106299574	2370	789	89.00	10.18	−0.055	95.64	46.62	12,604	112.8	42
BoCNGC24	Bo1g119320		LOC106326537	2232	743	85.06	10.06	−0.103	90.93	53.01	11,990	114.5	37
BoCNGC25	Bo5g122750		LOC106293228	2346	781	89.78	10.11	−0.176	90.73	47.88	12,686	115.0	34.5
BoCNGC26	Bo5g122740		LOC106293974	2235	744	85.72	9.3	−0.231	85.44	53.09	12,036	115.2	21.5

with an average of 717 amino acids. The ProtParam tool revealed that there was a considerable range in BoCNGC residue weight (112.795–116.128 g/mol) and molecular weight (63.938–89.775 kDa) depending on the number of atoms present. The computed average *pI* of majority of BoCNGC proteins was relatively high (range 8.23 to 10.18), signifying that these proteins are localized to membranes, and will supposedly participate in basic buffers. The BoCNGC19, which had *pI* than 7.4, indicate that this protein likely participate in the acidic buffers. Approximately one third of BoCNGC proteins had a low net charge (<17), while other are composed of more charged amino acids. Nearly all BoCNGC were hydrophilic, with BoCNGC17 and BoCNGC22 being slightly hydrophobic, which endorses its multifaceted role in cellular membrane transport. According to the instability index (II), only two proteins were stable in test tubes, namely BoCNGC2 and BoCNGC3. Aliphatic index showed that most BoCNGC proteins were thermostable at a wide temperature ranges, similar to other globular proteins.

Phylogenetic analysis of *BoCNGC* genes

Multiple sequence alignments and a maximum likelihood phylogenetic tree constructed between BoCNGCs and AtCNGCs were used to determine the similarity and homology between the *B. oleracea* and *A. thaliana CNGC* families. To strengthen the phylogenetic analysis, we identified and included 29 *CNGC* homolog genes from sister specie *Brassica rapa* (*BrCNGCs*) in current analysis. The sequence alignment revealed high similarity between the amino acid sequences of the three species, especially at the conserved domain regions (Additional file 2). The topology of the inferred maximum likelihood scoring tree revealed that the *BoCNGC* gene family can be divided into four major groups (i.e., Groups I–IV), which are based on the *A. thaliana* groups (Fig. 1) [5]. Groups I–III are monophyletic, while Group IV is sub-divided into two distinct clades (i.e., Groups IV-a and IV-b). Group IV contains 12 *BoCNGC* genes, while the other groups contain three to six members. Moreover, individual phylogenetic trees that were constructed based on the aligned *B. oleracea* and *A. thaliana* CNGC proteins produced similar clustering patterns (Additional files 3 and 4).

Chromosomal distribution and diversification of *BoCNGC* genes

The 26 *BoCNGC* genes were mapped onto *B. oleracea* chromosomes, and the position of each locus was determined. These genes were randomly distributed across the genome, and were detected on eight of nine chromosomes (i.e., C1–5 and C7–C9). The *BoCNGC* genes were unevenly distributed, with some chromosomes (i.e., C1 and C5) carrying five genes, while the rest had fewer genes (e.g. C7). Chromosome 6 did not carry any of the *BoCNGC* genes (Fig. 2a).

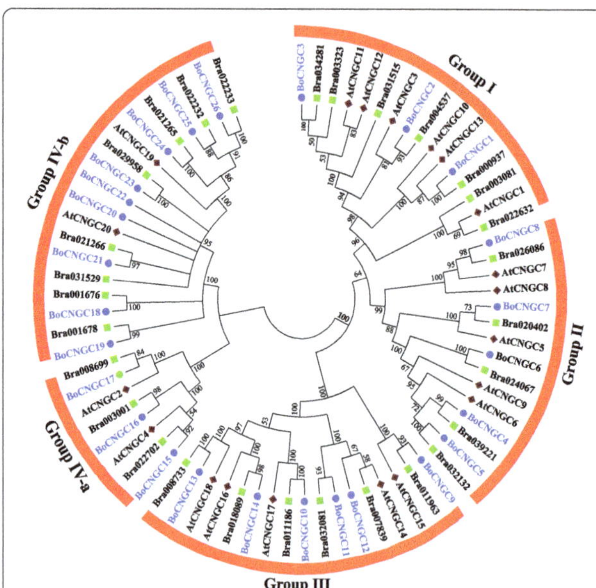

Fig. 1 Phylogenetic tree of *Brassica oleracea*, *Arabidopsis thaliana*, and *Brassica rapa* CNGC proteins. A maximum likelihood phylogenetic tree was created with MEGA 6.0, using the Jones–Taylor–Thornton model. The bootstrap values from 1000 replications are provided at each node. The BoCNGC proteins identified in this study are indicated with blue circles, while the AtCNGCs and BrCNGCs are indicated with maroon diamonds and green rectangles, respectively

Gene duplication events

Gene family expansion occurs via the following three mechanisms: tandem duplication, segmental duplication, and whole-genome duplication [25]. We investigated gene duplication events to clarify the genome expansion mechanism of the *B. oleracea BoCNGC* superfamily. An evaluation of the physical distance between *BoCNGC* gene loci revealed that eight genes (i.e., *BoCNGC18/BoCNGC19*, *BoCNGC21/BoCNGC22/BoCNGC24*, and *BoCNGC20/BoCNGC25/BoCNGC26*) were tandemly duplicated. These genes were detected on C3, C1, and C5, respectively. The data obtained from the Plant Genome Duplication Database revealed that 13 *BoCNGC* genes distributed across the *B. oleracea* genome were associated with segmental duplications (Fig. 2b). The *BoCNGC* gene clusters likely formed via tandem and segmental duplication events may have expanded and enhanced the functional diversity of the gene family.

Comparative syntenic and evolutionary analyses of orthologous *CNGC* gene pairs

The *B. oleracea* and *B. rapa* genomes are currently divided into three sub-genomes, namely LF (least fractionated), MF-I (most fractionated), and MF-II [26]. We

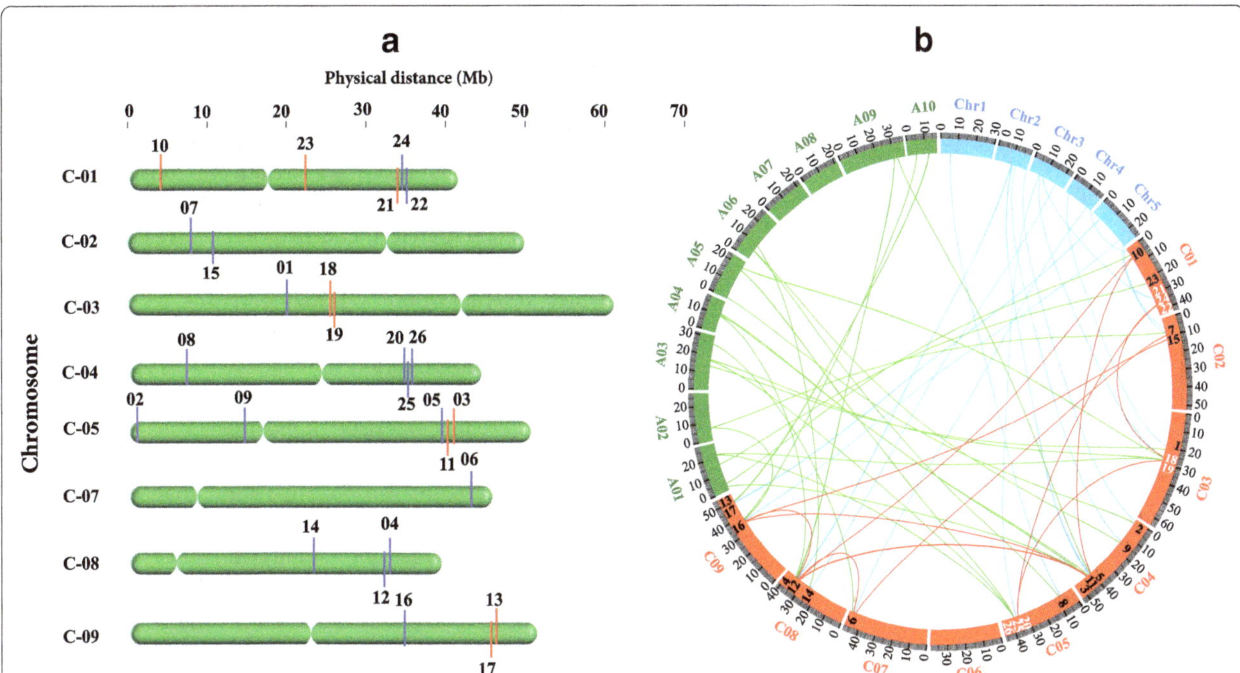

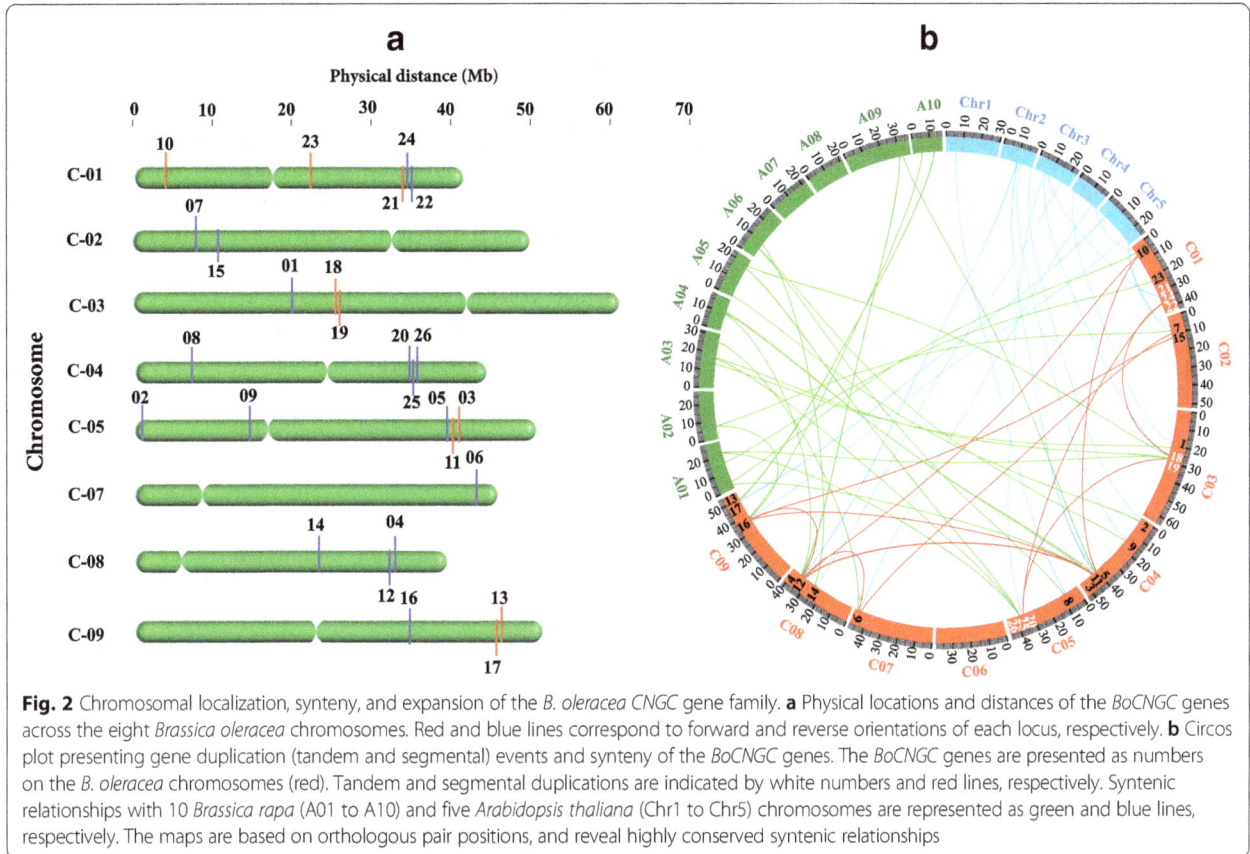

Fig. 2 Chromosomal localization, synteny, and expansion of the *B. oleracea CNGC* gene family. **a** Physical locations and distances of the *BoCNGC* genes across the eight *Brassica oleracea* chromosomes. Red and blue lines correspond to forward and reverse orientations of each locus, respectively. **b** Circos plot presenting gene duplication (tandem and segmental) events and synteny of the *BoCNGC* genes. The *BoCNGC* genes are presented as numbers on the *B. oleracea* chromosomes (red). Tandem and segmental duplications are indicated by white numbers and red lines, respectively. Syntenic relationships with 10 *Brassica rapa* (A01 to A10) and five *Arabidopsis thaliana* (Chr1 to Chr5) chromosomes are represented as green and blue lines, respectively. The maps are based on orthologous pair positions, and reveal highly conserved syntenic relationships

observed that the *B. oleracea* LF sub-genome contains the most *BoCNGC* genes (14), followed by sub-genomes MF-I (10) and MF-II (2) (Additional file 5). Because of a *Brassica*-lineage specific whole-genome triplication (WGT) [27], each *A. thaliana CNGC* gene was expected to generate three *Brassica* copies. However, there were 20 *A. thaliana CNGC* genes, 26 *B. oleracea CNGC* genes, and 29 *B. rapa CNGC* genes. To detect the retention or loss of *CNGC* genes after a WGT event, the syntenic map of *BoCNGC* genes with the model *A. thaliana* and *B. rapa CNGC* genes provided markers for defining the regions of conserved synteny among the three genomes (Fig. 2b). Compared with the ancestral *Brassicaceae* blocks (A to X) in *A. thaliana*, the synteny of 15 *AtCNGC* genes was preserved in *Brassica* species, based on the number of corresponding genes. Ten of the 20 *AtCNGC* genes were retained as a single copy in the equivalent blocks of both *Brassica* species. Three *AtCNGC* genes (i.e., AT2G23980, AT2G24610, and AT5G54250) located on the I and W syntenic blocks, were preserved as two copies in *Brassica* genomes, which were asymmetrically fractionated into three sub-genomes. Two *AtCNGC* genes (i.e., AT3G17690 and AT3G17700) in the F syntenic block were retained as three copies in each species. Two extra gene copies (i.e., *BoCNGC20* and *BoCNGC22*) were located on potential overlap/tandem repeat regions of the *B. oleracea* genome, thus producing phylogenetic cluster IV-b. Approximately 25 *B. oleracea CNGC* genes and 24 *B. rapa CNGC* genes exhibited clear syntenic relationships among the three species. Two gene pairs (i.e., *BoCNGC3* and *BoCNGC23*; Bra034281 and Bra029958) were not part of an *A. thaliana* syntenic block (Additional file 6), suggesting that these genes originated after the divergence from *A. thaliana*. The remaining four *B. rapa* genes were likely generated after the speciation event. In addition, 11 *BoCNGC* genes exhibited strong syntenic relationships with the genes from other plant species, implying this gene family is important for plant growth, development, and stress resistance (Additional file 6).

The orthologous *CNGC* gene pairs between the *B. oleracea* and *A. thaliana* genomes were used to estimate the Ka, Ks, and Ka/Ks values (Table 2). The mean Ka/Ks value of all orthologous gene pairs in the *B. oleracea CNGC* gene family was 1.98. Most of the *BoCNGC* genes had Ka/Ks ratios greater than 1. Additionally, the minimum and maximum Ka/Ks ratios were 1.05 (*BoCNGC26*) and 7.7 (*BoCNGC6*), respectively. These findings indicate that the *BoCNGC* gene family is under positive selection pressure, and might preferentially conserve functions and structures under this selective pressure.

Table 2 Comparative analysis of Ka, Ks and Ka/Ks values for *CNGC* gene pairs between *B. oleracea* compared to *A. thaliana*. Ka/Ks ratio greater than 1 indicates positive selection, a ratio less than 1 indicates functional constraint, and a Ka/Ks ratio equal to 1 indicates neutral selection

A. thaliana genes	*B. oleracea* genes	KA	KS	KA/KS
AtCNGC13	BoCNGC1	0.137	0.032	4.303387233
AtCNGC3	BoCNGC2	0.147	0.029	5.117236906
AtCNGC6	BoCNGC4	0.136	0.052	2.615208996
	BoCNGC5	0.167	0.048	3.468082016
AtCNGC9	BoCNGC6	0.188	0.024	7.768521972
AtCNGC5	BoCNGC7	0.124	0.040	3.098687155
AtCNGC7	BoCNGC8	0.111	0.033	3.40004813
AtCNGC15	BoCNGC9	0.147	0.060	2.456018066
AtCNGC17	BoCNGC10	0.113	0.034	3.357834045
AtCNGC14	BoCNGC11	0.133	0.031	4.245278743
	BoCNGC12	0.145	0.040	3.651430365
AtCNGC18	BoCNGC13	0.094	0.048	1.949453718
AtCNGC16	BoCNGC14	0.111	0.066	1.676233706
AtCNGC4	BoCNGC15	0.101	0.025	4.039237878
	BoCNGC16	0.103	0.034	3.056103924
AtCNGC2	BoCNGC17	0.118	0.029	4.091069466
AtCNGC19	BoCNGC18	0.246	0.126	1.950211367
AtCNGC20	BoCNGC19	0.202	0.178	1.133449904
	BoCNGC20	0.136	0.049	2.79662626
	BoCNGC21	0.099	0.041	2.44174101
AtCNGC19	BoCNGC22	0.202	0.119931313	1.680530313
	BoCNGC24	0.146	0.081	1.792823624
	BoCNGC25	0.146	0.081	1.792823624
AtCNGC20	BoCNGC26	0.131	0.125	1.054276842

Domain architecture and alignment of BoCNGC proteins

Domain composition analyses revealed that BoCNGC proteins contain two primary domains, namely CNBD and TM (Additional file 7). The sequence alignment of 26 BoCNGCs indicated that the two most conserved regions within the CNBD domain are a PBC, and an adjacent hinge region (Fig. 3; Additional file 8). The following highly conserved consensus motif was identified: [LI]-X(2)-[GSE]-X-[VFIY]-X-G-X(0,1)-[DE]-L-L-X-W-X-[LQ]-X(10,20)-S-X-[SAR]-X(7)-[VTI]-E-[AG]-F-X-L. This sequence can be used to classify newly annotated or un-annotated candidate sequences as *Brassica* CNGCs. Additionally, there was a relatively conserved IQ domain and a less conserved CaMBD adjacent to a CNBD present in 24 of the 26 BoCNGC proteins. Two proteins (i.e., BoCNGC18 and BoCNGC19) were observed to lack the CaMBD and IQ domains because their sequences are truncated at the C-terminal end of the CNBD. A high sequence divergence was noted among different groups, particularly between members of Sub-groups IV-a and IV-b. For example, the CaMBD [FLY[–X(10,12)-[AFI]-R-[FY](0,1), was not particularly conserved between Group IV-b and the other groups. However, the IQ motif [IV]-Q-X-X-W-R-X-X-X-[RKQ] was relatively conserved among the BoCNGC proteins (Fig. 3). Alignments between BoCNGCs, AtCNGCs, and BrCNGCs revealed a high sequence divergence at the C-terminal of the CNBD, in which several Group IV-b members lack the CaMBD and IQ motif (Additional files 9 and 10). Overall, our in silico analyses suggest that ion transport and CNBDs along with the PBC and hinge region are conserved in all three species, and are characteristic of plant CNGCs.

Gene structure and motif composition analysis

To characterise the structural diversity of the *BoCNGC* family members, we analysed the exon–intron organization of individual *BoCNGC* genes. The majority of the *BoCNGC* genes from phylogenetic Groups I–III contained six or seven exons, while the Group IV members had 8–11 exons (Fig. 4). Closely clustered *BoCNGC* genes in the same clades were similar regarding the number of exons and intron lengths. Most of the introns in *BoCNGC* genes were phase 0 introns, which occur in between complete codons. Fifty-four phase 2 introns (i.e., located between the second and third nucleotides of a codon) were observed in the *BoCNGC* family, in which the genes carried two phase 2 introns. The exceptions were *BoCNGC1* and *BoCNGC2*, which contained three phase 2 introns. Only the members of phylogenetic Group IV-b had single phase 1 introns at the terminal end of their sequences. A comparison between the exon–intron organizations of *BoCNGC* genes and the *AtCNGC* genes clustered in the same phylogenetic groups revealed several differences (Additional file 11). Most of the phase 1 introns were present in *AtCNGC* genes, implying that intron loss during evolution resulted in a decrease in the number of introns in *BoCNGC* genes, particularly those in Groups I–III and IV-a.

The BoCNGC protein sequences were used for domain or motif structure analyses with the Multiple Expectation Maximization for Motif Elicitation suite [28]. Ten conserved motifs were identified. According to Pfam codes [29] and WebLogo, only seven motifs (i.e., 1–5, 7, and 10) encode domains with known functions (Fig. 4; Additional files 12 and 13). Motif 2 was the biggest motif encoding a conserved domain, which is probably associated with peptidase_C50, putative aminopeptidase, or DNA polymerase III subunit tau_4. Motifs 1 and 5, which encode a CNBD and an ion transport domain, respectively, were conserved among all BoCNGC family members. The ion transport domain had the most motifs, including motifs 4,

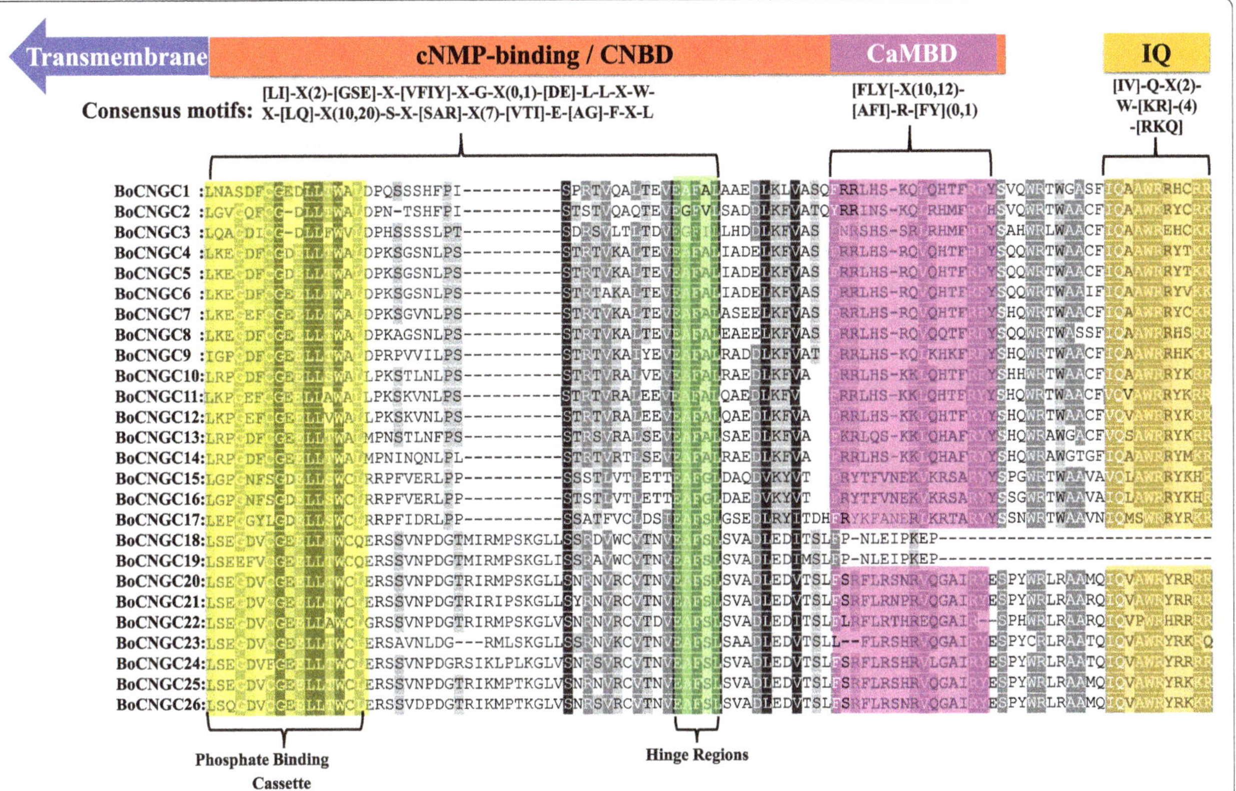

Fig. 3 Multiple sequence alignment of BoCNGC-specific domains, and a three-dimensional model of BoCNGC1. Cartoon model with characteristic CNGC domains provided on top. The BoCNGC-specific consensus motif keys are listed below the cartoon. Amino acids allowed in a specific position are presented in square brackets. X represents any amino acid, while numbers in round brackets indicate the number of amino acids. The multiple sequence alignment of BoCNGC proteins is presented with the CNBD, CaMBD, and IQ domain indicated with different colours. The CNBD domain includes a conserved PBC and hinge region, followed by the CaMBD. Residues shaded in black and grey indicate 100% and >50% similarity among the 26 BoCNGCs

5, 7, and 10. The IQ CaM-binding motif (PF00612) was conserved among BoCNGC family members, with the exception of BoCNGC18, 19, and 22. Group IV proteins contained the fewest functionally annotated motifs, suggesting that the closely related proteins in each group have similar motifs and are also probably functionally similar. The functions of the remaining motifs (i.e., 6, 8, and 9) remain to be determined.

Post-translational modification and phosphorylation of BoCNGC proteins

When BoCNGC protein sequences were analysed using ScanProsite [30], multiple putative phosphorylation sites were revealed. These sites may act as substrates for several kinases, including casein kinase II, protein kinase C, tyrosine kinase, and cAMP/cGMP kinases. Protein kinase C, a family of ten isoenzymes that play a vital role in cellular signal transduction [31], were the most abundant, with 16 sites in BoCNGC4, BoCNGC5, BoCNGC8, and BoCNGC12. Casein kinase II sites, which were the most abundant in Group IV members, are reported to influence different developmental and stress responsive pathways in Arabidopsis [32]. All BoCNGC proteins had multiple N-myristoylation/N-glycosylation motif sites, which are highly conserved compared with the other PTMs. The lipid modification by N-myristoylation might controls the redox disproportions originating from different stresses in plants [33], while glycosylation is crucial for correct growth [34]. The BoCNGC5 and BoCNGC18 proteins contained the most N-myristoylation (11) and N-glycosylation (10) sites, respectively. Other PTM sites, such as those for amidations, tyrosine kinase, serine- and glutamic acid- rich regions, cell attachment sequences, and leucine zipper patterns, were less conserved and randomly distributed (Table 3). Such phosphorylations deliver effective means to regulate most physiological activities, including metabolism, transcription, DNA replication and repair, cell proliferation [35].

Prediction of functional association network of BoCNGC proteins

To explore the relationships among different BoCNGC proteins, a hypothetical protein–protein interaction network was in silico predicted with the STRING program (accessed in April 2016) [36] and AtPID (*Arabidopsis thaliana* Protein Interactome Database), using using

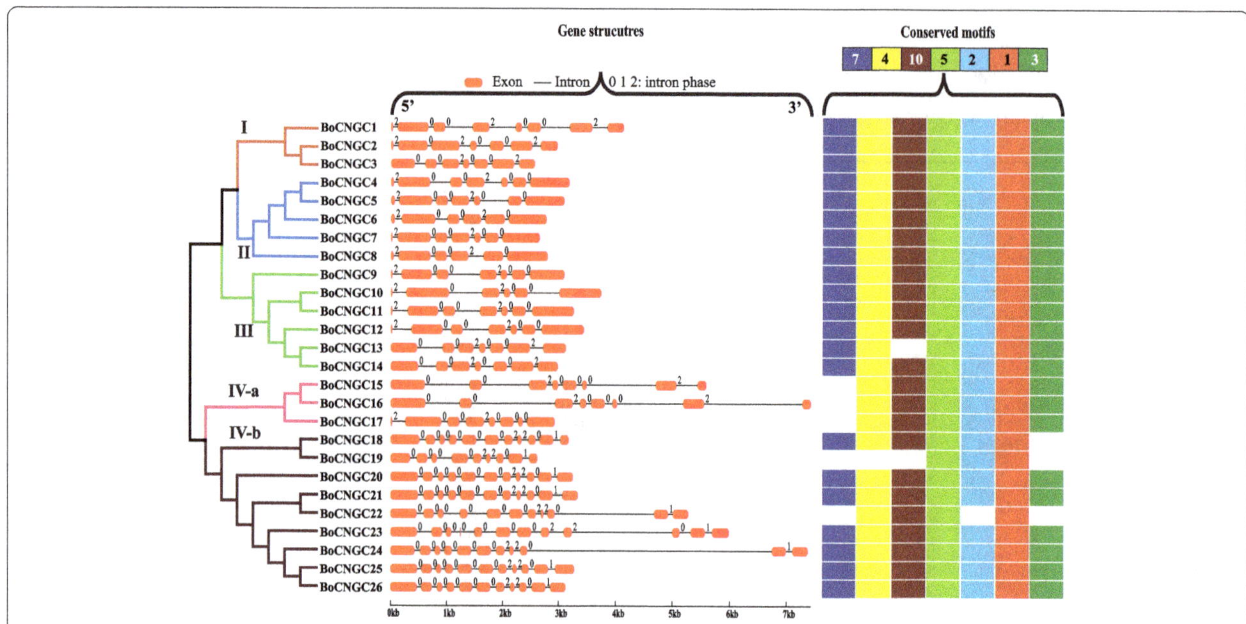

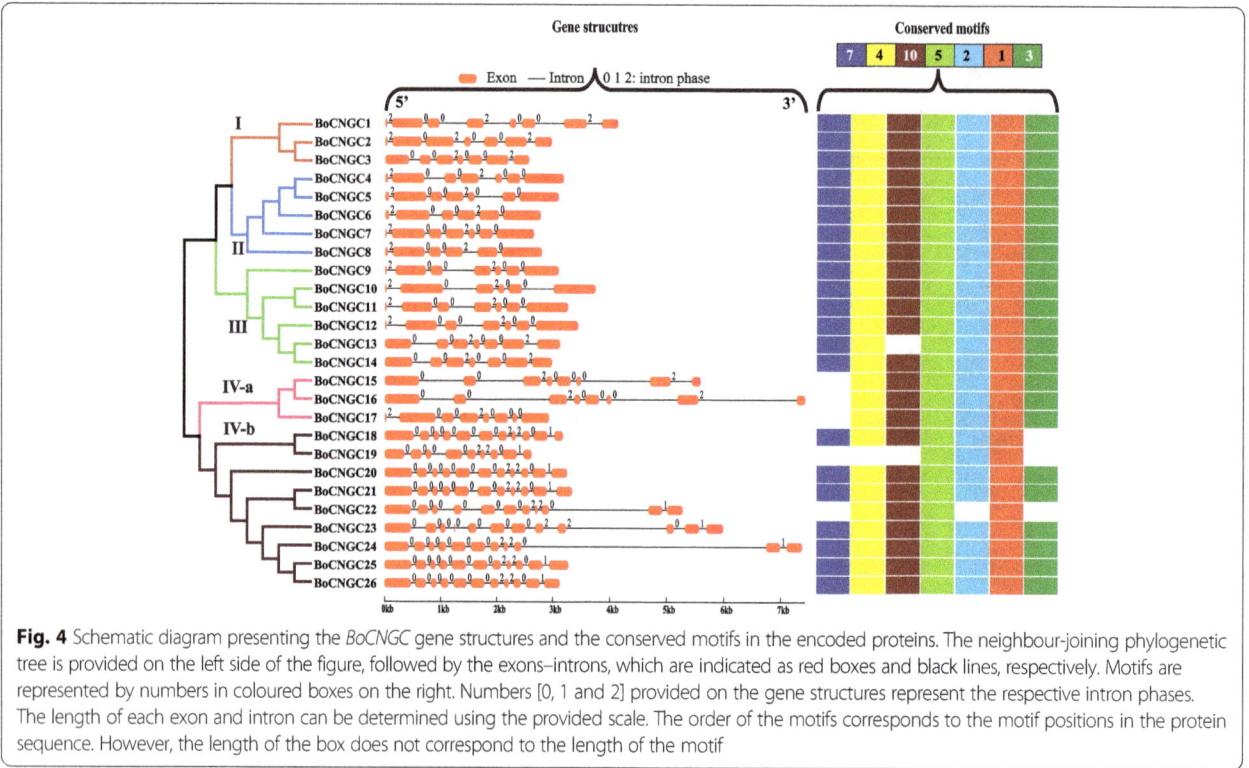

Fig. 4 Schematic diagram presenting the *BoCNGC* gene structures and the conserved motifs in the encoded proteins. The neighbour-joining phylogenetic tree is provided on the left side of the figure, followed by the exons–introns, which are indicated as red boxes and black lines, respectively. Motifs are represented by numbers in coloured boxes on the right. Numbers [0, 1 and 2] provided on the gene structures represent the respective intron phases. The length of each exon and intron can be determined using the provided scale. The order of the motifs corresponds to the motif positions in the protein sequence. However, the length of the box does not correspond to the length of the motif

orthologous AtCNGCs as query. The STRING interaction network for the first shell of interactors of AtCNGC proteins, supported by confidence score, is presented in Fig. 5a. Fourteen AtCNGCs, having 24 orthologs in *B. oleracea*, interact with flagellin-sensitive 2 (i.e., FLS2 or MPL12.8), represented by association in curated databases (confidence score: 0.8). This association was traced to manually curated plant–pathogen interaction pathway imported from the Kyoto Encyclopedia of Genes and Genomes database (Additional file 14). Supported by principal component analysis, a positive interaction (confidence score: 0.154) was observed between BoCNGC10 and BoCNGC13, which are the orthologues of AtCNGC17 and AtCNGC18, respectively. In another interaction network, BoCNGC1 interacts with BoCNGC2 and BoCNGC18–26, which are orthologues of AtCNGC13, 2, 19 and 20 respectively. This interaction is based on protein homology, association in curated human pathways (http://www.reactome.org/), or genes encoding these proteins have correlated expression levels. We also observed that the Group IV proteins are associated with constitutive photomorphogenic 1 and CaM proteins (i.e., CaM4, CaM6, and CaM7) (Fig. 5a).

Using orthologous *Arabidopsis* CNGCs as query in the AtPID uncover more potential interactions between CNGCs, and to other proteins, which are validated by experimental data from different assays (Fig. 5b; Additional file 15). The results exhibited strong interactions of co-expression and gene fusion between CNGC functional partners belonging to similar clades. For example, AtCNGC10 interacted with AtCNGC1, 3 and 13, while AtCNGC17 interacted with AtCNGC18 as mentioned earlier. AtCNGC10 interacted with more CNGCs than other proteins. In addition, some CNGCs (AtCNGC1, 5, 6, 9, 10, 13, 17, 18 and 19) interacted with many important signaling and stress related regulatory proteins, including calmodulins. These interactions are supported by data from yeast two-hybrid, and Affinity Capture-MS assays. Five *CNGC* genes (AtCNGC 1–4, and 11) were found to have available phenotypes of mutant data from seedlings, leaves and embryos, showing that these genes play important roles in hypersensitivity, pathogen and abiotic stress resistance (Additional file 15).

Additional evidence from experimental/biochemical data detected by protein kinase (MI:0424) and anti tag coimmunoprecipitation (MI:0007) assays in human putative homologs (i.e., Potassium voltage-gated channel 2 and Leucine rich repeat containing 47/Per-Arnt-Sim domain kinase) suggest a functional link between CNGCs and FLS2 [37, 38]. The experimental details and LC-MS/MS, yeast two-hybrid and phosphorylation of peptide arrays of human interacting KCNH2 and LRRC47/PASK proteins can be found in supplementary material of Behrends et al. [38]. Using Mating-Based Split Ubiquitin Assays in *A. thaliana*, Chen et al. [39] reported strong, positive (in both 500 μM methionine

Table 3 The number of predicted post-translational modification sites in BoCNGC protein sequences

Protein ID	CAMP	CK2	AMD	PKC	ASN	TYR	MYR	RGD	LEU	SER	GLU	ATP
BoCNGC1		7	2	8	7	1	4					
BoCNGC2	2	6		7	4	1	7					
BoCNGC3	2	7		10	4	1	3					
BoCNGC4	4	3		16	5	1	8					
BoCNGC5	3	4		16	4	1	10					
BoCNGC6	2	8		14	6	1	11					
BoCNGC7	1	5	1	9	5	2	7		1			
BoCNGC8	3	6		16	3	1	7					1
BoCNGC9	1	4		12	4	2	8					
BoCNGC10	1	6		12	4	2	7					
BoCNGC11	1	7		15	2	1	5		3			
BoCNGC12	2	9		16	2	1	5					
BoCNGC13	2	8	1	11	7		9		3			
BoCNGC14	1	8		13	5	1	8					
BoCNGC15	1	12	1	8	4		9		1		1	
BoCNGC16	2	13		10	5		8				1	
BoCNGC17	2	9		6	3	1	6					
BoCNGC18	2	11	1	9	10	1	5		1			
BoCNGC19	1	10		9	7	1	3	1	1			
BoCNGC20		13	1	9	2	1	5					
BoCNGC21		12	1	8	2		7	1				
BoCNGC22		11		6	4		5			1		
BoCNGC23	2	11	1	14	3		7					
BoCNGC24		14		15	6		8					
BoCNGC25		13		13	5		6					
BoCNGC26	1	14		6	7		4					

cAMP/cGMP cAMP/cGMP-binding motif profile, *SER* serine-rich region profile, *GLU* glutamic acid-rich region profile, *CAMP* cAMP- and cGMP-dependent protein kinase phosphorylation site; *CK2* casein kinase II phosphorylation site, *AMD* amidation site, *PKC* protein kinase C phosphorylation site, *ASN* N-glycosylation site, *TYR* tyrosine kinase phosphorylation site, *MYR* N-myristoylation site, *RGD* cell attachment sequence, *LEU* leucine zipper pattern, *ATP* ATP/GTP-binding site motif A (P-loop). Numbers given in each cell refer the total count of PTM sites found in each protein

and at least one 150 μM methionine conditions), and statistically significant interaction between these protein pairs, which are required for polarized tip growth of pollen tube [40]. In another interaction network, BoCNGC1 interacts with BoCNGC2 and BoCNGC18–26, which are orthologues of AtCNGC13, 2, 19 and 20 respectively. Additionally, we observed a weak interaction (confidence score: 0.151) between AtCNGC13 (i.e., orthologues of BoCNGC1) and BRI-associated receptor kinase 1 (BAK1), which was previously observed between AtCNGC17 and BAK1 [41]. Though, it is reported that evidence transfer from one model organism to the other seems feasible approach to study interaction conservation, and it has been implemented in several frameworks already [42]. However, these experimental proofs are essential to support this analysis in *B. oleracea*.

Identification of microRNA target sites

Identifying the targets of the predicted microRNAs (miRNAs) may provide insights into the biological functions of miRNAs influencing plant development, signal transduction, and stress adaptations [43]. We searched for potential miRNA targets in a set of identified *BoCNGC* transcripts using the plant small-RNA target analysis server (psRNATarget) [44]. Using a cut-off threshold of 5 for the search parameters, we identified 14 miRNAs with target sites in 17 *BoCNGC* transcripts, with expectation scores of 1.5–5 (Additional file 16). To decrease the number of false positive predictions, small-RNA/target site pairs with an expectation score and cut-off threshold of 3 were considered. Consequently, five miRNAs with target sites in nine *BoCNGC* genes were identified (Table 4). These miRNAs were localized to the 3′ arm of the stem-loop hairpin structure. Unlike bol-miR838d, which has

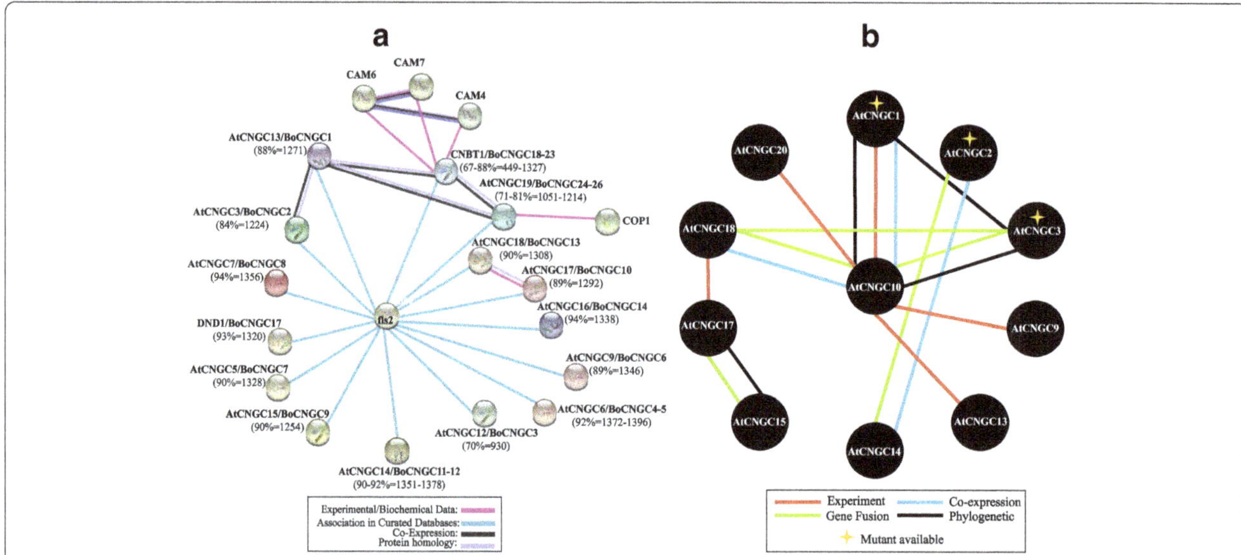

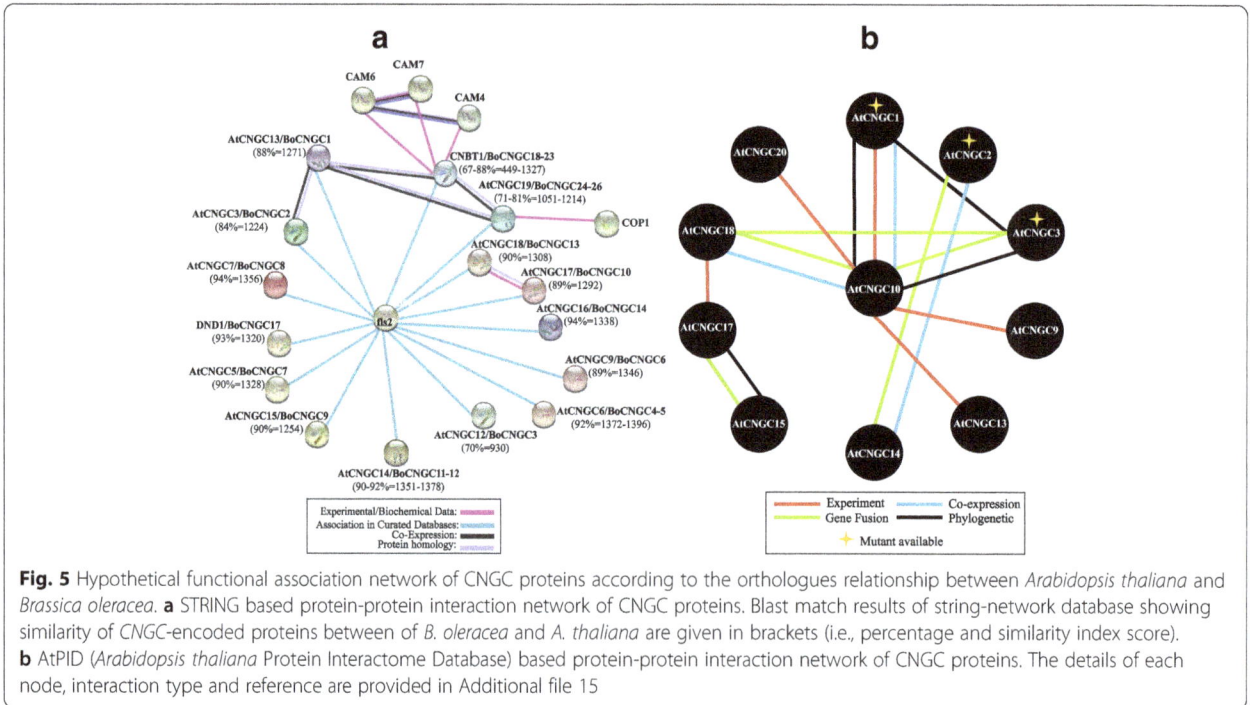

Fig. 5 Hypothetical functional association network of CNGC proteins according to the orthologues relationship between *Arabidopsis thaliana* and *Brassica oleracea*. **a** STRING based protein-protein interaction network of CNGC proteins. Blast match results of string-network database showing similarity of *CNGC*-encoded proteins between of *B. oleracea* and *A. thaliana* are given in brackets (i.e., percentage and similarity index score). **b** AtPID (*Arabidopsis thaliana* Protein Interactome Database) based protein-protein interaction network of CNGC proteins. The details of each node, interaction type and reference are provided in Additional file 15

five target genes, the remaining miRNAs have only one target gene. Moreover, only bol-miR838d has multiple target sites (i.e., complementary regions) on *BoCNGC15* and *BoCNGC16* transcripts. The accessibility of the target site varied from 2.883 (bol-miR838d) to 16.4 (bol-miR5021), where lower values correspond to a greater possibility of contact between the miRNA and target site. Four miRNAs were determined to be involved in cleaving the target transcript, while two miRNAs were predicted to inhibit the translation of target genes.

Gene ontology enrichment analysis

Using Blast2GO (v.3.3.5), we assigned 31 gene ontology (GO) classes to 26 *BoCNGC* genes with BLAST matches to known proteins in the InterPro database. The majority of the genes were assigned to biological process (22), followed by molecular function (7) and cellular components (3). All genes encoded integral membrane components associated with ion channel activity for transmembrane transport. Notably, *BoCNGC1* was associated with salicylic acid biosynthesis, negative regulation of defence responses, regulation of plant-type hypersensitive responses, and responses to chitin. Additionally, *BoCNGC6* was associated with DNA-mediated transformation (Additional file 17).

The level 2 GO enrichment analysis revealed that all 26 BoCNGC proteins are integral cell membrane components, with four proteins (i.e., BoCNGC1, BoCNGC4, BoCNGC5, and BoCNGC17) forming cell parts, and two proteins (i.e., BoCNGC4 and BoCNGC5) forming macromolecular complexes (Additional files 18-a and 19). These proteins are involved in cellular processes associated with transport, binding, and transduction (Additional files 18-b and 19). The biological process category at GO level 2 indicated that BoCNGC1 and BoCNGC17 are associated with cell death and immune responses to stimuli, while another eight CNGCs, including BoCNGC19, are associated with localization (Additional files 18-c and 19). Moreover, we mapped the 26 annotated sequences to reference pathways in the Kyoto Encyclopedia of Genes and Genomes database [45]. Twenty-four of these genes were defined as "cyclic nucleotide gated channels", and assigned to the "plant-pathogen interaction" pathway (Additional files 14 and 20).

Expression patterns in different plant parts

We investigated the steady-state *B. oleracea BoCNGC* expression patterns in seven tissues (i.e., leaf, stem, callus, root, silique, flower, and bud) using Illumina RNA-sequencing data from the Gene Expression Omnibus database. Of the 26 *BoCNGCs*, 19 were expressed at relatively high levels (fragments per kilobase of exon model per million mapped reads value >1) in at least one tissue, including 15 in the roots and siliques, 16 in leaves, and 17 in stems, buds, and flowers. The 19 genes were also expressed in calli (Fig. 6a). Some of the syntenic duplicates have diverged in expression patterns indicating sunfunctionalization. For example, BoCNGC26 and BoCNGC19 have very similar expression patterns. But their

Table 4 Putative microRNA targets predicted in 26 *BoCNGC* transcripts

miRNA Acc.	Target Acc.	Expectation	Target Accessibility	Alignment	Inhibition	Multiplicity
bol-miR5021a/j	BoCNGC4	2.5	16.412	miRNA 20 AGAAGAAGAAGAAGAAGAAU 1 Target 60 CUUUCCUCUUCUUCUUCUUA 79	Cleavage	1
bol-miR838d	BoCNGC5	2.5	15.037	miRNA 20 CUUGUUCUUCUUCUUCUUCU 1 Target 2884 GAAGAGGAAGAAGAGGAGGA 2903	Cleavage	1
bol-miR838d	BoCNGC6	2.0	15.121	miRNA 20 CUUGUUCUUCUUCUUCUUCU 1 Target 2593 GAAGAAGAGGAGGAAGAAGA 2612	Cleavage	1
bol-miR414b	BoCNGC8	3.0	16.071	miRNA 21 ACUUCUACUUCUACUUCUACU 1 Target 2589 UGAAGAUGAGUAUGAUGAUGA 2609	Translation	1
bol-miR838d	BoCNGC10	2.0	8.16	miRNA 22 CACUUGUUCUUCUUCUUCUUCU 1 Target 3520 GUGAUGAAGAAGAAGAAGAAGA 3541	Cleavage	1
bol-miR4234	BoCNGC12	3.0	15.182	miRNA 22 UGACGGUUGAUCAAAAUUCAAC 1 Target 2630 AUUUUCAAUUGGUUUUGAGUUG 2651	Cleavage	1
bol-miR838d	BoCNGC15	1.5	6.045	miRNA 20 CUUGUUCUUCUUCUUCUUCU 1 Target 103 GAAGAAGAGGAAGAAGAAGA 122	Cleavage	2
bol-miR838d	BoCNGC15	2.5	8.952	miRNA 21 ACUUGUUCUUCUUCUUCUUCU 1 Target 135 UGAGAAAGAUGAAGAAGAAGA 155	Cleavage	2
bol-miR838d	BoCNGC16	3.0	6.192	miRNA 20 CUUGUUCUUCUUCUUCUUCU 1 Target 94 GAAGAGGAGGACGAAGAAGA 113	Translation	2
bol-miR838d	BoCNGC16	1.5	2.883	miRNA 20 CUUGUUCUUCUUCUUCUUCU 1 Target 124 GAACAAGAGGAAGAGGAGGA 143	Cleavage	2
bol-miR_new2	BoCNGC26	3.0	11.462	miRNA 20 UGGGAUUUAGUAUUUAGGAU 1 Target 639 ACCUGGAAUCAUAAAUCCUC 658	Cleavage	1

duplicates BoCNGC21 and BoCNGC20 now have different expression patterns. An additional investigation revealed that *BoCNGC17* and *BoCNGC16* were the most highly expressed genes, especially in flowers, implying they may be important for *Brassica* species development. Among the other genes, *BoCNGC3* was highly expressed in roots, while *BoCNGC2* was highly expressed in siliques and calli, suggesting that the expression of this genes is induced by wounding. Most of the Group III and IV genes were expressed at low levels in the leaves, stems, calli, roots, and siliques, while *BoCNGC26* was not expressed in any tissue.

A review of the reported expression profiles of orthologus *Arabidopsis CNGCs* in the tissues of wild and mutant plants suggest that a) the mRNAs of this gene family are expressed in all plant tissues, b) expression in leaves is greater than in roots, stem and flower, c) group-I, II and IV *CNGCs* are highly expressed in flowers and apex of *Arabidopsis* mutants (Additional file 21) [46]. Some of these observations have been confirmed during earlier investigation of *CNGC* mutants in *Arabidopsis* plants, for example *AtCNGC1* [47]. Moreover, the expression patterns of *BoCNGC1* and *BoCNGC7* were consistent with their orthologs (*ATCNGC10* and *ATCNGC5*), which are highly expressed in roots than leaves [7]. Our results are also corroborated by the findings of Borsics et al. [6], showing that *AtCNGC10* mutant plants exhibited reduced mRNA levels in flower than its closest related member *AtCNGC13* and WT plants.

Expression patterns in response to abiotic and biotic stresses

Based on the *BoCNGC* expression patterns in different tissues, we attempted to determine whether these genes were associated with plant defence responses, especially against race- and species-specific *Brassica* pathogens. Therefore, we analysed the *BoCNGC* expression profiles in the shoots of 25-day-old *Brassica* plants infiltrated with *Xanthomonas campestris* pv. *campestris* (Xcc). The *BoCNGC* expression levels at 24 h post-inoculation are presented in Fig. 6b. The pathogen induced considerable changes to *BoCNGC*

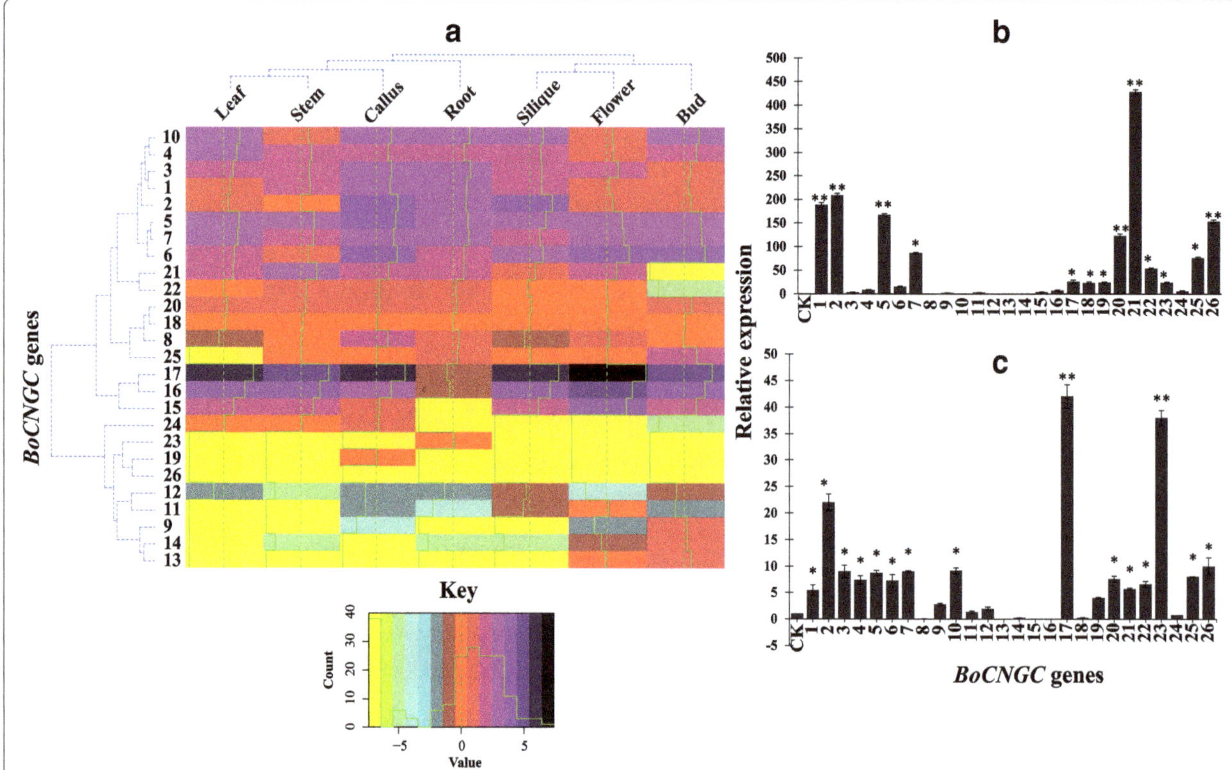

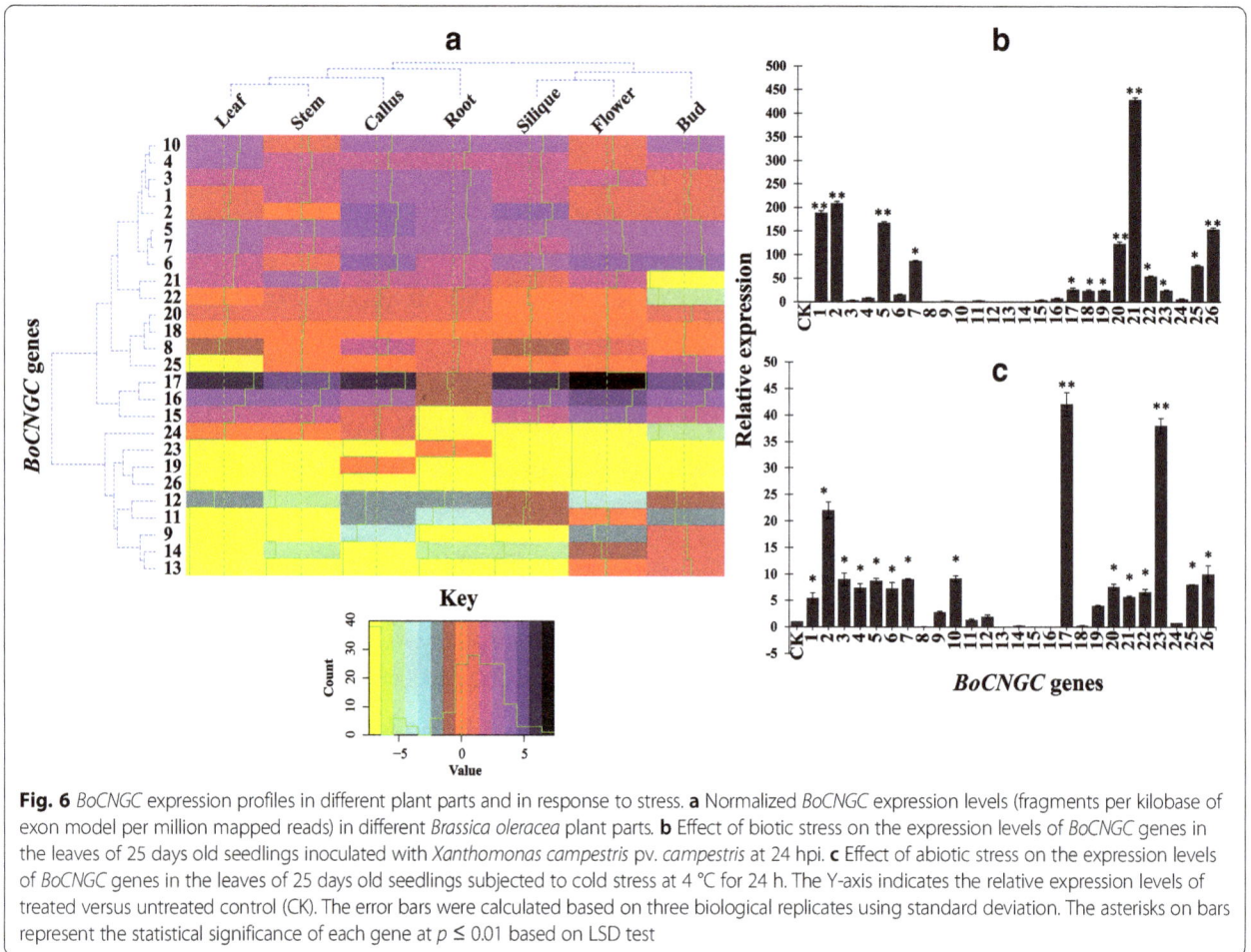

Fig. 6 *BoCNGC* expression profiles in different plant parts and in response to stress. **a** Normalized *BoCNGC* expression levels (fragments per kilobase of exon model per million mapped reads) in different *Brassica oleracea* plant parts. **b** Effect of biotic stress on the expression levels of *BoCNGC* genes in the leaves of 25 days old seedlings inoculated with *Xanthomonas campestris* pv. *campestris* at 24 hpi. **c** Effect of abiotic stress on the expression levels of *BoCNGC* genes in the leaves of 25 days old seedlings subjected to cold stress at 4 °C for 24 h. The Y-axis indicates the relative expression levels of treated versus untreated control (CK). The error bars were calculated based on three biological replicates using standard deviation. The asterisks on bars represent the statistical significance of each gene at $p \leq 0.01$ based on LSD test

expression levels, including the up-regulation of the expression of 10 *BoCNGC* genes in infiltrated seedlings, with the highest levels observed for *BoCNGC21*. This was followed by *BoCNGC2* and *BoCNGC1* from Group I, *BoCNGC5* and *BoCNGC7* from Group II, and *BoCNGC26* and *BoCNGC20* from Group IV-b. Interestingly, none of the Group III and IV-a genes were affected.

We also examined the *BoCNGC* expression levels under cold conditions. The expression of 13 of the 26 *BoCNGC* genes was up-regulated in cold-stressed plants, although the expression levels were lower than the levels induced by Xcc (i.e., biotic stress) (Fig. 6c). The expression levels of genes from Groups I, II, and IV were significantly induced by cold stress, with the highest levels observed for *BoCNGC17* and *BoCNGC23*. In contrast, the Group III *BoCNGCs* were expressed at low levels or not at all under cold conditions. Moreover, most of the duplicated gene pairs and genes encoding interacting proteins produced similar expression patterns (especially in response to Xcc). The exception was *BoCNGC24* whose expression was not significantly up-regulated like its duplicates (i.e., *BoCNGC21* and *BoCNGC22*).

The expression patterns of many *BoCNGCs* under pathogen stress were consistent with the expression patterns of their *Arabidopsis* orthologs obtained from the AtGenExpress project (Additional file 22) [46]. The involvement of group-IV CNGCs in disease resistance and hyper-sensitivity has been documented earlier [21, 22]. However, the cumulative profiles of group-I and IV CNGCs in *Arabidopsis* seedlings showed apposite trend of down-regulation by cold stress at 4 °C for 24 h, showing specie-specific divergence of expression pattern.

Discussion

The *CNGC* gene family has been reported for many agriculturally important plants [17, 18, 20]. However, a genome-wide identification and annotation of *CNGC* genes has not been reported for *B. oleracea*. In this study, we identified 26 *B. oleracea CNGC* genes, and determined that the *BoCNGC* gene family is larger than the *CNGC* families of most of the reported crops [4]. The isoelectric point (*pI*) and charge of a protein is important for solubility, subcellular localization, and interaction, depending on both insertion and deletions between orthologs, and the ecology of the organism [48]. It is reported that proteins in cytoplasm possess an acidic *pI* (*pI* < 7.4), nuclear proteins have more neutral *pI* (7.4 < *pI* < 8.1), while those in membrane have more basic *pI* [48], where basic residues located on either side of membrane spanning region play a role in the stabilization of the protein in membrane [49]. The net charge of a protein is a fundamental physical property, and its value directly influences the solubility, aggregation, and crystallization of the protein [50]. The 26 BoCNGCs were localized to membranes, greatly varied in physicochemical properties, and will theoretically participate in basic buffers. These variations reflects the changes in protein composition, and their effects on association of receptors with charged ligands, folding and stability, solubilization and precipitation, and selective transport of ions in protein channels [50].

Homologous genes within the same taxonomic group are assumed to exhibit similar structural, functional, and evolutionary properties, which may help clarify the role(s) of *B. oleracea CNGC* genes. Because of the close relationship between *B. oleracea* and *A. thaliana*, the *BoCNGC* genes were highly similar (>90%) to the corresponding *AtCNGC* genes regarding plant CNGC-specific domains, amino acid compositions, gene structures, and phylogenetic classifications. Interestingly, we revealed the absence of the CaMBD and IQ domain in BoCNGC18 and BoCNGC19, which raised the possibility that these were abnormal CNGC proteins. However, we found that many of their homologs in *A. thaliana*, pear and *B. rapa* reportedly lack the CaMBD [18]. Similar to other CNGCs, these proteins have regular 3D structural and membrane topologies, with conserved binding sites for cGMP/cAMP. Furthermore, the presence of conserved nickel- and zinc-binding sites suggests that BoCNGC18 and BoCNGC19 may have lost their secondary domains during evolution, but gained functional diversity. Additional research is required to clarify this point.

Proteins undergo post-translational modifications (PTMs), which increase the range of their functions through different mechanisms [51]. The associated PTMs likely affected protein function, localization, and stability, as well as the dynamic interactions with other molecules [52]. Following gene annotations and phylogenetic analyses, we predicted the presence of multiple PTM sites in BoCNGCs. Apart from evolutionarily conserved PTMs, other types of modification sites were detected in BoCNGCs, which diversified the functions and underlying mechanisms of CNGC-specific PTMs. Protein–protein interaction networks provide a base for systematic understanding of cellular processes that can be used to filter and assess the functional genomics data and provide an instinctive platform to annotate the structures, functions and evolutionary properties of proteins [53]. Using two different approaches, and orthologous *Arabidopsis* CNGCs as a reference, we found that most CNGCs were associated with various protein–protein interaction networks involving CNGCs and other proteins related to light signalling [54], regulation of enzyme activities [55] and cellular processes [56], brassinosteroid signal transduction [57], and resistance against pathogens [58]. These aanalyses can offer new information for future experimental research and provide cross-species predictions for efficient interaction mapping [53]. Additionally, of the 26 *BoCNGC* genes, nine included target sites for diverse groups of novel and conserved miRNAs. These miRNA families are highly conserved in *Brassicaceae* species, where they are expressed in leaves, siliques, and flowers. These miRNAs are reported to function in regulation of genes related to growth (miR157/171/824) [59], *Brassica*-specific stomatal organization (miR824), pollen development (miR824) [60], abiotic stress tolerance, and plant–pathogen interactions (miR398) [61].

Gene duplications during evolution increase the genomic content and expand gene functions to optimise the adaptability of plants [25]. *Brassica oleracea* is an ancient polyploid, whose genome underwent a WGT event approximately 16 million years ago, after diverging from *A. thaliana*, followed by large-scale chromosomal rearrangements (i.e., re-diploidisation). As a member of the classical triangle of U [62], the assembled genome of *B. oleracea* (540 Mb) is larger than that of its sister species, *B. rapa* (312 Mb) [63] that diverged from a common ancestor nearly 4 million years ago [64]. The less number of *CNGC* genes in *Brassica* genomes suggest that most of the duplicated gene copies were lost post-polyploidization. Reversion of the few duplicated *CNGC* genes to single copy might be due to neutral loss of unnecessary duplicates over time. Another possible explanation could be that CNGC proteins participate in dosage sensitive interactions that is affected by the copy number of each protein subunit (gene balance hypothesis) [24]. Synteny analysis revealed that more than 80% of the *BoCNGC* genes are located in conserved syntenic blocks, which lost and gained some genes. These results are consistent with the findings of Liang et al. [65]. We presume that functionally redundant gene copies are reportedly lost after genome duplication events, while some copies of functionally important genes are kept [51]. Our findings suggest that the WGT and segmental duplication events were important for the expansion of the *B. oleracea CNGC* family, where tandem duplications only affected the expansion of Group IV-b. Altogether, the conservation of *CNGC* genes after substantial genome reshuffling suggests that these genes are crucial for plant development [66]. Finally, the detailed analyses of gene expression in different tissues and under stress conditions further supported the importance of various *CNGC* genes for *B. oleracea* growth, development, and survival. To the best of our knowledge, this manuscript is the first to describe a comprehensive and systematic analysis of the *B. oleracea CNGC* gene family. The generated data may be useful for constructing protein–protein interaction networks and experimentally validating the miRNA targets, which regulate the development of *B. oleracea*. Besides, our results might help in understanding the functions of BoCNGCs related to the regulation of signal transduction pathways, and elucidate the expression profiles of the corresponding genes during plant development and stress responses. The results of the bioinformatics and comparative genomic analyses are also valuable for studying CNGC protein functions, with potential implications for the economic, agronomic, and ecological enhancement of *B. oleracea* and other *Brassica* species.

Conclusions

In conclusion, this work is the first comprehensive and systematic analyses of *CNGC* gene family in *B. oleracea*. There are 26 *CNGC* genes in *B. oleracea*, which are classified into 4 groups (I-IV) and fractionated into three subgenomes; this gene family appears to have expanded through WGT, segmental and tandem duplication events; the BoCNGC gene family is under positive selection pressure. All the BoCNGC protein sequences contain a CNGC specific domain CNBD that comprises a PBC and a "hinge" region, featured by a stringent motif: LI]-X(2)-[GSE]-X-[VFIY]-X-G-X(0,1)-[DE]-L-L-X-W-X-[LQ]-X(10, 20)-S-X-[SAR]-X(7)-[VTI]-E-[AG]-F-X-L. This study provided comprehensive information about domain structure, exon-intron structure, and the phylogenetic tree and expression analysis of *CNGC* genes in Chinese cabbage. These data are useful to construct protein-protein interaction network and experimentally validate the miRNA targets, which regulates and induces multiple responses in *B. oleracea*. The bioinformatics analysis and comparative genomic analysis also provides valuable information in the study of CNGC protein functions for the improvement of the economic, agronomic, and ecological benefits of Chinese cabbage. Furthermore, this study assists to elucidate the functions of differentially expressed candidate genes in the regulation of signal transduction pathway, plant development and stress resistance in *B. oleracea*.

Methods

Identification of *Brassica oleracea CNGC* genes

To identify the *B. oleracea CNGC* genes, 20 *Arabidopsis* CNGC protein sequences obtained from TAIR10 (https://www.arabidopsis.org/) [67] were used as queries to perform a homology-based search of the Ensembl Plants database (genome version v.2.1) [68]. This search was conducted with the default parameters of the BLASTP program. All non-redundant protein sequences were retrieved, and their domains were analysed with online servers: Simple Modular Architecture Research Tool (SMART) (http://smart.embl-heidelberg.de/) [69] and the Conserved Domains Database (CDD) (http://www.ncbi.nlm.nih.gov/Structure/cdd/wrpsb.cgi) [70]. The analyses were completed with the default cut-off parameters. Sequences containing the cNMP/CNBD (IPR000595) and transmembrane/ion transport protein (PF00520) domains as well as a plant CNGC-specific motif in the PBC and hinge region within the CNBD were recognized as CNGC proteins. The identified *BoCNGC* genes were named according to their positions in the phylogenetic tree.

Protein characterisation and amino acid properties

Details regarding gene and protein lengths as well as chromosomal locations were obtained from the Ensembl Plants database. Amino acid properties, including charge, molecular weight (kDa), aliphatic and instability indices, isoelectric points (pI), and grand average of hydropathy (GRAVY), were determined using the online available ProtParam tool (http://web.expasy.org/protparam/) [71]. The PTM sites were predicted with the ScanProsite web server (http://prosite.expasy.org/scanprosite/) [30].

Multiple sequence alignment and phylogenetic analysis

The identified CNGC proteins were aligned using the default settings of the ClustalX 2.0 program [72]. The conserved CNGC-specific domains were manually checked and shaded with the DNAMAN program (version 6.0.3.40; Lynnon Corporation, Quebec, Canada). The BoCNGC protein sequences were also aligned with CNGC sequences from *A. thaliana* and *B. rapa* (downloaded from the *Brassica* database; http://brassicadb.org/brad/) [73] using the default settings of the ClustalX 2.0 program. The alignments were viewed with the GeneDoc program [74]. A phylogenetic tree was constructed using the maximum likelihood method of MEGA 6.0 (1000 bootstrap replications) [75].

Chromosomal locations and gene duplication events

Details regarding the chromosomal locations of the *BoCNGC* genes were obtained from the Ensembl Plants database. The Plant Genome Duplication Database [76] was searched to identify segmentally duplicated genes. *BoCNGC* genes were defined as tandemly duplicated if the distance between the homologous loci was <50 kb [65]. The syntenic relationships among *BoCNGCs*, *AtCNGCs*, and *BrCNGCs* were evaluated using the Search Syntenic Genes tool in Bolbase [77].

Gene structure, motif composition, and prediction of three-dimensional models

Gene exon/intron structures were predicted with the Gene Structure Display Server (version 2.0) [78], with genomic and coding sequences as the input data. The conserved motifs in the CNGC sequences were identified using the Multiple Expectation Maximization for Motif Elicitation suite and the Motif Alignment and Search Tool [28] with the following parameters: optimal motif width: 6–200; maximum number of different motifs: 10. The detected motifs were annotated with Pfam [29]. Gene ontology enrichment analysis was performed using Blast2GO (v.3.3.5) [79].

Analysis of microRNA target sites and protein–protein interactions

The *B. oleracea* miRNA sequences obtained from the miRBase database at http://mirbase.org/ [80]. To detect potential miRNA target sites within the *BoCNGC* genes, the obtained miRNAs were analysed with the psRNATarget server (http://plantgrn.noble.org/psRNATarget/) [44] The information about protein-protein interaction, and available mutant information for *Arabidopsis CNGC*-encoded proteins was obtained from STRING (v10) [36] and AtPID (http://www.megabionet.org/atpid/webfile/query.php).

Analysis of *BoCNGC* transcriptome data

To investigate the *BoCNGC* expression profiles, we used the Illumina RNA-sequencing data available in the Gene Expression Omnibus database (accession number GSE42891) [24]. Transcript abundance was calculated as fragments per kilobase of exon model per million mapped reads, and the resulting values were $\log_2$ transformed. A hierarchical cluster was created and a heat map was generated with R language program [81].

Experimental conditions and quantitative real-time polymerase chain reaction assay

We used a quantitative real-time polymerase chain reaction (qRT-PCR) to quantify the *BoCNGC* expression levels in response to biotic (bacterial pathogen) and abiotic (cold) stresses. Cabbage (*B. oleracea* var. *capitata* L.) seedlings were grown for 25 days in a greenhouse at 23 ± 2 °C under natural light. For the cold stress treatment, seedlings were incubated at 4 °C for 24 h. For the bacterial infection, Xcc was first cultured in medium B [82] at 28 °C. Cells were collected by centrifugation, re-suspended in sterilized distilled water, and adjusted to an optical density at 600 nm of 0.1. The midvein of the first fully opened leaf (just above the petiole) was inoculated with the Xcc suspension using a 1-ml syringe. Sterilized ddH_2O was used as the control solution. The treated plants were returned to the greenhouse and sampled 24 h later. The extraction of RNA and synthesis of cDNA were completed as previously described [20]. Gene-specific primers were designed with Primer 5.0 (Additional file 23). The qRT-PCR was conducted using a StepOne Real-Time PCR System (Applied Biosystems, USA) and SYBR Premix Ex Taq reagents (TAKARA, Japan) as described by Kabouw et al. [83]. Finally, the $2^{-\Delta\Delta Ct}$ method [84] was used to calculate the relative gene expression values, which were subsequently transformed to $\log_2$- expression ratios and plotted in figures. Each experiment was performed with three technical replicates. The *Actin* gene (AF044573) was used as an endogenous control.

Statistical analysis

The RT-qPCR expression data was subjected to analysis of variance (ANOVA) using computer statistical package (SAS software SAS Institute, Cary, NC). Least significant difference (LSD) test at $p \leq 0.01$ was used to check the significant differences between the expression levels of different *BoCNGC* genes compared to control.

Additional files

Additional file 1: List of truncated gene accessions discarded during preliminary investigation. (XLSX 9 kb)

Additional file 2: Multiple sequence alignment of CNGC proteins from *B. oleracea*, *B. rapa* and *A. thaliana*. (PDF 2222 kb)

Additional file 3: Phylogenetic tree of CNGC proteins from *B. oleracea* (encoded by *BoCNGCs*). A multiple sequence alignment was performed using ClustalX 2.0 program with default settings. Maximum likelihood (ML) tree was create with MEGA 6.0, under the Jones-Taylor-Thornton (JTT) model. The bootstrap values from 1000 resampling are given at each node. (PDF 201 kb)

Additional file 4: Phylogenetic tree of *CNGC* genes from *Arabidopsis* (*AtCNGCs*). A multiple sequence alignment was performed using ClustalX 2.0 program with default settings. Maximum likelihood (ML) tree was create with MEGA 6.0, under the Jones-Taylor-Thornton (JTT) model. The bootstrap values from 1000 resampling are given at each node. (PDF 180 kb)

Additional file 5: Syntenic ancestral block structure between *A. thaliana* and three sub-genomes of *B. oleracea* and *B. rapa*. (XLSX 10 kb)

Additional file 6: Synteny of *BoCNGC* in other plant species. (XLSX 18 kb)

Additional file 7: Primary domain architecture of BoCNGC proteins. Information about domain annotation is obtained from SMART database. (PDF 339 kb)

Additional file 8: Multiple sequence alignment of BoCNGC proteins. Multiple sequence alignment was performed by clustalX2and viewed by GeneDoc software package. (PDF 767 kb)

Additional file 9: Multiple sequence alignment of *CNGC*-encoded proteins of *Arabidopsis* and *B. oleracea*. Multiple sequence alignment was performed by clustal X2 and viewed by GeneDoc software package. (PDF 1208 kb)

Additional file 10: Multiple sequence alignment of CNGC-encoded proteins of *B. oleracea* and *B. rapa*. Multiple sequence alignment was performed by clustal X2 and viewed by GeneDoc software package. (PDF 1436 kb)

Additional file 11: Schematic diagram showing the structures of *Arabidopsis CNGC* family genes. The exons-introns indicated as red boxes and black lines respectively, and the intron phases are displayed as numbers [0, 1 and 2]. The lengths of each exon and intron can be mapped to the scale given in the bottom. (PDF 154 kb)

Additional file 12: Functional annotation of the identified conserved MEME motifs. (XLSX 10 kb)

Additional file 13: Web logos of MEME-identified conserved functional motifs in BoCNGC proteins. The heights of the amino acids indicates the degree of conservation. (PDF 423 kb)

Additional file 14: KO pathway associated with plant-pathogen interaction (K05391). The pathway map was obtained from http://www.kegg.jp/kegg/kegg1.html. Details of *BoCNGC* genes allocated to K05391 pathway are given in Additional file 20. (PDF 126 kb)

Additional file 15: Table showing the details of protein-protein interaction, and available mutant information for *Arabidopsis CNGC*-encoded proteins. The information was obtained from AtPID. (XLSX 37 kb)

Additional file 16: The potential miRNA targets in the set of 26 *BoCNGC* transcripts using cut-off threshold of 5 in the search parameters. (XLSX 14 kb)

Additional file 17: GO term enrichment analysis of *BoCNGC* genes for Molecular function (MF), Biological process (BP) and Cellular component (CC). (XLSX 17 kb)

Additional file 18: Distribution of *BoCNGC* genes in major functional terms (GO terms Level 2) for categories Molecular Function **(a)**, Biological Process **(b)** and Cellular Component **(c)**. The details are given in Additional file 19. (PDF 97 kb)

Additional file 19: GO term enrichment analysis at level 2 for category: P: Biological process, F: Molecular function and C: Cellular component. (XLSX 9 kb)

Additional file 20: Reference KO pathway associated with *BoCNGC* genes. The pathway map was obtained from http://www.kegg.jp/kegg/kegg1.html. (XLSX 10 kb)

Additional file 21: Cumulative values of expression for *Arabidopsis CNGC* genes in different developmental samples. The expression data for 21 days old of wild type and mutant plants was obtained from Schmid et al. [48]. The information about different genotype mutants is given below the figures. (PDF 225 kb)

Additional file 22: Cumulative values of expression for *Arabidopsis CNGC* genes in response to pathogen (biotic) and cold (Abiotic) stress. The expression data for 21 days old of wild type and mutant plants was obtained from Schmid et al. [48]. (PDF 294 kb)

Additional file 23: List of primers used for gene expression via qRT-PCR. (XLSX 12 kb)

Abbreviations

AMD: Amidation; ASN: N-glycosylation; CaM: calmodulin; CaMBD: CaM-binding domain; cAMP: cyclic adenosine monophosphate; cGMP: cyclic guanosine monophosphate; CK2: Casein kinase II; CNBD: cNMP-binding domain; CNGC: Cyclic nucleotide-gated ion channel; IQ: Isoleucine-glutamine; IT: Ion transport; LEU: Leucine zipper pattern; MYR: N-myristoylation; PBC: Phosphate-binding cassette; p*I*: Isoelectric point; PKC: Protein kinase C; PTM: Post-translational modification; RGD: Cell attachment sequence; TYR: Tyrosine kinase; WGD: Whole-genome duplication; Xcc: *Xanthomonas campestris* pv. *campestris*

Acknowledgements

We thank Prof. Qing-yao Shu for his critical inputs and assistance during this study.

Funding

This research was financially supported by the Ministry of Science and Technology, China (grant No.: SQ2015IM3600010). The funders had no role in study design, data collection and analysis, decision to publish, or preparation of the manuscript.

Authors' contributions

KUK and ZN designed the study and conceptualised the methodology. KK and EA collected the data and completed the bioinformatics analyses along with KUK and ZN. RU and AA conducted the qRT-PCR experiments with assistance from KK. KUK and ZN analysed the data and wrote the manuscript with critical inputs from X-LR and Q-YS. X-LR and Q-YS revised the whole manuscript according to suggestions by referees, and supervised this study. All authors reviewed the manuscript at every stage. All authors read and approved the final manuscript.

Competing interests

The authors declare that they have no competing interests.

Author details

[1]State Key Laboratory of Rice Biology, Institute of Crop Science, Zhejiang University, Hangzhou 310058, China. [2]Molecular Genetics Key Laboratory of China Tobacco, Guizhou Academy of Tobacco Science, Guiyang 550081, China. [3]Wuxi Hupper Bioseed Technology Academy Ltd., Wuxi 214000, China. [4]Department of Biotechnology, BUITEMS, Quetta, Pakistan. [5]Department of Biological sciences, College of Education and Science, Albaydaa University, Rada'a, Yemen. [6]Department of Environmental Sciences, Quaid –i- Azam University, Islamabad, Pakistan. [7]Institute of Crop Sciences, Zhejiang University, 866 Yuhangtang Road, Hangzhou 310029, China. [8]Guizhou Academy of Tobacco Science, Longtanba Road No. 29, Guanshanhu District, Guiyang (550081), Guizhou, People's Republic of China.

References

1. Wu M, Li Y, Chen D, Liu H, Zhu D, Xiang Y. Genome-wide identification and expression analysis of the IQD gene family in moso bamboo (Phyllostachys Edulis). Sci Rep. 2016;6:24520.
2. Takáč T, Vadovič P, Pechan T, Luptovčiak I, Šamajová O, Šamaj J. Comparative proteomic study of Arabidopsis mutants mpk4 and mpk6. Sci Rep. 2016;6:28306.
3. DeFalco TA, Marshall CB, Munro K, Kang H-G, Moeder W, Ikura M, Snedden WA, Yoshioka K. Multiple Calmodulin-binding sites positively and negatively regulate Arabidopsis Cyclic Nucleotide-gated Channel12. Plant Cell. 2016;28(7): 1738–51.
4. Saand MA, Xu Y-P, Munyampundu J-P, Li W, Zhang X-R, Cai X-Z. Phylogeny and evolution of plant cyclic nucleotide-gated ion channel (CNGC) gene family and functional analyses of tomato CNGCs. DNA Res. 2015;22(6):471–83.
5. Mäser P, Thomine S, Schroeder JI, Ward JM, Hirschi K, Sze H, Talke IN, Amtmann A, Maathuis FJM, Sanders D. Phylogenetic relationships within cation transporter families of Arabidopsis. Plant Physiol. 2001;126(4): 1646–67.
6. Borsics T, Webb D, Andeme-Ondzighi C, Staehelin LA, Christopher DA. The cyclic nucleotide-gated calmodulin-binding channel AtCNGC10 localizes to the plasma membrane and influences numerous growth responses and starch accumulation in Arabidopsis Thaliana. Planta. 2007;225(3):563–73.
7. Christopher DA, Borsics T, Yuen CY, Ullmer W, Andème-Ondzighi C, Andres MA, Kang B-H, Staehelin LA. The cyclic nucleotide gated cation channel AtCNGC10 traffics from the ER via Golgi vesicles to the plasma membrane of Arabidopsis root and leaf cells. BMC Plant Biol. 2007;7(1):48.
8. Yuen CC, Christopher DA. The group IV-A cyclic nucleotide-gated channels, CNGC19 and CNGC20, localize to the vacuole membrane in Arabidopsis Thaliana. AoB Plants. 2013;5:plt012.
9. Charpentier M, Sun J, Martins TV, Radhakrishnan GV, Findlay K, Soumpourou E, Thouin J, Véry A-A, Sanders D, Morris RJ. Nuclear-localized cyclic nucleotide–gated channels mediate symbiotic calcium oscillations. Science. 2016;352(6289):1102–5.
10. Newton RP, Smith CJ. Cyclic nucleotides. Phytochemistry. 2004;65(17):2423–37.
11. Kaplan B, Sherman T, Fromm H. Cyclic nucleotide-gated channels in plants. FEBS Lett. 2007;581(12):2237–46.

12. Ma W, Berkowitz GA. Cyclic nucleotide gated channel and Ca2+–mediated signal transduction during plant senescence signaling. Plant Signal Behav. 2011;6(3):413–5.
13. Ma W, Smigel A, Walker RK, Moeder W, Yoshioka K, Berkowitz GA. Leaf senescence signaling: the Ca2+–conducting Arabidopsis cyclic nucleotide gated channel2 acts through nitric oxide to repress senescence programming. Plant Physiol. 2010;154(2):733–43.
14. Zelman AK, Dawe A, Gehring C, Berkowitz GA. Evolutionary and structural perspectives of plant cyclic nucleotide-gated cation channels. Front Plant Sci. 2012;3(195):95.
15. Guo KM, Babourina O, Christopher DA, Borsics T, Rengel Z. The cyclic nucleotide-gated channel, AtCNGC10, influences salt tolerance in Arabidopsis. Physiol Plant. 2008;134(3):499–507.
16. Nawaz Z, Kakar KU, Saand MA, Shu Q-Y. Cyclic nucleotide-gated ion channel gene family in rice, identification, characterization and experimental analysis of expression response to plant hormones, biotic and abiotic stresses. BMC Genomics. 2014;15(1):853.
17. Saand MA, Xu Y-P, Li W, Wang J-P, Cai X-Z. Cyclic nucleotide gated channel gene family in tomato: genome-wide identification and functional analyses in disease resistance. Front Plant Sci. 2015;6:303.
18. Chen J, Yin H, Gu J, Li L, Liu Z, Jiang X, Zhou H, Wei S, Zhang S, Wu J. Genomic characterization, phylogenetic comparison and differential expression of the cyclic nucleotide-gated channels gene family in pear (Pyrus bretchneideri Rehd.). Genomics. 2015;105(1):39–52.
19. Zelman AK, Dawe A, Berkowitz GA. Identification of cyclic nucleotide gated channels using regular expressions. In: Gehring C, editor. Cyclic nucleotide signaling in plants: methods and protocols. Totowa: Humana Press; 2013. p. 207–24.
20. Almoneafy AA, Kakar KU, Nawaz Z, Li B, Chun-lan Y, Xie G-L. Tomato plant growth promotion and antibacterial related-mechanisms of four rhizobacterial bacillus strains against Ralstonia solanacearum. Symbiosis. 2014;63(2):59–70.
21. Chin K, DeFalco TA, Moeder W, Yoshioka K. The Arabidopsis cyclic nucleotide-gated ion channels AtCNGC2 and AtCNGC4 work in the same signaling pathway to regulate pathogen defense and floral transition. Plant Physiol. 2013;163(2):611–24.
22. Finka A, Cuendet AFH, Maathuis FJ, Saidi Y, Goloubinoff P. Plasma membrane cyclic nucleotide gated calcium channels control land plant thermal sensing and acquired thermotolerance. Plant Cell. 2012;24(8):3333–48.
23. Warwick SI, Francis A, Al-Shehbaz IA. Brassicaceae: species checklist and database on CD-Rom. Plant Syst Evol. 2006;259(2–4):249–58.
24. Liu S, Liu Y, Yang X, Tong C, Edwards D, Parkin IAP, Zhao M, Ma J, Yu J, Huang S, et al. The Brassica Oleracea genome reveals the asymmetrical evolution of polyploid genomes. Nat Commun. 2014;5:3930.
25. Xu G, Guo C, Shan H, Kong H. Divergence of duplicate genes in exon–intron structure. Proc Natl Acad Sci. 2012;109(4):1187–92.
26. Parkin IAP, Koh C, Tang H, Robinson SJ, Kagale S, Clarke WE, Town CD, Nixon J, Krishnakumar V, Bidwell SL, et al. Transcriptome and methylome profiling reveals relics of genome dominance in the mesopolyploid Brassica Oleracea. Genome Bio. 2014;15(6):R77.
27. Lysak MA, Koch MA, Pecinka A, Schubert I. Chromosome triplication found across the tribe Brassiceae. Genome Res. 2005;15(4):516–25.
28. Bailey TL, Boden M, Buske FA, Frith M, Grant CE, Clementi L, Ren J, Li WW, Noble WS. MEME SUITE: tools for motif discovery and searching. Nucleic Acids Res. 2009;37(suppl_2):W202–8.
29. Finn RD, Bateman A, Clements J, Coggill P, Eberhardt RY, Eddy SR, Heger A, Hetherington K, Holm L, Mistry J, et al. Pfam: the protein families database. Nucleic Acids Res. 2014;42(D1):D222–30.
30. De Castro E, Sigrist CJA, Gattiker A, Bulliard V, Langendijk-Genevaux PS, Gasteiger E, Bairoch A, Hulo N. ScanProsite: detection of PROSITE signature matches and ProRule-associated functional and structural residues in proteins. Nucleic Acids Res. 2006;34(suppl 2):W362–5.
31. Leppänen T, Tuominen RK, Moilanen E. Protein Kinase C and its inhibitors in the regulation of inflammation: inducible nitric oxide Synthase as an example. Basic Clin Pharmacol Toxicol. 2014;114(1):37–43.
32. Mulekar JJ, Bu Q, Chen F, Huq E. Casein kinase II α subunits affect multiple developmental and stress-responsive pathways in Arabidopsis. Plant J. 2012; 69(2):343–54.
33. Traverso JA, Meinnel T, Giglione C. Expanded impact of protein N-myristoylation in plants. Plant Signal Behav. 2008;3(7):501–2.
34. Strasser R. Plant protein glycosylation. Glycobiology. 2016;26(9):926 39.
35. Lai S, Pelech S. Regulatory roles of conserved phosphorylation sites in the activation T-loop of the MAP kinase ERK1. Mol Biol Cell. 2016;27(6):1040–50.
36. Szklarczyk D, Franceschini A, Wyder S, Forslund K, Heller D, Huerta-Cepas J, Simonovic M, Roth A, Santos A, Tsafou KP. STRING v10: protein–protein interaction networks, integrated over the tree of life. Nucleic Acids Res. 2014;43(D1):D447–52.
37. Schläfli P, Tröger J, Eckhardt K, Borter E, Spielmann P, Wenger RH. Substrate preference and phosphatidylinositol monophosphate inhibition of the catalytic domain of the per-Arnt-Sim domain kinase PASKIN. FEBS J. 2011; 278(10):1757–68.
38. Behrends C, Sowa ME, Gygi SP, Harper JW. Network organization of the human autophagy system. Nature. 2010;466(7302):68–76.
39. Chen J, Lalonde S, Obrdlik P, Noorani Vatani A, Parsa SA, Vilarino C, Revuelta JL, Frommer WB, Rhee SY. Uncovering Arabidopsis membrane protein Interactome enriched in transporters using mating-based split Ubiquitin assays and classification models. Front Plant Sci. 2012;3:124.
40. Frietsch S, Wang Y-F, Sladek C, Poulsen LR, Romanowsky SM, Schroeder JI, Harper JF. A cyclic nucleotide-gated channel is essential for polarized tip growth of pollen. Proc Natl Acad Sci U S A. 2007;104(36):14531–6.
41. Ladwig F, Dahlke RI, Stührwohldt N, Hartmann J, Harter K, Sauter M. Phytosulfokine regulates growth in Arabidopsis through a response module at the plasma membrane that includes CYCLIC NUCLEOTIDE-GATED CHANNEL17, H+–ATPase, and BAK1. Plant Cell. 2015;27(6):1718–29.
42. Franceschini A, Szklarczyk D, Frankild S, Kuhn M, Simonovic M, Roth A, Lin J, Minguez P, Bork P, von Mering C, et al. STRING v9.1: protein-protein interaction networks, with increased coverage and integration. Nucleic Acids Res. 2013;41(Database issue):D808–15.
43. Witkos TM, Koscianska E, Krzyzosiak WJ. Practical aspects of microRNA target prediction. Curr Mol Med. 2011;11(2):93–109.
44. Dai X, Zhao PX. psRNATarget: a plant small RNA target analysis server. Nucleic Acids Res. 2011;39(suppl 2):W155–9.
45. Kanehisa M, Sato Y, Kawashima M, Furumichi M, Tanabe M. KEGG as a reference resource for gene and protein annotation. Nucleic Acids Res. 2015;44(D1):D457–62.
46. Schmid M, Davison TS, Henz SR, Pape UJ, Demar M, Vingron M, Scholkopf B, Weigel D, Lohmann JU. A gene expression map of Arabidopsis Thaliana development. Nat Genet. 2005;37(5):501–6.
47. Ma W, Ali R, Berkowitz GA. Characterization of plant phenotypes associated with loss-of-function of AtCNGC1, a plant cyclic nucleotide gated cation channel. Plant Physiol Biochem. 2006;44(7):494–505.
48. Khaldi N, Shields DC. Shift in the isoelectric-point of milk proteins as a consequence of adaptive divergence between the milks of mammalian species. Biol Direct. 2011;6(1):40.
49. Schwartz R, Ting CS, King J. Whole proteome pl values correlate with subcellular localizations of proteins for organisms within the three domains of life. Genome Res. 2001;11(5):703–9.
50. Gitlin I, Carbeck JD, Whitesides GM. Why are proteins charged? Networks of charge–charge interactions in proteins measured by charge ladders and capillary electrophoresis. Angew Chem Int Ed. 2006;45(19):3022–60.
51. Duan G, Walther D. The roles of post-translational modifications in the context of protein interaction networks. PLoS Comput Biol. 2015;11(2):e1004049.
52. Webster DE, Thomas MC. Post-translational modification of plant-made foreign proteins; glycosylation and beyond. Biotechnol Adv. 2012;30(2):410–8.
53. Schwartz AS, Yu J, Gardenour KR, Finley RL, Ideker T. Cost effective strategies for completing the Interactome. Nat Methods. 2009;6(1):55–61.
54. Bauer D, Viczián A, Kircher S, Nobis T, Nitschke R, Kunkel T, Panigrahi KCS, Ádám É, Fejes E, Schäfer E. Constitutive photomorphogenesis 1 and multiple photoreceptors control degradation of phytochrome interacting factor 3, a transcription factor required for light signaling in Arabidopsis. Plant Cell. 2004;16(6):1433–45.
55. Cohen P. Control of enzyme activity, illustrated edn. Berlin: Springer Science & Business Media; 2013.
56. Banerjee J, Magnani R, Nair M, Dirk LM, DeBolt S, Maiti IB, Houtz RL. Calmodulin-mediated signal transduction pathways in Arabidopsis are fine-tuned by methylation. Plant Cell. 2013;25(11):4493–511.
57. Sun Y, Li L, Macho AP, Han Z, Hu Z, Zipfel C, Zhou J-M, Chai J. Structural basis for flg22-induced activation of the Arabidopsis FLS2-BAK1 immune complex. Science. 2013;342(6158):624–8.
58. Murata Y, Mori IC, Munemasa S. Diverse stomatal signaling and the signal integration mechanism. Annu Rev Plant Biol. 2015;66:369–92.

59. Lukasik A, Pietrykowska H, Paczek L, Szweykowska-Kulinska Z, Zielenkiewicz P. High-throughput sequencing identification of novel and conserved miRNAs in the Brassica Oleracea leaves. BMC Genomics. 2013;14(1):801.
60. Song JH, Yang J, Pan F, Jin B. Differential expression of microRNAs may regulate pollen development in Brassica Oleracea. Gen Mol Res. 2015;14(4):15024–34.
61. He X-F, Fang Y-Y, Feng L, Guo H-S. Characterization of conserved and novel microRNAs and their targets, including a TuMV-induced TIR–NBS–LRR class R gene-derived novel miRNA in Brassica. FEBS Lett. 2008;582(16):2445–52.
62. Nagaharu U. Genome analysis in Brassica with special reference to the experimental formation of B. Napus and peculiar mode of fertilization. Jpn J Bot. 1935;7:389–452.
63. Chalhoub B, Denoeud F, Liu S, Parkin IAP, Tang H, Wang X, Chiquet J, Belcram H, Tong C, Samans B. Early allopolyploid evolution in the post-Neolithic Brassica Napus oilseed genome. Science. 2014;345(6199):950–3.
64. Liu S, Liu Y, Yang X, Tong C, Edwards D, Parkin IAP, Zhao M, Ma J, Yu J, Huang S. The Brassica Oleracea genome reveals the asymmetrical evolution of polyploid genomes. Nat Commun. 2014;5:3930.
65. Liang Y, Xiong Z, Zheng J, Xu D, Zhu Z, Xiang J, Gan J, Raboanatahiry N, Yin Y, Li M. Genome-wide identification, structural analysis and new insights into late embryogenesis abundant (LEA) gene family formation pattern in Brassica Napus. Sci Rep. 2016;6:24265.
66. Cheng F, Mandáková T, Wu J, Xie Q, Lysak MA, Wang X. Deciphering the diploid ancestral genome of the mesohexaploid Brassica Rapa. Plant Cell. 2013;25(5):1541–54.
67. Lamesch P, Berardini TZ, Li D, Swarbreck D, Wilks C, Sasidharan R, Muller R, Dreher K, Alexander DL, Garcia-Hernandez M. The Arabidopsis information resource (TAIR): improved gene annotation and new tools. Nucleic Acids Res. 2012;40(D1):D1202–10.
68. Kersey PJ, Allen JE, Armean I, Boddu S, Bolt BJ, Carvalho-Silva D, Christensen M, Davis P, Falin LJ, Grabmueller C. Ensembl genomes 2016: more genomes, more complexity. Nucleic Acids Res. 2016;44(D1):D574–80.
69. Letunic I, Doerks T, Bork P. SMART: recent updates, new developments and status in 2015. Nucleic Acids Res. 2015;43(D1):D257–60.
70. Marchler-Bauer A, Derbyshire MK, Gonzales NR, Lu S, Chitsaz F, Geer LY, Geer RC, He J, Gwadz M, Hurwitz DI. CDD: NCBI's conserved domain database. Nucleic Acids Res. 2014;43(D1):D222–6.
71. Gasteiger E, Hoogland C, Gattiker A, Duvaud Se, Wilkins MR, Appel RD, Bairoch A. Protein identification and analysis tools on the ExPASy server. Totowa: Humana Press; 2005.
72. Larkin MA, Blackshields G, Brown NP, Chenna R, McGettigan PA, McWilliam H, Valentin F, Wallace IM, Wilm A, Lopez R. Clustal W and Clustal X version 2.0. Bioinformatics. 2007;23(21):2947–8.
73. Cheng F, Liu S, Wu J, Fang L, Sun S, Liu B, Li P, Hua W, Wang X. BRAD, the genetics and genomics database for Brassica plants. BMC Plant Biol. 2011; 11(1):1.
74. Nicholas KB, Nicholas HBJ: GeneDoc: a tool for editing and annotating multiple sequence alignments. Distributed by the author; 1997.
75. Tamura K, Stecher G, Peterson D, Filipski A, Kumar S. MEGA6: molecular evolutionary genetics analysis version 6.0. Mol Biol Evol. 2013;30(12):2725–9.
76. Lee T-H, Tang H, Wang X, Paterson AH. PGDD: a database of gene and genome duplication in plants. Nucleic Acids Res. 2013;41(D1):D1152–8.
77. Yu J, Zhao M, Wang X, Tong C, Huang S, Tehrim S, Liu Y, Hua W, Liu S. Bolbase: a comprehensive genomics database for Brassica Oleracea. BMC Genomics. 2013;14(1):1.
78. Hu B, Jin J, Guo A-Y, Zhang H, Luo J, Gao G. GSDS 2.0: an upgraded gene feature visualization server. Bioinformatics. 2015;31(8):1296–7.
79. Conesa A, Götz S. Blast2GO: a comprehensive suite for functional analysis in plant genomics. Int J Plant Genomics. 2008;2008:12.
80. Kozomara A, Griffiths-Jones S. miRBase: annotating high confidence microRNAs using deep sequencing data. Nucleic Acids Res. 2014;42(D1): D68–73.
81. RCoreTeam. R: a language and environment for statistical computing. Vienna: R Foundation for Statistical Computing; 2014.
82. King EO, Ward MK, Raney DE. Two simple media for the demonstration of pyocyanin and fluorescin. J Lab Clin Med. 1954;44(2):301–7.
83. Kabouw P, Biere A, van der Putten WH, van Dam NM. Intra-specific differences in root and shoot Glucosinolate profiles among white cabbage (Brassica Oleracea Var. Capitata) cultivars. J Agric Food Chem. 2010;58(1):411–7.
84. Livak KJ, Schmittgen TD. Analysis of relative gene expression data using real-time quantitative PCR and the 2– ΔΔCT method. Methods. 2001;25(4): 402–8.

12

Comparative genomics of *Coniophora olivacea* reveals different patterns of genome expansion in Boletales

Raúl Castanera[1], Gúmer Pérez[1], Leticia López-Varas[1], Joëlle Amselem[2], Kurt LaButti[3], Vasanth Singan[3], Anna Lipzen[3], Sajeet Haridas[3], Kerrie Barry[3], Igor V. Grigoriev[3], Antonio G. Pisabarro[1] and Lucía Ramírez[1*]

Abstract

Background: *Coniophora olivacea* is a basidiomycete fungus belonging to the order Boletales that produces brown-rot decay on dead wood of conifers. The Boletales order comprises a diverse group of species including saprotrophs and ectomycorrhizal fungi that show important differences in genome size.

Results: In this study we report the 39.07-megabase (Mb) draft genome assembly and annotation of *C. olivacea*. A total of 14,928 genes were annotated, including 470 putatively secreted proteins enriched in functions involved in lignocellulose degradation. Using similarity clustering and protein structure prediction we identified a new family of 10 putative lytic polysaccharide monooxygenase genes. This family is conserved in basidiomycota and lacks of previous functional annotation. Further analyses showed that *C. olivacea* has a low repetitive genome, with 2.91% of repeats and a restrained content of transposable elements (TEs). The annotation of TEs in four related Boletales yielded important differences in repeat content, ranging from 3.94 to 41.17% of the genome size. The distribution of insertion ages of LTR-retrotransposons showed that differential expansions of these repetitive elements have shaped the genome architecture of Boletales over the last 60 million years.

Conclusions: *Coniophora olivacea* has a small, compact genome that shows macrosynteny with *Coniophora puteana*. The functional annotation revealed the enzymatic signature of a canonical brown-rot. The annotation and comparative genomics of transposable elements uncovered their particular contraction in the *Coniophora* genera, highlighting their role in the differential genome expansions found in Boletales species.

Keywords: Boletales, Brown-rot, Basidiomycete, Genome, Annotation, Transposable elements, Retrotransposon

Background

Coniophora olivacea is a basidiomycete fungus belonging to the order Boletales. *C. olivacea* produces brown-rot decay on dead wood of conifers (softwood) and, less frequently, on hardwood species. In addition, *C. olivacea* also damages wood buildings or construction materials. The genome sequence of its sister species *C. puteana* was made public in 2012 [1] and contributed to the understanding of genomic differences between brown and white-rot fungi. White-rot fungi are efficient lignin degraders, whereas brown-rot fungi attack cell wall carbohydrates leaving lignin undigested. The main responsible of this behavior are lignin-degrader peroxidases, which are abundant in white-rot species and particularly contracted in brown-rot and mycorrhizal fungi [2]. The Boletales order comprises a diverse group of species including saprotrophs and ectomycorrhizal species such as *Suillus sp.* or *Pisolithus sp.* During the last 6 years, up to 12 Boletales genomes have been sequenced and annotated [1, 3, 4]. Information that emerged from these studies showed important differences in genomic characteristics between the species belonging to this group, whose predicted common ancestor was dated 84 million years ago. Evolution from this boletales ancestor (supposed to be a brown-rot saprotroph) lead to the diversification and the appearance of ectomycorrhizae, which shows a particular

* Correspondence: lramirez@unavarra.es
[1]Genetics and Microbiology Research Group, Department of Agrarian Production, Public University of Navarre, 31006 Pamplona, Navarre, Spain
Full list of author information is available at the end of the article

contraction of the number of plant cell wall-degrading enzymes coding genes (PCWDE) [4, 5]. In addition, Boletales show important differences in their genome size and gene content. For example, the smallest assembled Boletales genome spans 38.2 Mb and has 13,270 annotated genes (*Hydnomerulius pinastri*), but the largest (*Pisolithus tinctorius*) spans 71.0 Mb and has 22,701 genes [4]. Previous studies in saprophytic basidiomycetes have shown that species with higher genome sizes tend to have more transposable elements [6]. Also, it has been described that species associated with plants (pathogenic and symbiotic) have genomes with expanded TE families [1, 7], although this trend varies between the three basidiomycete phyla [8]. In this paper, we describe the draft genome sequence and annotation of the brown-rot *C. olivacea*, and we compare it with the genomes of *C. puteana* as well as with that of three other Boletales showing important differences in genome sizes (*Serpula lacrymans*, *Pisolithus tinctorius* and *Hydnomerulius pinastri*). The results show that *C. olivacea* displays enzymatic machinery characteristic of brown-rot fungi encoded in a compact genome, carrying a small number of repetitive sequences. The comparative analysis with other Boletales shows that both ancient and modern LTR-retrotransposon amplification events have greatly contributed to the genome expansion along the evolution of Boletales.

Methods

Fungal strains and culture conditions

Coniophora olivacea MUCL 20566 was obtained from the Spanish Type Culture Collection and was cultured in SMY submerged fermentation (10 g of sucrose, 10 g of malt extract and 4 g of yeast extract per litre).

Nucleic acid extraction

Mycelia were harvested, frozen, and ground in a sterile mortar in the presence of liquid nitrogen. High molecular weight DNA was extracted using the phenol-chloroform protocol described previously [9]. DNA sample concentrations were measured using a Qubit® 2.0 Fluorometer (Life Technologies, Madrid, Spain), and DNA purity was measured using a NanoDrop™ 2000 (Thermo-Scientific, Wilmington, DE, USA). DNA quality was verified by electrophoresis in 0.7% agarose gels. Total RNA was extracted from 200 mg of deep-frozen tissue using Fungal RNA E.Z.N.A Kit (Omega Bio-Tek, Norcross, GA, USA), and its integrity was verified using the Agilent 2100 Bioanalyzer system (Agilent Technologies, Santa Clara, CA, USA).

Genome and transcriptome sequencing and assembly

A detailed description is provided in Additional file 1: Text S1. Briefly, the *C. olivacea* MUCL 20566 genome was sequenced using Illumina HiSeq-1 TB Regular 2 × 151 bp 0.309 kb. Sequenced reads were QC filtered for artifact contamination using BBDuk from the BBMap package (https://sourceforge.net/projects/bbmap/) and subsequently assembled with Velvet 1.2.07 [10]. The result -pair library with an insert size of 3000 +/− 300 bp *in silico* that was then assembled together with the original Illumina library with AllPathsLG [11]. Raw sequences were deposited in SRA (Sequence Read Archive) NCBI database under accession number SRP086489. Strand-specific RNASeq libraries were created and quantified by qPCR. Sequencing was performed using an Illumina HiSeq-2500 instrument. Reads were filtered and trimmed to remove artifacts and low quality regions using BBDuk. Transcriptome was de novo assembled using Trinity [12] and used to assist annotation and assess the completeness of the corresponding genome assembly using alignments of at least 90% identity and 85% coverage.

Whole-genome alignment

The genome assemblies of *C. olivacea* MUCL 20566 and *C. puteana* (http://genome.jgi.doe.gov/Conpu1/Conpu1.home.html) were aligned using the Promer tool from the MUMmer 3.0 package [13]. Genome rearrangements were identified in the alignment with dnadiff tool from the same package.

Genome annotation

The annotation of the *C. olivacea* MUCL 20566 assembly was performed using the Joint Genome Institute pipeline [14] to predict and functionally annotate protein-coding genes and other features such as tRNAs or putative microRNA precursors. The SECRETOOL pipeline [15] was used to identify putatively secreted proteins, considering the presence of signal peptides, cleavage sites, transmembrane domains and the GPI (glycosylphosphatidylinositol) membrane anchor. Carbohydrate-active enzymes (CAZys) were annotated based on BLAST [16] and HMMER [17] searches against sequence libraries and HMM (Hidden Markov Models) profiles of the CAZy database [18] functional modules. Protein structure predictions were carried out with Phyre2 [19]. Raw sequencing reads, genome assembly, transcriptome assembly, gene predictions and functional annotations are publicly available in the *C. olivacea* genome portal of Mycocosm database (http://genome.jgi.doe.gov/Conol1/Conol1.home.html).

Annotation of transposable elements

Transposable elements (TEs) were identified and annotated in the *C. olivacea* assembly using REPET package [20, 21], as well as in the following boletales assemblies available in Mycocosm database (http://genome.jgi.doe.gov/programs/fungi/index.jsf): *Coniophora puteana* v1.0 (ID: Conpu1), *Hydnomerulius pinastri* v2.0 (ID: Hydpi2), *Serpula lacrymans* S7.3 v2.0 (ID: SerlaS7_3_2), *Pisolithus tinctorius* Marx 270 v1.0 (ID: Pistil). Briefly, de novo TE detection was carried out with the TEdenovo pipeline [21] and the elements were classified with PASTEC [22]. The resulting TE library was fed into TEannot pipeline [20] in two consecutive iterations: the first one with the full library, and the second with an improved library consisting on consensus elements carrying at least one full-length copy after manually discarding false positives (i.e., *C. olivacea* genes).

Insertion age of LTR-retrotransposons

Full-length LTR-retrotransposons were identified using LTRharvest [23] followed by BLASTX against Repbase [24]. Long Terminal Repeats were extracted and aligned with MUSCLE [25]. Alignments were trimmed using trimAl [26] and used to calculate Kimura's 2P distances. The insertion age was calculated following the approach described in [27] using the fungal substitution rate of 1.05×10^{-9} nucleotides per site per year [6, 28].

Identification of gene families

All-by-all BLASTP followed by MCL (Markov Cluster Algorithm) clustering [29] was carried out with *C. olivacea* protein models using a threshold value of e^{-5} and an inflation value of 2. We considered gene families carrying four or more genes for further analyses.

Phylogenetic analyses

The predicted proteomes of the following species were downloaded from Mycocosm database (Mycocosm ID in parenthesis):

Agaricus bisporus var. *bisporus* H97 v2.0 (Agabi_varbisH97_2), *Boletus edulis* v1.0 (Boled1), *Coniophora olivacea* MUCL 20566 v1.0 (Conol1), *Coniophora puteana* v1.0 (Conpu1), *Cryptococcus neoformans* var. *grubii* H99 (Cryne_H99_1), *Fomitopsis pinicola* FP-58527 SS1 v3.0 (Fompi3), *Gyrodon lividus* BX v1.0 (Gyrli1), *Hydnomerulius pinastri* v2.0 (Hydpi2), *Leucogyrophana mollusca* KUC20120723A-06 v1.0 (Leumo1), *Paxillus involutus* ATCC 200175 v1.0 (Paxin1), *Phanerochaete chrysosporium* RP-78 v2.2 (Phchr2), *Pisolithus tinctorius* Marx 270 v1.0 (Pisti1), *Pleurotus ostreatus* PC15 v2.0 (PleosPC15_2), *Rhizopogon vinicolor* AM-OR11–026 v1.0 (Rhivi1), *Scleroderma citrinum* Foug A v1.0 (Sclci1), *Serpula lacrymans* S7.3 v2.0 (SerlaS7_3_2), *Suillus luteus* UH-Slu-Lm8-n1 v2.0 (Suilu3), *Trametes versicolor* v1.0 (Trave1). Species phylogeny was constructed as follows: all-by-all BLASTP followed by MCL clustering was carried out with a dataset containing the proteomes of all the species. The clusters carrying only one protein per species were identified, and the proteins were aligned using MAFFT [30]. The alignments were concatenated after discarding poorly aligned positions with Gblocks [31]. The phylogeny was constructed using RaxML [32] with 100 rapid bootstraps under PROTGAMMAWAGF substitution model. Phylogenetic reconstruction of Gypsy reverse-transcriptases was carried out as follows: Reverse transcriptase RV1 domains were extracted from LTR-retrotransposons of the TE consensus library using Exonerate [33] and aligned with MUSCLE. The alignments were trimmed using trimAl with the default parameters, and an approximate maximum likelihood tree was constructed using FastTree [34].

Results

C. olivacea assembly and annotation

The nuclear genome of *C. olivacea* was sequenced with 137 X coverage and assembled into 863 scaffolds accounting for 39.07 Mb, 90.3% of the genome size estimation based on k-mer spectrum (43.28 Mb). The mitochondrial genome was assembled into two contigs accounting for 78.54 kb. The assembly completeness was 99.78% according to the Core Eukaryotic Genes Mapping Approach (CEGMA [35]), with only one missing accession (KOG1322, GDP-mannose pyrophosphorylase). We assembled 66,567 transcripts (mean lenght = 2,744 nt, median = 2,154 nt) of which 97.8% could be mapped to the genome. The *C. olivacea* assembled genome was more fragmented than its close relative *C. puteana* (Table 1). The total repeat content was 2.91% of which 2.15% corresponded to transposable elements, 0.64% to simple repeats, and 0.12% to low complexity regions. The estimation of repeat content from low-coverage Illumina data (3.8X) yielded 6% of the genome size covered by transposable elements (Additional file 2: Table S1). We used transcriptomic information, *ab initio* predictions and similarity searches to predict a total of 14,928 genes—84.5% of them having a strong transcriptome support (spanning more than 75% of the gene length). In addition, 88.3% of the annotated genes had significant similarity to proteins from the NCBI nr database and 46.6% to the manually curated proteins from the Swiss-Prot database (cutoff e^{-05}) [36]. A total of 7,841 predicted proteins (52.3%) carried Pfam domains and 1,471 (9.8%) carried signal peptide, of which 470 were predicted to be secreted using the more stringent SECRETOOL pipeline.

The multigene phylogeny based on 1,677 conserved single copy genes displayed different classes, orders and families in branches congruent with previous phylogenetic data [37] and with very high support. *C. olivacea* was placed in a branch next to its sequenced closer species *C. puteana* representing the Coniophoraceae family in the order Boletales (Fig. 1).

The whole-genome protein-based alignment between the two Coniophoraceae species spanned 52.7% of the *C. olivacea* and 48.0% of *C. puteana* assemblies. It shows evidence of macrosynteny between the two species (Fig. 2a, Additional file 3: Fig. S1), with an average similarity of

Table 1 Summary of *C. olivacea* genome assembly and annotation

Feature	*C. olivacea*	*C. puteana*
Genome assembly size (Mb)	39.07	42.97
Sequencing coverage depth	137.7×	49.5×
Number of scaffolds	863	210
Scaffold N50 [a]	80	7
Scaffold L50 (Mb) [b]	0.14	2.40
N° scaffold gaps	127	412
Genome assembly gaps (%)	0.24	2.57
Assembly completeness (%)	99.78	Unknown
Repeat content (%) [c]	2.91	4.68
GC content (%)	52.82	52.4
Number of genes	14,928	13,761
Gene density (genes/Mb)	382.07	320.26
Predicted secreted proteins	470 (3.1%)	504 (3.7%)

[a] N50 indicates the number of scaffolds that account for 50% of the total assembled sequence
[b] L50 indicates that 50% of the total sequence is assembled in scaffolds larger than this size
[c] Includes TE, simple repeats and low complexity regions

78.4% in the aligned regions (Fig. 2b) and numerous inversions (1,027 regions). The good conservation between both genomes in protein coding regions was evidenced by the amount of orthologous genes obtained using the reciprocal best hit approach (7,468 genes with more than 70% identity over 50% of protein sequences) and by the number of *C. olivacea* proteins yielding significant tBLASTN hits against the *C. puteana* genome (13,572 genes, cutoff e-5, Fig. 2c). For the remaining 1,352 *C. olivacea*-specific (orphan) genes, only 48 could be functionally annotated based on KOG (Eukaryotic Orthologous Groups), KEGG (Kyoto Encyclopedia of Genes and Genomes), GO (Gene Ontology) or InterPro databases.

Carbohydrate-active enzymes of *C. olivacea*

The annotated proteome was screened for the presence of carbohydrate-active enzymes (CAZy). A total of 397 proteins were annotated and classified into different CAZy classes and associated modules. The CAZyme profile of *C. olivacea* was very similar to that of *C. puteana* although small differences were found in the glycoside hydrolases (GH, Additional file 4: Table S2). Some families such as GH5, GH18 or GH31 were smaller than in *C. puteana*. Similar to other brown-rot basidiomycetes, *C. olivacea* lacked Class II peroxidases (Auxiliar Activities AA2) and displayed a reduced set of other cellulolytic enzymes such GH6 (1), GH7 (1) and CBM1 (2) and AA9 (6).

Functional characteristics of *C. olivacea* predicted secretome

Using SECRETOOL pipeline we predicted 470 putatively secreted proteins in *C. olivacea* and 504 in *C. puteana*. An enrichment analysis of gene ontology (GO) terms was performed to determine what gene functions were over-represented in the secreted proteins. Thirty GO terms were significantly enriched including 24 corresponding to molecular functions, four to biological processes and two to cellular components (Table 2). The most enriched molecular function was "feruloyl esterase activity," which is responsible for plant cell-wall degradation. "Polysaccharide catabolic process" was the most enriched GO term within the biological processes, and "extracellular region" within the cellular components (Table 2).

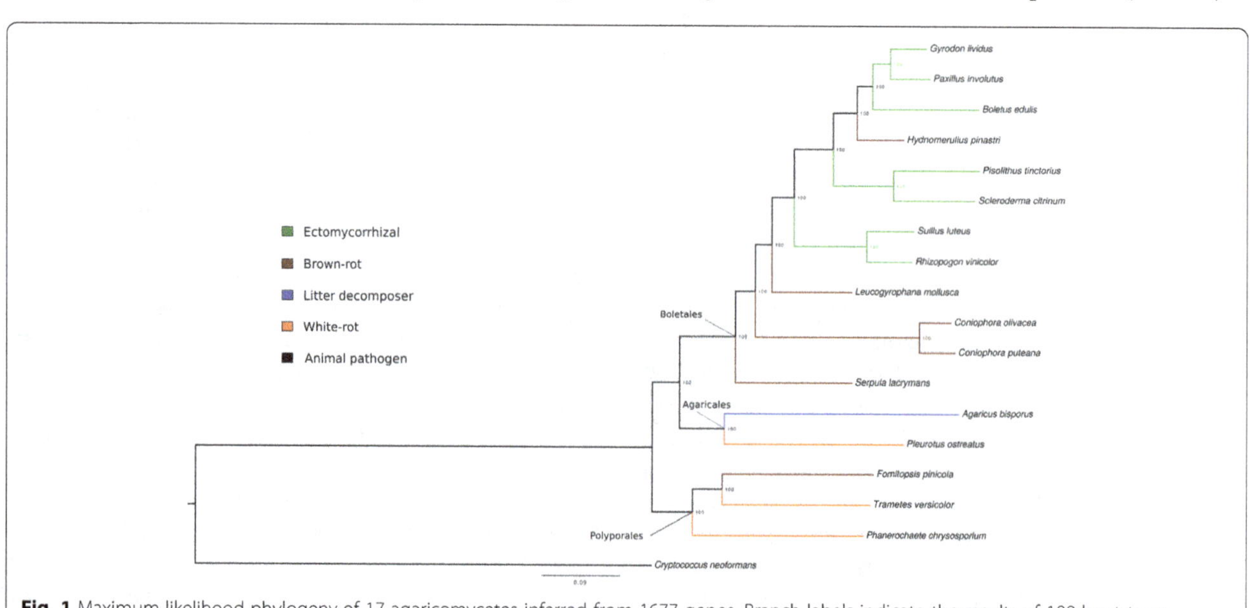

Fig. 1 Maximum-likelihood phylogeny of 17 agaricomycetes inferred from 1677 genes. Branch labels indicate the results of 100 bootstraps

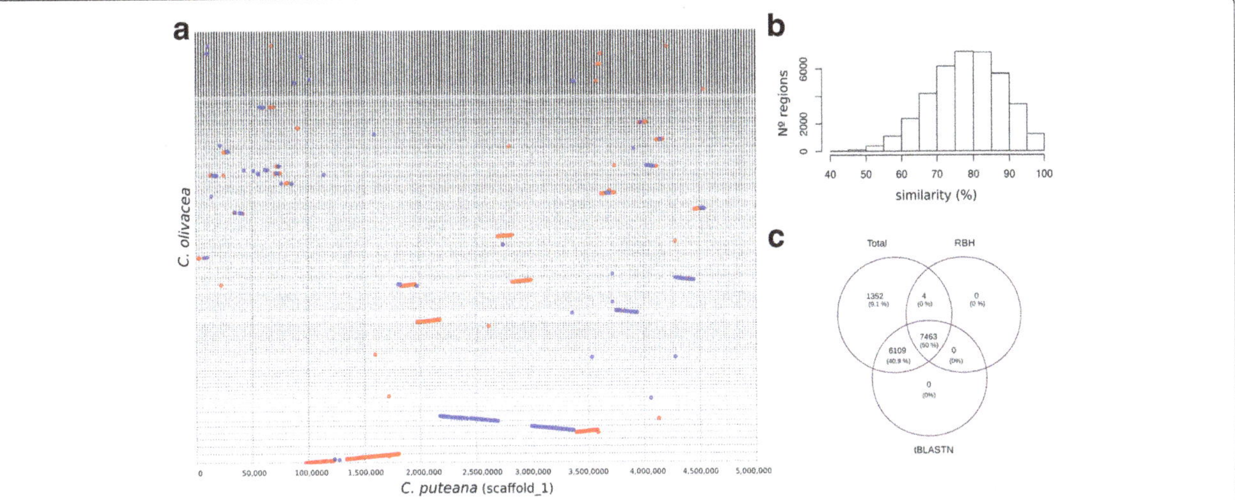

Fig. 2 a Synteny dot plot showing a fraction of the whole-genome alignment between *C. puteana* and *C. olivacea*. Every grid line in the y-axes represents the end of one scaffold and the beginning of the next. Forward matches are displayed in red, while reverse matches are displayed in blue. **b** Histogram of similarity of the 39,506 aligned regions. **c** Venn diagram summarizing the amount of genes shared by the two genomes based on reciprocal best hit (RBH) and tBLASTN is shown in panel C

Analysis of putatively secreted multigene families

Using all-by-all BLASTP followed by MCL we clustered by similarity the 1,471 proteins carrying signal peptides in *C. olivacea*. We used all proteins carrying signal peptides rather than only SECRETOOL predictions in order to obtain larger protein clusters. Up to 60% of the 1,471 proteins grouped in clusters were formed by 2 to 59 genes (Additional file 5: Table S3), showing the same distribution as the whole proteome ($p = 0.6032$, Wilcoxon test, 61% of the 14,928 predicted genes were found in clusters containing 2 to 157 members). For further analysis of the secreted genes found in clusters, we focused on the 70 clusters (families) formed by four or more gene members. Using the KOG, KEGG, InterPro and GO databases, we could assign functions to 45 out of the 70 gene families (Table 3). Cytochrome P450, hydrophobins and aspartic-peptidases were the largest gene families. In addition, 17 CAZys clusters were found including glycoside hydrolases (GH), carbohydrate esterases (CE), carbohydrate-binding modules (CBMs) and redox enzymes classified as auxiliary activities (AA). 25 clusters lacked functional annotation, and some of them had a high number of genes (clusters 2, 6 and 7 in Table 3). All of these genes belonging to families with unknown function were further analyzed with Phyre2 to predict their protein structure and used for PSI-BLAST (Position-Specific Iterated BLAST) analysis. Using this approach, two gene families were functionally annotated with high confidence (96.3–97.4% confidence for individual protein predictions): one as a copper-dependent lytic polysaccharide monooxygenase (LPMO, also known as AA9; cluster 16), and the other as thaumatin-lyke xylanase inhibitor (*tlxi*, cluster 48). The Cluster16 containing putative LPMOs was particularly interesting. This was formed by 10 genes coding for small proteins ranging from 130 to 162 amino acids with three exons (with the exception of protein ID839457 that shows only two). All these genes coded for proteins that have a signal peptide but lack of known conserved functional domains. Six were confidently annotated as LPMOs by Phyre2, and four of them were predicted to be secreted by SECRETOOL. In addition, this family of unknown proteins is conserved in all the agaricomycetes shown in Fig. 1. Interestingly, four members of this family appear as a tandem located in *C. olivacea* scaffold_124 (scaffold_426:4800–12,000).

Impact of repeat content on *C. olivacea* genome size and other Boletales

To study the role that TEs have played in the evolution of the Boletales genomes, we annotated and quantified the TE content in five species showing important differences in genome size: *C. olivacea* (39.1 Mb), *C. puteana* (42.9 Mb) [1], *Hydnomerulius pinastri* (38.2 Mb) [4], *Serpula lacrymans* (47.0 Mb) [3] and *Pisolithus tinctorius* (71.0 Mb) [4] (Additional file 6: Dataset S1, Additional file 7: Dataset S2, Additional file 8: Dataset S3, Additional file 9: Dataset S4, Additional file 10: Dataset S5). TEs were de novo identified and annotated using pipelines of the REPET package. The results yielded major differences in TE content between the five species, with *C. olivacea*, *C. puteana* and *H. pinastri* having low TE content (2.15%, 3.94% and 6.54% of their corresponding genome sizes), and *S. lacrymans* and *P. tinctorius* having up to 29.45% and 41.17% of their genomes occupied

Table 2 GO terms significantly enriched in the predicted secretome of *C. olivacea*

Molecular Function	Description	GO/Secretome	GO/Genome	*p* value[a]
GO:0030600	Feruloyl esterase activity	6/470	9/14,928	0.000171
GO:0042500	Aspartic endopeptidase activity, intra membrane cleaving	11/470	20/14,928	0.000192
GO:0008843	Endochitinase activity	8/470	14/14,928	0.000194
GO:0004568	Chitinase activity	8/470	14/14,928	0.000194
GO:0004650	Polygalacturonase activity	11/470	15/14,928	0.000354
GO:0004806	Triglyceridelipase activity	11/470	29/14,928	0.000376
GO:0016160	Amylase activity	25/470	40/14,928	0.000737
GO:0008933	Lytic transglycosylase activity	25/470	40/14,928	0.000737
GO:0015927	Trehalase activity	25/470	40/14,928	0.000737
GO:0015925	Galactosidase activity	25/470	40/14,928	0.000737
GO:0015924	Mannosyl-oligosaccharide mannosidase activity	25/470	40/14,928	0.000737
GO:0015929	Hexosaminidase activity	25/470	40/14,928	0.000737
GO:0015928	Fucosidase activity	25/470	40/14,928	0.000737
GO:0008810	Cellulase activity	9/470	11/14,928	0.00089
GO:0015926	Glucosidase activity	25/470	41/14,928	0.000948
GO:0015923	Mannosidase activity	25/470	41/14,928	0.000948
GO:0004620	Phospholipase activity	9/470	32/14,928	0.000968
GO:0004553	Hydrolase activity hydrolyzing O-glycosyl compounds	44/470	99/14,928	0.00105
GO:0004194	Obsolete pepsin A activity	17/470	42/14,928	0.00121
GO:0005199	Structural constituent of cell wall	16/470	33/14,928	0.00129
GO:0030246	Carbohydrate binding	9/470	25/14,928	0.00143
GO:0004190	Aspartic-type endopeptidaseactivity	20/470	44/14,928	0.00193
GO:0004099	Chitin deacetylase activity	5/470	9/14,928	0.00803
GO:0004185	Serine-type carboxypeptidase activity	5/470	12/14,928	0.0467
Biological Process				
GO:0000272	Polysaccharide catabolic process	5/470	6/14,928	0.000414
GO:0006508	Proteolysis	43/470	189/14,928	0.00128
GO:0005975	Carbohydrate metabolic process	65/470	161/14,928	0.00176
GO:0006629	Lipid metabolic process	10/470	50/14,928	0.00674
Cellular Component				
GO:0005576	Extracellular region	7/470	15/14,928	0.000354
GO:0005618	Cell wall	18/470	35/14,928	0.00224

[a] Bonferroni corrected, Fisher *p*-value

by TEs, respectively (Fig. 3, Table 4). In addition to higher TE content, species with larger genome assembly size showed higher TE diversity as reflected by the higher number of TE families, which ranged between 43 in *C. olivacea* to 432 in *P. tinctorius*.

The TEs found belong to seven out of the nine TE orders described by Wicker *et al* [38]: LTR, DIRS (*Dictyostelium* Intermediate Repeat Sequences), PLE (Penelope-like Elements), LINE (Long Interspersed Nuclear Elements), SINE (Small Interspersed Nuclear Elements), TIR (Terminal Inverted Repeats) and Helitrons. Two of the orders (LTR and TIRS, which contain long terminal repeats or terminal inverted repeats, respectively) were present in the five species. Class I TEs were primarily responsible for the observed genome size differences—especially the elements belonging to LTR in the Gypsy superfamily, which accounted for more than 15% of the assembly in *S. lacrymans* and *P. tinctorius,* but less than 3% in *H. pinastri, C. olivacea* and *C. puteana.* Of all the LTR/Gypsy families detected by TEdenovo, we observed that those elements belonging to the *Chromoviridae* group (carrying a Chromatin organization domain, PF00385, in the N-terminal region after the integrase, Fig. 4) were the most abundant LTR-retrotransposons in these five species, ranging from 44 to 83% of the total Gypsy coverage. LTR-retrotransposons in the Copia

Table 3 Size and functional annotation of *C. olivacea* predicted gene families targeted to the secretory pathway

Gene family	SignalP	SECRETOOL	Functional annotation
Cluster_1	59	3	Cytochrome P450
Cluster_2	33	0	Unknown
Cluster_3	32	17	Hydrophobin
Cluster_4	19	11	Aspartic peptidase
Cluster_5	18	12	Carboxylesterase
Cluster_6	17	0	Unknown
Cluster_7	15	0	Unknown
Cluster_8	14	12	Peptidase G1
Cluster_9	14	9	RlpA-likelipoprotein
Cluster_10	13	0	Pheromone mating factor, STE3
Cluster_11	13	0	Unknown
Cluster_12	12	3	Peptidase S8/S53
Cluster_13	11	9	Unknown
Cluster_14	10	0	CAZy:GH18
Cluster_15	10	9	Cytochrome P450
Cluster_16	10	6	Unknown/lytic polysaccharide monooxygenase (LPMO/ CAZy:AA9)
Cluster_17	9	5	Asparticpeptidase
Cluster_18	9	5	CAZy:CE4 Carbohydrate Esterase Family 4
Cluster_19	9	0	CAZy:GH16
Cluster_20	9	2	Peptidase S10
Cluster_21	9	5	Sugar transporter
Cluster_22	9	4	Unknown/putative lipoprotein
Cluster_23	8	6	Fungal lipase
Cluster_24	8	0	Isoprenylcysteinecarboxylmethyltransferase
Cluster_25	8	0	Monooxygenase, FAD-binding
Cluster_26	7	7	Ser-Thr-rich glycosyl-phosphatidyl-inositol-anchored membrane family
Cluster_27	7	0	Unknown
Cluster_28	7	1	Unknown
Cluster_29	6	5	CAZy:GH128
Cluster_30	6	0	CAZy:GH28
Cluster_31	6	3	CAZy:GH3
Cluster_32	6	2	Peptidase M28
Cluster_33	6	6	Thaumatin
Cluster_34	6	2	Unknown
Cluster_35	6	6	Unknown
Cluster_36	5	1	Aspartic peptidase
Cluster_37	5	2	CAZy:AA1_1
Cluster_38	5	4	CAZy:AA5_1
Cluster_39	5	5	CAZy:AA9
Cluster_40	5	1	CAZy:CBM5
Cluster_41	5	4	CAZy:GH12
Cluster_42	5	5	CAZy:GH30_3
Cluster_43	5	0	CAZy:GH47
Cluster_44	5	2	CAZy:GH71

Table 3 Size and functional annotation of *C. olivacea* predicted gene families targeted to the secretory pathway *(Continued)*

Gene family	SignalP	SECRETOOL	Functional annotation
Cluster_45	5	0	Monooxygenase
Cluster_46	5	2	Unknown
Cluster_47	5	0	Unknown
Cluster_48	5	5	Unknown/xylanase inhibitor tl-xi
Cluster_49	5	4	Unknown
Cluster_50	5	4	Unknown
Cluster_51	5	4	Unknown
Cluster_52	5	0	Unknown
Cluster_53	5	4	Unknown
Cluster_54	5	0	Unknown
Cluster_55	5	0	Unknown
Cluster_56	4	0	CAZy:GH18, CAZy:CBM5
Cluster_57	4	0	CAZy:GH31
Cluster_58	4	3	CAZy:GH55
Cluster_59	4	4	Flavin monooxygenase-like
Cluster_60	4	3	GOLD
Cluster_61	4	2	Histidine phosphatase superfamily, clade-2
Cluster_62	4	3	Lysophospholipase
Cluster_63	4	1	Peptidase S28
Cluster_64	4	0	Proteolipid membrane potential modulator
Cluster_65	4	3	RlpA-like, ceratoplatanin
Cluster_66	4	1	Thioredoxin-like fold
Cluster_67	4	3	Unknown
Cluster_68	4	3	Unknown
Cluster_69	4	3	Unknown
Cluster_70	4	0	Unknown

Protein IDs of each cluster are shown in Additional file 5: Table S3

superfamily were also particularly abundant in *S. lacrymans* and *P. tinctorius* (accounting for 2.4–6% of the total assembly size). Remarkably, non-coding LTR-retrotransposons such as TRIM (Terminal-repeat Retrotransposons In Miniature) and LARD (Large Retrotransposon Derivatives) were also found in three out of the five genomes, but in lower amounts (<1% of the genome, Table 4).

LINE, SINE, DIRS and PLE elements were also found in low copy numbers, but none of these were present in the five species. Regarding Class II transposons, TIR order was the most important in terms of abundance

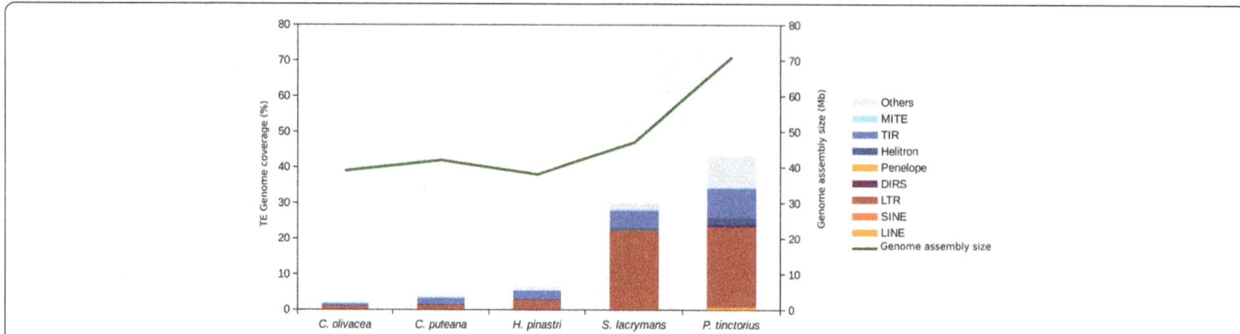

Fig. 3 TE content and genome size in five Boletales species. TE content is shown as a histogram, and genome size as a green line in panel A. Panel B shows a histogram representing the number of TE families found in each species

Table 4 Summary of TE content in four Boletales genome assemblies

Classification	*C. olivacea*			*C. puteana*			*H. pinastri*			*S. lacrymans*			*P. tinctorius*		
	(43) families			(108) families			(87) families			(230) families			(432) families		
	Copies	Full copies	Coverage (%)	Copies	Full copies	Coverage (%)	Copies	Full copies	Coverage (%)	Copies	Full copies	Coverage (%)	Copies	Full copies	Coverage (%)
Class I															
LINE	30	4	0.03	0	0	0.00	0	0	0.00	0	0	0.00	317	41	0.80
LINE (unknown)	29	5	0.02	11	3	0.01	23	3	0.02	0	0	0.00	14	1	0.01
SINE	0	0	0.00	6	2	0.00	0	0	0.00	0	0	0.00	9	1	0.00
LTR/Copia	36	7	0.09	441	27	0.83	267	10	0.72	3,773	86	6.04	1,617	101	2.43
LTR/Gypsy	394	13	0.93	299	28	0.54	767	42	2.13	6,949	268	16.27	8,434	575	19.28
LTR/LARD	0	0	0.00	60	8	0.08	0	0	0.00	0	0	0.00	361	2	0.53
LTR/TRIM	15	4	0.02	136	4	0.08	0	0	0.00	0	0	0.00	576	93	0.20
DIRS	0	0	0.00	0	0	0.00	0	0	0.00	0	0	0.00	361	36	0.58
Penelope	0	0	0.00	0	0	0.00	0	0	0.00	69	11	0.15	0	0	0.00
Class II															
Helitron	0	0	0.00	0	0	0.00	21	4	0.04	260	25	0.43	1,386	38	2.01
TIR/DDE	361	28	0.52	362	38	0.68	1,089	55	2.16	2,148	166	3.04	3,366	255	4.25
TIR (unknown)	143	34	0.18	720	115	1.10	323	49	0.36	736	67	1.55	1,115	40	1.85
MITE	410	85	0.30	702	264	0.56	121	43	0.12	539	98	0.62	1,102	227	0.59
Maverick (putative)	0	0	0.00	0	0	0.00	0	0	0.00	0	0	0.00	56	3	0.21
Unknown	67	4	0.07	99	9	0.06	865	133	0.99	1,138	167	1.34	8,611	708	8.44
TOTAL	1,485	184	2.15	2,836	498	3.94	3,476	339	6.54	15,612	888	29.45	27,325	2,121	41.17

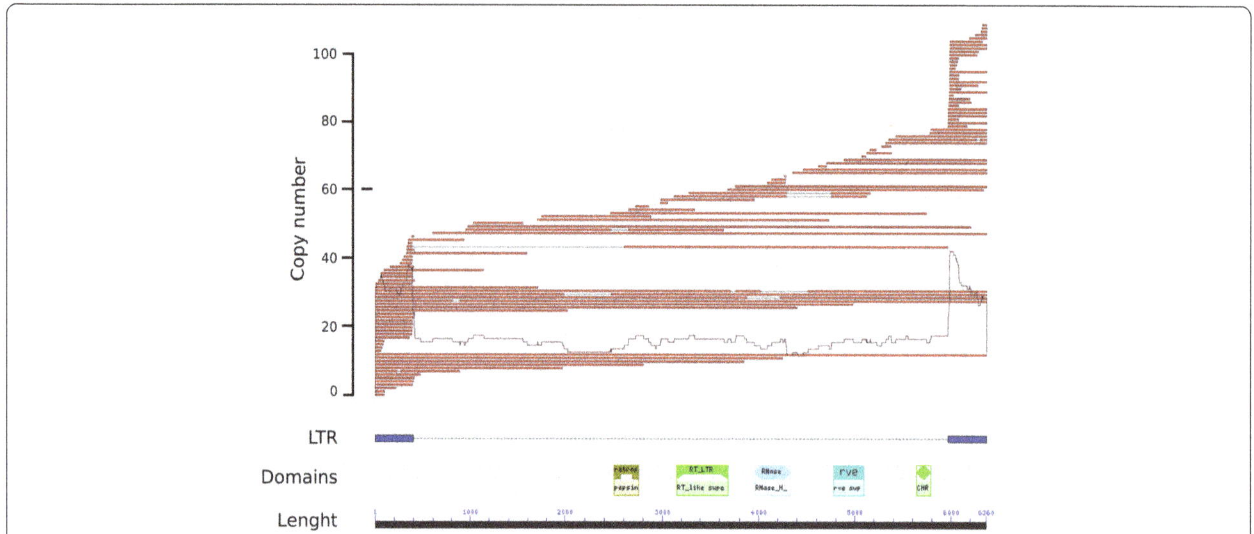

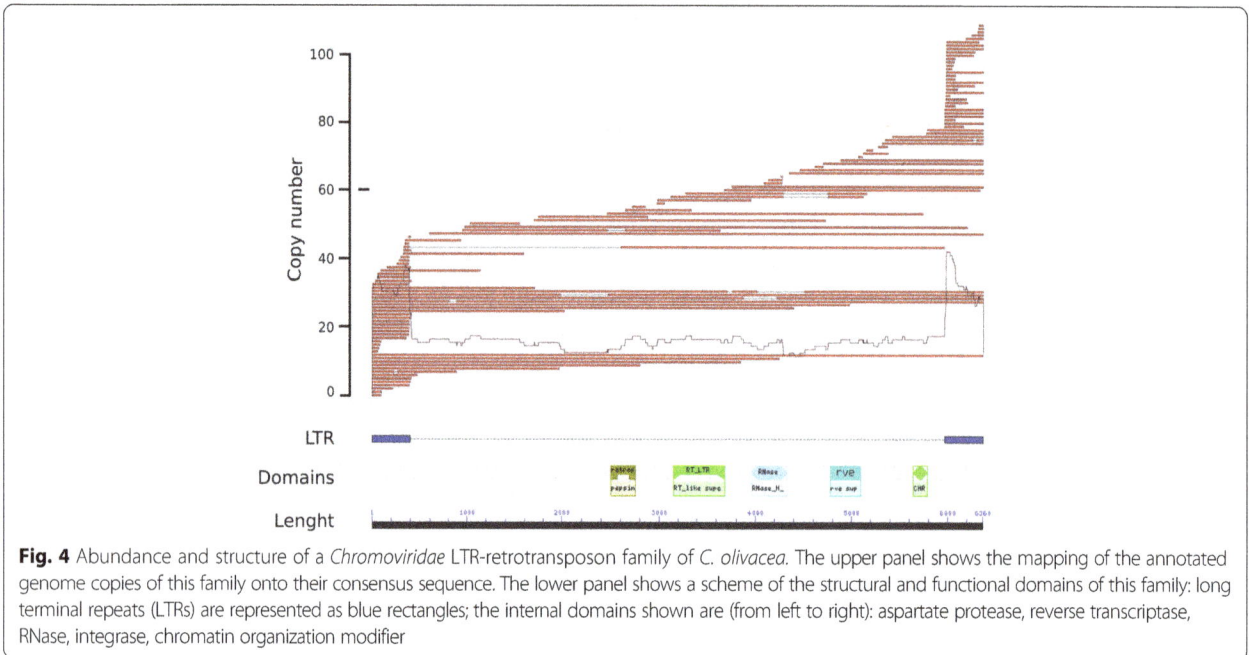

Fig. 4 Abundance and structure of a *Chromoviridae* LTR-retrotransposon family of *C. olivacea*. The upper panel shows the mapping of the annotated genome copies of this family onto their consensus sequence. The lower panel shows a scheme of the structural and functional domains of this family: long terminal repeats (LTRs) are represented as blue rectangles; the internal domains shown are (from left to right): aspartate protease, reverse transcriptase, RNase, integrase, chromatin organization modifier

and copy number with elements encoding DDE transposases present in the five species. The second most important were MITEs (Miniature Inverted–repeat Transposable Elements) and other non-coding elements carrying structural features (classified as TIR/unknown in Table 1). Rolling-circle helitrons were found in *H. pinastri*, *S. lacrymans* and *P. tinctorius*, while putative Mavericks were present only in this latter one.

Phylogenetic reconstruction of the LTR reverse-transcriptases

To understand the phylogenetic relationship between the LTR-retrotransposon familes in the five analyzed genomes, we inferred a maximum likelihood phylogeny of the LTR reverse-transcriptases of the Gypsy consensus sequences (Fig. 5). Three main clades were obtained (A, B and C). Clades A and B were formed, almost exclusively, by families found in the *P. tinctorius* genome. Moreover, while clade B is formed mostly by distantly related families, the profile of clade A suggests that an important fraction of the families underwent recent diversification. All LTR families found in the other four species grouped in clade C along with the remaining families of *P. tinctorius*. This clade contained several retrotransposon sub-clades sharing closely related families from three to five species.

Age of the LTR-retrotransposon amplification bursts in the Boletales

LTR-retrotransposons carrying conserved domains as well as intact Long Terminal Repeats (putative autonomous elements) were subjected to further study to investigate their amplification dynamics over the course of evolution. Based on the nucleotide divergence between the two LTRs, we estimated the time of insertion of each element using a substitution rate of 1.05×10^{-9} nucleotide substitutions per site per year. The number of intact, putative autonomous LTR-retrotransposons varied greatly in the five species ranging from 26 elements in *C. olivacea* to 944 in *P. tinctorius*. The LTR profiles of *C. olivacea*, *C. puteana* and *S. lacrymans* showed recent peaks of amplification with insertion dates at 0–5 million years (MY). LTR amplification in *H. pinastri* showed a peak at 10–15 MY ago, whereas the profile of *P. tinctorium* pointed to a much older amplification burst showing a maximum peak at 25–30 MY ago and few recent retrotransposition events (Fig. 6).

Discussion

Genomic and proteomic characteristics of *C. olivacea*

We report the 39.07 Mb draft genome assembly and annotation of brown-rot basidiomycete *C. olivacea*. In terms of genome size, this species is slightly smaller than *C. puteana*, but it falls in the range of other brown-rot basidiomycetes such as *Hydnomerulius pinastri* (38.3 Mb) [4] or *Serpuyla lacrymans* (47.0 Mb). As expected for closely related species, *C. olivacea* and *C. puteana* show macrosynteny, although due to the short scaffold lengths it is impossible to establish comparisons at a chromosome scale. We found very good conservation of protein-coding genes, although *C. olivacea* has up to 1,352 orphan genes—most of these are supported by structure and RNA evidence (i.e., no homology to any other known gene). In this sense, the higher number of

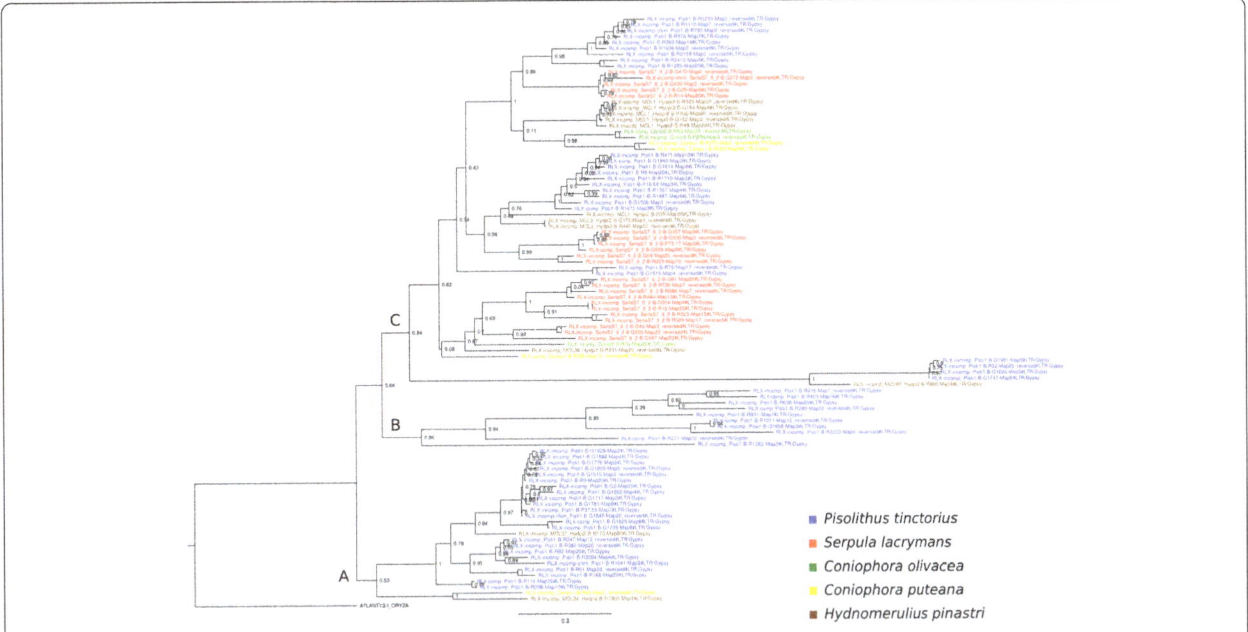

Fig. 5 Maximum likelihood phylogeny of the Gypsy reverse-transcriptases found in the *C. olivacea*, *C. puteana*, *S. lacrymans*, *H. pinastri* and *P. tinctorius* (blue) genomes. SH (Shimodaira-Hasegawa) local support values are shown in branches. The reverse-transcriptase from *Oryza sativa* ATLANTIS-I family consensus (Repbase) was used as outgroup

annotated genes in *C. olivacea* relative to *C. puteana* is probably related to the higher amount of assembled RNA contigs used to assist the annotation of the former (resulting from the higher RNAseq depth). The presence of about 10% of orphan genes is common in fungal genomes, and these genes often lack an *in silico* functional annotation as we found for *C. olivacea* [39, 40].

Wood-decaying species require a complex enzymatic machinery to degrade lignin and obtain nutrients. According to the CAZy enzymes identified in the genome, the *C. olivacea* proteome carries the main signatures of canonical brown-rot: (i) it completely lacks Class II peroxidases—enzymes primarily involved in lignin degradation [41], and (ii) it carries a reduced set of enzymes involved in degradation of crystalline cellulose. In fact, its profile is very similar to that of *C. puteana,* displaying only minor differences in several enzyme groups. As previously seen in other wood-degrading fungi, the *in silico* secretome of *C. olivacea* is enriched in functions related to lignocellulose degradation [42]. Our analysis showed that most intracellular and secreted proteins are members of multi-gene families of diverse size originating from gene duplications. The number of gene families that could not be functionally annotated by standard similarity-based methods was high, a phenomenon that is frequently observed in fungi.

To overcome this drawback, we used an alternative approach that combines similarity with structural information (Phyre-2). We then assigned a putative function to two multi-gene families conserved across the basidiomycete phylogeny but for which a putative function had not been previously proposed. Of special interest is the newly identified family of putative copper-dependent lytic polysaccharide monooxygenases (AA9, LPMO). The LPMOs are recently discovered enzymes used by microbes to digest crystalline polysaccharides [43]. They increase the saccharification yield of commercial enzyme cocktails [44]. Nevertheless, despite the promising results obtained *in silico*, experimental assays will be necessary to confirm the function of the members of this newly described gene family.

Impact of TEs in the evolution of Boletales genomes

The results of TE annotation in the five Boletales showed how different patterns of LTR-retrotransposon amplifications have shaped the architecture of their genomes. The expansion of LTR/Gypsy retrotransposons belonging to *Chromoviridae* occurred mainly in the species with large genomes, whereas the smaller genomes have a small amount of these families (ie, three families in *C. olivacea* and *C. puteana).* Chromoviruses are the most common LTR-retrotransposons in fungi [45], and the key to their success might be the presence of a chromo-integrase, which is thought to guide the integration of these elements into heterochromatic regions [46]. Heterochromatin is gene-poor, and it is silenced by epigenetic mechanisms such as DNA methylation and RNAi [47]. Thus, integration of these elements in such regions would allow them to skip purifying selection and increase their probability to persist in the genome. In fact, this could be the reason for the longer prevalence of *Gypsy* over *Copia*

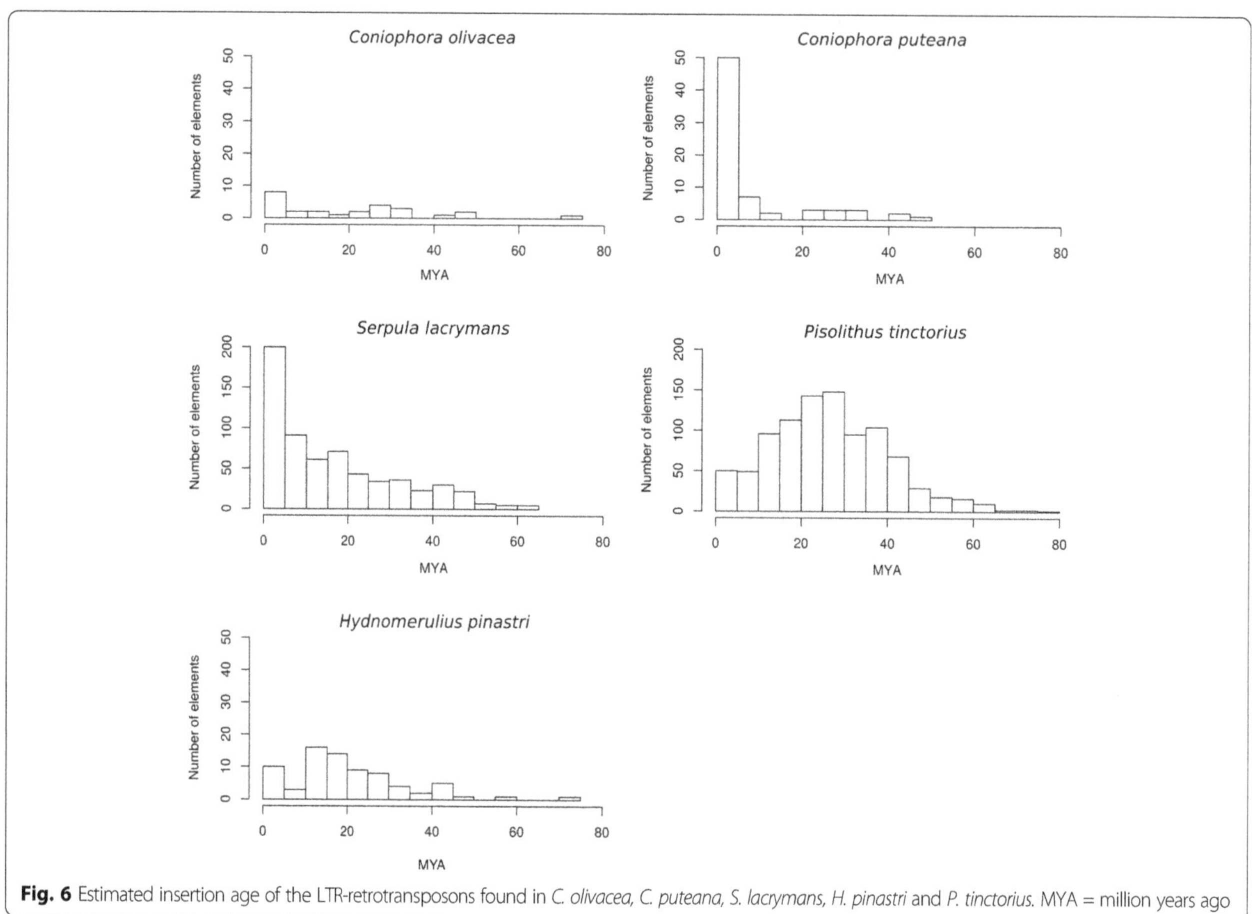

Fig. 6 Estimated insertion age of the LTR-retrotransposons found in *C. olivacea, C. puteana, S. lacrymans, H. pinastri* and *P. tinctorius.* MYA = million years ago

LTR-retrotransposons in most fungal species—the latter tend to integrate at random locations including euchromatic regions where transposon fixation is more difficult [48]. The LTR-retrotransposon amplification bursts of the Boletales indicate that elements from both *Coniophora* species are young and thus putatively active, and the profile of *S. lacrymans* also indicates a very strong activity of young copies with a progressive decrease in the amplification signals of older elements. Our findings suggest that the latter three species are currently in a period of genome expansion. Despite the different profile of *H. pinastri* and *P. tinctorius* we cannot rule out the same hypothesis, as both assemblies contain high gap content (7.7% and 13.3%, respectively). This fact usually leads to an underestimation in the amount of young retrotransposons [6], as they are difficult to assemble due to their repetitive nature and high sequence identity. In fact, we show that due to this reason the assembly-based TE quantification underestimated LTR content in *C. olivacea* in comparison to non-assembly based quantification (Additional file 2: Table S1). The profile of *P. tinctorius* is intriguing. This ectomycorrhizal (ECM) species undergoes a massive expansion of LTR-retrotransposons in the Gypsy superfamily (similar to that found for other symbiotic species in Agaricomycotina [7, 49]; however, the majority of elements are very old (20–40 MY) and still carry structural and coding domains necessary for transposition. The phylogeny of Gypsy reverse-transcriptases suggests that many *P. tinctorius*-specific families are distantly related to the other four species. In fact, its impressive retrotransposon content might be partially explained by the amplification and diversification of ancestral families (giving rise to clades A and B in Fig. 5). Our phylogenetic reconstruction suggests that such ancestral families were also present in other boletales but didn't proliferate in the genome (ie, *H. pinastri or C. puteana*). Whether genome defense mechanisms or lifestyle constraints are responsible of this phenomenon is still to be demonstrated. In this regards, it is interesting to note that the LTR-mediated genome amplification of *P. tinctorius* roughly coincides with the estimated origins of ECM symbiosis in Boletales [4]. Of the four Class I TE orders found, only the LTR elements were present in the five species. The most plausible scenario is that the elements from the other three orders (DIRS, LINE, and PLE) were lost by random drift in some of the species. Alternatively, they might be present in some genomes but in the form of very ancient and degenerated copies that are not detectable. Similarly, this patchy distribution was also found in class II elements (ie, helitrons were absent in the *Coniophora* genus and present in the remaining three species). Previous studies have shown that besides the conserved presence of LTR and TIR orders, the remaining TE groups tend to be present in variable amounts in basidiomycetes [6].

Conclusions

In this study we present the draft genome sequence and annotation of the brown-rot fungi *Coniophora olivacea*, along with a comparative analysis with *C. puteana* and other members of the Boletales order. Our results show evidence of macrosynteny and conservation in the protein coding genes of the two species. The functional analysis of *C. olivacea* secretome showed that it displays the main signatures of a canonical brown-rot, and uncovered a new family of putative LPMOs widely conserved in basidiomycota. The annotation of transposable elements revealed a particular contraction in these two species in comparison to other Boletales, mainly due to the differential expansion of *Chromoviridae* LTR-retrotransposons. By analyzing the distribution of insertion ages and phylogenetic relationships of these elements we show that these LTR-retrotransposons have played a key role in the genome expansion experienced by certain species in the Boletales order.

Additional files

Additional file 1: Text S1. Supplementary methods. (DOCX 16 kb)

Additional file 2: Table S1. Comparison of TE content estimation from REPET and Repeatexplorer. (XLSX 9 kb)

Additional file 3: Figure S1. Snapshot of synteny dot plot between *C. olivacea* and *C. puteana. (TIFF 582 kb)*

Additional file 4: Table S2. Comparison of CAZy proteins annotated in *C. olivacea* and *C. puteana. (XLSX 7 kb)*

Additional file 5: Table S3. Protein IDs of genes belonging to the 70 gene families with more than four members. (XLSX 9 kb)

Additional file 6: Dataset S1. TE annotation in *C. olivacea*. Classification information at the order level is included in the output format of PASTEC. *(GFF3 1342 kb)*

Additional file 7: Dataset S2. TE annotation in *C. puteana*. Classification information at the order level is included in the output format of PASTEC. *(GFF3 1273 kb)*

Additional file 8: Dataset S3. TE annotation in *H. pinastri*. Classification information at the order level is included in the output format of PASTEC. *(GFF3 1674 kb)*

Additional file 9: Dataset S4. TE annotation in *S. lacrymans*. Classification information at the order level is included in the output format of PASTEC. *(GFF3 8198 kb)*

Additional file 10: Dataset S5. TE annotation in *P. tinctorius*. Classification information at the order level is included in the output format of PASTEC. *(GFF3 12,724 kb)*

Abbreviations

AA: Auxiliar activity; CAZYs: Carbohydrate-active enzymes; CBM: Carbohydrate-binding modules; CE: Carbohydrate esterases; CEGMA: Core Eukaryotic Genes Mapping Approach; DIRS: Dictyostelium intermediate repeat sequence; ECM: Ectomycorrhizal; GH: Glycoside hydrolase; GO: Gene Ontology; GPI: Glycosylphosphatidylinositol; HMM: Hidden Markov Models; Kb: Kilobase; KEGG: Kyoto Encyclopedia of Genes and Genomes; KOG: Eukaryotic Orthologous Groups; LARD: Large retrotransposon derivative; LINE: Long interspersed nuclear elements; LPMO: Lytic polysaccharide monooxygenases; LTR: Long Terminal Repeats; Mb: Megabase; MITE: Miniature inverted-repeat transposable elements; MY: Million years; PCWDE: Plant cell wall-degrading enzymes; PLE: Penelope-like elements; PSI: Position-Specific Iterated; RBH: Reciprocal best hit; RNAi: RNA interference; RV: Reverse-transcriptase; SH: Shimodaira-Hasegawa; SMY: Sucrose, malt, yeast; SRA: Sequence Read Archive; TEs: Transposable elements; TIR: Terminal inverted repeats; TRIM: Terminal-repeat retrotransposon in miniature; tRNA: transfer RNA

Acknowledgments

The authors want to thank the helpful advices of the URGI team and especially Véronique Jamilloux on the use of REPET package, as well as Francis Martin, Joey Spatafora and In-Geol Choi for allowing the use of unpublished genome data and Bernard Henrissat for the annotation of CAZYmes.

Funding

This work was supported by Spanish National Research Plan (Project AGL2014-55971-R) and FEDER funds; Public University of Navarre. The work of RC is supported by an FPI grant from the Spanish Ministry of Economy and Competitiveness. The work conducted by the U.S. Department of Energy Joint Genome Institute, a DOE Office of Science User Facility, is supported by the Office of Science of the U.S. Department of Energy under Contract No. DE-AC02-05CH11231. Funders had no role in the design of the study and collection, analysis, and interpretation of data and in writing the manuscript.

Authors' contributions

All authors have read and approved the manuscript. Conceived the project: RC, AGP and LR. Wrote the article: RC. Contributed to material and data acquisition: RC, SH GP, LLV, JA, KL, AL, VS, KB. Analyzed the data: RC, SH, KL, AL, VS. Designed the experiments: RC, GP, LLV, JA, KL, VS, AL, SH, KB, IVG, AGP, LR. Critically revised the manuscript: RC, GP, LLV, JA, KL, VS, AL, SH, KB, IVG, AGP, LR. Lead the project: IGV, AGP and LR.

Competing interests

The authors declare that they have no competing interests.

Author details

[1]Genetics and Microbiology Research Group, Department of Agrarian Production, Public University of Navarre, 31006 Pamplona, Navarre, Spain. [2]URGI, INRA, Université Paris-Saclay, 78026 Versailles, France. [3]U.S.Department of Energy Joint Genome Institute, Walnut Creek, CA 94598, USA.

References

1. Floudas D, Binder M, Riley R, Barry K, Blanchette RA, Henrissat B, et al. The paleozoic origin of enzymatic lignin decomposition reconstructed from 31 fungal genomes. Science. 2012;336:1715–9.
2. Riley R, Salamov AA, Brown DW, Nagy LG, Floudas D, Held BW, et al. Extensive sampling of basidiomycete genomes demonstrates inadequacy of the white-rot/brown-rot paradigm for wood decay fungi. Proc Natl Acad Sci U S A. 2014;111:9923–8.

3. Eastwood DC, Floudas D, Binder M, Majcherczyk A, Schneider P, Aerts A, et al. The plant cell wall-decomposing machinery underlies the functional diversity of forest fungi. Science. 2011;333:762–5.
4. Kohler A, Kuo A, Nagy LG, Morin E, Barry KW, Buscot F, et al. Convergent losses of decay mechanisms and rapid turnover of symbiosis genes in mycorrhizal mutualists. Nat Genet. 2015;47:410–5.
5. Martin F, Kohler A, Murat C, Veneault-Fourrey C, Hibbett DS. Unearthing the roots of ectomycorrhizal symbioses. Nat Rev Microbiol. 2016;14:760–73.
6. Castanera R, López-Varas L, Borgognone A, LaButti K, Lapidus A, Schmutz J, et al. Transposable elements versus the fungal genome: impact on whole-genome architecture and transcriptional profiles. PLoS Genet. 2016;12
7. Hess J, Skrede I, Wolfe BE, LaButti K, Ohm RA, Grigoriev IV, et al. Transposable element dynamics among asymbiotic and ectomycorrhizal *Amanita* fungi. Genome Biol Evol. 2014;6:1564–78.
8. Castanera R, Borgognone A, Pisabarro AG, Ramírez L. Biology, dynamics, and applications of transposable elements in basidiomycete fungi. Appl Microbiol Biotechnol. 2017:1–14.
9. Larraya LM, Perez G, Penas MM, Baars JJP, Mikosch TSP, Pisabarro AG, et al. Molecular Karyotype of the white rot fungus *Pleurotus ostreatus*. Appl Envir Microbiol. 1999;65:3413–7.
10. Zerbino DR, Birney E. Velvet: algorithms for de novo short read assembly using de Bruijn graphs. Genome Res. 2008;18:821–9.
11. Gnerre S, MacCallum I, Przybylski D, Ribeiro FJ, Burton JN, Walker BJ, et al. High-quality draft assemblies of mammalian genomes from massively parallel sequence data. Proc Natl Acad Sci. 2010;108:1513–8.
12. Grabherr MG, Haas BJ, Yassour M, Levin JZ, Thompson DA, Amit I, et al. Full-length transcriptome assembly from RNA-Seq data without a reference genome. Nat Biotechnol. 2011;29:644–52.
13. Kurtz S, Phillippy A, Delcher AL, Smoot M, Shumway M, Antonescu C, et al. Versatile and open software for comparing large genomes. Genome Biol. 2004;5:R12.
14. Grigoriev IV, Nikitin R, Haridas S, Kuo A, Ohm R, Otillar R, et al. MycoCosm portal: gearing up for 1000 fungal genomes. Nucleic Acids Res. 2014;42:D699–704.
15. Cortázar AR, Aransay AM, Alfaro M, Oguiza JA, Lavín JLSECRETOOL. Integrated secretome analysis tool for fungi. Amino Acids. 2014;46:471–3.
16. Altschul SF, Gish W, Miller W, Myers EW, Lipman DJ. Basic local alignment search tool. J Mol Biol. 1990;215:403–10.
17. Johnson LS, Eddy SR, Portugaly E. Hidden Markov model speed heuristic and iterative HMM search procedure. BMC Bioinformatics. 2010;11:431.
18. Cantarel BL, Coutinho PM, Rancurel C, Bernard T, Lombard V, Henrissat B. The carbohydrate-active EnZymes databse (CAZY): an expert resource for Glycogenomics. Nucleic Acids Res. 2009;37(Database issue):D233–8.
19. Kelley LA, Mezulis S, Yates CM, Wass MN, Sternberg MJE. The Phyre2 web portal for protein modeling, prediction and analysis. Nat Protoc. 2015;10:845–58.
20. Quesneville H, Bergman CM, Andrieu O, Autard D, Nouaud D, Ashburner M, et al. Combined evidence annotation of transposable elements in genome sequences. PLoS Comput Biol. 2005;1:166–75.
21. Flutre T, Duprat E, Feuillet C, Quesneville H. Considering transposable element diversification in de novo annotation approaches. PLoS One. 2011;6:e16526.
22. Hoede C, Arnoux S, Moisset M, Chaumier T, Inizan O, Jamilloux V, et al. PASTEC: an automatic transposable element classification tool. PLoS One. 2014;9:e91929.
23. Ellinghaus D, Kurtz S, Willhoeft U. LTRharvest, an efficient and flexible software for de novo detection of LTR retrotransposons. BMC Bioinformatics. 2008;9:18.
24. Jurka J. Repbase update - a database and an electronic journal of repetitive elements. Trends Genet. 2000;16:418–20.
25. Edgar RC. MUSCLE: multiple sequence alignment with high accuracy and high throughput. Nucleic Acids Res. 2004;32:1792–7.
26. Capella-Gutierrez S, Silla-Martinez JM, Gabaldon T. trimAl: a tool for automated alignment trimming in large-scale phylogenetic analyses. Bioinformatics. 2009;25:1972–3.
27. SanMiguel P, Gaut BS, Tikhonov A, Nakajima Y, Bennetzen JL. The paleontology of intergene retrotransposons of maize. Nat Genet. 1998;20:43–5.
28. Dhillon B, Gill N, Hamelin RC, Goodwin SB. The landscape of transposable elements in the finished genome of the fungal wheat pathogen *Mycosphaerella graminicola*. BMC Genomics. 2014;15:1132.
29. Enright AJ, Van Dongen S, Ouzounis CA. An efficient algorithm for large-scale detection of protein families. Nucleic Acids Res. 2002;30:1575–84.
30. Katoh K, Misawa K, Kuma K, Miyata T. MAFFT: a novel method for rapid multiple sequence alignment based on fast Fourier transform. Nucleic Acids Res. 2002;30:3059–66.
31. Talavera G, Castresana J. Improvement of phylogenies after removing divergent and ambiguously aligned blocks from protein sequence alignments. Syst Biol. 2007;56:564–77.
32. Stamatakis A. RAxML version 8: a tool for phylogenetic analysis and post-analysis of large phylogenies. Bioinformatics. 2014;30:1312–3.
33. Slater GS, Birney E. Automated generation of heuristics for biological sequence comparison. BMC Bioinformatics. 2005;6:31.
34. Price MN, Dehal PS, Arkin AP. FastTree: computing large minimum evolution trees with profiles instead of a distance matrix. Mol Biol Evol. 2009;26:1641–50.
35. Parra G, Bradnam K, Korf I. CEGMA: a pipeline to accurately annotate core genes in eukaryotic genomes. Bioinformatics. 2007;23:1061–7.
36. Bairoch A, Boeckmann B. The SWISS-PROT protein sequence data bank. Nucleic Acids Res. 1991;19(Suppl):2247–9.
37. Hibbett DS, Binder M, Bischoff JF, Blackwell M, Cannon PF, Eriksson OE, et al. A higher-level phylogenetic classification of the fungi. Mycol Res. 2007;111:509–47.
38. Wicker T, Sabot F, Hua-Van A, Bennetzen JL, Capy P, Chalhoub B, et al. A unified classification system for eukaryotic transposable elements. Nat Rev Genet. 2007;8:973–82.
39. Grandaubert J, Bhattacharyya A, Stukenbrock EH. RNA-seq-based gene annotation and comparative genomics of four fungal grass pathogens in the genus *Zymoseptoria* identify novel orphan genes and species-specific invasions of transposable elements. G3 (Bethesda). 2015;5:1323–33.
40. Nagy LG, Riley R, Tritt A, Adam C, Daum C, Floudas D, et al. Comparative genomics of early-diverging mushroom-forming fungi provides insights into the origins of Lignocellulose decay capabilities. Mol Biol Evol. 2015;3:msv337.
41. Fernández-Fueyo E, Ruiz-Dueñas FJ, Martínez MJ, Romero A, Hammel KE, Medrano FJ, et al. Ligninolytic peroxidase genes in the oyster mushroom genome: heterologous expression, molecular structure, catalytic and stability properties, and lignin-degrading ability. Biotechnol Biofuels [Internet]. 2014;7:2.
42. Alfaro M, Oguiza JA, Ramírez L, Pisabarro AG. Comparative analysis of secretomes in basidiomycete fungi. J Proteome. 2014;102:28–43.
43. Vaaje-Kolstad G, Westereng B, Horn SJ, Liu Z, Zhai H, Sorlie M, et al. An oxidative enzyme boosting the enzymatic conversion of recalcitrant polysaccharides. Science. 2010;330:219–22.
44. Müller G, Várnai A, Johansen KS, Eijsink VGH, Horn SJ. Harnessing the potential of LPMO-containing cellulase cocktails poses new demands on processing conditions. Biotechnol Biofuels. 2015;8:187.
45. Muszewska A, Hoffman-Sommer M, Grynberg M. LTR retrotransposons in fungi. PLoS One. 2011;6:e29425.
46. Gao X, Hou Y, Ebina H, Levin HL, Voytas DF. Chromodomains direct integration of retrotransposons to heterochromatin. Genome Res. 2008;18:359–69.
47. Lippman Z, Martienssen R. The role of RNA interference in heterochromatic silencing. Nature. 2004;431:364–70.
48. Pereira V. Insertion bias and purifying selection of retrotransposons in the *Arabidopsisthaliana* genome. Genome Biol. 2004;5:R79.
49. Labbe J, Murat C, Morin E, Tuskan GA, Le Tacon F, Martin F. Characterization of transposable elements in the ectomycorrhizal fungus *Laccaria bicolor*. PLoS One. 2012;7:e40197.

PERMISSIONS

All chapters in this book were first published in GENOMICS, by BioMed Central; hereby published with permission under the Creative Commons Attribution License or equivalent. Every chapter published in this book has been scrutinized by our experts. Their significance has been extensively debated. The topics covered herein carry significant findings which will fuel the growth of the discipline. They may even be implemented as practical applications or may be referred to as a beginning point for another development.

The contributors of this book come from diverse backgrounds, making this book a truly international effort. This book will bring forth new frontiers with its revolutionizing research information and detailed analysis of the nascent developments around the world.

We would like to thank all the contributing authors for lending their expertise to make the book truly unique. They have played a crucial role in the development of this book. Without their invaluable contributions this book wouldn't have been possible. They have made vital efforts to compile up to date information on the varied aspects of this subject to make this book a valuable addition to the collection of many professionals and students.

This book was conceptualized with the vision of imparting up-to-date information and advanced data in this field. To ensure the same, a matchless editorial board was set up. Every individual on the board went through rigorous rounds of assessment to prove their worth. After which they invested a large part of their time researching and compiling the most relevant data for our readers.

The editorial board has been involved in producing this book since its inception. They have spent rigorous hours researching and exploring the diverse topics which have resulted in the successful publishing of this book. They have passed on their knowledge of decades through this book. To expedite this challenging task, the publisher supported the team at every step. A small team of assistant editors was also appointed to further simplify the editing procedure and attain best results for the readers.

Apart from the editorial board, the designing team has also invested a significant amount of their time in understanding the subject and creating the most relevant covers. They scrutinized every image to scout for the most suitable representation of the subject and create an appropriate cover for the book.

The publishing team has been an ardent support to the editorial, designing and production team. Their endless efforts to recruit the best for this project, has resulted in the accomplishment of this book. They are a veteran in the field of academics and their pool of knowledge is as vast as their experience in printing. Their expertise and guidance has proved useful at every step. Their uncompromising quality standards have made this book an exceptional effort. Their encouragement from time to time has been an inspiration for everyone.

The publisher and the editorial board hope that this book will prove to be a valuable piece of knowledge for researchers, students, practitioners and scholars across the globe.

LIST OF CONTRIBUTORS

Birgit S. Gruben
Fungal Physiology, Westerdijk Fungal Biodiversity Institute, Uppsalalaan 8, 3584, CT, Utrecht, The Netherlands
Microbiology, Utrecht University, Padualaan 8, 3584, CH, Utrecht, The Netherlands

Miia R. Mäkelä
Fungal Physiology, Westerdijk Fungal Biodiversity Institute, Uppsalalaan 8, 3584, CT, Utrecht, The Netherlands
Fungal Molecular Physiology, Utrecht University, Uppsalalaan 8, 3584, CT, Utrecht, The Netherlands
Department of Food and Environmental Sciences, Division of Microbiology and Biotechnology, Viikki Biocenter 1, University of Helsinki,Helsinki, Finland

Joanna E. Kowalczyk
Fungal Physiology, Westerdijk Fungal Biodiversity Institute, Uppsalalaan 8,3584, CT, Utrecht, The Netherlands
Fungal Molecular Physiology, Utrecht University, Uppsalalaan 8, 3584, CT, Utrecht, The Netherlands

Miaomiao Zhou
Fungal Physiology, Westerdijk Fungal Biodiversity Institute, Uppsalalaan 8,3584, CT, Utrecht, The Netherlands
Current affiliation: ATGM, Avans University of Applied Sciences, Lovensdijkstraat 61-63, 4818, AJ, Breda, The Netherlands

Isabelle Benoit-Gelber
Fungal Physiology, Westerdijk Fungal Biodiversity Institute, Uppsalalaan 8,3584, CT, Utrecht, The Netherlands
Microbiology, Utrecht University,Padualaan 8, 3584, CH, Utrecht, The Netherlands
Fungal Molecular Physiology, Utrecht University, Uppsalalaan 8, 3584, CT, Utrecht, The Netherlands
Current affiliation: Center for Structural and Functional Genomics, Concordia University, 7141 Sherbrooke St. W, Montreal QC, Canada

Ronald P. De Vries
Fungal Physiology, Westerdijk Fungal Biodiversity Institute, Uppsalalaan 8, 3584, CT, Utrecht, The Netherlands
Microbiology, Utrecht University,Padualaan 8, 3584, CH, Utrecht, The Netherlands
Fungal Molecular Physiology, Utrecht University, Uppsalalaan 8, 3584, CT, Utrecht, The Netherlands

Weiying Zeng, Zudong Sun, Zhaoyan Cai, Huaizhu Chen, Zhenguang Lai, Shouzhen Yang, Xiangmin Tang
Guangxi Academy of Agricultural Sciences, Nanning, Guangxi 530007, China

Frances R. Thistlethwaite, Blaise Ratcliffe, Yousry A. El-Kassaby
Department of Forest and Conservation Sciences, Faculty of Forestry, The University of British Columbia, 2424 Main Mall, Vancouver, BC V6T 1Z4, Canada

Jaroslav Klápště
Department of Forest and Conservation Sciences, Faculty of Forestry, The University of British Columbia, 2424 Main Mall, Vancouver, BC V6T 1Z4, Canada
Scion (New Zealand Forest Research Institute Ltd.), 49 Sala Street,Whakarewarewa, Rotorua 3046, New Zealand
Department of Genetics and Physiology of Forest Trees, Faculty of Forestry and Wood Sciences, Czech University of Life Sciences Prague, Kamycka 129, 165 21 Praha 6, Czech Republic

Ilga Porth
Département des sciences du bois et de la forêt, Université Laval, QC, Québec G1V 0A6, Canada

Charles Chen
Department of Biochemistry and Molecular Biology, Oklahoma State University, Stillwater, OK 74078-3035, USA

Michael U. Stoehr
British Columbia Ministry of Forests, Lands and Natural Resource Operations, Victoria, BC V8W 9C2, Canada

Md. Abdul Kayum, Jong-In Park, Ujjal Kumar Nath, Gopal Saha, Manosh Kumar Biswas, Ill-Sup Nou
Department of Horticulture, Sunchon National University, 255 Jungang-ro, Suncheon, Jeonnam 57922,Republic of Korea

Hoy-Taek Kim
University-Industry Cooperation Foundation, Sunchon National University, 255 Jungang-ro, Suncheon, Jeonnam 57922, Republic of Korea

Xinke Zhang, Pengxiu Dai, Yongping Gao, Xiaowen Gong, Hao Cui, Yihua Zhang
The College of Veterinary Medicine of the Northwest Agriculture and Forestry University, No. 3 Taicheng Road, Yangling, Shaanxi, People's Republic of China

Yipeng Jin
Clinical Department, College of Veterinary Medicine,China Agricultural University, Beijing, People's Republic of China

Ji Li, Jian Xu, Qin-Wei Guo, Zhe Wu, Ting Zhang, Kai-Jing Zhang, Chun-yan Cheng, Pin-yu Zhu, Qun-Feng Lou, Jin-Feng Chen
State Key Laboratory of Crop Genetics and Germplasm Enhancement, Nanjing Agricultural University, Nanjing 210095, China

Márcia Carvalho, Eduardo Rosa
Centre for Research and Technology of Agro-Environmental and Biological Sciences (CITAB), University of Trás-os-Montes and Alto Douro (UTAD),5000-801 Vila Real, Portugal

María Muñoz-Amatriaín
Department of Botany and Plant Sciences,University of California Riverside, Riverside, CA 92521-0124, USA

Isaura Castro
Centre for Research and Technology of Agro-Environmental and Biological Sciences (CITAB), University of Trás-os-Montes and Alto Douro (UTAD),5000-801 Vila Real, Portugal
Department of Genetics and Biotechnology, University of Trás-os-Montes and Alto Douro (UTAD), 5000-801 Vila Real, Portugal

Teresa Lino-Neto
Biosystems & Integrative Sciences Institute (BioISI), Plant Functional Biology Center (CBFP), University of Minho, Campus de Gualtar, 4710-057 Braga, Portugal

Manuela Matos
Department of Genetics and Biotechnology, University of Trás-os-Montes and Alto Douro (UTAD), 5000- 801 Vila Real, Portugal
Biosystems & Integrative Sciences Institute (BioISI), Sciences Faculty, University of Lisbon, Campo Grande, 1749-016 Lisbon, Portugal

Marcos Egea-Cortines
Instituto de Biotecnología Vegetal, Universidad Politécnica de Cartagena, 30202 Cartagena, Spain

Timothy Close
Department of Botany and Plant Sciences, University of California Riverside, Riverside, CA 92521-0124, USA

Valdemar Carnide
Centre for Research and Technology of Agro-Environmental and Biological Sciences (CITAB), University of Trás-os-Montes and Alto Douro (UTAD), 5000-801 Vila Real, Portugal.
Department of Genetics and Biotechnology, University of Trás-os-Montes and Alto Douro (UTAD), 5000-801 Vila Real, Portugal

Michie Kobayashi, Yukie Hiraka, Akira Abe, Hiroki Yaegashi, Satoshi Natsume, Hideko Kikuchi, Hiroki Takagi
Iwate Biotechnology Research Center, Iwate, Japan

Hiromasa Saitoh
Iwate Biotechnology Research Center, Iwate, Japan
Department of Molecular Microbiology, Tokyo University of Agriculture, Tokyo, Japan

Joe Win
The Sainsbury Laboratory, Norwich, UK

Sophien Kamoun
The Sainsbury Laboratory, Norwich, UK

Ryohei Terauchi
Iwate Biotechnology Research Center, Iwate, Japan
Kyoto University, Kyoto, Japan

Joanna Puławska, Monika Kałużna, Wojciech Warabieda, Artur Mikiciński
Research Institute of Horticulture, ul. Konstytucji 3 Maja 1/3, 96-100 Skierniewice, Poland

Wenjia Wang, Lianfeng Gu, Shanwen Ye, Hangxiao Zhang, Changyang Cai, Mengqi Xiang, Yubang Gao, Qin Wang, Qiang Zhu
Basic Forestry and Proteomics Center (BFPC), Fujian Provincial Key Laboratory of Haixia Applied Plant Systems Biology, Haixia Institute of Science and Technology, Fujian Agriculture and Forestry University, Fujian 350002, China

Chentao Lin
Basic Forestry and Proteomics Center (BFPC), Fujian Provincial Key Laboratory of Haixia Applied Plant systems Biology, Haixia Institute of Science and Technology, Fujian Agriculture and Forestry University, Fujian 350002, China
Department of Molecular, Cell & Developmental Biology, University of California, Los Angeles, California 90095, USA

Kaleem U. Kakar
State Key Laboratory of Rice Biology, Institute of Crop Science, Zhejiang University, Hangzhou 310058, China
Molecular Genetics Key Laboratory of China Tobacco, Guizhou Academy of Tobacco Science, Guiyang 550081,China

Zarqa Nawaz
Molecular Genetics Key Laboratory of China Tobacco, Guizhou Academy of Tobacco Science, Guiyang 550081,China
Wuxi Hupper Bioseed Technology Academy Ltd., Wuxi 214000, China

Khadija Kakar
Department of Biotechnology, BUITEMS, Quetta, Pakistan

Essa Ali
State Key Laboratory of Rice Biology, Institute of Crop Science, Zhejiang University, Hangzhou 310058, China

Abdulwareth A. Almoneafy
Department of Biological sciences, College of Education and Science, Albaydaa University, Rada'a, Yemen

Raqeeb Ullah
Department of Environmental Sciences, Quaid -i-Azam University, Islamabad, Pakistan

Xue-liang Ren
Molecular Genetics Key Laboratory of China Tobacco, Guizhou Academy of Tobacco Science, Guiyang 550081, China
Guizhou Academy of Tobacco Science, Longtanba Road No. 29, Guanshanhu District,
Guiyang (550081), Guizhou, People's Republic of China

Qing-Yao Shu
State Key Laboratory of Rice Biology, Institute of Crop Science, Zhejiang University, Hangzhou 310058, China
Institute of Crop Sciences, Zhejiang University, 866 Yuhangtang Road, Hangzhou 310029, China

Raúl Castanera, Gúmer Pérez, Leticia López-Varas, Antonio G. Pisabarro, Lucía Ramírez
Genetics and Microbiology Research Group, Department of Agrarian
Production, Public University of Navarre, 31006 Pamplona, Navarre, Spain

Index

www.ingramcontent.com/pod-product-compliance
Lightning Source LLC
Chambersburg PA
CBHW082025190326
41458CB00010B/3279
* 9 7 8 1 6 8 2 8 6 7 1 2 9 *